THERMAL CONDUCTIVITY 18

A Continuation Order Plan is available for this series. A continuation order will bring delivery of each new volume immediately upon publication. Volumes are billed only upon actual shipment. For further information please contact the publisher.

THERMAL CONDUCTIVITY 18

Edited by

T. Ashworth

and

David R. Smith

South Dakota School of Mines and Technology
Rapid City, South Dakota

PLENUM PRESS • NEW YORK AND LONDON

ISBN-13: 978-1-4684-4918-1   e-ISBN-13: 978-1-4684-4916-7
DOI: 10.1007/978-1-4684-4916-7

Proceedings of the Eighteenth International Thermal Conductivity Conference,
hosted and organized by the South Dakota School of Mines and Technology,
sponsored by RE/SPEC Inc. of Rapid City and CINDAS/Purdue University,
held October 3–5, 1983, in Rapid City, South Dakota

© 1985 by Purdue Research Foundation
Softcover reprint of the hardcover 1st edition 1985

Plenum Press is
a Division of Plenum Publishing Corporation
233 Spring Street, New York, N.Y. 10013

All rights reserved

No part of this book may be reproduced, stored in a retrieval system, or transmitted,
in any form or by any means, electronic, mechanical, photocopying, microfilming,
recording, or otherwise, without written permission from the Publisher

This volume is dedicated to
the memory of
C. F. Lucks,
Chairman of the
First International Thermal Conductivity Conference.

FOREWORD

The International Thermal Conductivity Conference was started
in 1961 with the initiative of Mr. Charles F. Lucks, who passed
away on 8 July 1982 and to the memory of whom this volume is
dedicated.

These Conferences on thermal conductivity grew out of the
needs of researchers in the field. The Conferences were held annu-
ally from 1961 to 1973 and have been held biennially since 1975
when our Center for Information and Numerical Data Analysis and
Synthesis (CINDAS) of Purdue University became the Permanent
Sponsor of the Conferences. These Conferences provide a broadly
based forum for researchers actively working on the thermal conductivity and closely related properties to convene on a regular basis
to exchange their ideas and experiences and report their findings
and results.

The Conferences have been self-perpetuating and are an example
of how a technical community with a common purpose can transcend
the invisible, artificial barriers between disciplines and gather
together in increasing numbers without the need of national pub-
licity and continuing funding support, when they see something
worthwhile going on. It is believed that this series of Conferences
not only will grow stronger, but will set an example for researchers in other fields on how to jointly attack their own problem
areas.

Of the first thirteen Conferences, only four published formal
volumes of proceedings. However, effective with the Fourteenth
Conference, a policy of publishing formal volumes of proceedings
on a continuing and uniform basis has been established. Thus,
including the present volume, the following formal volumes of proceedings have been published:

Conference (Year)	Title of Volume	Publisher and Year
7th (1967)	THERMAL CONDUCTIVITY Proceedings of the Seventh Conference	U.S. Government Printing Office (1968)
8th (1968)	THERMAL CONDUCTIVITY Proceedings of the Eighth Conference	Plenum Press (1969)
9th (1969)	NINTH CONFERENCE ON THERMAL CONDUCTIVITY	U.S. Atomic Energy Commission (1970)
13th (1973)	ADVANCES IN THERMAL CONDUCTIVITY, Papers Presented at XIII International Conference on Thermal Conductivity	University of Missouri, Rolla (1974)
14th (1975)	THERMAL CONDUCTIVITY 14	Plenum Press (1976)
15th (1977)	THERMAL CONDUCTIVITY 15	Plenum Press (1978)
16th (1979)	THERMAL CONDUCTIVITY 16	Plenum Press (1982)
17th (1981)	THERMAL CONDUCTIVITY 17	Plenum Press (1982)
18th (1983)	THERMAL CONDUCTIVITY 18	Plenum Press (1984)

Dr. T. Ashworth, Chairman of the Eighteenth Conference is to be congratulated for his excellent leadership in conducting the Conference; the painstaking efforts of Dr. Ashworth and his co-editor, Dr. David R. Smith, made the present volume possible. CINDAS looks forward to working with future host institutions to ensure that future Conferences continue to produce high-quality volumes of proceedings in this important, specialized field.

C. Y. Ho
Director
Center for Information and
  Numerical Data Analysis
  and Synthesis
Purdue University

August 1984
West Lafayette, Indiana

PREFACE

The 18th International Thermal Conductivity Conference (ITCC)
was held at the Rushmore Plaza Civic Center, Rapid City, South
Dakota on 3 - 5 October, 1983. It was hosted by the South Dakota
School of Mines and Technology (SDSM&T); the conference chairman
was T. Ashworth. RE/SPEC Inc. of Rapid City was a sponsor of the
conference, with continuing sponsorship from CINDAS/Purdue
University. In addition to the chairman, the other local committee
members were

      D. R. Smith, SDSM&T
      E. Ashworth, SDSM&T
      W. B. Krause, RE/SPEC Inc.
      J. D. Patterson, SDSM&T
      R. D. Redin, SDSM&T
      A. H. Redin, Spouses Program.

Papers presented covered a wide range of aspects of thermal
conductivity; special sessions were organized on glassy materials,
insulations, the high-temperature flash-laser method, and geologic
materials. Following welcomes by D. L. McElroy, Chairman of the
International Thermal Conductivity Conferences, R. A. Schleusener,
President of SDSM&T, Arthur P. LaCroix, Mayor of Rapid City, and
Paul F. Gnirk, President of RE/SPEC Inc., the technical portion of
the conference began with a Keynote Address by M. H. Cohen (EXXON)
on "Metastable Liquids and the Nature of the Glass Transition".
This was followed by an invited paper by R. Berman (Oxford University)
entitled "Solid Helium: A Review". Additional invited papers
(G. K. White (CSIRO) on "Lattice Thermal Conductivity at Normal and
High Temperature"; G. A. Slack (GE) on "Pressure Effects on Thermal
Conductivity"; P. G. Klemens (University of Connecticut) on "Theory
of the Thermal Conductivity of Amorphous Solids"; R. Taylor (UMIST)
on "Thermal Diffusivity of Layered and Fibrous Composites";
G. W. Milton (Cornell University) on "Estimating the Thermal
Conductivity of Composites") were intermingled with 81 contributed
papers. This mixture gave an enjoyable and stimulating balance
between review and research papers. For the first time a poster

session was included. Eleven papers were presented in this session, which was well attended with a high level of interest; the experiment was deemed a substantial success.

Overall statistics for the meeting include 109 registered delegates (75 from the U.S., 9 from Canada, and 25 from 11 countries around the world), and 89 presentations involving 140 authors.

Additional activities included visits to the SDSM&T Physics Department Thermophysical Properties Laboratory and RE/SPEC's Thermomechanical Laboratories, a mid-conference tour in the Black Hills, including Mount Rushmore, culminating in an informal banquet at the Chute Roosters in Hill City, and a special, invited presentation by L. Korb (Rockwell International) on "The Space Shuttle Orbiter Thermal Protection System"; this presentation brought the formal business of the conference to an end.

At the banquet the Thermal Conductivity Award was presented jointly to M. L. Minges of the Air Force Materials Laboratory, Wright-Patterson AFB. and G. K. White of CSIRO, Australia. Also R. U. Acton and P. Wagner, both of Sandia National Laboratories, were elected to Fellowships of the ITCC.

On a personal note, as Chairman of the 18th ITCC, I would like to express my appreciation to all the people who helped to make this conference a success. In particular I would like to extend my thanks to Dave McElroy and Jerry Hust for their encouragement and guidance, to the session chairmen for their services at the conference and for their help in obtaining reviews of the papers submitted for these proceedings, to the invited speakers for their wonderful presentations, to the members of the local committee for their assistance, and especially to my three main helpers, my secretary Mabel Miessner, my co-editor David Smith and my wife, Eileen.

The next meeting in this series will be held at Tennessee Technological University, under the chairmanship of D. W. Yarbrough, October, 1985. I look forward to seeing you all then.

Rapid City, SD.
June, 1984

T. Ashworth,
Chairman, 18th ITCC.

CONTENTS

SESSION D

LIQUIDS AND GASES

SESSION E

SALTS, OXIDES, AND BINARY COMPOUNDS

SESSION F

DATA ACQUISITION SYSTEMS

SESSION G

METALS AND ALLOYS

SESSION H

MISCELLANEOUS MATERIALS

SESSION J

MEASUREMENT APPARATUS

SESSION P

INVITED PAPER

SESSION Q

CERAMICS

SESSION U

COMPOSITES

SESSION V

GEOLOGICAL MATERIALS

SESSION Y

POSTER SESSION

SESSION A

Keynote Address

SPEAKER

Morrel H. Cohen
Exxon Research & Engineering Co.
Annandale, NJ

SESSION CHAIRMAN

D. L. McElroy
ORNL
Oak Ridge, TN

METASTABLE LIQUIDS AND THE NATURE OF THE GLASS TRANSITION

Morrel H. Cohen and Gary S. Grest

Corporate Research Science Laboratories
Exxon Research and Engineering Company
Route 22 East, Annandale, NJ 08801

ABSTRACT

A review is given of the free-volume theory of supercooled liquids in metastable equilibrium. That theory is made the basis for a quantative treatment of atomic and molecular movement which leads in turn to a quantitative treatment of relaxation processes in liquids and glasses. Results for glass transition temperatures and the magnitudes, temperature dependences and dispersions of relaxation rates are in excellent agreement with the experiment. The same theory leads to an explicit model of low-temperature tunneling centers in glasses.

I. INTRODUCTION

Much progress has been made in recent years in the qualitative and quantitative understanding of supercooled liquids, the glass transition, and glasses[1-4]. We have been able to elucidate the structure and dynamics of dense, supercooled liquids in metastable equilibrium[5,6], to describe the relaxation processes occurring within such materials[7,8,9], to obtain quantitatively accurate results for glass transition temperatures and relaxation spectra[7,8,9], and to develop an explicit model of low-temperature tunneling centers[10,11]. In section II of this paper, we describe the theory of the metastable equilibria of supercooled liquids [5,6]; in section III, the relaxation phenomena in liquids and glasses[7,8,9]; and, in section IV, the nature of tunneling centers[10,11].

II. METASTABLE EQUILIBRIUM OF SUPERCOOLED LIQUIDS

We suppose that we are dealing with a liquid at temperatures T well below its melting temperature T_M and that it is possible to cool the liquid sufficiently rapidly through the temperature range below T_M within which the nucleation rate for crystallization is maximal to avoid nucleation[12,13]. The liquid is dense; the movement of each atom, molecule or rigid subunit of a flexible molecule is strongly constrained by its neighbors. As a consequence, two distinct time scales emerge in the atomic motion. The first, t_v, characterizes the vibrational motion about pseudoequilibrium positions, and the second, t_D, characterizes the change of these pseudoequilibrium positions by diffusion. As the temperature T decreases, the diffusional time scale t_D increases markedly relative to the vibrational time scale t_v; $t_D \gg t_v$. In equilibrium, the system passes through a sequence of structures Y_1, Y_2, Y_3 in which local changes occur on the time scale t_D. The typical structures Y_i have associated with them free energies which are approximately the same on the scale of $k_B T$, where k_B is the Boltzmann constant. However, the diffusion-induced changes in structure give rise to an additional entropy, the communal entropy S_c. Thus the total free energy F of the system can be decomposed as follows,

$$F = F(Y_i) - TS_c. \tag{1.}$$

The clean separation between $F(Y_i)$ and $-TS_c$ is possible because $t_D \gg t_v$.

We can say what we mean by the structures Y_i in detail only for monatomic liquids. However, the considerations that evolve from that answer are generally applicable to all liquids, when suitably extended.

The radial distribution function g(r) for liquid argon or a Lennard-Jones liquid near its melting point is sketched in Figure 1a. The value of g(r) at its first peak is typically 2.8 for $T \simeq T_M$. g(r) can be obtained via molecular dynamics from the paths followed by the individual molecules, $\vec{X}_i(t)$. Both the vibrational and diffusional motion are then mixed up together. Important features of the underlying structures Y_i are obscure by the vibrational motion. Instead, following Tanemura et al.[14] or Jacucci[15], one should construct time-averaged positions,

$$\langle \vec{X}_i \rangle_\tau = \frac{1}{\tau} \int_t^{t+\tau} \vec{X}_i(t')dt', \quad t_v \ll \tau \ll t_D . \tag{2.}$$

$\bar{g}(r)$ can then be recalculated from the $\langle \vec{X}_i \rangle_\tau$. As shown in Figure 1b, the structure in $\bar{g}(r)$ sharpens up considerably, the height of the first peak increases from 2.8 to 6, and a shoulder is present

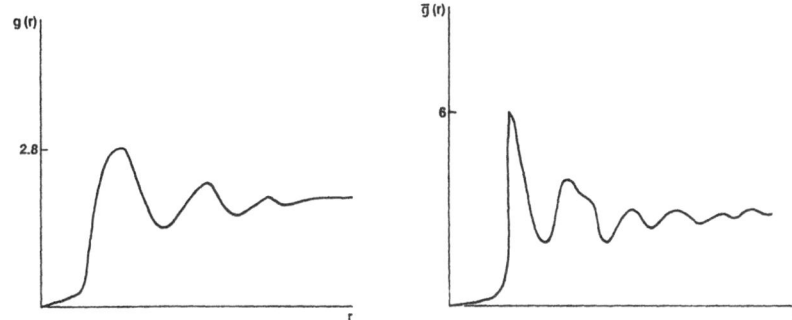

Figure 1.   a)   Sketch of the normal radial distribution function
             g(r) constructed from the separations between pairs of
             instantaneous positions at the melting point T_M.
             b)   Sketch of the radial distribution function $\bar{g}(r)$
             constructed from the separations between averaged posi-
             tions at the melting point T_M. Note the sharpening of
             the structure, the increase in the height of the first
             peak from 2.8 to 6 and the shoulder on the second peak.

in the second peak[15]. Apart from changes in the positions of the
peaks arising from thermal expansion, $\bar{g}(r)$ at T_M is essentially
identical to g(r) found for the simulated glass at much lower temperatures, which is quite similar to that of the dense random packing of hard spheres. A systematic investigation of this sharpening
of the structure has been carried out by Yonezawa and Kimura[16].

    The averaging process projects out the vibrational motion,
leaving a set of pseudoequilibrium positions. Such sets form the
structures Y_i, which are the various equivalent realizations of the
dense random packing of "soft" spheres.

    Dividing all space into cells, each containing one equilibrium
position, by for example a Voronoi tesselation, leads to a natural
generalization of the Lennard-Jones, Devonshire, Kirkwood cellular
model of the liquid state which incorporates the random characteristics of the arrangement of cells in a real liquid. In this generalized cellular model the structural part of the free energy
$F(Y_i)$ in (1.) is

$$F(Y_i) = F_{CELLULAR} + F_{CLUSTER} \tag{3.}$$

$F_{CELLULAR}$ in (3.) is obtained directly from the cellular model, but
$F_{CLUSTER}$ is new and will be discussed below. The communal entropy[17] term in (1.) was introduced earlier by Hirschfelder et
al.[18] and by Kirkwood[19], but we have greatly extended their
analysis.

The steps followed in obtaining $F_{CELLULAR}$ are to assume that Y_i has an infinite life. This assumption becomes accurate as t_D becomes much longer than t_v. Then assign to each atom, molecule, or rigid segment of a flexible molecule a volume coordinate v. A simple model for $F_{CELLULAR}$ is then constructed in analogy with the Lennard-Jones, Devonshire, Kirkwood theory,

$$F_{CELLULAR} = N \{ \int P(v)[f(v) + k_B T \ln P(v)] dv \}, \qquad (4.)$$

where N is the total number of cells and P(v) is the probability distribution of the values of v. The term $N \int P(v)f(v)dv$ is the sum of the free energies of the individual cells. f(v) contains the potential energy of interaction between the molecules and the entropy of the "vibrations" within the cells and must be interpreted as

$$\frac{\delta}{\delta P(v)} \{ F_{CELLULAR} - N k_B T \int P(v) \ln P(v) dv \} = f(v)$$

to avoid double counting of the interactions. The term $-N k_B \int P(v) \ln P(v)dv$ is the entropy of mixing arising from the distribution P(v) of cell volumes. We note that $F_{CELLULAR}$ does not distinguish the liquid from the crystal because it incorporates only quantitative information about the cell volume distribution but no other structural information. The communal entropy is associated with diffusive movement, which has not yet been introduced into the theory. Diffusion can be attacked with the aid of the free volume concept, introduction of which requires a more explicit description of f(v).

The single-cell free energy f(v) contains the work on the rest of the molecules required to make a cell of volume v, the interaction energy of the molecule within the cell with the rest of the molecules, and the entropy of "vibrational" movement within the cell. Consequently f(v) has two universal features: 1) a minimum at a volume v_0. The temperature dependence of v_0 gives rise to that component of the thermal expansion expected in the glassy state; and 2) a point of inflection at some $v_1 > v_0$. The importance of the point of inflection was stressed by Turnbull and Cohen[20], and we have made it the basis of the free-volume theory. We make a simplifying approximation to f(v):

$$f(v) = f_0 + 1/2 \kappa (v-v_0)^2 , \quad v < v_c \qquad (5.)$$

$$= f(v_c) + \gamma (v-v_c) , \quad v > v_c$$

$$v_0 < v_c < v_1$$

The parameters f_0, κ, v_0, γ and v_c should, in principle, be determined by the best fit of Eq (5) to the actual $f(v)$. The relevant features of $f(v)$ are sketched in Figure 2. It is the presence of the point of inflection which permits the neglect of the effects of the curvature of $f(v)$ for $v > v_c$ on a scale of $k_B T$ for sufficiently high temperatures.

We define $v_f = v - v_c$ as the free volume present in a cell of volume $v > v_c$. Cells with a quadratic dependence of $f(v)$ on v, $v < v_c$, are solid-like cells. Those with a linear dependence of $f(v)$ on v, $v > v_c$, are liquid-like. Only liquid-like cells have free volume. The fraction p of liquid-like cells is $p = \int_v P(v) dv$. The contribution per cell of the liquid-like cells to the total single-cell free energy is, $\int_v f(v) P(v) dv = p(f(v_c) + \overline{\gamma}_f)$ where $p \overline{v}_f = \int_v v_f P(v) dv$. Thus changes of $P(v)$ which leave p and $\overline{v}_f$ unchanged do not affect the single-cell free energy. Such changes of $P(v)$ amount to exchanges of free-volume among the liquid-like cells. The invariance of the single-cell free energy under exchange of free volume is our reason for terming $v_f = v - v_c$ the free volume.

A simple theory of diffusive movement in liquids and glasses was developed by Turnbull and Cohen [3] based on the idea that for diffusive movement of a molecule, a void of size at least v_M must open beside it and it does so by free exchange of free volume. These simple ideas led directly to the forms $\eta = \eta_0 \exp(v_m/v_f)$ and $D = D_0 \exp(-v_m/v_f)$ for the viscosity η and self-diffusion coefficient D. These forms, together with a highly simplified argument of how $\overline{v}_f$ depended on T led to the first satisfactory theoretical accounting of the phenomenological results of Fulcher[21], Vogel[22], and Doolittle[23] within the free volume model, as introduced by Fox and Flory[1] and by Williams, Landel and Ferry[2]. However, as additional, lower-temperature data accumulated for η, it became clear that the earlier arguments as to

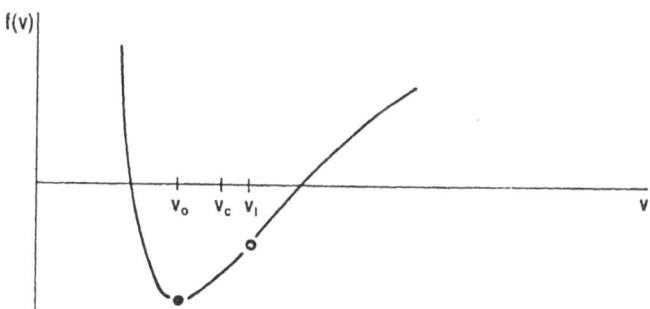

Figure 2.   The principal feature of the single-cell free energy, $f(v)$, as a function of the cell-volume v.

the temperature dependence of v_f were inadequate[5,6]. Moreover, the theory that was constructed by Turnbull and Cohen[3] to obtain the free-volume distribution did not make adequate connection with the thermodynamics of the system. Our further development of the free-volume theory[5,6] corrected these inadequacies.

When one considers that the cellular volume must be defined by partitioning processes as, e.g. in the Voronoi construction, it becomes clear that free volume can be exchanged only between two liquid-like cells which are nearest neighbors. Moreover, the volumes of some of the other nearest-neighbors of the pair in question must also change if the partitioning of volume between the pair is changed. This exchange of free volume can therefore be free in the sense that the sum of single-cell free energies is invariant only if a sufficient number of the nearest neighbors of each of the liquid-like cells are themselves liquid-like. We estimate that a liquid-like cell must have at least $z \cong \bar{c}/2$ liquid-like neighbors before it can participate in free exchanges of free volume, where $\bar{c}$ is the mean coordination number. This last statement defines a clustering criterion. Free exchange of free volume thus occurs only within clusters of liquid-like cells, or liquid-like clusters. The as yet undefined $F_{CLUSTER}$ simply adds the interfacial free energy and the shape entropy of the clusters to the total free energy F. These quantities can be obtained straightforwardly from Fisher's droplet theory of condensation[24,5,6].

Imposing the restriction that free exchange of free volume can occur only within liquid-like clusters and introducing $F_{CLUSTER}$ brings cell-cell interactions more explicitly into F. Also, it is $F_{CLUSTER}$ which distinguishes between the liquid and the crystal in this theory. In the model of a metastable liquid and, it will turn out, of a glass thus far constructed, we have built up a picture of clusters of liquid-like cells within a solid matrix. Precisely the same picture can be constructed for the crystal. In the former case, the solid matrix is amorphous, but in the latter it is crystalline. The interfacial free energy between cluster and matrix is small in the case of the metastable liquid or glass, because the structures of cluster and matrix are closely similar, differing only slightly in volume. However, the structure of the crystal and the cluster are very different, and the interfacial free energy

correspondingly large. Premelting phenomena are therefore suppressed in the crystal, and the melting point T_M is significantly higher than the glass transition temperature T_G.

If ν is the cluster size, then $\nu \bar{v}_f$ is the total free volume within the cluster. Voids of size exceeding v_M are required for diffusion. Such voids form by the redistribution of the free volume. Therefore, no diffusion can occur within a cluster unless ν

8

exceeds $\nu_M = v_M / \bar{v}_f$. We term clusters smaller than ν_M liquid-like clusters and those larger than ν_M liquid clusters or liquid drops. Diffusion occurs only within the latter.

The extra entropy associated with the diffusive motion is the communal entropy. A drop of size $\nu \geqslant \nu_M$ contributes $S_{c,\nu} = \nu\, k_B$ to the communal entropy S_c in a natural generalization of the formulation of Hirschfelder et al.[18] and of Kirkwood[19]. The communal entropy S_c is therefore $S_c = N_M\, k_B$, where N_M is the total number of cells in liquid drops,

$$N_M = N\, p\, a_z(p)\phi(\nu_M). \tag{6.}$$

In (6), $a_z(p)$ is the fraction of liquid-like cells in clusters, and $\phi(\nu_M)$ is the fraction of such cells which are in liquid clusters.

The final form for the free energy contains three contributions

$$F = F_{CELLULAR} + F_{CLUSTER} + F_{COMMUNAL} \tag{7.}$$

which depend only on p, P(v), and C_ν, the cluster size distribution,

$$F = F(p, P(v),\ C_\nu) \tag{8.}$$

F is thus a Landau-Ginsburg-Wilson free-energy functional from which a mean-field theory of the metastable-equilibrium state of a supercooled liquid can be constructed simply by minimizing F with respect to $C_\nu, P(v)$, and p.

The mean-field theory yields the following results:

1. Minimizing F with respect to C_ν yields a C_ν much like that encountered in ordinary percolation theory[25], but with C_ν much reduced for $\nu < \nu_M$. Clustering is random in ordinary percolation processes, without a driving force. In most known cases of non-random clustering, however, the driving forces towards clustering are almost invariably energetic in origin. The present case is unusual because the clustering is entropy driven.

2. Minimizing F with respect to P(v) yields a bimodal distribution of v. One peak corresponds to the solid-like cells, and the second peak to the liquid-like cells. The second peak would be absent were it not for the communal entropy, which greatly enhances the liquid-like portion of P(v).

3. Results 1. and 2. imply that two local configurations are present in the metastable liquid, with a clustering of the liquid-like configuration. This result derives from the flatness of the

configurational free energy on the scale of $k_B T$ in local regions of the configuration space, in this case the space of cell-volume values. It should, however, be completely general for all classes of liquids, including those for which cell-volume coordinates are incorrect or incomplete. The flatness derives from the existence of points of inflection which can be expected to be present for all liquids whatever the appropriate coordinates are.

4. Minimizing F with respect to p yields a first-order phase transition at a temperature T_p at which p falls from above to below the percolation threshold p_c. This first-order phase transition is unobserved, because the system falls out of metastable equilibrium at the glass transition temperature $T_G > T_p$.

5. There is a deep relationship between entropy and movement. Kauzmann[26] pointed out that there is a temperature T_s at which the entropy of the liquid appears, upon extrapolation to low temperatures, to fall below that of the crystal. The empirical formulas of Fulcher[21] and Vogel[22] or of Doolittle[23] for the viscosity appear to predict a divergence of the viscosity at a temperature T_η upon extrapolation from higher temperatures. Angell and coworkers[27] have shown that $T_s = T_\eta$ for many systems for which data are available. While these results only hold upon extrapolation, the implication is that the excess entropy of the liquid relative to the solid disappears when diffusive movement ceases and the liquid becomes rigid.

The free-volume theory described above has that feature built into it. The system becomes rigid when $\bar{v}_f$ vanishes. We[5,6] obtain the expression:

$$p = \frac{\bar{v}_f}{\bar{v}_f + a} \qquad (9.)$$

where a is slowly varying. Thus p vanishes when $\bar{v}_f$ vanishes. However, the extra entropy of the liquid over the solid derives entirely from the presence of the liquid-like cells and goes to zero when p vanishes. Thus, $T_s \simeq T_h$ is an automatic consequence of the theory as a strict equality.

6. The theory also yields an explicit expression[5,6] for $\bar{v}_f$:

$$\bar{v}_f = \frac{v_A}{T_0} \{(T-T_0) + [(T-T_0)^2 + T\,T_1]^{1/2}\} \qquad (10.)$$

When inserted into $\eta = \eta_0 \exp(v_m/\bar{v}_f)$, Eq (10) yields a fit to the experimental values of η to within experimental error for all the cases we have studied, which embrace a wide range of materials[5,6]. It encompasses the Fulcher-Vogel and Doolittle results at relatively higher temperatures as well as data at lower temperature which show departures from the former[5,6].

The theory for η, however, has four adjustable parameters. The fact of the excellent fit to the data is therefore less significant than the fact that we can extract the temperature dependent values of $v_M/\bar{v}_f$ from the data for η and establish values for the most sensitive parameters entering the thermodynamics.

IV. RELAXATION PHENOMENA

We can treat nonequilibrium phenomena such as the glass transition[7] because we now have a description of the metastable equilibrium state towards which the system relaxes. The variables of state are p, P(v), and C_v. P(v) relaxes on the vibrational time scale t_v. C_v is controlled by p. The latter is thus the dominant, slowly relaxing variable. It relaxes by transfer of cellular volume across the cluster-matrix interface by diffusional rearrangement of the cellular structure.

In single relaxation time approximation, the relaxation time has the form:

$$\tau = \tau_0 \, e^{v_M/\bar{v}_f},$$ (11.)

with

$$\tau_0^{-1} = a \left(\frac{S}{V}\right)\sqrt{\frac{k_B T}{M}}.$$ (12.)

In (12.) a is of order unity, (S/V) is the ratio of the total area of the clusters to the total cluster volume, and M is the molecular mass.

When the metastable liquid is cooled and the heat capacity C_p measured, a sharp drop in C_p is observed centered at a point of inflection at a temperature T_G which depends on the heating rate q'. On reheating, a sharp rise is observed at a different temperature T_G' which depends on q', as illustrated in Figure 3. Inverting the relationship between T_G or T_G' and $|q|$, one finds empirically a linear relationship between $\ln|q|$ and T_G or T_G'.

Using the metastable equilibrium theory of section II and Eqs (11.) and (12.), it is easy to simulate the relaxation processes occurring during cooling or heating for any material for which

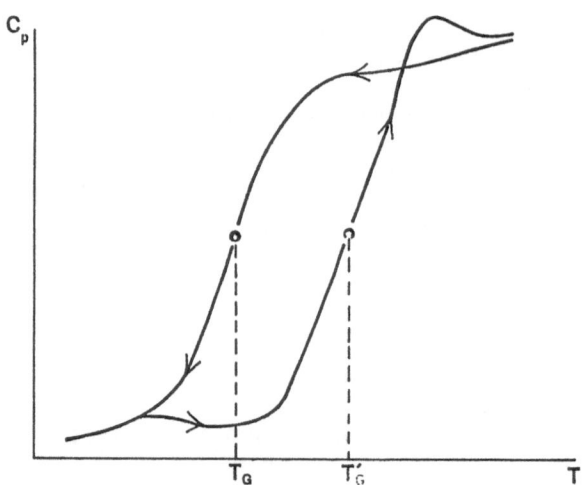

Figure 3.   The glass transition in the specific heat C_p occurs at T_G on cooling but at $T_G'>T_G$ on reheating.

adequate viscosity data are available[7].   We have studied, e.g., the ionic melt (0.6) $KNO_3 \cdot (0.4)Ca(NO_3)_2$ and find a linear relationship between $\ln|q|$ and T_G.   Our theoretical result[7] for the slope is 3.9K compared with the experimental result[23] of 3.8K. Both theory and experiment yield 10^{-10} sec for the magnitude of τ_0.

One can draw three conclusions from this excellent agreement between theory and experiment obtained without adjustable parameters:

1.   All relaxation times important for the glass transition are proportional to the viscosity.

2.   The free energy F is smooth around T_G.   The glass transition is not a smeared phase transition.   The system simply falls out of equilibrium when the relaxation times become too long for the system to follow temperature changes.

3.   The theory gives the correct temperature dependence and magnitudes of the relaxation times.   That the temperature dependence of τ was that of η was known before.   The capability of estimating τ_0 is new.

In writing Eqs. (11.) and (12.), we have supposed that all disjoint clusters and all parts of the infinite cluster relax together with (S/V) in τ_0 of (12.) the mean value for the entire interface.   This is patently incorrect[8].   The different clusters will relax independently, each with a relaxation rate given by:

$$\frac{1}{\tau_\nu} = a \left(\frac{S}{V}\right)_\nu \sqrt{\frac{kT}{M}} \, e^{-\nu_M/\bar{\nu}} f, \tag{13.}$$

when $(S/V)_\nu$ is the surface to volume ratio for a cluster of size ν. We suppose that the infinite cluster can be decomposed into parts of size ν which do not communicate in a relaxation time τ_ν also given by (13.) and that the probability distribution of the sizes of these parts is essentially the same as for the finite clusters.

$(S/V)_\nu$ depends on the cluster sizes as:

$$(S/V)_\nu \propto \nu^{-x}, \, 0 < x < 1/3, \tag{14.}$$

where the lower limit for x pertains to highly ramified clusters in which there is no distinction between surface and interior and the upper limit to the case of compact clusters. Comparing (13.) and (14.), we see that in general there is a dispersion of relaxation rates because of the dispersion in cluster sizes.

The resulting nonexponential relaxation function may be written as:

$$R(t) = \int P(\nu) \, e^{-t/\tau_\nu} \, d\nu \tag{15.}$$

where $P(\nu)$ the probability distribution for the sizes of the clusters and for pieces of the infinite cluster broken at their weak links can, according to our previous study of C_ν, be taken from percolation theory,

$$P(\nu) \propto e^{-A\nu^y}, \, 2/3 < y < 1. \tag{16.}$$

The integral in (15) is easily done either for relatively short or relatively long times. For short times we obtain:

$$R(t) \sim e^{-t/\tau}, \tag{17.}$$

with τ given by (11) and (12). For long times we obtain:

$$R(t) \sim e^{-(t/\tau')^\beta} \tag{18.}$$

where,

$$\beta = \frac{y}{x+y}, \, \frac{2}{3} < \beta < 1. \tag{19.}$$

13

Experimental results[29,30,31] for thermodynamic relaxation show precisely the asymptotic form given by Eq. (18.) with values of β in the range 0.67 to 0.87. Thus, we have succeeded in obtaining the magnitudes, temperature dependences and dispersion of the relaxation times in supercooled liquids and glasses.

If instead of the relaxation of a thermodynamic potential such as the enthalpy as considered just above, one studies dielectric or stress relaxation[31], one finds experimentally Eq. (18.) but with β in the range 0.45 to 0.75. The difference in β is readily understood. Consider dielectric relaxation in an ionic glass. The relevant relaxation time for a cluster of size ν and dimension L_ν is the time required for diffusion across the cluster, $\tau_\nu = L_\nu^2/4D$. Because, $L_\nu \propto \nu^x$ holds, it immediately follows that Eq. (18.) still holds but with a modified exponent,

$$\beta' = \frac{y}{2x+y} \ , \ 1/2 < \beta' < 1 \tag{20.}$$

in excellent agreement with experiment.

There are two physical situations in which the dispersion of relaxation times will not be observed[9]. The first case corresponds to ramified clusters so that x→0 and β and β' →1. In that event, the initial, purely exponential behavior of R(t) given by (17.) persists to very long times up to, say, t_1 with:

$$\frac{t_1}{\tau} \simeq \frac{\pi}{3} \ \frac{\beta^2}{(1-\beta)^2} \tag{21.}$$

The second case is nondispersive for all times. Here the relaxation process does not involve the interface between the liquid drop and the solid matrix. The relaxation time is then given by:

$$\tau = \tau_0 \ e^{v^*/\bar{v}_f}, \tag{22.}$$

where τ_0 is an appropriate preexponential and v^* is the change of volume required for the system to pass through a saddle configuration.

A beautiful example of this case was studied experimentally by Zeller[32]. He measured dielectric relaxation in fully oriented supercooled nematic liquids and glasses and found pure exponential relaxation over many orders of magnitude in R(t), with a relaxation time consistent with the form Eq. (22.). The relaxation process here is reorientation of a molecule by a 180° flip which can take place anywhere in the interior of a liquid drop, in our picture, and need not involve the interface.

To obtain the above excellent results for many aspects of relaxation processes in supercooled liquids and glasses, we have used primarily the notion of the clustering of one out of two local conditions plus percolation theory. The theory is thus very robust. Many details can be changed as the theory is improved or extended to other kinds of materials, but there will be little, if any, effect on the relation between theory and experiment.

IV. THE NATURE OF TUNNELING CENTERS

The specific heats[33] and thermal expansion coefficients[34] have an approximately linear dependence on temperature at very low temperatures[35]. There is anomalous temperature dependence in the thermal conductivity[39] and ultrasonic attenuation[39]. The latter exhibits saturation[39], and both phonon[38] and electric

echoes[39] have been observed. All of these diverse phenomena have been explained by supposing that there are two-level systems present at low temperatures in glasses with a statistical distribution of level separations which is finite at zero separation[40]. The two level systems are interpreted as tunneling centers[41]. The tunneling is multiparticle tunneling of a few atoms or molecules made possible because of the near degeneracy of equivalent structures of the glassy systems[41].

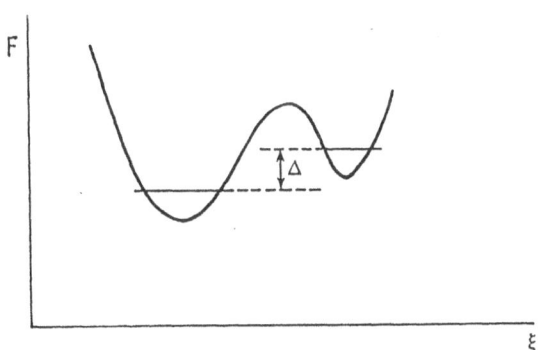

Figure 4. The free energy F as a function of a multiparticle configuration coordinate ξ. Δ is the separation of the vibrational ground states in each minimum before multiparticle tunneling is included.

In Figure 4 the free energy is sketched as a function of a multiparticle configuration coordinate. There are two minima giving rise to two ground vibrational levels separated by Δ. The energy splitting becomes:

$$E = \sqrt{\Delta^2 + \Delta_0{}^2} \qquad\qquad (23.)$$

when the tunneling energy Δ_0 is small and can be neglected when discussing the distribution of values of E, i.e., the density of states N(E). From heat capacity measurements, the density of tunneling levels N(0) is found to be about 10^{-6} per atom per K. Moreover, N(0) is found to be inversely proportional to T_G in two separate sets of experiments. Reynolds[42] inferred it from results for phonon mean free paths on a range of materials. Raychaudhuri and Pohl[43] also demonstrated it by measuring C_p in H_2O doped K-CaNO$_3$ glasses.

The free-volume theory affords interesting insights into the nature of tunneling centers which allow explanations both of the magnitude of N(0)[10] and its proportionality to $1/T_G$[11].

Thus far, we have supposed F to be flat on the scale of $k_B T$ in localized regions of configuration space for $T \gtrsim T_G$. This cannot remain true as T decreases. The curvature in F becomes significant at low temperatures. Minima, maxima and saddle points must emerge in F. With no detailed information about them, we can still be confident that their energy scale is $k_B T_G$.

As cooling proceeds, the liquid goes out of equilibrium at T_G. p and v_f become frozen apart from changes of $\bar{v}_f$ associated with ordinary thermal contraction. The system descends from minima in F to lower accessible minima in F as it cools. More concretely, the clusters freeze from the outside inward. Since $\bar{v}_f$ cannot relax, the free volume in each cluster initially of size $v > v_M$ is concentrated among fewer and fewer atoms. Since an amount of free volume of size v_M is all that is needed for structural rearrangement, the free volume will tend to be concentrated in chunks of size v_M within clusters significantly larger than v_M and within the infinite cluster. When this process is complete, there is one center per $(v_M/\bar{v}_f)_{T_G}$ of the molecules originally in liquid clusters. These centers have free volume v_M distributed over a few molecules and are quite mushy. That is, there are nearly equivalent configurations of the molecules around each center which are separated by saddle points for which the energy scale is $k_B T_G$.

16

Identifying these regions in which the free volume is concentrated as the tunneling centers[17] permits us to estimate $N(0)$. We obtain within a half-order of magnitude a value of 10^{-6} per atom per K as is observed. $N(E)$ is the total number of centers times the probability distribution of values of E. But E is dominated by Δ, and Δ in turn is dominated by the separation between adjacent minima in F in configuration space. Thus $N(0)$ is proportional to the probability distribution of the separation of adjacent minima in F for zero separation. However, there is only one energy scale in the problem, $k_B T_G$, so that the probability distribution and therefore $N(0)$ must be proportional to $(k_B T_G)^{-1}$, the total number of centers being weakly material-dependent, according to our theory[11].

References

1. T. G. Fox and P. J. Flory, J. Appl. Phys. 21; 581 (1950); J. Phys. Chem. 55: 221 (1951).
2. M. L. Williams, R. F. Landel and J. D. Ferry, J. Am. Chem. Soc. 77; 3701 (1955).
3. M. H. Cohen and D. Turnbull, J. Chem. Phys. 31; 1164 (1959); D. Turnbull and M. H. Cohen, ibid 34; 120 (1961).
4. P. B. Macedo and T. Litovitz, J. Chem. Phys. 42; 245 (1965).
5. G. S. Grest and M. H. Cohen, in Advances in Chemical Physics, ed. by I. Prigogine and S. A. Rice, Wiley, New York (1981), Vol. 48, p. 55.
6. M. H. Cohen and G. S. Grest, Phys. Rev. B20; 325 (1979); Phys. Rev. B26; 6313 (1982) Eratta.
7. G. S. Grest and M. H. Cohen, Phys. Rev. B21; 4113 (1980).
8. M. H. Cohen and G. S. Grest, Phys. Rev. B24; 4091 (1981).
9. M. H. Cohen and G. S. Grest, Phys. Rev. B26; 2664 (1982).
10. M. H. Cohen and G. S. Grest, Phys. Rev. Lett. 45; 1271 (1980).
11. M. H. Cohen and G. S. Grest, Solid State Comm. 39; 143 (1981).
12. D. Turnbull and M. H. Cohen, J. Chem. Phys. 29; 1049 (1958).
13. D. Turnbull and M. H. Cohen, in Modern Aspects of the Vitreous State, ed. J. D. Mackenzie, Butterworths, London (1960).
14. M. Tanemura et al., Prog. Theor. Phys. 58; 1079 (1977). See also A. C. Belch, S. A. Rice and M. G. Sceats, Chem. Phys. Lett. 77; 455 (1981); 84; 245 (1981); S. A. Rice, J. Phys. Chem. 85; 1108 (1981).
15. G. Jacucci (unpublished, see refs. 5 and 6).
16. M. Kimura and F. Yonezawa, Topological Disorder in Condensed Matter, eds. F. Yonezawa and T. Ninomiya, Springer-Verlag, Berlin (1983), p. 80; F. Yonezawa and M. Kimura, J. Non-Cryst. Solids 61 & 62; 761 (1984).
17. For a review, see T. L. Hill, Statistical Mechanics, McGraw-Hill, New York (1956).

18. J. O. Hirschfelder, D. P. Stevenson and H. Eyring, J. Chem. Phys. 5; 896 (1937).
19. J. G. Kirkwood, J. Chem. Phys. 18; 380 (1959).
20. D. Turnbull and M. H. Cohen, J. Chem. Phys. 34; 120 (1961).
21. G. S. Fulcher, J. Am. Ceramic Soc. 6; 339 (1925).
22. H. Vogel, Phys. Z., 22; 645 (1921).
23. A. K. Doolittle, J. Appl. Phys. 22; 1471 (1951).
24. M. E. Fisher, Physics 35; 394 (1975).
25. D. Stauffer, Phys. Repts. 54; 1 (1979).
26. W. Kauzmann, Chem. Rev. 43; 219 (1948).
27. C. A. Angell and K. J. Rao, J. Chem. Phys. 57; 470 (1972); C. A. Angell and J. C. Tucker, J. Phys. Chem. 78; 278 (1974).
28. M. A. DeBolt, A. J. Easteal, P. B. Macedo and C. T. Moynihan, J. Am. Ceram. Soc. 59; 16 (1976).
29. C. T. Moynihan et al., Ann. N. Y. Acad. Sci. 279; 15 (1976); C. T. Moynihan and A. V. Lesikan, ibid. 371; 151 (1981).
30. A. J. Kovacs, J. J. Aklonis, J. M. Hutchinson and A. R. Ramos, J. Polym. Sci. Polym. Ed. 17; 1097 (1979); A. J. Kovacs, Ann. N.Y. Acad. Sci. 371; 38 (1981).
31. J. Wong and C. A. Angell, "Glass; Structure by Spectroscopy", Dekker, New York (1976).
32. H. R. Zeller, Phys. Rev. Lett. 48; 334 (1982).
33. R. C. Zeller and R. O. Pohl, Phys. Rev. B4; 2029 (1971).
34. G. K. White, Phys. Rev. Lett. 34; 204 (1975).
35. For a review of the experimental results on glasses at low temperatures, see R. O. Pohl and G. L. Salinger, Ann. N.Y. Acad. Sci. 279; 150 (1976).
36. R. B. Stephens, Phys. Rev. B8; 2896 (1973); and J. C. Lasjaunias, A. Ravex, M. Vandorpe and S. Hunklinger, Solid State Commun. 17; 1045 (1975).
37. S. Hunklinger and W. Arnold, in Physical Acoustics, W. P. Mason and R. N. Thurston, Vol. 12, p. 155, Academic Press, New York (1976); and B. Golding, J. E. Graebner, B. I. Haperin and R. J. Schutz, Phys. Rev. Lett. 30; 223 (1973).
38. B. Golding and J. E. Graebner, Phys. Rev. Lett. 37; 852 (1976).
39. L. Bernald, L. Piche, G. Schumacher, J. Joffrin and J. Graebner, J. Phys. Lett. 39; L-126 (1978); and B. Golding, M. V. Schickfus, S. Hunklinger and K. Dransfeld, Phys. Rev. Lett. 43; 1817 (1979).
40. P. W. Anderson, B. I. Halperin and C. M. Varma, Phil. Mag. 25; 1 (1972); and W. A. Phillips, J. Low Temp. Phys. 7; 351 (1972).
41. J. L. Black, in Metallic Glasses, by H. J. Guntherodt, Springer, New York (1982).
42. C. L. Reynolds, Jr., J. Non-Cryst. Solids 30; 371 (1979); 37; 125 (1980).
43. A. K. Raychaudhuri and R. O. Pohl, Phys. Rev. B25; 1310 (1982).

SESSION B

Invited Paper

SPEAKER

R. Berman
University of Oxford
England

SESSION CHAIRMAN

G. K. White
CSIRO
Lindfield, NSW
Australia

SOLID HELIUM: A REVIEW

R. Berman

Clarendon Laboratory
Oxford University

A former President of the Royal Society has expressed his unease at the fact that authors of scientific papers are encouraged to present their work as a logical development and not to clutter up the literature with personal reminiscences, so that it is not clear what part chance and good fortune have played. Having recently retired, I am happy to admit that my theoretical and experimental defects have been largely patched over by a sequence of pieces of extremely good luck.

It was luck that brought into the conductivity field two theoreticians who have had a strong influence on the subject, and it was very rewarding to have been associated with them at the outset of their careers in the subject. When Paul Klemens arrived in Oxford the nuclear physics project on which he was to have worked had been assigned to another graduate student who had got to Oxford earlier. I had just observed the now well-known plateau in the thermal conductivity of glass and it was suggested that Klemens should investigate this theoretically. A few years later I was measuring the conductivity of diamond and it was obvious to me that length and surface effects in addition to those involved in the familiar boundary scattering regime were involved. I got a little way by using the analogy of gas flowing in a short pipe, the pipe having walls which were not completely rough (the Smoluchowski correction takes care of this for gases), but I could not do the proper calculation. John Ziman had just completed his doctorate in Oxford on a theoretical topic in magnetism and had an award for a further two years in Oxford, so it was suggested to him that he look into the effects of surface roughness and finite length (the latter being a feature of most diamonds available to experimentalists) on heat flow in the boundary scattering regime.

I cannot imagine that I would have made much progress in the subject if chance had not brought Klemens and Ziman into the field while they were in Oxford.

After some further skirmishing with glasses and with irradiation effects on crystals, I started to search for evidence of the exponential rise in conductivity with decreasing temperature predicted by Peierls in 1929 and not observed in a series of measurements under de Haas in Leiden in the 1930's (we now know that they were measuring the wrong crystals at the wrong temperatures). By 1956 the conductivity of nine very pure crystals had been measured at low temperatures by several authors and only two - helium and sapphire (Al_2O_3) - showed a temperature region where the conductivity could be described as varying exponentially with temperature. The similarity in the conductivity of specimens of KCl measured by Glen Slack and by me led us to think that something other than normal imperfections was masking the exponential and that this might be the presence of isotopes in the elements of which the crystal is composed (for vibrational properties, an atom of the wrong mass constitutes a defect). Slack did indeed mention this possibility at a Physical Society meeting and it is referred to in the corresponding Bulletin. A few months later while carrying out my wife's suggestion that I shouldn't waste any more time trying to make sense of our collected experimental results, but dig the garden instead, it occurred to me that a comparison of just the shapes of the conductivity curves (suitably normalized for temperature and magnitude) with the isotopic content of each material would indicate whether isotopes were indeed the cause of the departures from Peierls' predictions. The correlation was so good that I wrote to Ziman, who by then had moved to Cambridge. By the next piece of good luck he was writing his book 'Electrons and Phonons' and had reached the necessary chapter so that he could reply to my letter by return of post with some preliminary estimates confirming the effect (my best birthday present that year).

The next stage was obviously to measure specimens of one material in which the isotopic proportions were varied, and this was done by Geballe and Hull on a crystal grown from enriched germanium. The conductivity was indeed enhanced and the analysis of the results was the first application by Callaway of his method of taking three-phonon normal-(N-) processes into account. We chose to work on two systems: LiF and He, because of the availability of almost completely pure ^6Li, ^7Li, ^3He and ^4He (natural fluorine consists of only ^{19}F). By luck the person in the U.K. Atomic Energy Authority who could supply enriched lithium had been a wartime colleague. We later realized that our choice, made because of availability of material was fortunate in that LiF can be considered for our purposes a 'classical' crystal, while helium at low densities is the best example of a quantum crystal in which the zero point energy is comparable with the binding energy.

By using the Callaway method of analysis for our LiF results we found that the N-process relaxation rate could be pinned down quite closely (our frequency and temperature dependence and even the absolute magnitude was compatible with results of ultrasonic measurements by de Klerk and Klemens) and that the scattering by an isotopic defect was just about what would be calculated for scattering by a wrong mass.

1. ISOTOPE EFFECT IN HELIUM

We made series of measurements of $\lambda(T)$ for various concentrations of ^{3}He in ^{4}He for crystals of different densities grown at easily accessible pressures. Analysis showed that it was essential to choose the N-process relaxation correctly to fit all the $\lambda(T)$ curves for a given density and then the scattering by the isotopic defect was found to be several times greater than the mass-only value, this ratio decreasing with increasing density. This is related to the fact that helium gradually loses its quantum nature as the atoms get closer together; the zero point energy increases much more slowly with decreasing atomic separation than does the binding energy. Our obvious desire to go to still higher densities and so 'squeeze out' completely the quantum effects was tempered by our reluctance to learn proper high-pressure techniques. It was time for another piece of luck and it came in the form of a letter from Professor Swenson (Ames, Iowa) asking whether his graduate student Howard Sample could usefully spend a year in Oxford. This brought the necessary high-pressure expertise into our group and we made measurements on crystals under such pressure that their density was double the minimum helium crystal density. From analysis of the conductivity curves we deduced that the ratio of the scattering by an isotopic defect to the mass-only value had come down to about unity.

The reasonableness of these conclusions was tested in two ways. We did similar measurements on ^{3}He crystals containing small amounts of ^{4}He. The mass-only scattering at a given density is nearly double that for ^{4}He in ^{3}He (scattering proportional to $(\Delta M/M)^2$, i.e. to 1/9 and 1/16 respectively). If the quantum effect is about the same it will be relatively less important for ^{4}He in ^{3}He. This was confirmed very well. Lawson and Fairbank made much more difficult measurements of the effect of very small concentrations (10 and 15 ppm) of ^{3}He in very perfect ^{4}He single crystals. It turns out theoretically that for such small concentrations in very perfect crystals the analysis is very simple and at the density they used they showed that the isotope scattering was double the mass-only value; at the same density our ratio was 2.7 - good agreement in the field of thermal conductivity.

2. ANISOTROPY IN CONDUCTIVITY OF HCP HELIUM

Mezhov-Deglin noticed that the conductivity of very perfect
single crystals grown at the same density on different occasions
could be very different, and he suggested that this reflected an
anisotropy, 'allowed' for hcp crystals, the form for all the crystals discussed here. This suggestion was pursued by Hogan, Guyer
and Fairbank who seeded crystals on gold foil to obtain more preferential orientation of the specimen axis relative to the crystal
axis. Without measuring this orientation they deduced the $\lambda(T)$
dependence for the two principal directions (heat flow parallel
and perpendicular to the c-axis). The anisotropy reached a factor of about 20 and was enormously temperature dependent.

I thought it would be satisfying to measure the orientation
and conductivity simultaneously on each specimen, but was deterred
by the thought of the apparatus required. The next piece of luck
was that Professor Blaisse (Delft, Holland) had a graduate student
Hans Vos who wanted to spend a year in Oxford. As he had been
measuring the birefringence of hcp helium crystals I rapidly re-
plied and we used the technique he brought to measure $\lambda(T)$ curves
over a wide range of known orientations. We confirmed the
suggestions of Hogan et al.

The anisotropy occurs in the frequency of Umklapp-(U-) processes which are responsible for the Peierls exponential of the
form $\exp(\Theta*/T)$, where $k\Theta*$ is an energy related to the dimensions
of the Brillouin zone (and seems for most crystals to be about
$k\Theta_D/4$). The values of $\Theta*$ in different directions in a non-cubic
crystal can be different. Near the conductivity maximum $\Theta*/T$ is
found to be about 10 for pure crystals, so that a difference of
only 30% in $\Theta*$ would lead to the observed 20-(e^3) fold difference
in conductivity. The decreasing anisotropy of the exponential
at higher temperatures and the influence of boundary scattering at
lower temperatures tend to equalize the conductivities (the actual
processes in helium are more complicated than might be supposed
from this simple explanation).

In principle, the same sort of anisotropy should be observable
in other non-cubic crystals, but it has not been clearly seen in
anything except helium.

3. POISEUILLE FLOW OF PHONONS

The analogy is often made between the flow of phonons in a
crystal (heat flow) and the flow of gas molecules in a tube (particle flow). The (generally) low-temperature boundary scattering
regime of heat flow is a clear analogue of Knudsen flow of a gas at
very low pressures - the probability of collisions among the

'particles' is much less than that of collisions of a 'particle' with the walls. The flow under Knudsen conditions is proportional to r^3 (r is the radius) and for heat flow $\dot{Q}$ is also proportional to the specific heat (thus at the appropriate temperatures to T^3) and to the temperature gradient (dT/dx). As we for some reason like to define a conductivity, even when the conditions are quite inappropriate for such a definition, by $\dot{Q}/(\pi r^2 dT/dx)$, the 'conductivity' under Knudsen conditions must be proportional to rT^3, the well-known result.

For a gas at normal pressures the flow rate is determined by the combination of collisions with the tube walls and by collisions of the molecules with one another. We assume that the inter-molecular collisions are elastic (energy and momentum conserved) and that the drift velocity of the gas at the walls is zero, to obtain the Poiseuille equation for gas flow rate $G \propto r^4/\eta$ (where η is the viscosity). The role of intermolecular collisions is only to convey to the bulk of the gas knowledge of the stationary layer at the walls. These collisions would not by themselves introduce any resistance to flow in a tube which had perfectly smooth walls. If our analogy between heat flow and gas particle flow is valid we should observe a similar type of heat flow. This is almost always masked by the existence for phonons of processes in which the 'momentum' (wave-vector) is not conserved (Peierls' Umklapp processes and scattering by defects) even though energy is conserved. These occur regardless of the presence of crystal boundaries and lead to a heat flow $\dot{Q} \propto l_R r^2 \, dT/dx$, (where l_R is the phonon mean free path for processes which do not conserve momentum) from which a true conductivity can be defined, independent of the size or shape of the specimen. It is only in helium that the conditions for Poiseuille flow of phonons can be easily obtained.

The viscosity which enters the Poiseuille relation is proportional to the molecular mean free path, so that $G \propto r^4/l$. The mean free path appears in the denominator because the smaller it is, the shorter the distance over which information about the stationary walls is conveyed. The molecules nearer the centre of the tube are unaware of the walls and can have large drift velocity. The analogue of l for phonons is the mean free path for 'momentum' conserving processes, l_N, and for Poiseuille flow of phonons the frequency of such processes must enormously exceed the frequency both of 'momentum' non-conserving processes (U-processes and scattering by defects) and of collisions of phonons with the crystal walls. The corresponding heat flow is then $\dot{Q} \propto T^3 r^4/l_N \, dT/dx$, with the appropriate mean free path in the denominator. As $l_N \propto T^{-n}$, where n can be up to about 5, the 'conductivity' deduced $\lambda \propto T^m r^2$, with m around 7 or 8. Such a variation can be clearly observed over a very narrow temperature range within which the stringent conditions $l_N \ll r$, $l_N l_R \gg r^2$ hold.

Poiseuille flow has been studied by Mezhov - Deglin, Hogan, Guyer and Fairbank and more recently by Golub et al. [1]. Great care was taken by the last authors about the conditions under which the crystals were grown and they find that the 'conductivity' is indeed proportional to about T^7. The observation of Poiseuille flow provides a means of deriving values for l_N, and the temperature variation and absolute magnitude given by the last authors is remarkably close to the values we deduced from analysing the variation of conductivity with isotopic concentration.

4. HELIUM-NEON MIXED CRYSTALS

The last topic I want to discuss also owed its origin to a timely piece of luck. I had been wondering whether it would be possible to incorporate some impurity other than an isotope in a helium crystal. If one assumes that the concentration of another element (Ne or H_2, say) in the high-pressure fluid helium from which the crystal is to be grown is determined by its normal vapour pressure (measured in the standard way with the condensed phase only in contact with its own vapour), it would be necessary to grow crystals at a high starting pressure of helium in order that the corresponding freezing temperature would be high enough for the vapour pressure of Ne/H_2 to represent a high enough concentration in the mixture for an appreciable effect on the conductivity to be observable. The luck arose in the setting of a question in the 1970 Oxford Physics Final Examination papers. In one question, candidates were asked to show that the vapour pressure of a condensed phase is enhanced by the application of pressure. The actual process is more complicated than the students were expected to assume, but the effect is real and the question did draw my attention to such a phenomenon. Taking, for example, numbers applicable to our experiments: at 14.3 K the freezing pressure of helium is 10^8 Pa; the 'bare' vapour pressure of neon is 20 Pa, so that if the enhancement effect did not occur, the maximum concentration of neon in fluid helium at the freezing pressure for 14.3 K would be 0.2 ppm. However, the enhancement effect we determined [2] for the fluid under these conditions is ~ 5000, so that the partial pressure of neon is 10^5 Pa, giving a concentration of 1000 ppm.

We grew a series of crystals at several densities with different starting concentrations of neon and the main observations were:

1) As the starting neon concentration increased, the conductivity decreased, but a limiting conductivity was reached which we took to correspond to the limiting concentration of neon which could be in a helium crystal of a given density [3]. From these results some information can be derived about the He-Ne phase diagram at low neon concentrations [4].

2) There is a pronounced dip in the $\lambda(T)$ curves which we as-
cribe to scattering by the resonant mode formed when a heavy atom
is dissolved in a crystal composed of a light element. By a
Callaway analysis, the resonant frequency could be deduced and, just
as for an isotopic defect, the transition from a quantum to a classical value of this frequency could be followed. The same effect
was observed for Ne in ^{3}He with the appropriate change in numerical
values consequent to changing M from 4 to 3 and ΔM from 16 to 17
[3].

5. SUMMARY

From the experiments described, information has been derived of
interest both from the point of view of understanding phonon conduction and for understanding the behaviour of helium crystals:

1) Together with the work on LiF, we learned how to analyse
thermal conductivity results and showed the importance of normal
processes in determining the conductivity of not too imperfect
crystals. For both LiF and He the relaxation rates for N-processes
which we deduced are compatible with those deduced from rather
different experiments.

2) We showed quantitatively the effect of isotopes on conductivity and explained why it is difficult to observe the exponential
rise in conductivity with decreasing temperature, predicted by
Peierls.

3) We observed the progression from quantum-dominated properties of low-density helium to classical behaviour at high densities,
illustrated by the phonon scattering by an isotopic defect and by
the introduction of a heavy atom impurity.

4) We confirmed the existence of a new type of conductivity
anisotropy related to umklapp processes, which is enormously
temperature-dependent and is not observed in any other type of
crystal.

5) Although we have observed it, we have not done much work on
Poiseuille flow of phonons, which is a true analogue of Poiseuille
flow of a gas. As mentioned in 1) above, this regime enables the
relaxation rate for N-processes to be deduced.

6) We have shown that mixed He-Ne crystals can be grown; we
derived some information on the phase diagram and on the resonant
mode produced by insertion of the heavy atom, and observed the dependence of frequency on density. As mentioned in 3) above, this
gives further evidence for the transition of helium from a 'classical' to a 'quantum' crystal with increasing density.

References

References, except those given below can be found in 'Thermal Conduction in Solids' by the present author, Oxford University Press (1976).

1   Golub, A.A. and Svyatko, S.V., "Thermal Conductivity of Low-Density Solid ^{4}He", Fiz. Nizk. Temp. $\underline{6}$, 957-967 (1980) [Sov. J. Low Temp. Phys. $\underline{6}$, 465-470 (1980)]; Golub, A.A., Zuev, N.V. and Mikhaĭlov, G.A., "The Thermal Conductivity of Single Crystals of HCP ^{4}He Grown from Superfluid", Fiz. Nizk. Temp. $\underline{9}$, 453-459 (1983) [Sov. J. Low Temp. Phys. $\underline{9}$, 229-232 (1983)].

2   Berman, R., Chaves, F.A.B., Livesley, D.M. and Swartz, C.D. "The Solubility of Solid H_2 and Ne in High-Pressure He", J. Phys. C: Solid State Phys. $\underline{12}$, L777-780 (1979).

3   Berman, R. and Livesley D.M., "Thermal Conductivity of ^{3}He and ^{4}He Crystals Containing Neon Impurities", J. Phys. C: Solid State Phys. $\underline{14}$, L945-949 (1981).

4   Livesley, D.M. "Phonon Scattering and Phase Equilibrium in ^{3}He and ^{4}He Crystals Containing Neon Impurities", J. Phys. C: Solid State Phys. $\underline{16}$, 2881-2888 (1983); "Phase Separation in Freezing ^{3}He-Ne and ^{4}He-Ne Mixtures", ibid. $\underline{16}$ 2889-2895 (1983).

SESSION C

Theory and Modeling

SESSION CHAIRMAN

M. J. Laubitz
NRC
Ottawa, Canada

A PROBLEM OF THE HEAT CONDUCTION EQUATION WITH A MOVING BOUNDARY IN THREE-DIMENSIONAL SPACE, E^3

Lian-ke Dong and Ben-lian Zhou

Institute of Metal Research
Academia Sinica
Shenyang, China

ABSTRACT

It is well known that a parabolic partial differential equation for the case of a moving boundary can describe some moving-boundary problems in different fields of science and technology. In this paper, we give an analytical method to obtain the analytical solution of the heat conduction equation for a system with a boundary moving in three-dimensional space:

$$\frac{\partial T(p;t)}{\partial t} = \alpha \Delta_3\, T(p;t), \quad p \in \dot{D}, \quad t > 0$$

(I) $\qquad T(p;0) = f(p), \quad p \in \dot{D}$

$$T[R(t),\, \theta,\, z;\, t] = T_c,$$

where $T(p;t)$ is the temperature function, $\dot{D} \subset E^3$ is an open domain and $R(t)$ is a time dependent function describing the moving boundary and having a continuous first derivative.

In this paper, we suggest a compressing transformation with respect to space-coordinates, by use of which problem (I) transforms to a homogeneous parabolic-type equation with variable coefficients for a system with fixed boundaries and a moving heat source; hence the existence and uniqueness of the solution of problem (I) will be proved. By a further transformation for the homogeneous parabolic-type equation with variable coefficients, an analytical solution is given.

31

ON THE EFFECTIVE THERMAL DIFFUSIVITY OF

MACROSCOPIC HETEROGENEOUS TWO PHASE MATERIALS

B. Schulz

Kernforschungszentrum Karlsruhe
Institut für Material - und Festkörperforschung
Postfach 3640, D-7500 Karlsruhe

ABSTRACT

The idea of an effective diffusivity of a composite normally requires the assumption that the composite is quasihomogeneous, i.e.
the size of the individual particles of the phases is small compared
with the size of sample. Using the laser flash technique it is investigated to what extent this assumption has to be fulfilled in
defining an effective thermal diffusivity for heterogeneous materials.
Samples of the following geometry were examined: disc-shaped pores in
metals, copper spheres in Plexiglass, Plexiglass discs in metals and
molten materials in non-transparent capsules. The results are interpreted on the background of the model of the effective diffusivity,
where the encapsulated phase is characterized by a shape factor and
a definite orientation with respect to the heat flow in the sample.

INTRODUCTION

If "effective" properties of composites are discussed, it is
necessary first to define an "effective" medium and the special features of an "effective" property. This paper follows completely the
definition which was given in ref.1, i.e., media which are heterogeneous
in a microscale are preferred for a description with effective parameters and a treatment as "effective" medium for macro-properties.
With this definition it is assumed that the effective medium behaves
like a homogeneous one as long as macroproperties are discussed and
it is considered as a quasihomogeneous material. For such quasihomogeneous composites the effective thermal conductivity λ_e is well
defined and related to the effective diffusivity α_e in the following
way:

$$\alpha_e = \lambda_e/c_e \cdot \rho_e \qquad (1)$$

33

c_e and ρ_e are the effective heat capacities and densities given by the mixing rule applied to the pure phases, which build up the composite.

The assumption of quasihomogeneity is normally fulfilled, if the number of the individual particles of the phases is large and their size small. The fact, that it is difficult to point out what in this context a large number really is, makes it also difficult to draw a sharp line between quasihomogeneous and heterogeneous composites.

To test the range of applicability of the model of the "effective" diffusivity, specimens are examined with the laser flash technique which are doubtless heterogeneous in the above-mentioned sense.

The effective thermal conductivity of composites with matrix structure

The matrix type of structure of composites is characterized as a structure in which individual particles of one phase are inclusions - dispergents - in the matrix of the second phase (example: closed pores in a material).

A general equation based on the work of Niesel[6], Bruggemann[5] has been derived for the conductivity of such composites[2]:

$$1-C_D = (\lambda_M/\lambda)^m \frac{\lambda_D-\lambda}{\lambda_D-\lambda_M} \cdot \left(\frac{\lambda+n\lambda_D}{\lambda_M+n\lambda_D}\right)^q \qquad \lambda \equiv \lambda_e \qquad (2)$$

$$m = \frac{F(1-2F)}{1-\cos^2\alpha(1-F)-2F(1-\cos^2\alpha)} \qquad n = \frac{1-\cos^2\alpha(1-F)-2F(1-\cos^2\alpha)}{2F(1-\cos^2\alpha)+\cos^2\alpha(1-F)}$$

$$q = \frac{F(1-2F)}{1-\cos^2\alpha(1-F)-2F(1-\cos^2\alpha)} + \frac{(1-F)2F}{2F(1-\cos^2\alpha)+\cos^2\alpha(1-F)} - 1$$

Here λ_M, λ_D are the conductivities of the matrix and the dispersed phase, respectively; and c_D is the volume concentration of the dispersed phase, while F is the shape factor and $\cos^2\alpha$ the orientation factor of the dispersed particles. F is the well known de-electrification factor. It represents the perturbations in an originally homogeneous field, which are caused by the introduction of particles which can be approached as spheroids. F is a function of the ratio of the main axis to the subordinate axis of the spheroids z/x, if the orientation of the main axis, z, relative to the direction of the field is constant. The axis ratio is >1 for prolate spheroids, <1 for oblate spheroids, and unity for spheres (Fig.1). Any deviation from the spherical shape requires the consideration of the orientation of the particles with respect to the direction of the field. The orientation parameter is defined by the square of the cosine of the angles, α, between the axis of the ellipsoid, a,b, c, and the direction of the original field: $\cos^2\alpha_a$, $\cos^2\alpha_b$, $\cos^2\alpha_c$, with $\Sigma\cos^2\alpha = 1$. For b = c (that is a spheroid):

$$\cos^2\alpha_b + \cos^2\alpha_c = 1-\cos^2\alpha_a = 1-\cos^2\alpha$$

where α is the angle between the axis of revolution and the original direction of the field.

34

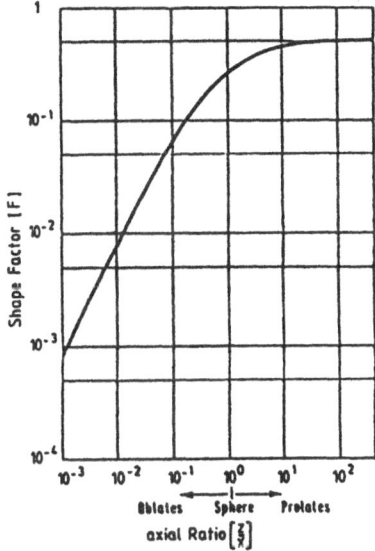

Figure 1　Shape factor F as a function of the axial ratio z/x of the spheroid, curve after /17/.

Table 1　Special cases of the general equasion 2 for the effective thermal conductivity.

Inclusion	λ_D	F	$\cos^2\alpha$	equation	
pores, bacelit H_2O, ethylenglycol in metal	$\to 0$	disc shape	1	2a	$\lambda_e/\lambda_M = (1-c_D)^{1/2F}$
	$\to 0$	sphere 1/3	1/3	2b	$\lambda_e/\lambda_M = (1-c_D)^{1.5}$
Cu disc in Epoxy	$\gg \lambda_M$	disc shape	1	2c	$\lambda_e/\lambda_M = (1-c_D)^3$
Cu sphere in Epoxy	$\gg \lambda_M$	sphere 1/3	1/3	2d	$\lambda_e/\lambda_M = (1-c_D)^{-1/1-2F}$
molten metal in stainless steel	comparabel to λ_M	disc-shape	1	2e	$(1-c_D) = (\lambda_M/\lambda_e)^{1-2F} \cdot \dfrac{\lambda_D - \lambda_e}{\lambda_D - \lambda_M}$

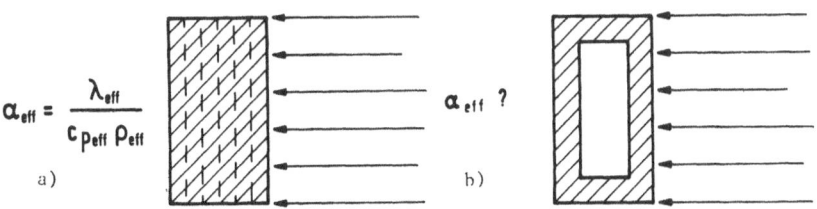

$$\alpha_{eff} = \frac{\lambda_{eff}}{c_{p_{eff}} \rho_{eff}}$$

α_{eff} ?

a)　　　　　　　　　　　　b)

Figure 2　Example for a heterogeneous material (b) being a unit cell of a quasihomogeneous material (a).

A proof of the validity of equation (2) has already been given for the thermal conductivity[2] and the electric conductivity[7]. It was applied to very extreme structures like powder aggregates and thermal insulators[8] and was extended to the viscosity of suspensions and high temperature creep of composites[9].

In contrast of composites, in the following we consider only one inclusion being surrounded by another material. In applying eq. (2) for special cases of the parameters λ_D/λ_M, F and $\cos^2\alpha$ of the incapsulated inclusions, one has to deal with the equations given in table 1.

Formal application of the effective medium concept to the diffusivity of heterogeneous samples

The measurement of the diffusivity with the laser flash technique is based on a temperature-time-function of the following way:

$$V = 1 + 2 \cdot \sum_{n=1}^{\infty} (-1)^n \exp\left(-\frac{n\pi^2\alpha t}{L^2}\right) \quad \text{(ref.10)} \tag{3}$$

This is valid for a cylindrical sample being heated entirely, uniformly and instantaneously over the front face of the sample to a depth g which is small compared to the thickness L of the sample; t is the time variable, V is the temperature at the rear face of the sample, normalized to the maximum temperature. No radiative heat losses and no finite pulse-time-effect have to be considered. Then

$$\alpha = 0,1388 \frac{L^2}{t_{1/2}} . \tag{4}$$

$t_{1/2}$ is the time at which the temperature reaches half its maximum. As pointed out in ref.10, knowing $t_{1/2}$ and thus an "ideal" diffusivity it is possible to calculate a theoretical time-temperature-function according to equation (7) and compare it with the experimental one.

If one now considers a heterogeneous sample as given in fig. 2b and compares it with a quasihomogeneous one (fig. 2a), both being heated by a flash over the whole face of the sample as indicated, one can state.

Seen only from the point of heating and absorbing the heat in the above-defined small depth g, the heterogeneous sample appears as nothing more than a unit cell of the quasihomogeneous composite. In this context the definition of an effective diffusivity for heterogeneous samples should be allowed. The comparison between the theoretical and experimental temperature rise should give information, whether one can derive an effective diffusivity α_e or not, assuming no heat losses and no finite pulse-time-effect are present. As an example a comparison between a heterogeneous sample (capsule no. 3 in table 3) and a homogeneous sample (SS 1.4970) is considered.

The variation of the data of various percentage temperature rise

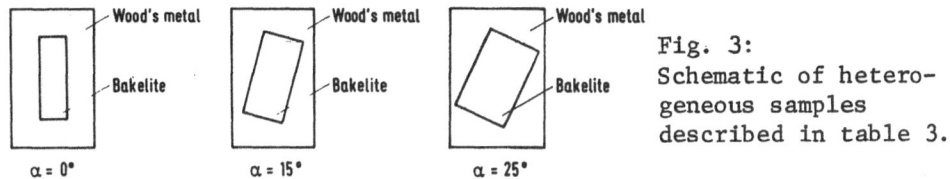

Fig. 3:
Schematic of hetero-
geneous samples
described in table 3.

Table 2 Data of temperature rise in homogeneous SS 1.4970 and SS 1.4970 capsule no. 3 with one disc shaped pore of 62% of the bulk volume. (Data for room temperature).

Temperature-rize (%)	α-SS 1.4970 (cm²/sec)	α-SS 1.4970 capsule no.3 (cm²/sec)
10	0.0379	0.003635
20	0.0395	0.003599
25	0.0403	0.003590
30	0.0408	0.003591
33.3	0.0409	0.003590
40	0.0401	0.003592
50	0.0423	0.003582
60	0.0422	0.003614
66	0.0418	0.003626
70	0.0419	0.003648
75	0.0436	0.003605
80	0.0440	0.003739
90	0.0445	0.003895

Table 3 Geometric and stereological data of samples . ρ_e – effective density of sample.

Matrix Inclusions	c_D (Vol.%)	ρ_e (g/cm³)	F	α	$\cos^2\alpha$	Exponent in Eq 2f
High-conducting Wood-metal –Bakelit	8.7 / 11.0 / 12.3	8.86 / 8.64 / 8.53	0.333 / 0.164 / 0.0857	25 / 15 / 0	0.821 / 0.933 / 1	1.51 / 2.92 / 5.83

	c_D (Vol.%)	c_D^{t} (Gew.%)	ρ_e (g/cm³)	F	$\cos^2\alpha$	Exponent in Eq 2b
low conducting Epoxy 1 coppersphere	7.8 / 13.9 / 23.3	39.2 / 55.2 / 69.8	1.78 / 2.25 / 2.98	0.333 / 0.333 / 0.333	0.333 / 0.333 / 0.333	– 3.0 / – 3.0 / – 3.0

						Exponent in Eq 2g
low-conducting Epoxy 1 copperdisc	19.4 / 14.1 / 9.8	67.0 / 59.0 / 45.0	2.67 / 2.19 / 1.93	0.0857 / 0.164 / 0.33	1.00 / 0.933 / 0.821	– 1.21 / – 1.80 / – 2.41

for this capsule at room temperature is given in table 2 and compared to that of a homogeneous one-phase sample (SS 1.4970). No principle difference in the variation of the individual α-values with % temperature rise can be detected between the two samples.

Another point which has to be discussed is the history of the temperature size at the rear face of the sample. The concept of quasi homogeneity requires an undisturbed temperature distribution in the sample seen from a point outside. The above mentioned pertubations of the originally homogeneous distribution of the isotherms introduced by the inclusion will diminish more and more if the distance from the inclusion increases ($\equiv$ increase in thickness of the sample), tending to zero at infinity. If one calculates the ratio of the external field strength E_a to the undisturbed field E_O one gets in the case of a non conducting sphere of radius a in a conducting matrix $E_a/E_O = 0.875$ at $r = 2a$ and 0.946 at $r = 2.5a$. For the opposite case of a conducting sphere in a nonconducting matrix we obtain $E_O/E_a = 1.25$ at $r = 2a$ and 1.13 at $r = 2.5a$. It is worthwhile to note, that the influence of the inclusion on the field in a nonconducting matrix is doubled compared to the opposite case. This restricts the application of the model of the effective diffusivity to lower sizes of the inclusion for the case, where the ratio of matrix to inclusion conduction λ_M/λ_D is <1. On the other hand the ratios E_a/E_O given above are calculated for the extreme cases of $\lambda_M/\lambda_D \to 0,\infty$.

Preparation of samples

The first group of samples is characterized in table 3. Low and high conducting discs as given in fig. 3 are surrounded by a matrix of wood's metal or epoxy respectively. The variation of $\cos^2\alpha$ is expressed for the effective conductivity by the following equations:

$$\frac{\lambda_e}{\lambda_M} = (1-c_D)^{\frac{1-\cos^2\alpha}{1-F} + \frac{\cos^2\alpha}{2F}} \quad (2f) \qquad \frac{\lambda_e}{\lambda_M} = (1-c_D)^{\frac{\cos^2\alpha-1}{F} - \frac{\cos^2\alpha}{1-2F}} \quad (2g)$$

$$\lambda_D/\lambda_M \ll 1 \qquad\qquad\qquad \lambda_D/\lambda_M \gg 1$$

Furthermore 1 copper sphere of 3 different radii, is prepared in a matrix of epoxy. A second group of samples is described in table 4,5. Here all inclusions have the shape of a disc, where the axis of revolution is oriented parallel to the heat flow. Capsule 6 Tab.4 was filled with H_2O and ethylene glycol.

Experimental technique

For all samples the effective diffusivity was measured with the laser flash technique as described in ref. 10. The temperature sensors were thermocouples (Ni-NiCr) pointed-welded to the metal surface and a special infrared sensor developed at the Nuclear Research Center, Karlsruhe[11]. The system was calibrated for the sensor with POCO

Table 4 Geometric data and stereometric factors of matal capsules used in the laser flash apparatus (all capsules electron beam welded). ($\phi_{a,i}$ - outer and inner diameter of the capsule; $H_{a,i}$ - outer and inner height of the capsule; $\rho_{e,o}$ effective and true density of the material).

Capsule Material No.	Inclusion Material	Capsule Geometry				$\frac{\rho_e}{\rho_o}$	Stereomatric Factor				Exponent in Eq 2c
		ϕ_a	H_a	ϕ_i	H_i		c_D	$z/x = \frac{4\,H_i}{\pi\,\phi_i}$	F	$\cos^2\alpha$	
				(cm)							
1) W	Vacuum He 1 bar	1.26_7	0.315	1.16_7	0.215	42	0.58	0.2354	0.133	1	2.69
2) Pt	Vacuum	1.27	0.535	1.17	0.415	31.6	0.684	0.452	0.2217	1	3.61
3) SS 1.4970	Vacuum	1.36_7	0.315	1.26_7	0.230	38	0.62	0.230	0.1385	1	3.61
4) SS 1.4970	Vacuum	1.36_7	0.322	1.26_7	0.121	67	0.33	0.122	0.0814	1	6.14
5) SS 1.4970	Vacuum	1.36_7	0.315	1.30_7	0.0503	84.5	0.146	0.0490	0.0361_8	1	13.82
6) SS 1.4970	Vacuum	1.36_8	0.365	1.26_8	0.190	48	0.52	0.191	0.1201	1	4.16

Table 5 Geometrical data for capsules containing liquids (capsules were laser beam welded).

Capsule Material	Inclusion Material	Capsule Geometry				c_D	F	$\cos^2\alpha$	Exponent in Eq 2b
		ϕ_a	H_a	ϕ_i	H_i				
SS 1.4970	Pb,Cd,Hg- molten	1.365	0.315	1.26	0.23	0.62	0.1393	1	3.59

Table 6a: Comparison of measured diffusivities of liquids and molten metals α_{De} with literature data. $\Delta = \dfrac{|\alpha_{ec} - \alpha_{em}|}{\alpha_{ec} + \alpha_{em}}$

Sample		α_{DLit} (cm^2/sec)	α_{De} (cm^2/sec)	Δ (%)
H_2O	300 K	$1.46 \cdot 10^{-3}$	$9.88 \cdot 10^{-4}$	18
ethylene glycol	300 K	$9.42 \cdot 10^{-4}$	$9.40 \cdot 10^{-4}$	1
Pb molten	625 K	0.0987	0.110	6
Cd molten	625 K	0.209	0.161	13
Hg molten	300 K	0.0448	0.0390	9
wood's metal molten	355 K	0.129	0.0967	6

graphite and shows good agreement with data summarized in ref. 12 over the whole temperature range from RT to 2000 K. The deviation between the results of POCO graphite with the sensor mentioned above and those from ref. 12 was less than 5%.

Comparison between measured and calculated effective diffusivities (α_e)

If one allows for the measured α_{em} a standard deviation ±5% and for the calculated α_{ec} ±7.5%, which arises from inaccuracies in the determination of the stereological factors, the value $|\alpha_{ec} - \alpha_{em}|/\alpha_{ec} + \alpha_{em}$ may be around 12.5% to support the model of the effective diffusivity for heterogeneous samples.

The calculation of the effective diffusivity was done as follows. The conductivity of the pure phases is taken from literature for Cd, Pb, Hg[13], H_2O, Ethylene glycol, He[15]. The conductivity of W at room temperature (1.25 W.cmK), Pt (0.71 W/cmK), Cu (3.93 W/cmK) SS 1.4970 (0.151 W/cmK at 300 K, 0.184 W/cmK at 355 K, 0.207 W/cmK at 625 K) and epoxy (0.00171 W/cmK) were determined in using own measurements of α of the pure phase. With these data and the stereological factors given in table 3-5, the effective conductivity λ_e for the various samples was calculated using the equations given in the tables. With the effective specific heat capacities and densities α_{ec} was calculated (eq. 1). Specific heat capacities of W, Pt, and Cu were taken from ref. 15; for Wood's metal from ref. 14; for Cd, Pb, and Hg from ref. 13; for H_2O, ethylene glycol and He from ref. 15; and epoxy from ref. 16. Tables 6a and 6 give the comparisons.

Those samples which are capsules filled with liquids or molten metals are of interest because these experiments could involve the possibility of a measurement of the thermal conductivity of molten substances with the laser flash technique. Therefore in this special case, from the measured effective diffusivity, the effective conductivity is calculated for the capsules with the liquid, and then the phase conductivity λ_D is calculated from what follows directly the phase diffusivity α_D of the liquid. In table 6a the diffusivity of liquids as known from literature α_{DLit} and those determined with the model of the effective diffusivity α_{De} are compared

Discussion and Summary

This paper should be considered as a first step to test the model of the effective diffusivity for heterogeneous materials. The comparison between calculated and measured values of the diffusivity encourages further work in this field. It is of interest that the results on the samples with copper spheres in EPOXY are outside of the allowed standard deviation. This is in agreement with the theory, if the sample thickness is not large enough to avoid systematic errors, due to perturbations arising from the embedded sphere. The experiments with liquids and molten metals are part of a test of a method to determine thermal conductivities of molten materials at very high temperatures

40

Table 6: Comparison between calculated and measured effective diffusivities for various heterogeneous samples. $\Delta = |\alpha_{ec} - \alpha_{em}|/\alpha_{ec} + \alpha_{em}$; T = 300 K

Sample	α_{ec} cm^2/sec	α_{em} cm^2/sec	Δ (%)
1 Cu disc in Epoxy			
$\alpha = 0^o$	$2.08 \cdot 10^{-3}$	$2.18 \cdot 10^{-3}$	2.5
$\alpha = 15^o$	2.70	3.22	9
$\alpha = 25^o$	2.78	3.31	9
1 Cu sphere Epoxy			
$c_D = 7.8$ Vol.%	$2.61 \cdot 10^{-3}$	$3.48 \cdot 10^{-3}$	14.4
$c_D = 13.9$ Vol.%	3.13	5.13	24
$c_D = 23.3$ Vol.%	3.57	($\sim 10^{-2}$)	Nonuniform heating
1 Bakelite disc in Wood's metal			
$\alpha = 0^o$	0.0504	0.0420	9
$\alpha = 15^o$	0.0740	0.0623	9
$\alpha = 25^o$	0.0970	0.0885	5
W Capsule			
$c_D = 58$ Vol.% Vacuum	$5.93 \cdot 10^{-2}$	$5.64 \cdot 10^{-2}$	3
$c_D = 58$ Vol.% Helium	5.93	5.71	2
Pt Capsule			
$c_D = 68.4$ Vol.% Vac.	$5.65 \cdot 10^{-2}$	$5.00 \cdot 10^{-2}$	6
SS Capsule, Vacuum			
$c_D = 62$ Vol.%	$3.28 \cdot 10^{-3}$	$3.72 \cdot 10^{-3}$	6.5
$c_D = 33$ Vol.%	5.23	4.62	6
$c_D = 14.6$ Vol.%	5.42	5.05	4
$c_D = 52$ Vol.%	5.18	4.18	11

with the laser flash technique under conditions where the well known more accurate methods will not work. The results of these experiments are not uniform. Obviously there is an influence due to the heat transfer between molten material and capsule, if their conductivities are comparable. On the other hand at very high temperatures the standard deviation of conductivity measurements will be much higher than 5% (50% for molten uranium dioxide[18]). This means, that under circumstances where the molten material needs a thermodynamically closed system for measurements of its thermal conductivity the laser flash technique together with the model of the effective diffusivity may be used.

ACKNOWLEDGMENT

The author wishes to thank Mr. Uwe Jauch for his interest and help in performing the experiments.

REFERENCES

1. E. Kröner, B. Schulz; Workshop on Composites 8 ETCP Sept. 1982 Baden-Baden High-Temperatures High Pressures, in print.
2. B. Schulz; Report Kernforschungszentrum Karlsruhe 1974 KfK 1988.
3. G. Ondracek; Report Kernforschungszentrum Karlsruhe 1978 KfK 2688.
4. C.J. Maxwell; A Treatise on Electricity and Magnetism, Oxford Clarenden Press (1904) Vol. 1.
5. D.A.G. Bruggemann; Ann. Phys. 24 (1935) 636.
6. W. Niesel; Ann. Phys. 10 (1952) 336.
7. G. Ondracek; Zeitschrift für Werkstofftechnik 8 (1974) 416.
8. B. Schulz; High-Temp. High Press. 13 (1981) 649.
9. W.D. Sältzer; Zur Theorie und Messung von Eigenschafts-kenngrössen zweiphasiger Werkstoffe im plastisch-viskosen Verformungsbereich. Thesis University Karlsruhe 1983 (KfK 3581) und W.D. Sältzer, B. Schulz; Proc. 4th RISØ; Symposium on Metallurgy and Materials Science 1983; Deformation on Multi-phase and particle containing Materials, p. 511-524 (in English).
10. R.E. Taylor; High Temp. High Press. 11 (1979) 43.
11. W. Rapp; Report Kernforschungszentrum Karlsruhe, (1983) KfK 3616.
12. R.E. Taylor; High Temp. High Press. 12 (1980) 147.
13. C.H. Smitheless; Metals Reference book Vol. III 4th Edition London, Butterworth 1967.
14. Gnelins Handbuch der anorganischen Chemie 8. Auflage Band Blei Teil 4C Verbindungen Syst. Nr. 47 Band 47 Verlag Chemie GmbH Weinheim/Bergstrasse - 1971.
15. D'Ans'Lax (Ed.); Taschenbuch für Chemiker und Physider Vol. 1 Springer Verlag Berlin 1967.
16. Ullmanns Enzyklopädie der technischen Chemie 3. Auflage Band 11 Urban and Schwarzenberg München-Berlin 1960.
17. U. Stille; Archiv Elektrotechnik 38 (1944) 91.
18. H.A. Tasman, D. Pel, J. Richter, H.E. Schmidt; 8th European Conf. on Thermophys. Properties, Baden-Baden 27.09.-01.10.1982, in High Temperature - High Pressure, in print.

THE USE OF NUMERICAL HEAT TRANSFER TECHNIQUES TO ANALYZE THERMAL

COMPARATOR CONDUCTIVITY MEASUREMENTS*

James N. Sweet,
Marvin Moss and
Carlton E. Sisson

Sandia National Laboratories
P. O. Box 5800
Albuquerque, NM   87185

ABSTRACT

In the comparative conductivity measurement technique, a uniform axial heat flow is assumed to exist in a stack composed of two reference disks with a sample disk sandwiched between them. The sample conductivity is found from measured temperature drops across and thicknesses of the stack elements. In practice, the heat flow in the stack is non-uniform and corrections must be made for this effect. We have made a detailed numerical heat transfer analysis of the commercially available Dynatech TCFCM comparator with the aid of a finite difference numerical heat transfer code (SINDA). The goal of this study has been to determine the effect of non-uniform axial heat flux on the measured conductivity and to define the magnitude of the errors likely to be observed in various experimental situations. Correction factors are given for correcting the measured conductivities of samples with true conductivities in the range ≈ 0.05 - 1 W/m-K when measured against Pyrex or Pyroceram. Estimated errors are given for measurements with various combinations of Pyrex and Pyroceram references and samples. Results are also given to show the utility of numerical calculations in situations where the sample diameter is significantly less than the reference diameter. Various approximate methods of correcting for nonuniform heat flow are compared to the numerical predictions for selected cases.

*This work performed at Sandia National Laboratories supported by the U.S. Department of Energy under Contract Number DE-AC04-76DP00789.

INTRODUCTION

Experimental determination of the thermal conductivity of a solid test specimen with a slab geometry is frequently performed by measuring the thermal gradient across the sample produced by a known axial heat flux. In the comparative or cut-bar method, the sample is sandwiched between references of known conductivity, as shown schematically in Fig. 1. These references act as heat flux gages. By measuring the temperature gradient across each reference, the flux may be calculated using the known reference conductivity. The selection of an experimental geometry and choice of reference materials are made so to as insure that the same axial heat flux passes through all the elements in the stack. In actual situations, the axial flux is not uniform throughout the stack and a correction must be applied to the measured conductivity value. The magnitude of this correction and its dependence on experimental parameters is the subject of this paper.

The comparative technique is widely used for measurement of conductivities of solids in the $\approx 0.5 - 50$ W/m-K range because it is fast and reasonably accurate. A commercial instrument is available, the Dynatech model TCFCM,[1] and suitable reference materials

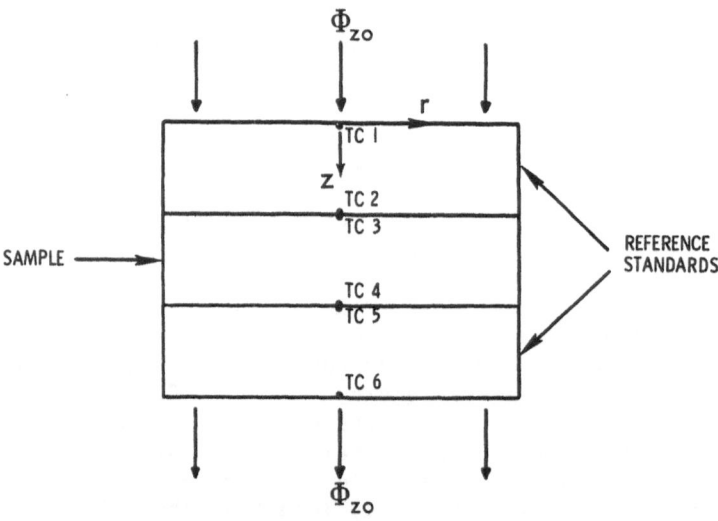

Fig. 1. Schematic of the comparative method of thermal conductivity measurement. A uniform axial heat flux Φ_{zo} flows through the stack consisting of two identical references and a sample, all with the same circular cross-section. Temperatures are measured with centerline thermocouples TC1-TC6.

are available in the 0.5 - 50 W/m-K conductivity range. The conductivity vs. temperature relations for these reference materials have been published by the National Bureau of Standards.[2] The operation of a Dynatech TCFCM system at Sandia National Laboratories has been discussed in detail by Moss, Koski, and Haseman.[3] The experimental reference sample-reference stack is placed between a main or high-temperature heating element and a low-temperature or auxiliary element. Thermocouples made from 0.010 in. wire are cemented into 0.015 in. grooves in the sample and references as described in Ref. 3. After the stack is in place with the thermocouples wired to terminal strips, an insulating powder is poured into the annular space between the stack and the guard. Upper and lower guard heaters are controlled to match the corresponding guard thermocouple voltages with thermocouple voltages in the main and auxiliary surface plates respectively. This method of control establishes an axial temperature profile at the outer boundary of the insulation which is approximately the same as that in the stack. Radial temperature gradients in the insulation are thus reduced and presumably the uniformity of the axial heat flux in the stack is enhanced.

THERMAL COMPARATOR ANALYSIS

In the usual method of data analysis for a comparator system the stack axial heat flux, Φ_z, is determined as the average of the heat flux in the top and bottom reference disks,

$$\Phi_z = \frac{1}{2} \left[\left(\lambda_r \frac{\Delta T}{\Delta z} \right)_{\text{top ref}} + \left(\lambda_r \frac{\Delta T}{\Delta z} \right)_{\text{bottom ref}} \right] \qquad (1)$$

In Eq. (1), λ_r is the reference conductivity and is evaluated at the average temperature for a given reference disk. ΔT and Δz are respectively the temperature drop across and thickness of the given reference. With the heat flux through the stack calculated from Eq. (1), the conductivity of the sample in the uniform heat flux approximation, λ_{so}, is found from Fourier's law,

$$\lambda_{so} = \Phi_z / (\Delta T / \Delta z)_{\text{sample}} \qquad (2)$$

The various sources of error in utilizing Eqs. (1) and (2) for reduction of comparator data have been discussed by Laubitz[4] and by Moss et al.[3] They can be grouped in two classes, fundamental errors and measurement errors. Fundamental errors arise from either a deviation from uniform axial heat flow, which was assumed in the derivation of Eq. (2), or from deviation of the actual λ_r for a specific set of references from the recommended values.[2] Measurement errors result from uncertainties in the measurement of any or all of the ΔT and Δz values or from uncertainties in the precise location of the thermocouples in the stack.

The major goal of the following numerical heat transfer analysis will be the calculation of the correction factor, C_f, required to convert the uniform heat flux conductivity, λ_{so}, to an actual experimental conductivity; i.e.,

$$\lambda_s = C_f \, \lambda_{so} \qquad (3)$$

The correction factor, C_f, will be a function of the stack and guard geometry, the thermal conductivities of the various elements, and the temperature profile of the guard. In any estimate of the precision of a thermal comparator measurement, the quantity $\delta\lambda_s/\lambda_s$ must be evaluated, where $\delta\lambda_s$ is the uncertainty in a λ_s measurement resulting from uncertainties in all of the experimental parameters. This uncertainty can be expressed symbolically as,

$$\delta\lambda_s/\lambda_s = \delta\lambda_{so}/\lambda_{so} + (1/C_f)\left[\sum_j\left(\frac{\partial C_f}{\partial \alpha_j}\right)\delta\alpha_j\right], \qquad (4)$$

where the α_j are the parameters in the thermal model and the $\delta\alpha_j$ are the uncertainties in these parameters. Estimates of the magnitudes of some of the terms in the summation will be made in the next section.

NUMERICAL HEAT TRANSFER ANALYSIS

In order to accurately determine the correction factor, C_f, and the uncertainty in this factor, a detailed numerical heat transfer analysis of the comparator system was performed.[5] The basic tool employed was the SINDA[6,7] (Systems Improved Numerical Differencing Analyzer) computer code which was utilized to solve Laplace's equation in the stack and insulation regions. An interactive preprocessing sequence was employed to create a SINDA thermal model. This was performed on a Digital Equipment Corp. VAX 11/780 mini-computer. The SINDA solution was then performed on a Control Data Corp. Cyber 76 mainframe computer and the results were post-processed on the VAX 11/780.

The actual comparator stack structure shown in Ref. 3 was approximated by the generalized structure shown in Fig. 2 for the purpose of making numerical calculations. This figure shows a cross section of the stack-insulation structure from the stack centerline to the Inconel sleeve which forms the guard inner boundary. The stack parts are the main and auxiliary heater blocks (Inconel), the top and bottom references, and the sample. The radii of all of these stack parts can be adjusted. The number of mesh intervals in the radial direction, IY1 to IY6, and in the axial direction, IX1 to IX5, can be varied in order to provide sufficient resolution for accurate isotherm plotting. The positions of the thermocouples on the stack centerline (H2, H4, H5, H7, H8, and H10) can be varied to

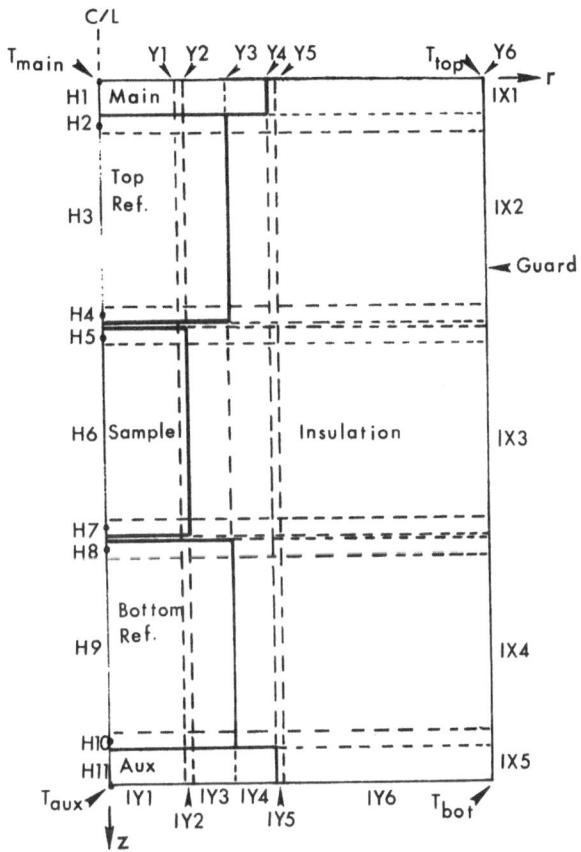

Fig. 2. Model cross-section of a comparator used for generation of node data. Y1-Y6 represent various radial dimensions and H1-H11 represent part thicknesses in the axial direction. The number of mesh intervals IX1-IX5 and IY1-IY6 may be varied in order to provide sufficient resolution for accurate solution and isotherm plotting.

study the effect of thermocouple misplacement on calculated conductivity.

In order to complete the problem specification, the materials properties must be given and the temperatures and/or heat flows at all of the boundary nodes must be specified. The boundary conditions approximate the effect of the rest of the apparatus on the portion of the system shown in Fig. 2. The top and bottom of the main and auxiliary heaters were assumed to be at constant temperatures, T_{main} and T_{aux}, respectively. The nodes at the top and bottom of the guard were assumed to be at temperatures T_{top} and T_{bot} respectively and the radial temperature profiles along the top and bottom of the insulation region were assumed to vary linearly between T_{main} and T_{top} or T_{aux} and T_{bot} respectively. The temperature profile along the guard, $T(r=Y6, z)$, could be specified arbitrarily by any profile varying from T_{top} at $z = 0$ to T_{bot} at $z = W$. The standard or "matched-linear guard" case is specified by: $T_{top} = T_{main}$, $T_{bot} = T_{aux}$ and a linear guard temperature variation between T_{top} and T_{bot}. Some test calculations involving deviations from this case will be discussed below.

The modeling code[5] allows temperature-dependent thermal conductivities to be input for all of the different elements in Fig. 2. For the reference materials, the standard NBS curves of λ vs T were used for Pyrex 7740, Pyroceram 9606, and Armco iron.[2,3] These conductivity values are felt to be accurate to approximately 5-10%, depending on the specific material and temperature range.[2] The λ vs T relation for another reference, Inconel 718, was determined by Dynatech.[3] The main and auxiliary heaters can be made of either Inconel or type 347 stainless steel. The λ vs T relations were obtained from the TPRC data series[8] but calculated results do not depend critically on the choice of conductivity for the heater surface plates. The insulation used in our systems is a diatomite filter material, Johns Manville Celite 560 powder. The thermal conductivity of this material was measured in air at ambient pressure as a function of temperature by the transient line source (thermal probe) technique. Measurements made at both the tap density (≈ 0.38 g/cm^3) and the pour density (≈ 0.31 g/cm^3) are shown in Table I. Each individual measurement was felt to be accurate to $\pm 15\%$. The effect of 50% variations in insulator conductivity will be discussed in the next section.

RESULTS OF NUMERICAL CALCULATIONS

The model and associated computer programs discussed in the preceding section were developed to aid in the analysis of data generated by the specific comparator systems used in our laboratory. However, model calculations have been performed which should aid in understanding the performance of any comparator system when standard reference materials and sample sizes are employed. The stack temperature drop, $\Delta T = T_{main} - T_{aux}$, was specified as 50°C for all calculations, in accordance with our usual experimental practice. (The dimensions utilized for our test calculations are shown in Table II).

The largest deviations from "ideal" behavior occur for low conductivity sample materials. Low conductivity samples increase the thermal resistance of the stack and thus increase the amount of "bypass" or leakage heat flux which flows out of the stack into the

Table I. Thermal conductivity of Celite 560 powder in air
          - Transient line source (probe) measurement

Tap Density = 0.38 g/cm^3		Pour Density = 0.31 g/cm^3	
T(°C)	k_l(W/m-K)	T(°C)	k_l(W/m-K)
30	0.080 ± 15%	30	0.06 ± 15%
255	0.100	425	0.11
430	0.125		
605	0.142		

insulation region. In practice, the thickness of low conductivity samples is usually decreased below its nominal value of 0.5 in. in order to decrease the temperature drop across the sample and the amount of leakage flux. However, decreasing the sample thickness increases the contribution of a given amount of sample thickness error to the uncertainty in the value of λ_{so}.

In the first series of test calculations, the dimensions given in Table II were employed with a $\Delta T = 50°C$ and matched linear guarding. Pyrex 7740 and Pyroceram 9606 were employed in all combinations of sample and reference stack elements and the temperature profiles were calculated at several average stack temperatures from 25-450°C. From the temperature distribution on the stack centerline, the apparent conductivity, λ_{so}, was calculated from Eqs. (1) and (2) and compared to the input assumed sample conductivity at the average sample temperature. The results are shown in Table III, where the maximum percent error and the approximate temperature at which it occurs are given. A negative % error means $\lambda_{so} > \lambda_s$. From this table it can be seen that the maximum predicted absolute error is about 0.7% for a case with Pyroceram references and a Pyrex sample. This is somewhat better than the estimated ± 2% precision associated with the experimental determination of λ_{so} from temperature, thickness, and thermocouple placement measurements.

Table II. Dimensions used for test calculations

Dimension	Symbol (Fig. 3)	Length (in.)
Reference, sample radii	Y2, Y3	1.00
Guard radius	Y6	2.30
Top reference thickness	H2 + H3 + H4	0.50
Bottom reference thickness	H8 + H9 + H10	0.50
Heater plate thickness	H1, H11	0.10
Sample thickness	H5 + H6 + H7	0.50

Table III. Maximum predicted % error for reference vs. reference measurements and temperature where the maximum deviation is predicted. % error = $100 \, (\lambda_s - \lambda_{so})/\lambda_s$

Reference	Sample	Max % Error/Temp (°C)
Pyrex 7740	Pyrex 7740	−0.34/450
Pyrex 7740	Pyroceram 9606	+0.19/25
Pyroceram 9606	Pyroceram 9606	−0.11/300−400
Pyroceram 9606	Pyrex 7740	−0.65/25

In the next set of calculations, a range of temperature independent sample conductivities between 0.05 W/m–K and 10 W/m–K was employed and values of C_f were calculated at average stack temperatures in the range 25–450°C. Results for the dependence of C_f on λ_{so} are shown in Figs. 3 and 4 for Pyrex and Pyroceram references respectively. For sample conductivities which are much lower than that of the references, the sample heat flux is

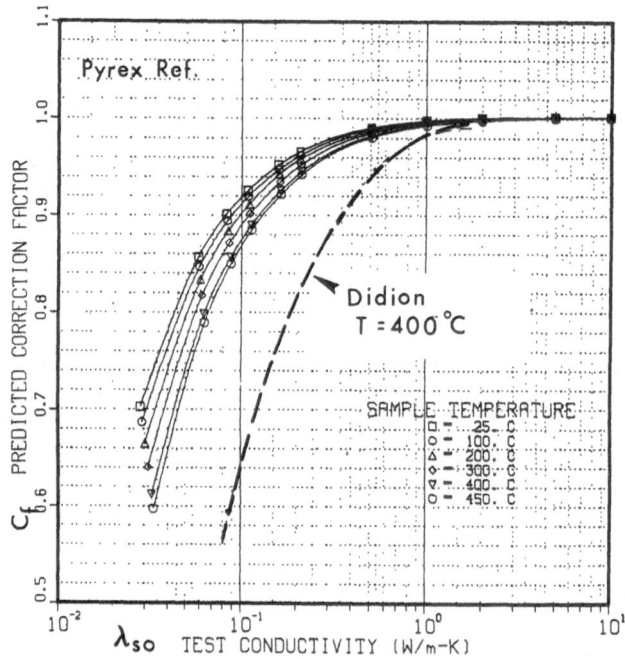

Fig. 3. Correction factor C_f vs uncorrected or test conductivity λ_{so} for measurements made with Pyrex references. Sample thickness is 0.5 in. and $\Delta T = 50°C$. The dashed line is from an approximate calculation of Didion as discussed in the text.

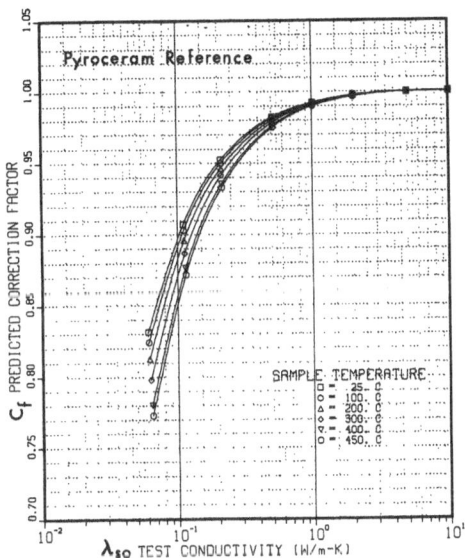

Fig. 4. Correction factor C_f vs λ_{so} for measurements with
Pyroceram references. Same conditions as for Fig. 3.

overestimated by Eq. (1) with the result that $\lambda_s < \lambda_{so}$ or $C_f < 1$.
It is interesting to observe that samples with conductivities as
low as 0.5 W/m-K can be measured to an uncorrected accuracy of $\approx 2\%$
with either type of reference. Comparison of Figs. 3 and 4 indicates that the correction factors for both types of references are
roughly equal, even though Pyroceram conductivities are 2-4 times
larger than equivalent temperature Pyrex conductivities. The curves
in these figures may be used to correct raw experimental data for
low conductivity samples in the standard geometry.

For an estimate of the precision with which the correction
factor, C_f, may be determined, calculations were performed with
some of the model parameters varied. An obvious parameter of importance is the thermal conductivity of the insulation material in
the annular space between the stack and the guard. To estimate the
effect of the insulator conductivity λ_i on C_f, runs were made
with standard conditions, a sample conductivity $\lambda_s = 0.1$ W/m-K,
and temperature-independent λ_i values of 0.05, 0.10, and 0.15
W/m-K. Fig. 5 shows a plot of the isotherms for the $\lambda_i = 0.1$
W/m-K case. The temperature difference between adjacent isotherms
is 5°C. Since most of the distortion in the isotherms occurs in
the insulation region, it is evident that λ_i will have an important effect on the amount of deviation from uniform axial heat flow.
Table IV shows the results for the fractional variation in C_f per

Table IV. Relative change in C_f, $\delta C_f/C_f$, produced by a relative change, $\delta\lambda_i/\lambda_i$ in insulator conductivity

$\lambda_s = 0.1$ W/m-K, $\lambda_i = 0.1$ W/m-K, $\delta\lambda_i = \pm 0.05$ W/m-K
$\Delta T = 50°C$, Pyrex references

$\overline{T}$ (°C)	$\delta\lambda_i = -0.05$ W/m-K $(\delta C_f/(C_f)/(\delta\lambda_i/\lambda_i)$	$\delta\lambda_i = +0.05$ W/m-K $(\delta C_f/C_f)/(\delta\lambda_i/\lambda_i)$
25	0.079	-0.067
100	0.081	-0.070
200	0.084	-0.073
300	0.086	-0.076
400	0.088	-0.077
450	0.089	-0.079

unit fractional variation in λ_i at six temperatures in the range 25–450°C. For an assumed ± 20% uncertainty in λ_i, consistent with the estimated experimental uncertainty in λ_i for Celite powder, the relative uncertainty in C_f is only about 2%.

Another important variable in comparator measurements is the temperature profile in the guard at the outer radius of the insulation region. The inevitable mismatch between vertical locations of thermocouples in the upper guard and main surface plate leads to a temperature mismatch at the top of the stack. In addition, the axial temperature profile in the guard may not be exactly linear. Since our measurements indicate that the guard temperature profile is reasonably linear, only deviations from matched conditions at the top of the stack were considered.

In Fig. 6 isotherm plots are shown for a system with $\lambda_s = 0.1$ W/m-K, Pyrex references, an average sample temperature of 25°C and upper guard temperatures 25°C greater than and less than the main surface plate temperature of 50°C. These plots may be compared with the one shown in Fig. 5 for the matched guard condition. The plots clearly indicate that most of the isotherm distortion produced by the guarding mismatch occurs outside the stack region and that only very small shifts in centerline temperatures are produced by the mismatch. An analysis of these calculations shows that 25°C upper guard mismatches produce a change of only 1-3% in the calculated correction factor. Experimentally, upper guard mismatches are estimated to be no larger than ~ ±10°C for our system and hence, the uncertainty in C_f produced by upper guard mismatches is estimated to be ≤ 1%.

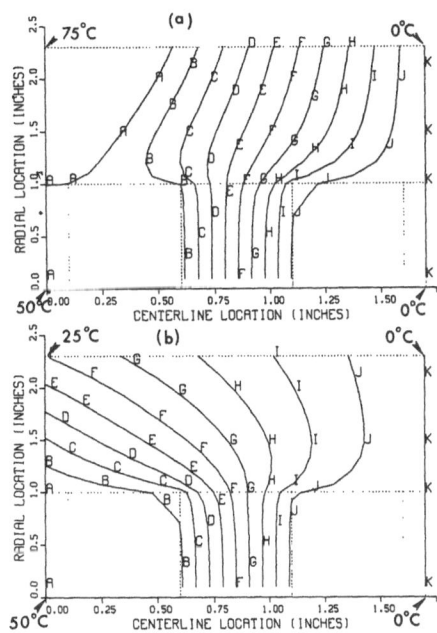

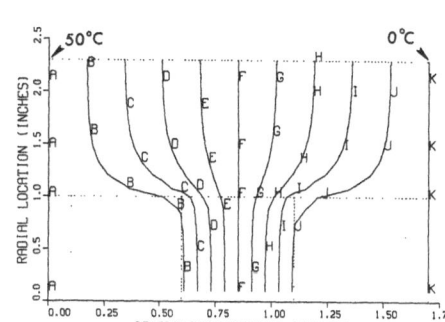

Fig. 5. Isotherms calculated for a stack with λ_i = λ_s = 0.1 W/m-K and Pyrex references at T = 50°C.

Fig. 6. Isotherms for the same conditions as in Fig. 6 but with upper guard mismatches of ± 25°C. (a) T_{top} = T_{main} + 25°C, (b) T_{top} = T_{main} - 25°C.

COMPARISON WITH PREVIOUS ANALYSIS

The heat flow and temperature distributions in simplified model comparator systems have been previously studied by Flynn[9] and Didion.[10] In these analyses, the heat flow in the annular insulation region of the comparator is analyzed under the assumption that the axial temperature profile at the outer stack radius is piecewise linear. The calculated temperature profile in the guard region, $T_i(r,z)$, is then used to calculate the radial heat flux which must flow from the stack to the insulation region to support this temperature field. Constant temperature independent reference, sample, and insulation conductivities (λ_r, λ_s, and λ_i respectively) were assumed.

The solution for the insulation temperature field is made by the standard Fourier series method. Details are given by Didion[10] who suggests that the correct axial heat fluxes to use in the conductivity determination are the average values for the reference and sample regions. We have made a comparison between Didion's method and the numerical method for the case with Pyrex references and an average sample temperature of 400°C with $\Delta T = 50°C$. The axial temperature profiles for the stack centerline, outer radius, and guard from the numerical calculation are shown in Fig. 7 for the case with $\lambda_s = 0.1$ W/m-K. It can be seen that the temperature profile at the outer stack radius is in fact approximately piecewise linear but that temperatures are different than centerline temperatures. The magnitude of the temperature difference between the stack outer

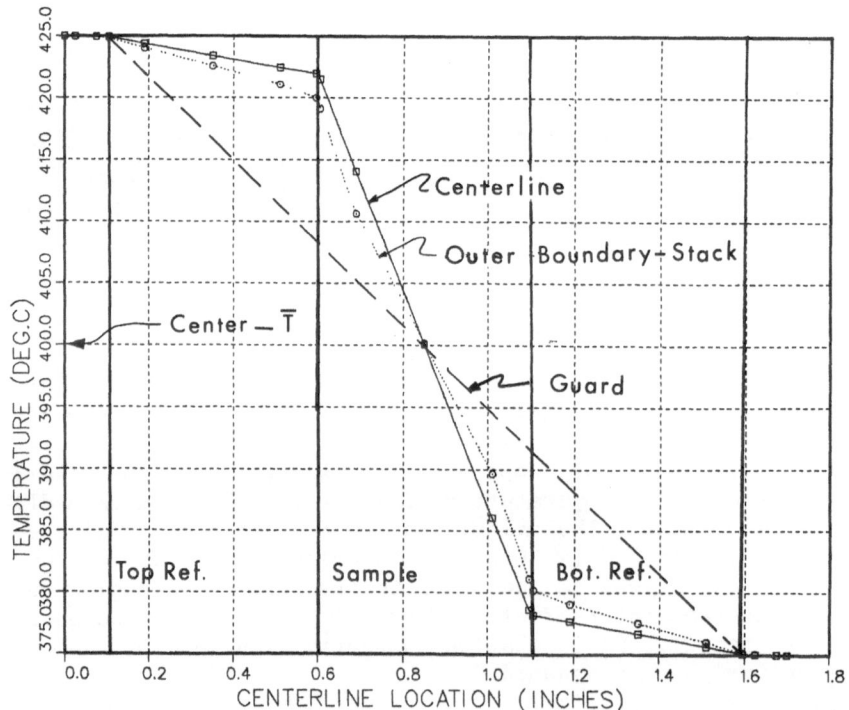

Fig. 7. Axial temperature profiles at stack centerline and outer radial boundary calculated for a Pyrex-sample-Pyrex stack with $k_s = 0.1$ W/m-K and $\Delta T = 50°C$ at $T = 400°C$. The linear guard profile is also shown.

radius and the guard is smaller than the temperature difference between the stack centerline and the guard. Hence, the use of centerline temperatures at the outer stack radius for the analytical calculation would be expected to overestimate the radial heat flux and lead to larger percentage corrections (smaller C_f values) than the numerical calculation. The results of a comparison of the two methods are shown in Fig. 3. At low λ_s values, the Didion method significantly overestimates the required correction, as predicted. However, the dependence of C_{fD}, the Didion correction factor, on λ_{so} is similar to that predicted numerically. In making the calculation for C_{fD}, an average reference conductivity for Pyrex at 400°C, $\lambda_s = 1.60$ W/m-K, was assumed.

The dependence of C_{fD} on insulator conductivity may also be derived from Didion's analysis. The results indicate that C_{fD} is more sensitive to changes in λ_i than is in fact the case. For example, Table IV shows that $(dC_f/C_f)/d\lambda_i/\lambda_i)$ is in the range -0.08 to -0.09 at T = 400°C with Pyrex standards and $\lambda_s = 0.1$ W/m-K, while the Didion analysis predicts $(dC_{fD}/C_{fD})/(d\lambda_i/\lambda_i) = -0.24$. From the above analysis, we conclude that the Didion-Flynn theory is not quantitatively accurate for very low conductivity samples. It can be utilized, however, for making rough qualitative predictions for situations in which the sample conductivity is relatively close to the reference conductivity.

Comparator systems are sometimes utilized to measure the conductivity of samples with diameters smaller than the reference diameter. This is usually due to the inability to obtain a sample with the standard 2 in. diameter. Instead of making new references with reduced diameters, it can be desirable to use the regular references and correct the measured λ_{so} for non-uniform heat flow effects. For the case in which the sample conductivity is much larger than the reference conductivity, the results of an analytical study by Kennedy[11] may be used to make this correction. Kennedy calculates the net thermal resistance of a cylindrical bar of radius b with a uniform heat flux injected into one end of the bar within a circular area of radius a $\leqslant$ b. If it is assumed that the sample with radius = a has high conductivity and supports an approximately uniform axial heat flux, then Kennedy's results may be used to calculate the corrected thermal resistance of the references.

The results in Ref. 11 may be used to cast the correction factor into the form,

$$C_{fK} = (z_r/a)/H_K(a/b, z_r/b), \qquad (5)$$

where z_r = reference thickness and H_K (a/b, z_r/b) is a dimensionless function given by Kennedy. A comparison was made between the predictions of Eq. (5) and those of a numerical calculation for a case with Pyrex references and a sample with $\lambda_s = 10$ W/m-K and

Table V. Correction factors found by the numerical method and by the numerical method and by the approximate method of Kennedy for reduced diameter high conductivity samples

λ_s = 10 W/m-K, Pyrex reference, $\overline{T}$ = 50°C,
ΔT = 50°C, b = 1.00 in.

Parameter \ Sample Radius, a(in.)	1.00	0.50	0.25
H_K	0.5	0.7	0.8
C_{fK}	1.00	1.42	2.50
C_f (numerical)	1.00	1.36	2.21
$(C_{fK} - C_f)/C_f$	1.00	0.044	0.132

reduced radii of 0.5 and 0.25 in. relative to a 1.0 in. reference radius. The results are given in Table V. The correction factors are quite large, but the Kennedy prediction is only in error about 4% for the 0.5 in. sample. Hence, for making approximate measurements on high conductivity samples with a > 0.5 in., Kennedy's method should be sufficiently accurate for finding the correction factor.

COMPARISON WITH EXPERIMENTAL DATA AND DISCUSSION

The results presented above facilitate the use of Dynatech comparator systems in the measurement of low conductivity samples or samples with non-standard dimensions. Another important application is the precise definition of expected errors when checking one reference against another, as shown in Table III. The results in Table III indicate that the maximum absolute error which should occur in a measurement with Pyrex or Pyroceram sample and references in any combination is < 0.5%. Mirkovich[12] has shown that excellent results could be obtained with Pyroceram 9606 references and samples of alumina Al-300 and fosterite "L". The agreement between the conductivity measured by comparator and by "absolute" methods appears to be within 5%. Recent use of our Dynatech comparator system for measuring the thermal conductivity of geologic materials in conjunction with radioactive waste disposal studies[13,14] has stimulated efforts to delineate more precisely the sources of error which can arise in these measurements.

Inspection of Pyrex and Pyroceram conductivity data indicates that the spread in the Pyroceram 9606 data is relatively small, < 5% at T > 125°C.[2] Below this temperature, the conductivity increases rapidly with decreasing temperature and the measurements appear less precise. In contrast to this case, the spread in the Pyrex data is much larger, > 10%.[2] On the basis of this comparison, it is tentatively concluded that Pyroceram 9606 is a "better" working standard than Pyrex. According to the results in Table III, in the absence of large experimental errors, measurements with Pyroceram references and a Pyrex sample should determine the Pyrex conductivity to an accuracy very close to that associated with the Pyroceram conductivity. Calibration data obtained from one of our comparator systems showed that, in a Pyrex vs Pyrex measurement the rms deviation from the NBS curve was ≈ 2%, in agreement with the estimated measurement error.[3] This measurement cannot detect "errors" in the NBS curve, it can only show that the sample and references are similar and that there are no anomalies in the system operation when a stack of three identical parts is utilized. In measurements with Pyroceram 9606 and fused silica samples with Pyrex references, there was a small but definite tendency for the measured conductivities to be higher than published data at high temperature and lower at low temperature.[3] In both cases, maximum observed deviations for average temperatures in the range 20–425°C were < 8%.

Recent measurements with a Pyrex sample and Pyroceram references indicate that there is a small but definite tendency for the experimental Pyrex conductivity to be higher than the NBS suggested conductivity at T < 300°C. The deviation of the 30°C experimental conductivity from the published value is about 8%. On the basis of the known accuracy of the Pyroceram 9606 conductivity and the calculated accuracy of the experiment from Table III, we have tentatively concluded that the conductivity of our Pyrex references has a small but definite deviation from the NBS suggested curve. Further experiments are necessary to confirm this hypothesis in detail.

CONCLUSIONS

In this analysis, the temperature distribution in a thermal comparator conductivity measurement system, operating at steady state, has been determined in detail for a number of cases of experimental interest. The computer codes developed in this investigation facilitate the determination of the dependence of measured sample conductivity on a number of experimental parameters.

One major result of this study has been the accurate determination of the correction factors needed to correct raw data for very low conductivity samples. As a result of this work, we feel that a Dynatech comparator with either Pyrex or Pyroceram references can be used to measure conductivities as low as 0.05 W/m-K with an accuracy ~ 15%. The major uncertainty in the derived correction factors is associated with variations in the thermal conductivity of the insulation used in the annular space between the stack and the guard. This uncertainty together with the small reference temperature drops associated with measurements on low conductivity materials determine the predicted measurement accuracy. Since comparator measurements are much easier to make than guarded hot-plate measurements and since they can be made with smaller samples, there is an incentive to use comparator systems for low conductivity measurements. Further experimental comparisons between comparators and guarded hot-plates are needed to confirm this predicted comparator accuracy.

Another major prediction of our numerical analysis is that measurements of the standard low conductivity reference materials, Pyrex 7740 and Pyroceram 9606, against each other do not require correction for non-uniformities in axial heat flux by more than 0.7%. Hence, the uncertainty in making conductivity measurements on these materials is dominated by measurement errors and errors in the assumed reference conductivities. This finding has led us to question the accuracy of the standard NBS Pyrex conductivity vs temperature relation, assuming the NBS Pyroceram conductivity curve is accurate at the 5% level. It is anticipated that further use of the numerical methods described in this paper will lead to additional understanding of comparator behavior and improvements in the accuracy achievable with these instruments.

ACKNOWLEDGMENTS

The authors would like to thank W. D. Drotning for measuring the thermal conductivity of Celite powder and G. M. Haseman for making the comparator thermal conductivity measurements discussed in this paper.

REFERENCES

1. The Dynatech Model TCFCM comparative thermal conductivity instrument is manufactured by Dynatech R/D Co., Cambridge, MA. Reference to a particular product or company implies neither a recommendation nor an endorsement by Sandia National Laboratories or the U. S. Department of Energy, nor a lack of suitable substitutes.
2. Powell, R. W.; Ho, C. Y.; Liley, P. E.; "Thermal Conductivity of Selected Materials," NBS publication NSRDS-NBS8, 1966.
3. Moss, M; Koski, J. A.; Haseman, G. M.; "Measurement of Thermal Conductivity by the Comparative Technique," Sandia National Laboratories report SAND82-0109, 1982.*

4. Laubitz, M. J., "Measurement of the Thermal Conductivity of Solids at High Temperatures by Using Steady State Linear and Quasi-Linear Heat Flow," in Thermal Conductivity Vol. I, R. P. Tye, Ed(Academic Press, NY, 1969) Chap. 3. See especially pp. 174-182.

5. Sisson, C. E., "An Automated 2-D Thermal Mode to Predict Temperatures in the Material Stack of a Thermal Comparator Test Device," Sandia National Laboratories report SAND83-1900, 1983.*

6. TRW Report, 14690-H001-R0-00, "SINDA User's Manual,"prepared under NASA Contract 9-10435, April, 1971.

7. TRW Report, 14690-H002-R0-00, "SINDA Engineering-Program Manual," prepared under NASA Contract 9-10435, June 1971.

8. Touloukian, Y. S., Ed, "TPRC Data Series - Vol. I, Thermal Conductivity of Metallic Solids," IFI/Plenum, NY, 1970; p. 1017, 1174.

9. Flynn, D. R., "Thermal Conductivity of Ceramics," in "Mechanical and Thermal Properties of Ceramics," Ed by J. B. Wachtman Jr., NBS Special Publication 303, May 1969, pp. 63-123.

10. Didion, D. A., "An Analysis and Design of a Linear Guarded Cut-Bar Apparatus for Thermal Conductivity Measurements," Tech Report No. 2 prepared under Contract NONR 2249(12) for ONR, Catholic Univ. of America, Wash., DC, 1968.

11. Kennedy, D. P., "Spreading Resistance in Cylindrical Semiconductor Devices," J. Appl. Phys. 31 1490 (1960).

12. Mirkovich, V. V., "Comparative Method and Choice of Standards for Thermal Conductivity Determinations," J. Am. Ceramic Soc. 48 387 (1965).

13. Sweet, J. N.; McCreight, J. E.; "Thermal Conductivity of Rocksalt and Other Geologic Materials from the Site of the Proposed Waste Isolation Power Plant," Thermal Conductivity 16, Ed by D. C. Larson, Plenum, NY, 1983, pp. 61-78.

14. Moss, M; Koski, J. A.; Lappin, A. R.; "Thermal Conductivity of Tuff: the Effects of Composition, Porosity, Bedding Plane Orientation, Water Content and a Joint," Proc. of ASME/JSME Thermal Engineering Joint Conference, Vol. II, Honolulu, March, 1983, p. 199.

*Sandia reports available from NTIS, U. S. Dept. of Commerce, 5285 Port Royal Road, Springfield, VA 22161.

A NUMERICAL METHOD FOR SOLVING THE INVERSE PROBLEM OF HEAT TRANSFER

AND ITS APPLICATION TO DETERMINATION OF THE THERMAL CONDUCTIVITY

Cangming Yu

Thermophysics Division, Dept. of Engineering Mechanics
Tsinghua University
Beijing 100084, China

ABSTRACT

The usual practice in the measurement of the thermal conductivity is as follows: a sample with specified shape is employed in
a given experimental apparatus; as the foundation of the experimental
measurement, the thermal conductivity is calculated by a formula
based on the assumption that the differential equation of heat conduction is a linear one, i.e. the thermal conductivity is constant.
In this paper, an approach is employed in which a numerical method
for solving the inverse problem of heat conduction is used to
determine the thermal conductivity. With this approach, the thermal
conductivity can be determined from the data obtained directly from
an object in its natural situation instead of from the data obtained
by a sample in an experimental apparatus. In order to show the
applicability of the suggested approach, the thermal conductivity
is measured first for a sample of stainless steel by using a method
of direct electric heating within the temperature region from 200°
to 800°C. The thermal conductivity and its dependence on temperature
are calculated by using Fourier's law. The thermal conductivity and
its temperature dependence are then calculated by using a finite-
difference method and a finite-element method for solving the inverse
problem of heat conduction. All these experimental and calculated
results are compared with the curve recommended by TPRC. The agreement is good (within $\pm 5\%$).

The approach is also employed to determine the thermal conductivity of engineering material. By means of the temperature history
measured directly at several points within a 4.5 ton vermicular-
graphite-cast-iron ingot mold, the thermal conductivity and its
temperature dependence within the region 0°-800°C are calculated by

using the finite-difference method for solving the inverse problem of nonlinear heat conduction. The calculated values of the thermal conductivity compared favorably with the experimental ones obtained by use of a sample of the same material as the ingot mold and of the direct electric heating method.

SESSION D

Liquids/Gases

SESSION CHAIRMAN

J. G. Hust
NBS
Boulder, CO

TOWARDS REFERENCE STANDARDS FOR THE THERMAL CONDUCTIVITY OF

LIQUIDS

[1]C.A. Nieto de Castro and [2]W.A. Wakeham

[1]Department de Química, FCUL
 Centro de Química Estrutural
[2]1096 Lisboa, Codex, Portugal
 Department of Chemical Engineering
 Imperial College, London SW7 2BY, England

ABSTRACT

In view of the common need to calibrate instruments for the measurement of liquid-phase thermal conductivities, standard reference values for this property are urgently required. However, the difficulties associated with the measurement of the thermal conductivity of liquids mean that there is no consensus among workers in the field regarding either the most suitable liquids or the reference values to be adopted despite several independent recommendations. In this paper we consider developments in the measurement of thermal conductivity in the last decade which have produced results with an uncertainty almost an order of magnitude lower than hitherto. These new results are used to assess the situation with regard to standard reference materials and to formulate suggestions for future experimental and theoretical work to establish suitable reference values.

1. INTRODUCTION

The thermal conductivity of fluids has proved to be one of the most difficult transport properties to measure with great accuracy. In general, this is because of the influence of the other mechanisms of energy transport which inevitably accompany the imposition of a temperature gradient in a fluid within the earth's gravitational field. It is therefore common practice to arrange that measurements of thermophysical properties are performed in a relative manner in which a parameter or group of parameters, characteristic of the instrument, are determined by means of a calibration using standard reference values for the property for one or more fluids. In the case

of the thermal conductivity of gases such standard reference values
may be generated from accurate, independent measurements of the viscosity of monatomic gases with the aid of exact relationships from
kinetic theory [1]. However, for liquids no such theory is available and reference values must be determined by direct measurement.
The difficulties of measurement have caused the results of different
authors, made with different instruments, to differ substantially
from each other for every liquid, even at ambient conditions. For
this reason there has been no consensus among workers in the field
either on the most suitable liquids or on the reference values to be
adopted.

The first liquid suggested as a standard reference material
for thermal conductivity was toluene [2]. Prior to this suggestion,
measurements had been performed by several authors, in different types
of instrument yielding results of poor precision and accuracy which
were often in disagreement [3]. Following Riedel's recommendation
[2], more precise measurements were performed. However, these new
measurements did not drastically affect the downward trend in the
thermal conductivity of toluene, at a particular temperature, which
is apparent when the values are plotted as a function of year of
measurement (see figure 1). This trend is an illustration of the

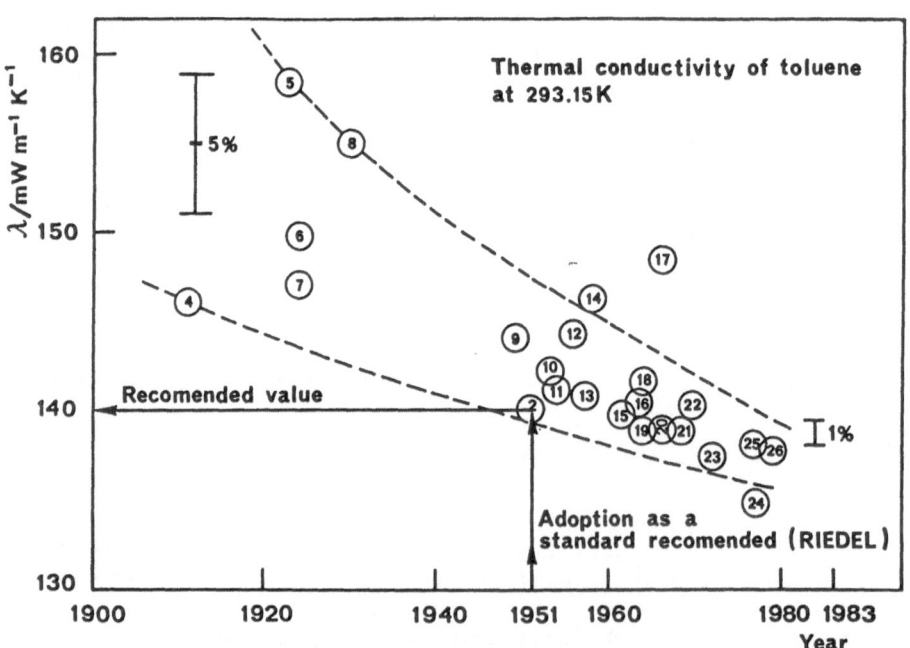

Fig. 1.   The thermal conductivity of toluene at 293.15 K measured
          by different authors as a function of the year of measure-
          ment, each one quoted by its reference number.

considerable developments there have been in the measurement of thermal conductivity in the last thirty years. In particular, the development of the transient hot wire technique in the last decade has brought about an improvement in the precision of measurement of about an order of magnitude [27-29]. It therefore seems that it is now appropriate to reconsider the establishment of a standard reference material for thermal conductivity, the instruments to be used for the measurement of this standard, and a critical analysis of the data produced. The present paper is concerned with the criteria that should guide the choice of standard reference materials for thermal conductivity of liquids as well as with the method of their determination. It is specifically not intended to provide definitive reference values, which will be an activity undertaken on a collaborative basis by a number of laboratories throughout the world.

2. EXPERIMENTAL METHOD

An instrument for the measurement of a thermophysical property must satisfy a number of conditions before the results obtained with it may be considered to have high accuracy. First, it is necessary that a complete working equation describing the instrument should be available together with a complete set of corrections. Second, the working equation should be of a form such that the sensitivity of the measured property to the principal variables does not magnify the random errors of management. Third, each of the principal variables which enter the working equation should be amenable to measurements of high precision.

For the particular case of thermal conductivity of fluids the available experimental methods may be divided into two groups: steady-state methods and transient methods. In any instrument of either type operating in a gravitational field, the imposition of the temperature gradient required for the measurement inevitably gives rise to convective and radiative heat transport in addition to pure conduction. In order to isolate pure conductive heat transport and retain a complete and accurate working equation, it is therefore necessary either to arrange the experimental conditions so as to render the effects of other modes of heat transfer negligible or, to have a mathematical description of the process which permits corrections to be applied to experimental data.

So far as convection is concerned there are no exact solutions of the equations of hydrodynamics to allow the latter course of action. On the other hand, extremely careful design of an instrument is necessary to ensure that the effects of convective flow upon the measurement are negligible. This is particularly true for instruments of the steady-state variety where the effects of convection are rendered small by the use of geometric arrangements of parallel plates [30] and coaxial cylinders [31] which require careful alignment and construction. In only a few cases has it proved possible to achieve these aims satisfactorily.

In transient methods it is possible to eliminate the effects of convection by making use of the inertia of the fluid. That is, if the measurement is performed on an appropriate time scale, elements of the fluid do not attain a velocity sufficient to unduly perturb the conductive heat flow. It has been demonstrated that it is always possible to find such a time scale for gases or liquids in transient hot wire instruments, so that the effect of convective heat transfer can be entirely eliminated from the measurement [27,32].

The radiative heat transport associated with any measurement of the thermal conductivity (steady-state or transient) is easily treated in the case where the fluid does not contribute directly or indirectly to the radiation process. This is because in such cases the radiative heat transport is not coupled to the conduction process and therefore the radiative and conductive heat fluxes are additive. A correction for the radiation heat flux is then rather simply evaluated either by calculation [31] or from experiment [29]. When the fluid absorbs and emits radiation the problem is considerably more complicated because then the conductive and radiative heat fluxes are coupled and the situation is described by an integro-partial differential equasion [33]. There has been a large number of attempts to treat the coupled problem in both steady-state and transient regimes which have, on occasions, produced conflicting results. Thus in steady-state methods it has been argued on the one hand [20] that radiation augments the process of conduction in a fluid and on the other hand [34] it reduces the net heat flux. The most complete treatment of the process in a parallel plate geometry, given by Viskanta and Grosh [35], has seldom been applied in practice because the necessary optical properties of the fluid and its bounding surfaces are not available and are not easily determined. Thus, the steady state techniques used with fluids contributing to the radiative transfer fail to satisfy the condition that a complete working equation and a complete set of corrections should be available.

In the case of transient methods of measurement the situation has, until recently, been the same. The only analyses available for the coupled conduction-radiation problem were presented in the form of numerical solutions which were often inconvenient to implement and required a detailed knowledge of the optical properties of the fluid and its boundaries. For these reasons it has been usual either to report thermal conductivity data contaminated with a ra - diation component or to apply an uncertain correction, and decrease the estimated accuracy of the final data correspondingly. The most recent analysis of the radiation problem for the transient hot wire method has made use of a numerical solution in a rather different way [36]. It has been shown that in the transient hot wire method the principal radiative contribution arises from the coupling between the conduction and the radiant emission by the fluid. This is seen to occur because in this particular technique the region where the spatial temperature gradients are very large is localized within a very small cylinder of fluid. On the other hand the region where radiation is absorbed by the fluid is distributed in a different, larger volume where the temperature gradients are small. These observations allow us to neglect all radiative terms in the governing equation other than those arising from the emission by the fluid. The resulting equation is then amenable to exact analytic solution [36] and yields the usual working equation for the transient hot-wire method together with a correction term.

The working equation for the transient hot-wire method for non conducting fluids relates the temperature rise of a thin heat source in the fluid to the time of heating in the form

$$\Delta T = \frac{q}{4\pi\lambda} \ln \frac{4\alpha t}{a^2 C} \; . \tag{1}$$

Here ΔT, is the temperature rise of the wire, corrected for the effect of the finite thermal properties of the wire and for the finite extent of the fluid, q represents the rate of heat generation in the wire per unit length, a represents the wire radius, λ is the thermal conductivity of the fluid at a suitable reference temperature and density and α is the thermal diffusivity, $\lambda/\rho c_p$. The linearity between ΔT and $\ln t$ given by equation (1) allows the thermal conductivity to be evaluated by means of a straight forward linear regression analysis of experimental data.

When allowance is made for the radiative contribution of the fluid the equation takes the form

$$\Delta T = \frac{q}{4\pi\lambda} B + \frac{q}{4\pi\lambda} (1+B) \ln \frac{4\alpha t}{a^2 c} - \frac{q}{\pi\rho c_p} \frac{B}{a^2} t \; . \tag{2}$$

Here, B is a dimensionless number given by

$$B = \frac{4 \, \varepsilon_f n \, \sigma \, T_o^3 \, a^2}{\lambda} \; ; \tag{3}$$

n is the refractive index of the fluid, σ the Stefan-Boltzmann constant, T_o is the equilibrium temperature of the fluid prior to the heat generation and ε_f is a property of the fluid which is defined by the condition that

$$Q_e = - 4 \, \varepsilon_f E_i \; . \tag{4}$$

Here Q_e represents the gradient of the radiant heat flux emitted by a volume element of fluid, and E_i is the radiant flux from a black body at the same temperature[36, 37].

Because the thermal conductivity of the fluid is determined from the scope of the ΔT vs. ℓnt [27] the occurrence of the time-independent term in equation (2) is of no consequence to the measurement. However, the third term in the equation indicates that, in principle, the line ΔT vs ℓnt should be curved and concave to the ℓnt axis. Finally, there is a modification to the term linear in ℓnt, related to the magnitude of the radiant emission, characterized by B.

Equation (2) can readily by rewritten in the form usually employed for the working equation of the hot wire method

$$\Delta T = \frac{q}{4\pi\lambda} \ell n \frac{4\alpha t}{a^2 C} + \Delta T_{rad}, \qquad (5)$$

where the correction

$$\Delta T_{rad} = - \frac{4 \varepsilon_f n^2 \sigma T_o^3 q}{\rho c_p \pi \lambda} t. \qquad (6)$$

It is not possible to determine the quantity ε_f in an independent optical measurement because it is a quantity particular to the transient hot wire experiment itself. Nevertheless, the form of equation (2) shows that if the effect of radiation is significant its value may be derived from the experimental data directly, by an examination of the line ΔT vs. ℓnt. The derived value may then be employed in the evaluation of the correction ΔT_{rad}.

In all experiments carried out in liquids so far the radiative contribution has been found to be negligible as is shown by the fact that the line ΔT vs ℓnt has displayed no curvature whatsoever. This is illustrated in Figure 2 which contains deviations of a set of ΔT vs ℓnt data for a measurement carried out in toluene at 27.15^oC from the linear fit. There is clearly no evidence of any curvature. On the basis of this and many other similar plots as well as the detailed analysis [36] we conclude that radiation contributes <u>insignificantly</u> to the measurement of the thermal conductivity of toluene in the transient hot-wire instrument.

It must be emphasized that the preceding review of the effects of radiation in thermal conductivity measurements is restricted to the <u>transient hot wire method</u>. It is not proven that a similar simplification of the problem is possible for the steady-state techniques. The accumulated evidence set out here indicates that only the transient hot wire method is suitable for the measurement of the thermal conductivity for standard reference purposes. It is corollary that the results obtained in such instruments may be used to determine the contribution of radiation in other types of equipment.

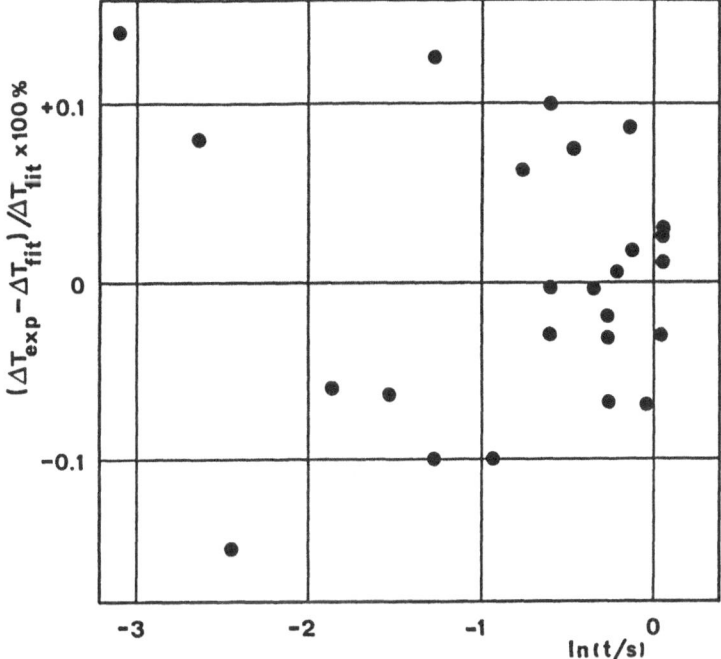

Fig. 2. - Scattering diagram for a run with toulene at 27.15°C

3. STANDARD REFERENCE MATERIALS

In the past it has been argued, owing to the absence of a reliable analysis of the contribution of radiation to the thermal conductivity measurements, that fluids to be selected as standard reference materials should be those for which the radiative contribution is very small. In the light of the discussion in the preceding section it may be that this is no longer a necessary condition, although it remains a desirable one. The reference materials should therefore be selected following a series of transient hot wire measurements which will reveal whether radiation plays a significant role or not. The measurements carried out to date, on normal alkanes [25, 29, 38] toluene and dimethylphthalate [25] and carbon disulfide [42] indicate that, judged on this criterion alone, any of these materials could be suitable. However, there are several other conditions to be fulfilled by standard materials for liquid thermal conductivity which were originally set out by Ziebland [15] :

a) The temperature interval between the freezing point and the normal boiling point, i.e. the extent of the liquid range, must be large and include some thermometric fixed point.

b) The liquid should be non-toxic and non-corrosive with respect to the usual engineering materials.

c) It should be obtainable at reasonably low cost and guaranteed high purity.

To these conditions we would add two further, more general conditions.

d) A series of liquids covering as wide a range of thermal conductivity as possible should be selected.

e) It will not generally be possible to operate a transient hot-wire instrument exactly at the conditions of saturation most suitable for a reference state. Consequently, it will be necessary to extrapolate to saturation conditions from higher pressures so that reliable values of liquid densities for saturated and compressed liquids are required.

These considerations lead us to propose the following set of liquids for standard reference materials for liquid thermal conductivity:

Table 1: Proposed Thermal Conductivity Standards

Fluid	Thermal conductivity at 298 K mW m^{-1}K^{-1}
iso-octane	98
heptane	123
toluene	130
methanol	197
water	600

The table indicates that there is a substantial gap in thermal conductivity values between methanol and water. Although there is no pure fluid which is available to bridge this gap, it could be filled by a mixture of methanol and water if it were deemed necessary.

4. FUTURE WORK

It is hoped that a systematic series of careful measurements of the thermal conductivity of the proposed materials in a number of different laboratories throughout the world can be arranged with the aim of obtaining agreement upon the reference values. All measure - ments will be carried out using transient hot wire instruments. The agreed standard values should provide the basis for improvement in the accuracy of future measurements of the thermal conductivity of liquids.

ACKNOWLEDGEMENTS

This work was partially financed by Science Research Council (UK) and NATO Grant 1984. CANC would like to acknowledge financial assistance from his department permitting him present at the Conference.

REFERENCES

1. Maitland, G.C., Rigby, Smith, M., Smith, E.B., Wakeham, W.A., "Intermolecular Forces - Their Origin and Determination", Clarendon Press, Oxford (1981).

2. Riedel, L., Chem. Ing. Tech., 23, 321 (1951).

3. Jamieson, D.T., Irving, J.B., Tudhope, J.S., "Liquid Thermal Conductivity - A data survey to 1973", HMSO, Edinburg(1975)

4. Goldschmidt, R.,Physik. Z., 12, 417 (1911).

5. Bridgman, P.W., Proc. Amer. Acad. Arts. Sci.,59, 141 (1923).

6. Rice, C.W., Phys.Rev. 23, 306 (1924).

7. Davis, A.H., Phil. Mag., 47, 972 (1924).

8. Smith, J.F.D., Ind. Eng. Chem., 22, 1246 (1930).

9. Abas-Zade, A.K., Dokl. Akad. Nauk. Azerb. SSR, 3, 3 (1947).

10. Filipov, L.P., Vestnik Mosk.gos.Univ., 8, 109 (1953).

11. Schmidt, E., Leidenfrost, W., Chem.Ingr.Tech., 26, 35 (1954).

12. Challoner, A.R., Powell, R.W., Proc.Roy.Soc. 238A, 90 (1956).

13. Vargaftik, N.B., Proc.Conf.Thermodynamic and Transport Properties of Fluids, IME, London (1957), pp.142.

14. Schrock, V.E., Starkman, E.S., Rev.Scient.Inst.,29, 625(1958).

15. Ziebland, Int.J.Heat Mass Transfer, 2, 273 (1961).

16. Horrocks, J.K., McLaughlin, E., Ubbelohde, A.R., Trans.Farad. Soc., 59, 1110 (1963).

17. Tufeu,R., Le Neindre,B., Johannin,P. C.R.Acad.Sci. Paris, 262B, 229 (1966).

18. Venart, J.E.S., J. Scient. Inst. $\underline{41}$, 727 (1964).

19. Mukhamedzyanov, G. Kh., Usmanov, A.G., Tarzimanov, A.A., Izv. Vyssh Ucheb.Zaved. Neft'i Gaz., $\underline{7}$, 70 (1964).

20. Poltz,H., Jugel, R., Int. J.Heat Mass Transfer,$\underline{10}$,1075(1967).

21. Pitmann, J.F.T., Ph.D Thesis, ICST, London (1968).

22. Rastorguev, Yu.L., Pugach, V.V., Izv.Vyssh. Ucheb. Zaved., Neft'i Gaz., $\underline{13}$, 69 (1970).

23. Mani, N., Venart, J.E.S., Proc. 12th Conf. Thermal Conductivity, Birmingham, Alabama (1972) pp 166.

24. Trump,W.N., Luebke, H.W., Fowler, L., Emery, E.M., Rev.Sci. Inst., $\underline{48}$, 47 (1977).

25. Nieto de Castro, C.A., Calado, J.C.G., Wakeham, W.A., Proc. 7th Symp. Therm. Prop. ed ASME (1978) pp 730.

26. Nagasaka, Y. and Nagashima, A., Proc. 1st Japanese Symposium on Thermophysical Properties, Tokyo (1980).

27. Healy, J.J., de Groot, J.J., Kestin, J. Physica,$\underline{82}$, 392(1976).

28. Nieto de Castro, C.A., Calado, J.C.G., Wakeham, W.A. Dix, M., J. Phys. E., Sci. Inst., $\underline{9}$, 1073 (1976).

29. Menashe, J. Wakeham, W.A., Ber. Bunsenges, Phys.Chem. $\underline{85}$, 340 (1981).

30. Michels, A., Sengers, J.V., Van der Gulik, P.S., Physica,$\underline{28}$, 1201 (1962).

31. Le Neindre, Thesis, Université Paris VI (1969).

32. Nieto de Castro, C.A., Wakeham, W.A., "Thermal Conductivity 15", Plenum Press, (1978), p. 236.

33. Menashe, J., Wakeham, W.A., Int.J. Heat Mass Transfer,$\underline{25}$, 661 (1982).

34. Leidenfrost, W., Int. J. Heat Mass Transfer $\underline{7}$, 447 (1964).

35. Viskanta, R., Grosh, R., J. Heat Transfer, Trans. ASME, 84C, 63 (1962).

36. Li, S.F.Y., Nieto de Castro, C.A., Wakeham, W.A., Int. J. Therm. (in press).

37. Hottel, H.C., Sarafim, A.F.,"Radiative Transfer", McGraw Hill (1967).

38. Nieto de Castro, C.A., Calado, J.C.G., Wakeham, W.A., High Temperatures-High Pressure, $\underline{11}$, 551 (1979).

39. Menashe, J., Wakeham, W.A., Ber. Bunsenges. Phys. Chem., $\underline{86}$, 541 (1982).

40. Mustafa,M., Sage, M., Wakeham, W.A.,Int.J.Therm.,$\underline{3}$,217(1982).

41. Calado, J.C.G. Fareleira, J.M.N.A., Nieto de Castro, C.A., Wakeham, W.A., Int. J. Therm., $\underline{4}$, 193 (1983).

42. Mardolcar, U.V., Wakeham, W.A., Int. J. Therm., $\underline{4}$, 1 (1983).

ORGANIC LIQUID MIXTURES: MEASUREMENT AND

ESTIMATION AT DIFFERENT TEMPERATURES

C. Baroncini, G. Latini and P. Pierpaoli

Istituto di Energetica, Facoltà di Ingegneria
Università di Ancona
Via della Montagnola n. 30, 60100 Ancona - ITALY

EXTENDED ABSTRACT

The value of the thermal conductivity λ_m of organic liquid
mixtures is required in several engineering problems, but experimental data exist for only a small portion of the enormous variety
of the possible mixtures, and the measurements generally are carried
out at only one value of temperature. A general correlation [1,2],
proposed for pure organic and inorganic liquids, in the reduced
temperature range 0.3 to 0.8 and at atmospheric pressure or along
the saturation line, was generalized in the form [3]:

$$\lambda_m = [A_1 x_1^2 + A_2 x_2^2 + n \sqrt{A_1 A_2} \, x_1 x_2] \; \frac{[1 - T_{rm}]^{0.38}}{T_{rm}^{1/6}} \qquad (1)$$

in order to predict the thermal conductivity λ_m of the organic
liquid binary mixtures. In equation [1] A_1 and A_2 are factors
characteristic of the mixture's components, x_1 and x_2 their respective molar concentrations, n a pure number independent of x_1 and
x_2 ranging from 1.0 to 2.0, and T_{rm} the "critical" temperature of
the mixture, obtained by a linear mole - fraction average. The
mean and maximum deviation between predicted and experimental thermal conductivity data for 30 binary mixtures were 1.8% and 8.3%
respectively.

The advantages of correlation [1] are the following:
1) equation [1] does not need values of the thermal conductivity
   of the components in the mixture, but contains the temperature-
   independent factors A_1 and A_2;

2) equation [1], if A_1, A_2 |1,4| and n are known, can be used at every temperature.

The subject of the present work is the pure number n which changes from mixture to mixture.

A preceding investigation [3] leads one to suppose that n is independent of temperature and does not give any indication of its value when experimental thermal conductivity data λ_m are not available; thus the objects of the present investigation are two:

1) n is effectively a constant (independent of temperature and of concentration) characteristic of each given mixture;
2) accurate indications are given in order to predict the value of n for binary mixtures.

The method used to obtain the above results is based on a critical study of experimental thermal conductivity data obtained at Ancona University on 25 organic liquid binary mixtures at different temperatures and concentrations of the components (250 experimental λ_m data).

The mixtures investigated are listed in Table I. The correlation [1] contains parameters easily obtained by means of the most important physical properties.

Table-1 Investigated Mixtures

ACETONE - n-BUTYL ALCOHOL
ACETONE - sec-BUTYL ALCOHOL
n-HEXYL ALCOHOL - ACETONE
n-BUTYL ACETATE - sec-BUTYL ALCOHOL
n-BUTYL ACETATE - DIETHYL ETHER
n-BUTYL ACETATE - ETHYLMETHYL KETONE
n-BUTYL ACETATE - TOLUENE
n-BUTYL ALCOHOL - DIETHYL KETONE
n-BUTYL ALCOHOL - n-HEXYL ALCOHOL
sec-BUTYL ALCOHOL - TOLUENE
DI-n-BUTYL ETHER - METHYL ALCOHOL
CARBON TETRACHLORIDE - CYCLOHEXANE
CARBON TETRACHLORIDE - ETHYL ALCOHOL
CARBON TETRACHLORIDE - n-HEXYL ALCOHOL
CARBON TETRACHLORIDE - n-PROPYL ALCOHOL
CHLOROFORM - DI-n-BUTYL ETHER
CHLOROFORM - DIETHYL KETONE
CHLOROFORM - ETHYLMETHYL KETONE
CHLOROFORM - METHYL ALCOHOL
CYCLOHEXANE - ETHYL ALCOHOL
CYCLOHEXANE - n-PROPYL ALCOHOL
ETHYL ALCOHOL - METHYL ALCOHOL
ETHYL ALCOHOL - n-PROPYL ALCOHOL
ETHYLMETHYL KETONE - TOLUENE
n-PROPYL ALCOHOL - TOLUENE

REFERENCES

1. Baroncini, C., Di Filippo, P., Latini, G., and Pacetti, M.,
    Int. J. Thermophys. 2: 21 (1981).
2. Baroncini, C., and Latini, G., "Symposium on Transport
    Properties of Fluids". Lisbon (Portugal) 25-26 March 1982.
3. Baroncini, C., Di Filippo, P., Latini, G., and Pacetti, M.,
    "Thermal Conductivity,"Vol. 17 (Plenum Press, N.Y. 1983).
4. Baroncini, C., Di Filippo, P., and Latini, G., Int. J. Re-
    frigeration, 6 (1): 60 (1983).

THERMAL CONDUCTIVITY OF METHANE AND ETHANE

R. C. Prasad

University of New Brunswick
Saint John, N.B., Canada

J. E. S. Venart

University of New Brunswick
Fredericton, N.B., Canada

and

N. Mani

Private Consultant
Calgary, Alberta, Canada

ABSTRACT

The transient line-source technique has been utilized to obtain thermal conductivity data for methane and ethane. Measurements are obtained for ethane in a temperature range from 295 K to 600 K at pressures to 70 MPa. Measurements are obtained for methane in the range of 100 K to 400 K and pressures from 2 to 70 MPa.

Correlations for the thermal conductivity of methane and ethane at low pressure as well as general correlations of the experimental data are presented. The thermal conductivity anomaly is noted along 295 K, 318 K, 350 K and 398 K isotherms for ethane. A simple model is recommended to estimate the anomalous conductivity. Some evidence of critical anomaly is seen along the 200 K and 240 K isotherms in the case of methane.

It is estimated that the measurements on ethane are obtained with a maximum uncertainty of about 1.5% in the dense region and

about 2.5% in the low-density and critical regions. The correlation
has an estimated uncertainty of less than 2.5 percent. In the case
of methane, the maximum uncertainties are estimated to be 1 percent
and 3 percent for the measurements and the correlation respectively.

KEYWORDS

Methane, ethane, thermal conductivity, heat conductivity,
transport property, thermophysical property, conductivity measurement.

INTRODUCTION

A survey of experimental measurements of the thermal conductivity of methane[1,2] and ethane[3,4] indicate the need for additional
and more accurate measurements.

In this paper, general correlations for the experimental data
of the thermal conductivity of methane and ethane in the temperature
range of 120-400 K and 295-600 K respectively at pressures to 700
bars are reported. The correlation for methane is accurate to within
3 percent. For ethane, the correlation is estimated accurate to
within 2.5 percent in general and 5 percent in the anomalous region.

EXPERIMENTAL

The correlations are based on experimental data obtained with
a transient hot-wire instrument[5,6] utilizing a thin platinum wire.
Potential leads were welded onto the main wire to provide a test
section of 8 to 9 centimeters long.

The platinum wire was stretched vertically in the test fluid
and electrically heated with a step current. To a first approximation, the wire simulated an infinite line source with constant heat
dissipation. The conductivity was calculated[7] from

$$\Delta T_w(t_1,t_2) = \frac{\dot{Q}_\ell}{4\pi\lambda} \ln(t_2/t_1) \tag{1}$$

Corrections for finite heat capacity[8] of the wire and non-
constant heat dissipation were applied. The apparent thermal
conductivities at increasing times were computed and a true conductivity estimated by extrapolation to zero time with the
condition[5]

$$\frac{d\lambda}{dt}\bigg|_{t=0} = 0 \qquad\qquad (2)$$

The transient response of the wire was recorded by an on-line computerized data-acquisition system[5,6].

RESULTS

The thermal conductivity of methane and ethane were measured along fifteen nominal isotherms from 120 K to 400 K and six nominal isotherms from 295 K to 600 K respectively. These data are reported elsewhere[2,4].

CORRELATION

A correlation was attempted according to the model[9](Figure-1):

$$\lambda = \lambda_{bg} + \Delta\lambda_{cr} \qquad\qquad (3)$$

$$\lambda_{bg} = \lambda_1 + \Delta\lambda_e \qquad\qquad (4)$$

Thermal Conductivity at Low Pressure

Thermal conductivity of ethane vapor at low pressure ($\rho \to 0$, $P \cong 1$ bar) was obtained[4] by correlating the measurements at low and moderate pressures in terms of density for a particular isotherm. For methane, the experimental data reported by other investigators[10-18] were used.

A correlation of λ_1 was obtained in terms of temperature with[9]

$$\lambda_1(T) = \sqrt{T_r} \Big/ \sum_{k=o}^{n} (a_k/T_r^k) \qquad\qquad (5)$$

and least square estimates of the parameters a_k were obtained[2,4] (Tables-1a,2a).

The Excess Thermal Conductivity ($\Delta\lambda_e$)

The measurements in high density regions indicated an increase in the excess thermal conductivity along any isotherm. Anomalous increases were observed in the critical region[2,4] and near the saturation line[4]. However, data remote from the anomalous region was used to develop a correlation. In the case of methane, other

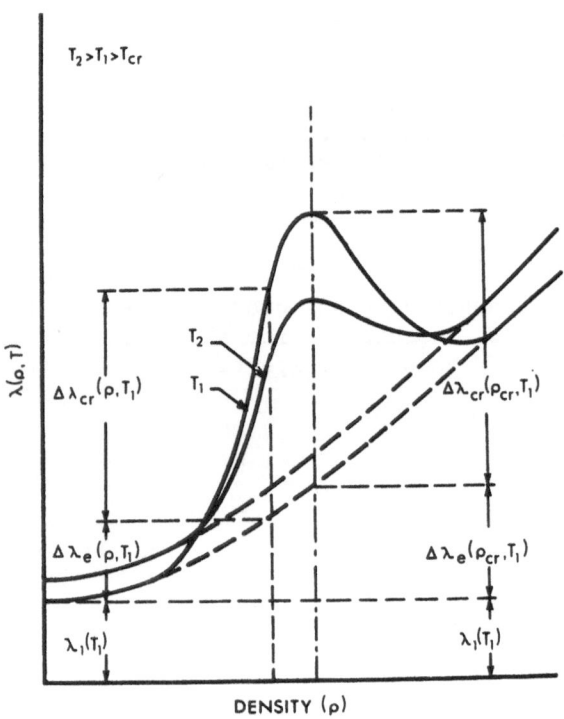

Figure-1   The thermal conductivity model

Table-1   Estimates of the parameters of the correlation model
for the thermal conductivity of methane.

(a) Correlation for $\lambda_1(T)$ in the range of 120 K to 622 K
with equation-5 :

$$
\begin{aligned}
n=3 \qquad a_0 &= -0.47313668 \times 10^{-2} \\
a_1 &= 9.35882659 \times 10^{-2} \\
a_2 &= -5.37104001 \times 10^{-2} \\
a_3 &= 1.32157337 \times 10^{-2}
\end{aligned}
$$

(b) Correlation for $\Delta\lambda_e$ as a function of density with
equation-6 :

$$
\begin{aligned}
k=1,2,5 \qquad b_1 &= 15.94004983 \\
b_2 &= 10.20614596 \\
b_5 &= 0.47065428
\end{aligned}
$$

Table-2 Estimates of the parameters of the correlation model
for the thermal conductivity of ethane.

(a) Correlation for $\lambda_1(T)$ in the range 295 K $\leq$ T $\leq$ 600 K
with equation-5 :

$$n=2 \qquad \begin{aligned} a_0 &= -0.284425 \times 10^{-3} \\ a_1 &= 2.862191 \times 10^{-2} \\ a_2 &= 1.786388 \times 10^{-2} \end{aligned}$$

(b) Correlation of $\Delta\lambda_e$ as a function of density and temperature
with equation-7 :

m=2 , n=3

| | b_{ij} | | |
| | | j | |
i	0	1	2	3
0	0.0	-14.158643	15.995975	1.917451
1	-1.396130	62.724318	-54.924712	17.953475
2	1.000647	-26.933607	27.389732	-9.696567

(c) Correlation of $\Delta\lambda_{cr}$ as a function of density and
temperature with equation-8 :

$$\begin{aligned} c_1 &= 0.40594041 \\ c_2 &= 0.01903895 \\ c_3 &= 0.14378686 \\ c_4 &= 1.86698399 \end{aligned}$$

---

experimental data [10-17,19,20] in the region $\rho < 0.06$ g.cm^{-3} were
included in the data set. Correlations in terms of density alone
with equation

$$\Delta\lambda_e(\rho) = \sum_k b_k \rho_r^k \quad , \quad b_o = 0 \qquad (6)$$

as well as in terms of density and temperature with equation

$$\Delta\lambda_e(\rho,T) = \sum_{i=o}^{m} \sum_{j=o}^{n} b_{ij} T_r^i \rho_r^j \quad , \quad b_{oo} = 0 \qquad (7)$$

were attempted[2,4,21].

For methane, a correlation of the data to within 2 to 3 percent and those of other's data within 3 to 4 percent was obtained[2] with equation-5 (n=3) and equation-6 (k=1,2,5). Least-squares estimates of b_k are listed in Table-1b and Figure-2 shows the deviation plot.

For ethane, a correlation within 2.5 percent was obtained[4,21] with equation-5 (n=2) and equation-7 (m=2,n=3). Least-squares estimates of the parameters b_{ij} are listed in Table-2b and the deviation plot is shown in Figure-3.

The Anomalous Thermal Conductivity ($\Delta\lambda_{cr}$)

The anomalous conductivity values were estimated by subtracting the background thermal conductivity, λ_{bg}, from the experimental values in the anomalous region.

A correlation of $\Delta\lambda_{cr}$ in terms of density and temperature with[22,4]

$$\Delta\lambda_{cr}(\rho,T) = A e^{-x^2} \tag{8a}$$

$$A = c_1/[(\Delta T^*)^2 + c_2] \tag{8b}$$

$$x = c_4[\Delta\rho^* - c_3(\Delta T^*)^n] \tag{8c}$$

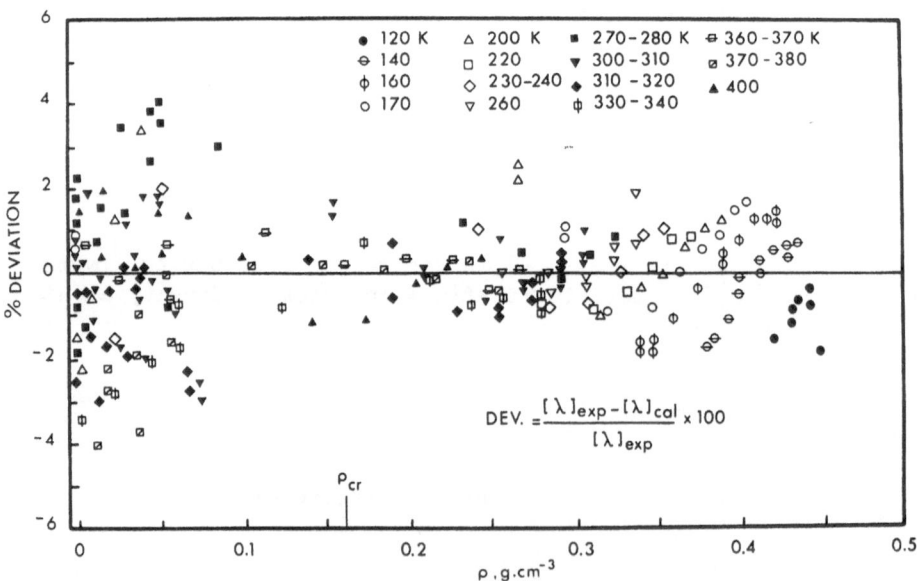

Figure-2   Deviation plot - correlation of the thermal conductivity of methane.

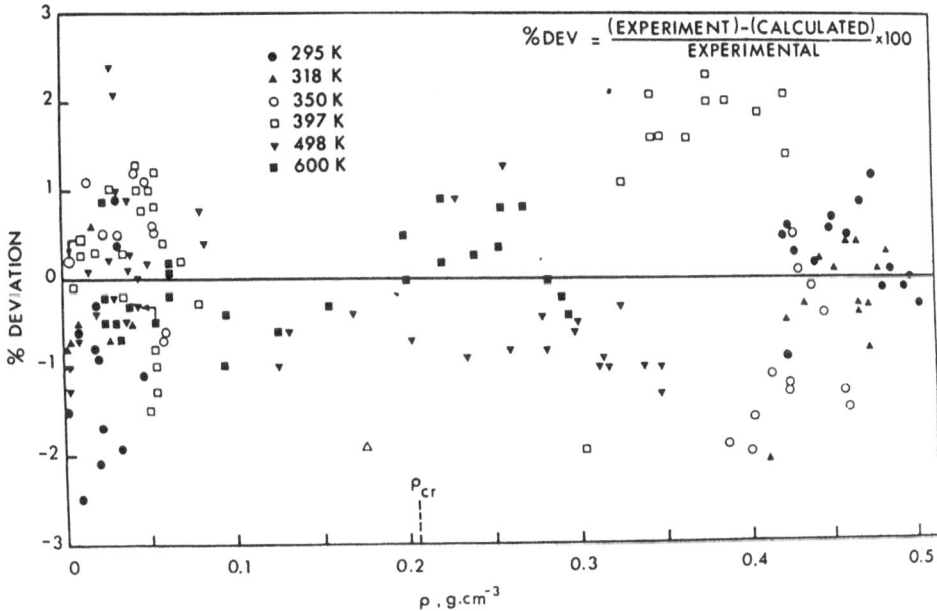

Figure-3 Deviation plot — correlation of thermal conductivity
of ethane (Equations-5,7).

was obtained[4] within 5 percent with n=0 for ethane. Estimates of
the parameter c_1 to c_4 are listed in Table-2c and the deviation
plot is shown in Figure-4. A similar correlation for methane was,
however, not possible due to lack of experimental data in the
anomalous region.

DISCUSSION

Based on the correlations for thermal conductivity of
methane[2] and ethane[4], smoothed values were computed. These
values are shown in the form of $\lambda \sim P$ isotherms (Figures-5 and 6).

The correlation reported for $\Delta \lambda_{cr}$ of ethane is based on
measurements in the extended critical region[22] and, thus, recommended in this region. In the critical region proper where

$$\left| \Delta T^* \right| \leqq 0.03 \quad \text{and} \quad \left| \Delta \rho^* \right| \leqq 0.25 , \tag{9}$$

the method suggested by Sengers et al[9] should be employed.

In absence of a correlation for $\Delta \lambda_{cr}$ of methane, only the
background thermal conductivity is determined in the anomalous
region and total conductivity outside this region.

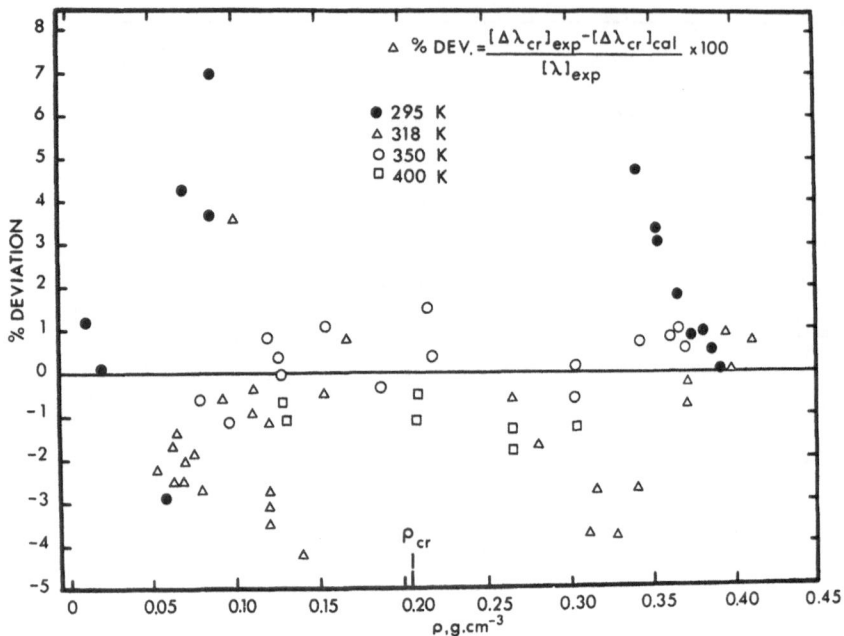

Figure-4 Deviation plot - correlation of anomalous thermal
conductivity $\Delta\lambda_{cr}$ of ethane (Equation-8).

CONCLUSION

Correlations for the thermal conductivity of methane and ethane
in the temperature range of 120-400 K and 290-600 K respectively at
pressures to 700 bars are reported.

The correlation for methane is estimated accurate to within
about 3 percent outside the anomalous region. Accurate measurements
are needed in this region to develop an accurate statistical model
for the anomalous conductivity.

For ethane, this correlation is estimated accurate to within
2.5 percent in general and 5 percent in the anomalous region.

NOMENCLATURE

a_k
b_{ij} Parameters of regression model: k=0 to n, i=0 to m,
b_k j=0 to n
c_i

P Pressure, MPa (or bar)
$\dot{Q}_\ell$ Heat flux per unit length, mW.m^{-1}

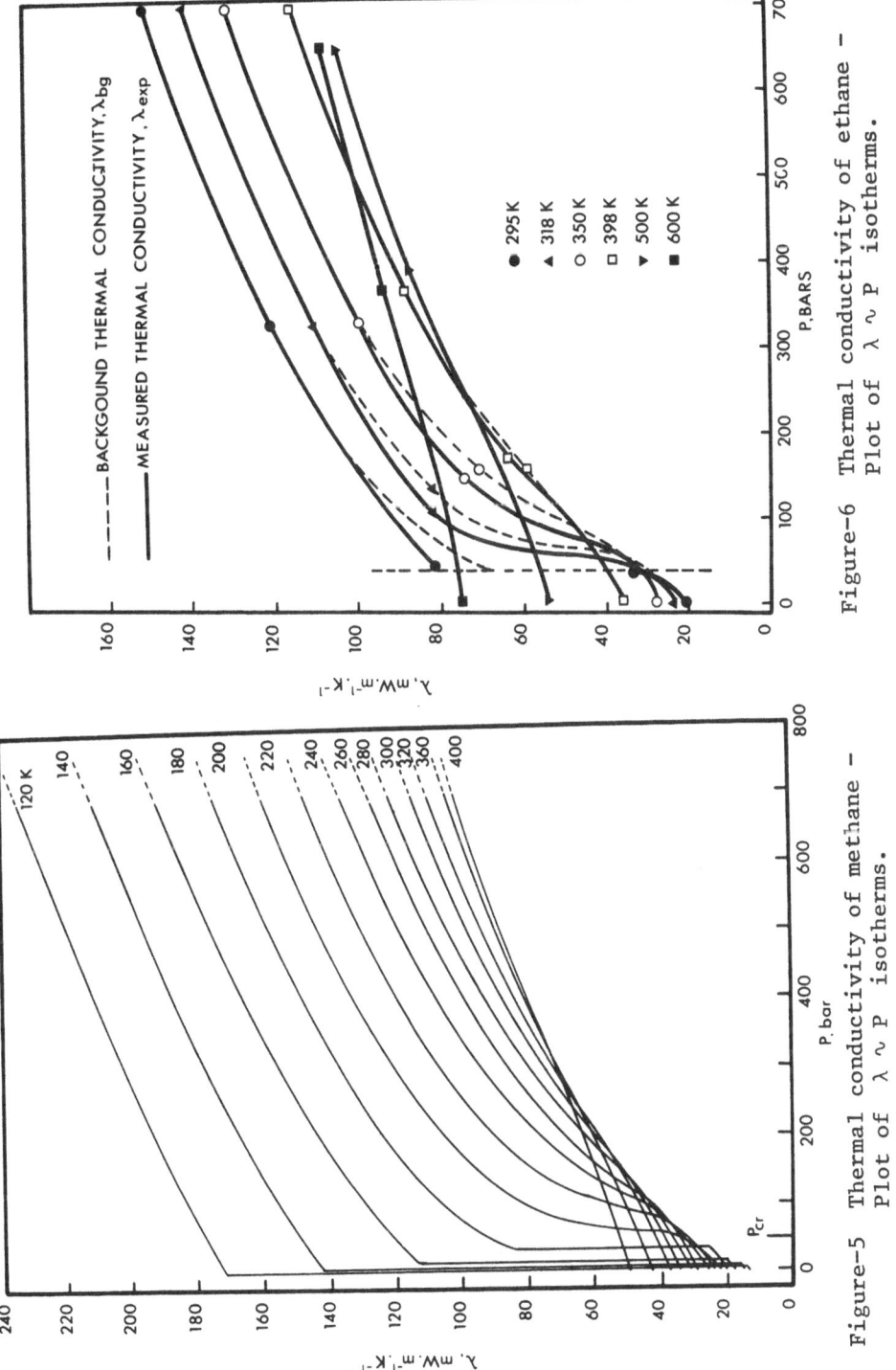

Figure-6 Thermal conductivity of ethane –
Plot of $\lambda \sim P$ isotherms.

Figure-5 Thermal conductivity of methane –
Plot of $\lambda \sim P$ isotherms.

89

t	Time, second
T	Temperature, K
T_{cr}	Critical Temperature, K
T_r	Reduced temperature = T/T_{cr}
ΔT_w	Temperature rise of wire between time t_1 and t_2
ΔT^*	Reduced temperature difference = $(T - T_{cr})/T_{cr}$
λ	Thermal conductivity, $mW.m^{-1}.K^{-1}$
λ_1	Thermal conductivity at low pressure ($P \cong 1$ bar), $mW.m^{-1}.K^{-1}$
λ_{bg}	Background thermal conductivity, $mW.m^{-1}.K^{-1}$
$\Delta\lambda_{cr}$	Thermal conductivity anomaly, $mW.m^{-1}.K^{-1}$
$\Delta\lambda_e$	Excess thermal conductivity, $mW.m^{-1}.K^{-1}$
ρ	Density, $g.cm^{-3}$
ρ_{cr}	Critical density, $g.cm^{-3}$
ρ_r	Reduced density = ρ/ρ_{cr}
$\Delta\rho^*$	Reduced density difference = $(\rho-\rho_{cr})/\rho_{cr}$

ACKNOWLEDGEMENTS

This work was performed under a program of studies funded by the Natural Sciences and Engineering Research Council, Canada under Grant no. A8859.

REFERENCES

[1] Hanley, H. J. M., Gubbins, K. E. and Murad, S., J. Phys. Chem. Ref. Data 16, 1167-1180 (1977).

[2] Prasad, R. C., Mani, N. and Venart, J. E. S.,"The thermal conductivity of methane", to be published in Int. J. Thermophys.

[3] Hanley, H. J. M., Haynes, W. M. and McCarty, R. D.,"The viscosity and thermal conductivity coefficients for dense gaseous and liquid methane", J. Phys. Chem. Ref. Data 6, 597-609(1977).

[4] Prasad, R. C. and Venart, J. E. S., "The thermal conductivity of ethane", to be published in Int. J. Thermophys.

[5] Mani, N.,"Precise determination of the thermal conductivity of fluids using absolute transient hot-wire technique", Ph.D. dissertation (1971), Univ. of Calgary, Canada.

[6] Prasad, R. C. and Venart, J. E. S., "An apparatus to measure the thermal conductivity of fluids using the transient line-source technique", to be published.

[7] van der Held, E. F. M. and van Drunen, F. G., "Method of measuring the thermal conductivity of fluids", Physica 15, 865-881 (1949).

[8] Carslaw, H. S. and Jaeger, J. C., "Conduction of Heat in Solids", Oxford University Press (1959).

[9] Sengers, J. V., Basu, R. S. and Sengers, J. M. H. L., "Representative equations for the thermodynamic and transport properties of fluids near the gas-liquid critical point", NASA Contractor Report 3424 (1981).

[10] Johnston, H. L. and Grilly, E. R., "Thermal conductivity of eight common gases between 80 K and 380 K", J. Chem. Phys. 14, 233 (1946).

[11] Keyes, F. G., "Thermal conductivity of gases", Trans. ASME, July (1954).

[12] Golubev, I. F., "A calorimeter for determining the thermal conductivity of gases and liquids at high pressures and various temperatures", Teploenergetica 12, 78 (1963).

[13] Carmichael, L. T., Reamer, H. H. and Sage, B. H., "Thermal conductivity of fluids. Methane II", J. Chem. Engg. Data 11, 52 (1966).

[14] Misic, D. and Thodos, G., "Thermal conductivity measurements of methane in the dense gaseous state", Physica 32, 885(1966).

[15] Le Neindre, B., Tufeu, R., Bury, P., Johannin, P. and Vodar, B., "Experimental study of thermal conductivity coefficients of methane and ethane between 25 C and 450 C at pressures up to 1000 bars", Laboratoire des Hautes Pressions - C.N.R.S.-1, Place a. Briand-Bellevue, France (1969).

[16] Eucken, A., Z. Physik 14, 342 (1913), referred from [18].

[17] Weber, S., Ann. d. Physik 54, 437 (1917), referred from [18].

[18] Mann, W. B. and Dickins, B. G., "The thermal conductivities of the saturated hydrocarbons in the gaseous state", Proc. Roy. Soc. (London) A134, 77-96 (1931).

[19] Ikenberry, L. D. and Rice, S. A., "On the kinetic theory of dense fluids. XIV. Experimental and theoretical studies of thermal conductivity in liquid Ar, Kr, Xe and CH_4", J. Chem. Phys. 39, 1561 (1963).

[20] Rosenbaum, B. H. and Thodos, G., "Thermal conductivity of mixtures in the dense gaseous state. The methane-carbon tetrafluoride system", Physica 37, 442 (1967).

[21] Prasad, R. C. and Venart, J. E. S., "The thermal conductivity of ethane and ethylene", Proc. 8th Symposium on Thermophysical Properties, J. V. Sengers (Editor), ASME, New York, 263-268 (1982).

[22] Roder, H. M., "The thermal conductivity of oxygen", J. of Research of N.B.S. 87, 279 (1982).

SESSION E

Salts, Oxides and Binary Compounds

SESSION CHAIRMAN

P. J. Klemens
University of Connecticut
Storrs, CT.

THERMAL DIFFUSIVITY MEASUREMENT OF MOLTEN FLUORIDE SALT

CONTAINING ThF_4 (IMPROVEMENT OF THE SIMPLE CERAMIC CELL)

Yoshio Kato

Department of Fuels and Materials Research
Japan Atomic Energy Research Institute
Tokai, Ibaraki-Ken,319-11, Japan

Nobuyuki Araki, Kiyosi Kobayasi and Atsushi Makino

Department of Mechanical Engineering
Shizuoka University
3-5-1, Johoku, Hamamatsu, Shizuoka, Japan

ABSTRACT

Design conditions of a cylindrical ceramic cell were estimated. One can use it to measure the absolute value of thermal diffusivity of molten salts by applying the stepwise heating method. For a thickness ℓ of the salt specimen, the thickness of the upper and lower parallel ceramic base plates which hold the specimen between them should be greater than 5ℓ and the radius of the base plates should be greater than 15ℓ. This satisfies the theoretical conditions of a semi-infinite solid base plate to sufficient accuracy. Using a ceramic cell which has a large enough value of ℓ, the thermal diffusivity, a, of molten fluoride salt: $LiF-BeF_2-ThF_4$ (67-18 -15 mol%) was measured. The value a = $2.34 \times 10^{-7} + 6.10 \times 10^{-11}(T-788)$ $m^2 s^{-1}$ was obtained in the temperature range 823 $\leqslant$ T $\leqslant$ 983 K.

INTRODUCTION

Molten salt is expected to be used in nuclear systems such as the Molten-Salt Reactor, the Accelerator Molten-Salt Breeder, the Fusion Reactor Blanket Coolant, the Fuel Reprocessing System, and so on. Many candidate salts have been recommended for that purpose but there is little data on their thermal properties. Therefore a simple ceramic cell has been developed to measure conventionally

the thermal diffusivity of molten salts over a wide range of
temperatures. The apparatus for measuring the thermal diffusivity
of molten salt is required in general to satisfy the following
conditions.   (i) It should be designed to prevent the onset of
convective flow of the sample salt. (ii) The materials contacting
the molten salt should have high resistance to corrosion. (iii) It
should be easy to remove the bubbles from the specimen salt.

In the previous report[3] , the structure and the size of a
ceramic cell had been experimentally determined to satisfy approximately the theoretical boundary conditions. In this report, the
geometrical design conditions of the cell are analysed and the
feasibility of the ceramic cell is proven by measuring molten
fluoride salt(LiF-BeF$_2$-ThF$_4$) up to a temperature of about 980 K..
The salt is one of the candidate target salts of the Accelerator
Molten-Salt Breeder.

THEORY

A model of the ceramic cell is shown in Fig.1. In this figure,
the layer 2 indicates the sample salt and the layers 1 and 3 also
indicate the parallel ceramic base plates. A plane,infinitely thin
heat source is assumed to exist at the boundary of the layers 1 and
2.   A resistance thermometer which measures the transient
temperature at the bottom of layer 2 is located at the boundary of
the layers 2 and 3.   The physical properties of each layers are
distinguished by suffixes.

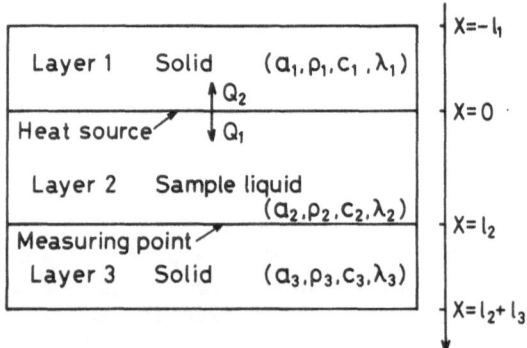

Fig. 1 Model of the cell

To simplify the problem, it is assumed at first that layers 1
and 3 are infinitely thick and wide semi-infinite solids. When the
sample salt is heated stepwise by the plane heat source, the
temperature rise $\theta_2(\ell_2,t)$ at the resistance thermometer is obtained
[1,2] as

$$\theta_2(\ell_2,t)=2\left[\frac{Q\ell_2}{\lambda_2}\right]\left[1+\frac{\lambda_1}{\lambda_2}\sqrt{\frac{a_2}{a_1}}\right]^{-1}(1-\delta)\sqrt{Fo_2}\{\sum_0^\infty(\delta)^{2n}\mathrm{ierfc}[(2n+1)/2\cdot\sqrt{Fo_2}]\},(1)$$

96

where material of the layer 1 and layer 3 is assumed to be the same and

$$ierfc(y) = \frac{1}{\sqrt{\pi}} \exp(-y) - y \cdot erfc(y), \qquad (2)$$

$$Fo_2 = \alpha_2 t / \ell_2^2 \qquad (3)$$

$$\delta = (1-\phi)/(1+\phi), \qquad \phi = \sqrt{\lambda_2 \rho_2 C_2 / \lambda_1 \rho_1 C_1},$$

α : thermal diffusivity $\quad \lambda$: thermal conductivity
C : specific heat, $\qquad\qquad \rho$: density
Q : generating heat flux ($Q = Q_1 + Q_2$; Q is step function)

The ratio of the temperature rise at two points in time, t and 2t, defined by

$$V_2 = \theta_2(\ell_2, 2t)/\theta_2(\ell_2, t), \qquad (4)$$

is calculated as a function of the Fourier number Fo_2. It was found that in the region of $Fo_2 < 0.3$, the value of V_2 did not depend on the parameter δ which describes the relative thermal properties of the layers 1 and 3 [1,2,3].

The absolute value of thermal diffusivity α_2 can be obtained as follows: from a measured temperature rise curve, the value of V is obtained and the corresponding value of Fo_2 is calculated theoretically. As ℓ_2 and t are the known values, α_2 is determined from eq. (3).

Next we consider the effect of finite size of the cell to obtain the conditions from which eq.(1) can be applied approximately. It is still assumed in Fig.1 that heat flow from the plane heat source is one-dimensional and the adiabatic condition is satisfied at the surface of the cell. When the sample salt is heated stepwise, the temperature rise at ℓ_2 becomes as follows:

$$\theta_2 = \frac{2\ell_{1/2}}{3(c\rho)_{3/2} a_{1/2}}$$

$$\times \left[\frac{3\{1+(\ell_{1/3}^2/a_{1/3})+(2a_{1/2}Fo_2/\ell_{3/2}^2)\}(X_1\omega_1+X_2\omega_2+X_3\omega_3+X_4\omega_4)-(X_1\omega_1^3+X_2\omega_2^3+X_3\omega_3^3+X_4\omega_4^3)}{(X_1\omega_1+X_2\omega_2+X_3\omega_3+X_4\omega_4)^2} \right.$$

$$\left. - \sum_{\kappa=1}^{\infty} \frac{12}{\gamma_\kappa^2} \cdot \frac{\cos(\ell_{1/3}\gamma_\kappa^2/(\sqrt{a})_{1/3}) \cdot \cos(\gamma_\kappa) \cdot \exp(-\gamma_\kappa^2 a_{3/2} Fo_2/\ell_{3/2}^2)}{X_1\omega_1\cos(\omega_1\gamma_\kappa)+X_2\omega_2\cos(\omega_2\gamma_\kappa)+X_3\omega_3\cos(\omega_3\gamma_\kappa)+X_4\omega_4\cos(\omega_4\gamma_\kappa)} \right], \qquad (5)$$

where

$X_1 = (\sqrt{a})_{1/3}(c\rho)_{1/3}+(\sqrt{a})_{1/2}(c\rho)_{1/2}+(\sqrt{a})_{2/3}(c\rho)_{2/3}+1, \quad \omega_1 = \ell_{1/3}/(\sqrt{a})_{1/3}+\ell_{2/3}/(\sqrt{a})_{2/3}+1$

$X_2 = (\sqrt{a})_{1/3}(c\rho)_{1/3}-(\sqrt{a})_{1/2}(c\rho)_{1/2}+(\sqrt{a})_{2/3}(c\rho)_{2/3}-1, \quad \omega_2 = \ell_{1/3}/(\sqrt{a})_{1/3}+\ell_{2/3}/(\sqrt{a})_{2/3}-1$

$X_3 = (\sqrt{a})_{1/3}(c\rho)_{1/3}-(\sqrt{a})_{1/2}(c\rho)_{1/2}-(\sqrt{a})_{2/3}(c\rho)_{2/3}+1, \quad \omega_3 = \ell_{1/3}/(\sqrt{a})_{1/3}-\ell_{2/3}/(\sqrt{a})_{2/3}+1$

$X_4 = (\sqrt{a})_{1/3}(c\rho)_{1/3}+(\sqrt{a})_{1/2}(c\rho)_{1/2}-(\sqrt{a})_{2/3}(c\rho)_{2/3}-1, \quad \omega_4 = \ell_{1/3}/(\sqrt{a})_{1/3}-\ell_{2/3}/(\sqrt{a})_{2/3}-1$

$\theta_2 = \theta_2(\ell_2, t)/(Q\ell_2/\lambda_2), \quad Fo_2 = a_2 t/\ell_2^2.$

and γ_k is the K'th positive root of the equation

$$X_1\sin(\omega_1\gamma)+X_2\sin(\omega_2\gamma)+X_3\sin(\omega_3\gamma)+X_4\sin(\omega_4\gamma)=0. \quad (6)$$

In eq.(5) the suffix 3/2, for example, means the ratio of the physical property of layer 3 to layer 2. An example of a numerical calculation of eq.(5) is shown in Fig.2. The meaning of $\ell_{1/2}=0$, for example, is that the layer 1 does not exist; its thickness is zero.

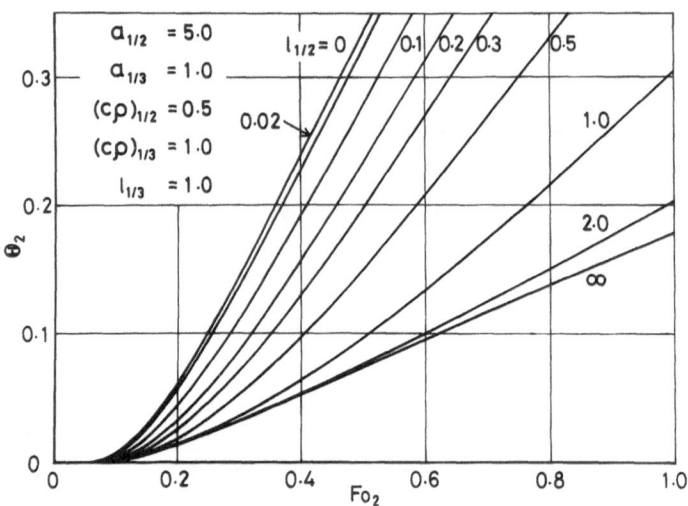

Fig. 2   Theoretical curve of dimensionless temperature rise.

The ratio of temperature rise V_2 given from eq.(5) as a function of Fo_2 is shown in Fig.3.   It is clear from Fig.3 that V_2 depends largely on the values of $\ell_{1/2}$.   The relations of V_2 and $\ell_{1/2}$ are shown in Fig. 4. As seen from this figure, V_2 becomes constant in the region $5 \leqslant \ell_{1/2} \leqslant \infty$ for $Fo_2=0.3$. It means that the cell should be designed to $\ell_{1/2}$(or $\ell_{3/2}$) $\geqslant$ 5 at least, and Fo_2 should be determined in the region $Fo_2 < 0.3$ in order to apply the semi-infinite approximation of eq.(1)

However the structure of the real ceramic cell is cylindrical and heat loss takes place from all surfaces of the cell by the temperature rise due to the stepwise heating. These effects are numerically analyzed for the model shown in Fig.5. The results are shown in Fig.6. In Fig.6, the X-axis denotes the ratio of the radius Ro of the cell to the thickness ℓ_2 of the sample salt, and the Y-axis indicates the error in the Fourier number Fo_2 determined by the temperature rise ratio V_2, where Fo_2' and Fo_2 correspond to the cases when a heat loss does or does not occur.

It is clear from Fig.6 that in the region Ro/ℓ_2 > 10, the error becomes constant and heat flow in the cell is considered to be one-dimensional.

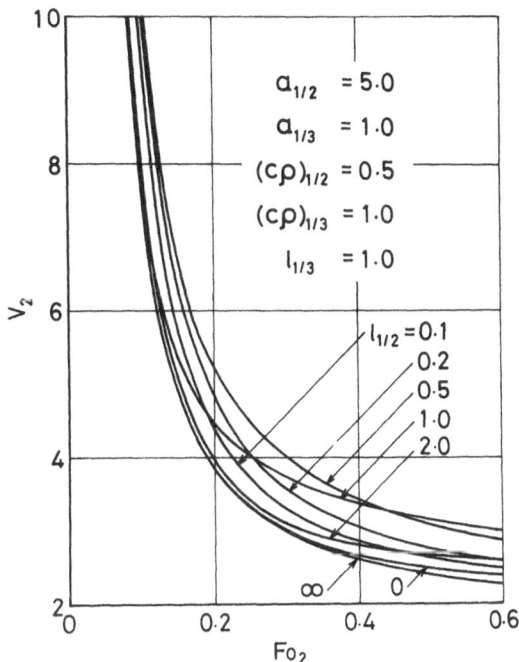

Fig. 3 Effect of the parameter $\ell_{1/2}$ on the theoretical curve
of the temperature rise ratio versus Fo_2.

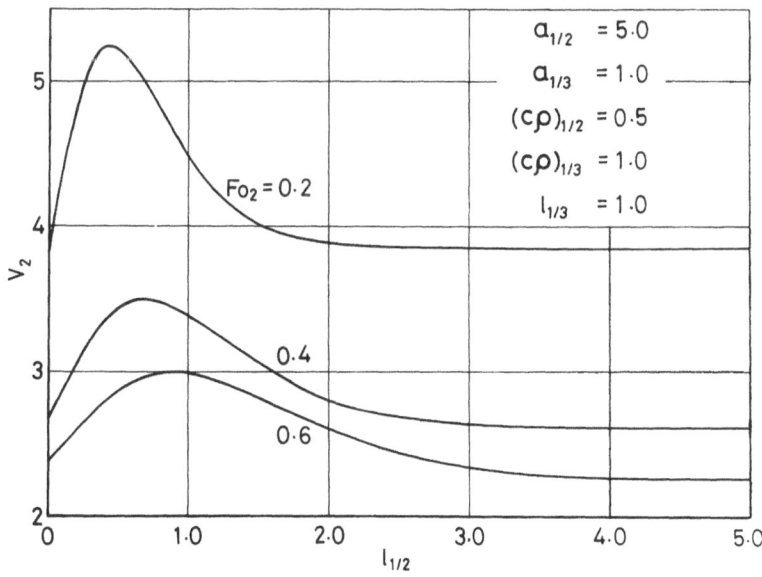

Fig. 4 Theoretical curve of the temperature rise ratio
versus $\ell_{1/2}$.

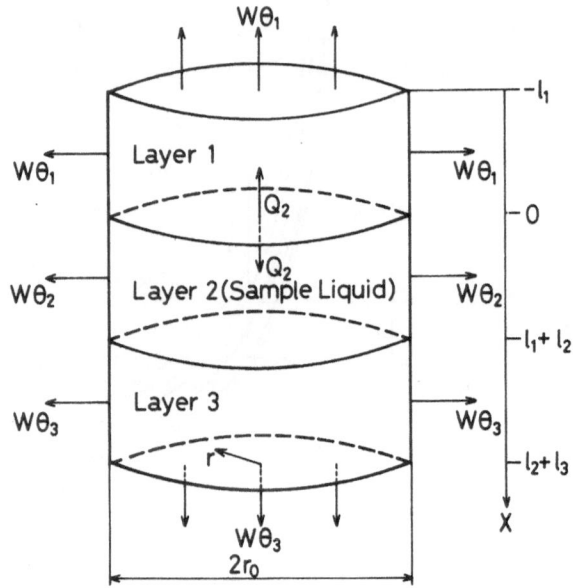

Fig. 5 Model of the cell with heat loss.

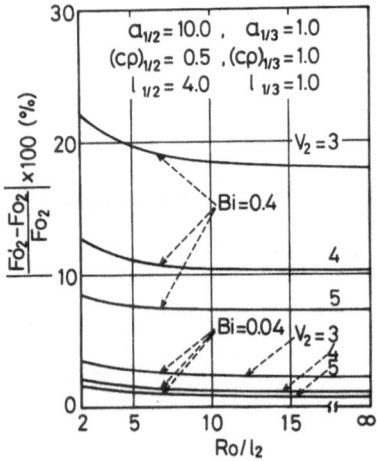

Fig. 6 Error caused by heat loss.

The error still remaines at $Ro/\ell_2 = \infty$. It is caused by the effect of heat loss from upper and lower surfaces of the cell. The error should decrease as the thickness increases. The Biot number is defined by $Bi = W\ell_2/\lambda$, where W means the equivalent heat transfer coefficient.

EXPERIMENTAL APPARATUS AND PROCEDURE

The ceramic cell used in this experiment was the same in size and structure as that in the previous report[2,3]. However the technique of a thin alumina coating on the metallized layers or the materials for adhesion of each ceramic part of the cell have been improved. The improved ceramic cell was expected to have better compatibility with the molten fluoride salts. A view of the ceramic cell and each alumina coating on the surfaces of the plane heater and the resistance thermometer are shown in Fig.7.

The thickness and the radius of both the upper and the lower alumina base plates which correspond to the layer 1 and 3 in Fig.1 are 30mm and 45mm respectively. Thus it is confirmed by the above theoretical analysis that the cell has a sufficient size for eq.(1) to apply.

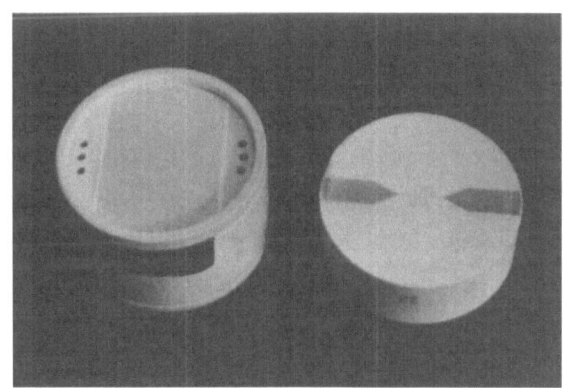

(a)   (b)
Fig. 7 Ceramic cell
(a) A view of the ceramic cell.
(b) Upper and lower base plates of the cell(sample).
    The metallized plane heat source and the resistance
    thermometer (right) are shown. Each metallized layer
    ($\sim$20μm)is coated by thin alumina layers ($\sim$20μm).

101

The cell is set in a cylindrical container as shown in Fig.8. In this figure, two sets of electrodes are for the heat source and for the resistance thermometer. Each electrode is used as a cover gas inlet tube or also as a guide for thermocouples. The upper part of the container has a water jacket. The temperature of three points on the surface of the cell (upper, middle and bottom) are controlled within ±1 K by a cylindrical furnace and auxiliary heater in the container.

A specimen salt LiF-BeF$_2$-ThF$_4$(67-18-15 mol %) was prepared as follows. At first LiF-BeF$_2$(70-30 mol % :17.49g), LiF(5.86g) and ThF$_4$(41.76g) were mixed. Then the mixture was heated up to 723 K to dry before melting and was kept under reduced gas pressure of about 10^{-3} Torr for 15 hours. Next,the salt temperature was increased for the melt to about 923 K and was kept for about 5 hours. He gas was bubbled in the molten salt to remove H$_2$O from the salt. Afterwards, the salt was cooled and the melting point was determined to be 788 K from the time-temperature cooling curve.

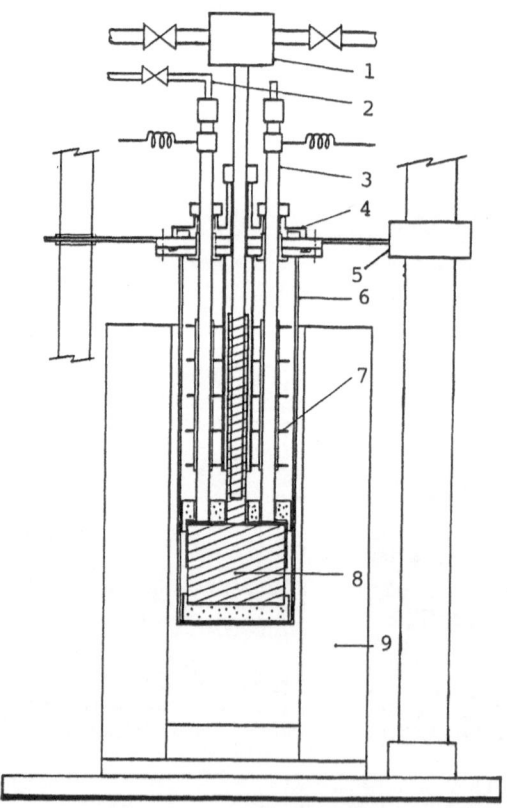

Fig. 8   Measuring apparatus

1: Vapour trap, 2: Cover gas inlet, 3: Electrode tube,
4: Water jacket, 5: Container lifter, 6: Container,
7: Reflector,   8: Ceramic cell,   9: Furnace.

RESULTS AND CONCLUSION

To test for reliable operation of the improved ceramic cell, the thermal diffusivities of distilled water were first measured and the results agreed well with the standard values. The thermal diffusivities of molten LiF-BeF$_2$-ThF$_4$(67-18-15 mol %) were then measured. The experimental results are shown in Fig.9. A slight temperature dependence was observed. The solid line was obtained by a least squares method as follows

$$a = 2.34 \times 10^{-7} + 6.10 \times 10^{-11} (T-788) \quad m^2 s^{-1},$$

$$823 \leqslant T \leqslant 983 \quad K.$$

The measurements for the salt were made in a He atmosphere at a pressure 5 Torr higher than atmospheric. The reproducibility depends on the initial thermal stability of the salt. In this case, the initial rate of temperature drift was kept below).04 K min^{-1}.

In conclusion, the design conditions of the simple ceramic cell have become clear on the basis of theoretical considerations and the feasibility of the cell has been proven in this experiment.

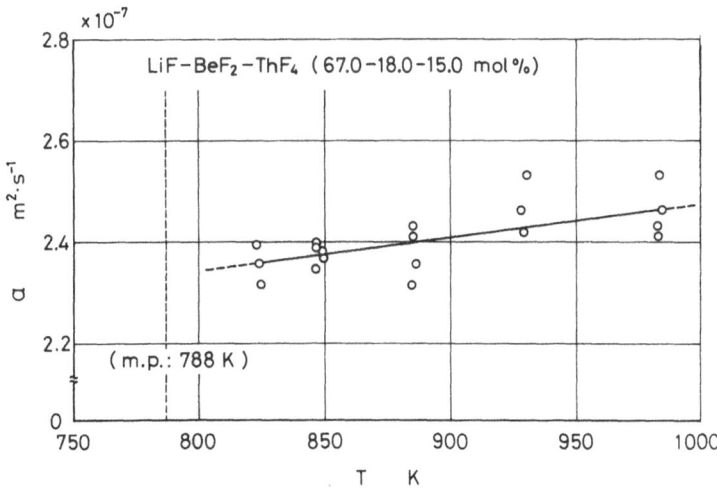

Fig. 9   Thermal diffusivity of molten LiF-BeF$_2$-ThF$_4$(67-18-15 mol %)

ACKNOWLEDGEMENT

The authors wish to thank Mr. Yutaka Sasahara of Sizuoka Univ. for his sincere co-operation in the experiment.

REFERENCES

[1]   Kobayasi, K. and Araki, N.,"Measuring Method of Thermal Diffusivity of Liquid by Stepwise Heating of Thin Layer", Proc. 5th Int. Heat Transfer Conf. Vol. V, (The Science Council of Japan), 247-251 (1974).

[2]   Kato, Y. and Furukawa, K.,"Thermal Diffusivity Mesurement of Molten Salts by Simple Apparatus", Proc. 1st Int. Symp. on Molten Salt Chemistry and Technology, (Molten Salt Committee of The Electrochemical Society of Japan), 269-272 (1983).

[3]   Kato, Y., Furukawa, K., Araki, N. and Kobayasi, K., "Thermal diffusivity measurement of molten salts by use of a simple ceramic cell", High Temp.-High Press., 15, 191-198 (1983).

THE THERMAL CONDUCTIVITY AND EXPANSION ENHANCEMENT ASSOCIATED

WITH FORMATION OF THE SUPERIONIC STATE IN $SrCl_2$*

J. P. Moore
Martin Marietta Energy Systems, Inc.
P. O. Box Y
Oak Ridge, Tennessee 37831

F. J. Weaver, R. S. Graves, and D. L. McElroy
Martin Marietta Energy Systems, Inc.
Metals and Ceramics Division
Oak Ridge National Laboratory
Oak Ridge, Tennessee 37831

A second-order phase transition in $SrCl_2$ near 1000 K produces superionic conduction and is often called the Bredig transition. This anion disordering occurs in other technologically important materials, such as UO_2, but at very high temperatures where measurements are difficult. Property studies on $SrCl_2$ at low temperatures may aid future high temperature studies.

Fine-grained, 99% dense, $SrCl_2$ samples containing three volume percent TiO_2 to reduce radiant transport were used to measure the differential thermal expansion coefficient (α) by push-rod dilatometry and the thermal conductivity (λ) by a radial heat flow method. Both properties show maxima near the Bredig transition. The peak α-value is over $75 \times 10^{-6} K^{-1}$. X-ray data in the literature show no evidence for this increase, and it is presumably due to the formation of localized defects. This large expansion coefficient corresponds to a volumetric increase of over 0.5%. Thus, push-rod dilatometry is a powerful diagnostic tool to explore the disordering process that occurs in many important AB_2 and A_2B compounds.

---

*Research sponsored by the Division of Materials Science, U.S. Department of Energy, under contract DE-AC05-84OR21400 with the Martin Marietta Energy Systems, Inc.

The data obtained from the radial heat flow method show good agreement with earlier tests at low temperatures, and the high temperature results show a local enhancement of λ of about 0.06 W/m·K. This enhancement cannot be attributed to electronic or radiant transport, but is due to a new mechanism of thermal diffusion of vacancy-anion interstitial pairs in the superionic state. The α and λ values used by the U.S. Nuclear Regulatory Commission for UO_2 fuel do not include the effects of this transition.

INTRODUCTION

Nearly 20 years ago ORNL scientists[1] reported the existence of a diffuse transition in solid ionic compounds (AB_2) having the fluorite (CaF_2) structure. The transition corresponds to a change of the B ion arrangement from an ordered array to a 3 to 15% state of disorder, while the A ions remain fixed. Subsequent research has shown this cooperative phenomenon occurs in many AB_2 and A_2B compounds at 80 to 90% of the melting temperature and is accompanied by a rapid increase of heat content,[2] a maximum in the specific heat, and development of super-ionic conductivity with increases in conductivity from 0.5 $\Omega^{-1}m^{-1}$ (characteristic of normal salt at its melting temperature) to 100 $\Omega^{-1}m^{-1}$ (characteristic of an ionic melt).[3]

The influence of this disordering on thermal conductivity has never been measured. For $SrCl_2$ the transition is near 1000 K (T_{mpt} = 1146 K, $T_t/T_{mpt} \sim 0.87$), which lies within the range of our capacity to measure thermal expansion (α) and thermal conductivity (λ). Obtaining and understanding such data can help define expectations for technologically important materials, such as, UO_2, ThO_2, and CaF_2.

SAMPLE PREPARATION AND MEASUREMENT TECHNIQUES

Fine-grained, 99% dense $SrCl_2$ specimen disks (containing 3 v/o TiO_2 to reduce radiant transport) were prepared for radial heat flow measurements of λ and for fused-quartz dilatometer measurements of α. The fabrication procedure used to produce the fine-grained disks has been described.[4] Because $SrCl_2$ is hygroscopic, the disks were machined under a paraffin oil to final dimensions with grooves and holes for thermocouples. The radial heat flow specimen included two central disks and each contained six thermocouple holes, 0.158 cm diam and 1.9 cm deep, with three at an inside diameter of 1.43 cm and three at an outside diameter of 4.32 cm. Table 1 indicates the height of the assembled disks

used in the two radial heat flow runs. This same machining procedure was used to produce two thermal expansion specimens, one from
a radial heat flow disk (SP2) and one from a melt of reagent grade
$SrCl_2$ that was slow cooled in a tapered graphite crucible (SP1).
A thin disk of the fine-grained $SrCl_2$ containing 3 v/o TiO_2 was
used to measure the percent transmission in the range 300 to
1000 K by optical spectrometry.

Table 1. $SrCl_2$ Specimens and Measurements

Thermal Expansion by fused Quartz Push Rod Dilatometry

	Nominal Length and Diameter (cm)	Temperature Range Studied (K)
SP1 - Cast $SrCl_2$	4.96 x 0.98	300 to 1000 300 to 1025
SP2 - Hot pressed $SrCl_2$ with 3 v/o TiO_2	4.20 x 0.99	300 to 1020

Thermal Conductivity by a Radial Heat Flow Technique

	Disk Dia. and Stack Hgt. (cm)	Temp. Range Studied (K)	Data Points
RHF1 - Hot pressed $SrCl_2$ with 3 v/o TiO_2	4.95 x 23	300 to 900	6
RHF2 - Hot pressed $SrCl_2$ with 3 v/o TiO_2	4.95 x 23	300 to 1020	12

Transmission by Spectrometry

	Disk Thickness (cm)	Measurement Temperatures (K)
T1 - Hot pressed $SrCl_2$ with 3 v/o TiO_2	0.0434	300, 540, 860, 1000

RESULTS

Thermal Expansion

The fused quartz push-rod dilatometer[5,6] was used to measure the temperature dependence of the differential length change, δL, between the quartz support tube and the specimen plus push rod. The length change is measured at a series of 24 hour holds at constant temperature, and the resulting δL versus T data are used to calculate the coefficient of thermal expansion. A correction is applied for the expansion of the quartz and dilatometer components based on data for SRM 739 fused silica. The equilibrium temperature holds were not equally spaced but selected to study the details of the transformation near 1000 K. Data were obtained for temperature intervals from 2 to 100 K between 300 and 1025 K for the specimens listed in Table 1. In general, the

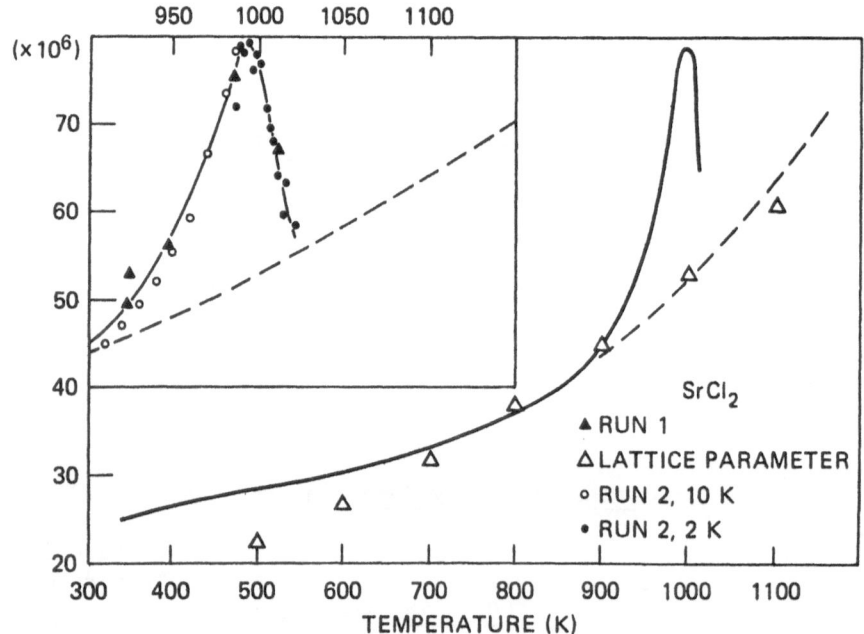

Figure 1: Differential thermal expansion coefficient of $SrCl_2$ as a function of temperature. Bold line represents smoothed values from a fused quartz dilatometer. Values indicated by (Δ) are from lattice parameter results of X-ray powder diffractometry.

Table 2. Smoothed values of the differential
thermal expansion coefficient*, α, of $SrCl_2$ and $SrCl_2$ with
3 v/o TiO_2 as a function of temperature

Temperature (K)	Differential Thermal Expansion Coefficient x $10^6 K^{-1}$	
	$SrCl_2$	$SrCl_2$ with 3 v/o TiO_2
300	24.2	23.1
400	26.2	25.1
500	28.4	27.2
600	30.8	29.55
700	33.5	32.1
800	37.25	35.4
900	43.3	42.3
920	47.2	47.3
940	51.5	51.5
960	60.3	60.3
970	66.3	66.3
980	75.0	70.6
990	79.4	71.8
1000	75.2	66.7
1010	67.3	58.3
1020	57.6	48.3

$$*\alpha = \frac{L(T_1)-L(T_2)}{L_{20}(T_1-T_2)} \quad , \text{ assigned to } T = (T_1+T_2)/2.$$

behavior of the specimens as a function of temperature was very
similar. Smoothed values of the differential thermal expansion
coefficient α are given in Table 2 and Figure 1. The α-values
rise from about 24 x $10^{-6} K^{-1}$ near 300 K to a peak value of over
75 x$10^{-6} K^{-1}$ near 1000 K and then decrease. Literature values[7]
of α derived from the lattice parameter of $SrCl_2$ as a function of
temperature are smooth, lower than the dilatometry results at low
temperatures and show no anomalous increase in α in the transition
region "such as might be expected, for example, if chlorine ions
were transferring sites of tetrahedral coordination to sites of
octohedral coordination."[7]

The volume increase from 300 to 1025 K is 7.9% with approximately 0.5% due to the transition. The α-value maximum (997 K) is
within 4 degrees of the specific heat maxima (1001 K) reported by
others for $SrCl_2$[2,8]. The current expansion results illustrate
a valuable tool for exploring this phenomenon in other compounds.

Thermal Conductivity

Two runs were made using the radial heat flow technique[9,10,11] to measure the temperature dependence of λ for $SrCl_2$ from 300 to 1020 K. The first run was terminated after obtaining data to 900 K because the thermal expansion of the $SrCl_2$ specimen caused open circuits to occur in the measuring thermocouples. The thermocouple assembly for the second run allowed for the thermal expansion of the specimen and yielded data to 1020 K. The measured λ values from Run 2 were within 1% of the Run 1 values to 900 K, and the average λ for the two measuring disks are given in Table 3.

The fraction of radiation transmitted by a 0.0434 cm thick $SrCl_2 + 3$ v/o TiO_2 specimen measured between 500 and 4000 wave numbers and 300 and 1000 K corresponded to about 3% at the wave length maximum for a black body.[12] This yields an absorption coefficient of 8070 m^{-1}, which was used with the Ginzel equation to calculate the radiative component of thermal conductivity indicated in Table 3. The resulting phonon contribution was corrected for the TiO_2 contribution using the Eucken relation[13] and values of λ for TiO_2 recommended by TPRC.[14] The previously described results on $SrCl_2$[4] were treated in the same way and the combined

Table 3. Measured and corrected thermal
conductivity values for $SrCl_2$ from 300 to 1020 K

T (K)	Thermal Conductivity W/m•K			Thermal Resistivity cm•K/W
	Two Plane Average	Radiation Correction	Corrected for TiO_2	
299.01	2.306	0.003	2.2074	45.30
399.04	1.659	0.006	1.5765	63.43
700.51	0.9872	0.035	0.9044	110.57
896.01	0.8523	0.0776	0.7386	135.39
902.18	0.8681	0.0746	0.7546	132.52
920.71	0.8561	0.0797	0.7378	135.54
940.15	0.8608	0.0850	0.7376	135.57
941.43	0.8341	0.0850	0.7115	140.55
961.71	0.8158	0.0905	0.6886	145.22
979.91	0.7730	0.0963	0.6415	155.88
1000.00	0.7648	0.1023	0.6280	159.23
1020.68	0.7325	0.1086	0.5907	169.29

thermal resistivity data set was fitted to a quadratic equation[15] by the least squares method. The λ^{-1} results from 87 K through 700 K were described to $1 \frac{cm \cdot K}{W}$ by

$$\lambda^{-1} = -1.535 + 0.1588T + 0.211 \cdot 10^{-5} T^2 \left(\frac{cm \cdot K}{W} \right) \qquad (1)$$

From 700 K to 1020 K the λ values deduced from the measurements are greater than the λ values predicted by Eq. 1 as shown in Figure 2. This enhanced thermal conductivity occurs between 860 and 1000 K, which coincides with the region of anomalous thermal expansion. The difference between the extrapolated and the measured values reaches a maximum value of about 0.06 W/m·K near

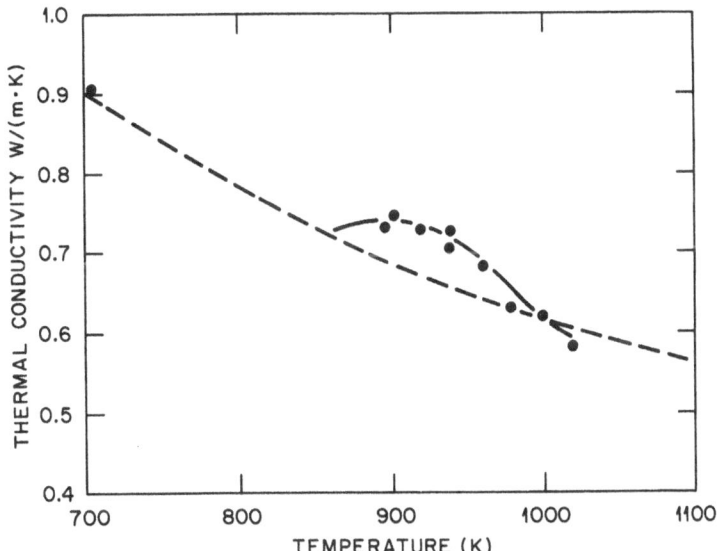

Figure 2: The thermal conductivity of SrCl$_2$ from 700 to 1020 K. Dashed line is an extrapolation based on results from 100 to 700 K. Values indicated by dots (·) are from a radial heat flow technique.

111

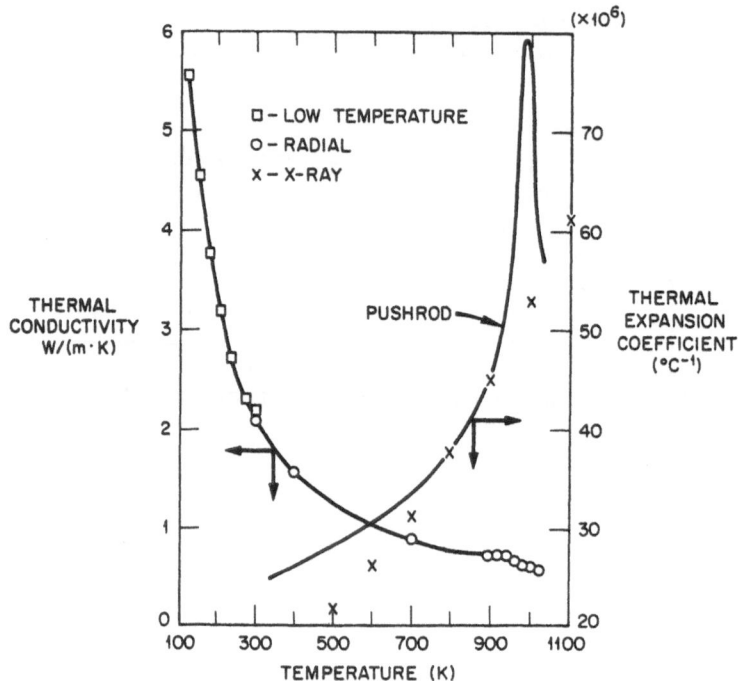

Figure 3: Comparision of the expansion coefficient and thermal conductivity of $SrCl_2$ as a function of temperature, showing the enhancement of values by the Bredig transition near 1000 K.

920 K. Figure 3 provides a comparison of both properties as a function of temperature. The thermal expansion coefficient maximum and the thermal conductivity excess maximum do not coincide. A theoretical explanation[15] of this has been offered and does not require coincidence of the maxima. The proposed conduction mechanism is similar to ambipolar diffusion of electron-hole pairs in an intrinsic semiconductor, but for the transition in $SrCl_2$ assumes that vacancies and interstitials diffuse together. The predicted excess thermal conductivity depends on the steady-state flux of vacancy-interstitial pairs, which in turn depends on temperature, and the heats of transport. The predicted excess is about 0.02 W/m·K.

Implications

The current results indicate that the Bredig transition in $SrCl_2$ is accompanied by enhanced values of the thermal expansion coefficient and thermal conductivity. Neither enhancement had been previously reported and both are significant since $SrCl_2$ is an analog to UO_2, ThO_2, and CaF_2, which are AB_2 compounds with the calcium fluorite structure. Thermal conductivity values for UO_2 recommended by Argonne National Laboratory[16] include an effect due to the transition in UO_2 at 2670 K. The thermal conductivity values for UO_2 used by the U.S. Nuclear Regulatory Commission[17] do not include a transition effect. Neither[16,17] include transition effects on thermal expansion. The current results suggest that λ-values derived from thermal diffusivity (λ/volumetric heat capacity) tests are questionable near this type transition.[18] These differences, the existence of this transition in other fluorite type compounds, the proposed conduction mechanisms, and the lack of knowledge of the effect of this transition on other properties are invitations for further study of AB_2 and A_2B compounds.

ACKNOWLEDGMENTS

The authors are grateful for reviews by D. W. Yarbrough and J. V. Cathcart and for preparation of the manuscript by Brenda Hickey and Carolyn Whitus. We would like to thank R. K. Williams for preparing SP-1, the cast $SrCl_2$ specimen.

REFERENCES

1. A. S. Dworkin and M. A. Bredig, J. Chem. Eng. Data 67, 697–8, (1963).
2. W. Schroter and J. Nolting, "Specific Heats of Crystals with the Fluorite Structure," Journal de Physique C6 41(7), 66 (1980).
3. A. V. Chadwick, F. G. Kirkwood, and R. Saghafian, "Point Defect Parameters for Strontium Chloride from Ionic Conductivity Studies," Journal de Physique C6, supplement to Vol. 41, page C6-216 (1980).
4. J. P. Moore, F. J. Weaver and D. L. McElroy, "The Thermal Conductivities of $SrCl_2$ and SrF_2 from 85 to 400 K," in Proceedings of the 18th Thermal Conductivity Conference, T. Ashworth, ed., Plenum Press, New York (in press).
5. T. G. Kollie, D. L. McElroy, J. T. Hutton and W. M. Ewing, "A Computer Operated Fused Quartz Differential Dilatometer," p. 129–146 in Thermal Expansion — 1973, AIP Conference Proceedings 17, R. E. Taylor and G. L. Denman, eds., American Institute of Physics, New York (1974).
6. T. G. Kollie, "Measurement of the Thermal Expansion Coefficient of Nickel from 300 to 1000 K and Determination of the Power-Law Constants Near the Curie Temperature," Phys. Rev. B 16(11), 4872–4881 (1977).
7. M. Shand, R. C. Hanson, C. E. Derrington, and M. O'Keefe, Solid State Communications 18, 769–772 (1976).

8. W. Schroter, "Anomale Spezifische Wärmen vom Kristallen Mit Fluoritstruktur," PhD Dissertation, Gottingen (1979).

9. T. G. Godfrey, W. Fulkerson, T. G. Kollie, J. P. Moore and D. L. McElroy, "Thermal Conductivity of Uranium Dioxide from -57° to 1100°C by a Radial Heat Flow Technique," J. Amer. Ceram. Soc. 48(6), 297—305 (1965).

10. W. Fulkerson, J. P. Moore, and D. L. McElroy, "Comparison of the Thermal Conductivity, Electrical Resistivity and Seebeck Coefficient of a High-Purity Iron and an Armco Iron to 1000°C," J. Appl. Phys. 37(7), 2639—53 (1966).

11. W. Fulkerson, J. P. Moore, R. K. Williams, R. S. Graves and D. L. McElroy, "Thermal Conductivity, Electrical Resistivity, and Seebeck Coefficient of Silicon from 100°C to 1300°K," Phys. Rev. 167(3), 765—82 (1968).

12. Private Communication from J. B. Bates, Oak Ridge National Laboratory, October 1982.

13. A. Eucken, Forsch. Gebiete Ingenieur. B3, Forschungshaft 353 (1932); Ceram. Abs. 11(11), 576 (1932); Ceram. Abs. 12(6), 231 (1933).

14. R. W. Powell, C. Y. Ho, and P. E. Liley, "Thermal Conductivity of Selected Materials," NSRDS-NBS 8, (1966).

15. R. K. Williams, "Enhanced Thermal Conduction Associated with Formation of the Superionic State," Phys. Rev. B 26(10), 5983—5986 (1982).

16. J. K. Fink, M. G. Chasanov, and L. Leibowitz, "Thermophysical Properties of Uranium Dioxide," J. of Nucl. Materials, 102, 17—25 (1981).

17. "Matpro — Version II," A handbook of materials properties for use in the analysis of light water reactor fuel rod behavior, compiled and edited by Donald L. Hagrman and Gregory A. Rexmann, NUREG/CR-0497, (1979).

18. Private communication from Roy Taylor, U.M.I.S.T., Manchester England, July 25, 1983.

THE THERMAL CONDUCTIVITIES OF $SrCl_2$ AND SrF_2 FROM 85 TO 400 K*

J. P. Moore
Martin Marietta Energy Systems, Inc.
P. O. Box Y
Oak Ridge, Tennessee 37831

F. J. Weaver, R. S. Graves, and D. L. McElroy
Martin Marietta Energy Systems, Inc.
Metals and Ceramics Division
Oak Ridge National Laboratory
Oak Ridge, Tennessee 37831

Measurements of thermal conductivity, λ, for the alkaline earth compounds $SrCl_2$ and SrF_2 indicate that the λ of $SrCl_2$ is less than that of SrF_2 by a factor of 4 to 5 over the temperature range from 85 to 400 K. This difference can be related to the differences in other parameters of the two compounds. At the lower end of the temperature range, the thermal resistivities, λ^{-1}, of both materials are linear; but large positive deviations occur at the higher end of the range. Calculated values using an equation for acoustic phonon conduction agree with the experimental values to within 30% over the entire measurement range. The positive deviations from linearity appear to be due to the effect of lattice dilation on the Debye temperatures and Grüneisen constants.

INTRODUCTION

Although the thermal conductivity, λ, of SrF_2 has been measured at intermediate temperatures,[1,2] that of $SrCl_2$ has never been determined. This is partly due to the hygroscopic nature of $SrCl_2$ which limits its technological usefulness and makes its fabrication and handling difficult. This paper describes the

---

*Research sponsored by the Division of Materials Sciences, U.S. Department of Energy under contract DE-AC05-840R21400 with the Martin Marietta Energy Systems, Inc.

experimental λ of SrF_2 and $SrCl_2$ over a temperature range from 85 to 400 K, a range where their Debye temperatures indicate that the λ of SrF_2 should exceed that of $SrCl_2$ by a factor of 3. At high temperatures (~1000 K), an order-disorder transformation in $SrCl_2$ causes a peak in the specific heat and may alter other properties such as thermal expansion and thermal conductivity. This behavior, which is possibly caused by disordering of the anions, is similar to that which occurs in many other AB_2 compounds[3-5] including UO_2. High temperature transport properties of $SrCl_2$ will be the subject of later papers.

MEASUREMENT TECHNIQUES AND EXPERIMENTAL ERRORS

The thermal conductivities were measured from 85 to 400 K using a guarded longitudinal technique described previously and tested extensively by comparison of results to those of others on common specimens.[6] This technique has been used on electrically-insulating low λ solids,[7] but a correction for heat shunting must be made to the data at the upper end of the measurement range.[8] The maximum values of this correction were -8% and -1.8% at 400 K for the $SrCl_2$ and SrF_2, respectively. Neither data set has been corrected for infrared radiation because it was negligible compared to the conducted heat. This is because of the high λ of the SrF_2 and because of an opacifier added to the $SrCl_2$ specimen. The most probable determinate error for the $SrCl_2$ data was $\pm 4.7\%$ at 400 K and $\pm 3.3\%$ below 200 K. The error for the SrF_2 data was below $\pm 1.5\%$ at all temperatures. These errors are due to uncertainties related to determination of dimensions and temperature differences, correction for heat exchange with the guard, and correction for the presence of TiO_2 ($SrCl_2$ only).

SPECIMEN DESCRIPTION

The $SrCl_2$ specimen was cut from a 50 mm diameter by 25 mm thick disk which had been fabricated primarily for high temperature studies. Since infrared radiation could possibly have an adverse effect, the specimen contained 3 vol % TiO_2 to assist in opacification. The first step in disk preparation was to separately dry reagent grade powders of $SrCl_2$ and TiO_2 under vacuum at temperatures between 120 and 160°C. The powders were then blended and the combination was dried under the same conditions. This mixture was then placed in a graphite die which was then installed in a vacuum hot press. Following evacuation, this powder was heated to a temperature of 900 to 1000 K and pressed at 27.5 MPa. The resulting composite had a theoretical density of 3.088 Mg/m^3. The dominant impurities in the specimen and their levels in atomic ppm were Al(974), Ca(4372), Si(624), and S(820). Metallographic analysis indicated uniformly spaced TiO_2 particles and grain sizes between 10 and 20 μm. The SrF_2 specimen was a rod cut from a single crystal disk obtained from Optovac, Inc., of North Brookfield, Massachusetts. The nominal level of purity for this disk was 99.99%.

RESULTS

Smoothed values of the measured thermal conductivites and the thermal resistivities (λ^{-1}) are given in Table 1 as a function of temperature. The thermal conductivity of the $SrCl_2$, λ_s, was calculated from the measured thermal conductivity of the $SrCl_2$ + 3 vol % TiO_2, λ_c, using the equation[9]

$$\lambda_c = \lambda_s \left[\frac{1 - \rho + \rho \left(\frac{3\lambda_T}{2\lambda_s + \lambda_T} \right)}{1 - \rho + \rho \left(\frac{3\lambda_s}{2\lambda_s + \lambda_T} \right)} \right] , \tag{1}$$

where

ρ = vol fraction of the TiO_2 (0.03) and

λ_T = thermal conductivity of the TiO_2.

Values for λ_T were obtained from Charvat and Kingery[10] and an iterative procedure was used on Eq. (1) to calculate λ_s. The difference between λ_c and λ_s was 4 to 6% depending on the measurement temperature. The sample temperature during the brief time required for hot pressing could have caused some of the TiO_2 to dissolve in the $SrCl_2$. Since the solubility of TiO_2 in liquid $SrCl_2$ is only $9-23 \times 10^{-6}$ mole fraction,[11] the presence of TiO_2 in the $SrCl_2$ lattice should not influence the lattice conductivity.

Comparison of the thermal conductivities and resistivities of SrF_2 and $SrCl_2$ in Figures 1 and 2 shows that the λ of SrF_2 is 4–5 times greater than that of $SrCl_2$ over this temperature range. The authors are unaware of any other thermal conductivity results in $SrCl_2$, but Parfenreva et al.[1] and Mogilevskii and Tumpurova[2] have measured the thermal conductivity of SrF_2. The data from Parfenreva et al.[1] agree with the present results to within 1% at 100 K but are 5% higher at 300 K, whereas, those from Mogilevskii and Tumpurova[2] are about 20% above the present results from 100 to 300 K.

Table 1. Smoothed Values of the Measured Thermal Conductivity and Thermal Resistivity of SrF_2 and $SrCl_2$

T (K)	SrF_2[a] λ (W/m·K)	SrF_2[a] W x 10^2 (m·K/W)	$SrCl_2$[b] λ (W/m·K)	$SrCl_2$[b] W x 10^2 (m·K/W)
85	44.1	2.27	–	–
90	39.9	2.50	7.17	13.9
100	34.0	2.95	6.56	15.2
120	26.1	3.83	5.59	17.9
140	21.2	4.72	4.86	20.6
160	17.8	5.63	4.27	23.4
180	15.3	6.54	3.78	26.5
200	13.4	7.47	3.37	29.7
220	11.88	8.42	3.03	33.0
240	10.66	9.38	2.74	36.5
260	9.65	10.36	2.49	40.2
280	8.81	11.36	2.27	44.1
300	8.07	12.39	2.11	47.4
320	7.45	13.42	1.92	52.1
340	6.93	14.42	1.74	57.5
360	6.48	15.43	1.58	63.3
380	6.08	16.44	1.43	69.9
400	5.74	17.43	1.30	76.9

[a]A least squares fit showed that all W data were (with the exception of one value) within ±0.4% of the equation W = −1.37 + 0.041629 T + 0.1374 x 10^{-4} T^2. These smoothed values are from the equation.
[b]These values were obtained by first fitting the data to a low order polynomial and then plotting the deviations between the individual datum and the smoothing polynomial. A smooth curve was then drawn through the data and values obtained at even temperature intervals.

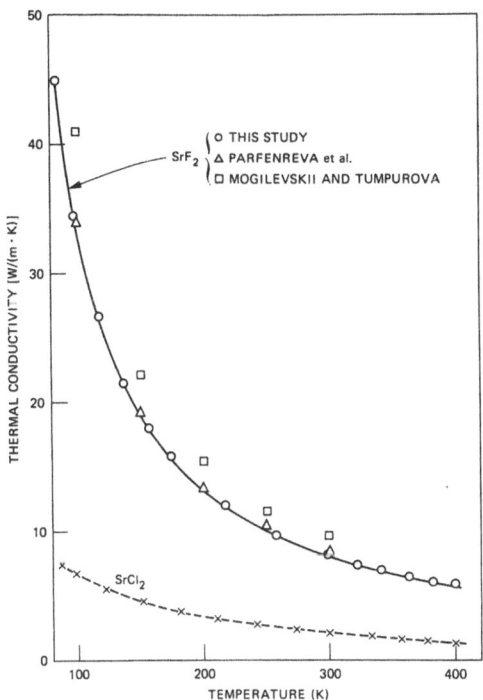

Figure 1: Comparison of the thermal conductivity of SrF_2 to previous results and to that of $SrCl_2$.

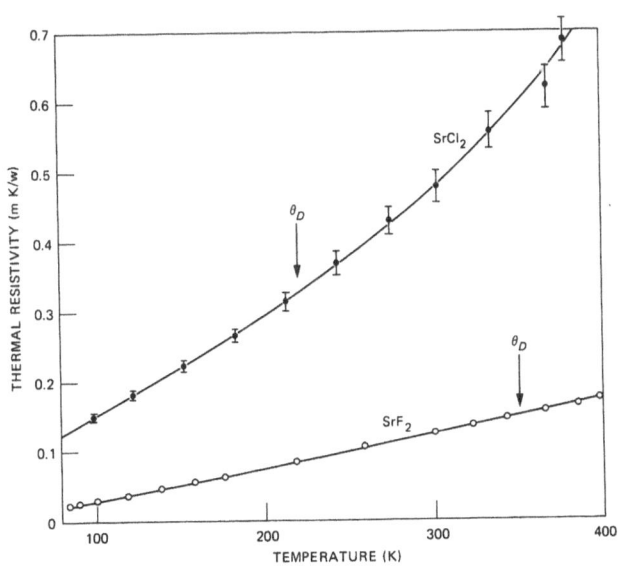

Figure 2: Experimental thermal resistivities of SrF_2 and $SrCl_2$ versus temperature. Error bars for the SrF_2 data are too small to be seen on this scale.

DISCUSSIONS OF RESULTS

The thermal resistivity, W, of $SrCl_2$ has been corrected for the impurities using an approach described by Abeles [12]. These corrected W values, which would be 0.0256 m·K/W lower than the measured values tabulated in Table 1, are compared to the W of SrF_2 in Figure 3. At the lower end of the measurement range, the W of both SrF_2 and $SrCl_2$ are linear, but positive deviations from linearity occur at the upper temperatures. These positive deviations are similar to those observed in the rubidium halides.[7] The ratio of the corrected W values of $SrCl_2$ to those of SrF_2 is 4.3 at 100 K and 3.6 at 200 K. Most of this large difference between $SrCl_2$ and SrF_2 can be related to differences in other parameters as shown below.

The acoustic phonon modes are dominant in heat conduction in these two materials since the group velocities of the optic modes are small. When scattering is by a three-phonon Umklapp process and $T > \theta_D/4$ (where θ_D is the Debye temperature) it has been shown that the thermal resistivity[13] can be written as

$$W(T) = \frac{5.57 \times 10^{-32} \, \gamma^2 \, T}{n^{1/3} M \, \delta \, \theta_0^3}, \tag{2}$$

where

γ = Gruneisen constant

M = average mass per atom (kg/atom)

δ = cube root of the atomic volume (m)

T = absolute temperature (K)

n = number of atoms/molecule (3 in this case)

and θ_0 is the Debye temperature for the acoustic branch at $T = 0°K$. The latter is given by

$$\theta_0 = \theta_D \, n^{-1/3} \tag{3}$$

where θ_D is the customary value of the Debye temperature. Equation (2) indicates that the thermal resistivity should increase linearly with temperature. Klemens and Ecsedy[14] have shown that an additional term should be used to account for the effects of lattice dilation. Equation (2) then becomes

$$W(T) = \frac{5.57 \times 10^{-32} \gamma^2 T}{n^{1/3} M \delta \theta_0^3} \left[1 + \left(3\gamma + \frac{2}{3} + \frac{2\gamma'}{\gamma} \right) 3 \alpha \Delta T \right] \quad (4)$$

where

α = the linear expansion coefficient (K^{-1})

γ' = the derivative of γ with thermal dilation

$\Delta T = T - T_1$,

and where T_1 is assumed to be the lower end of the measurement range (85 K).

Values of the various parameters of Eq. (2) are given in Table 2. The Debye temperatures of these two compounds were calculated in three ways. The first method[15] employed the equation

$$\theta_D = \frac{h}{K} \left[\frac{3n}{4\pi} \frac{N\rho}{M.W.} \right]^{1/3} \nu_m \quad , \quad (5)$$

where

h = Planck's constant

K = Boltzmann's constant

n = number of atoms/molecule

M.W. = molecular weight

N = Avogadro's number

ρ = density

ν_m = the mean sound velocity.

The latter was calculated[13] using

$$\nu_m = \left[\frac{1}{3} \left(\nu_T^{2/3} + \nu_L^{1/3} \right) \right]^{-1/3} , \quad (6)$$

where v_T and v_L are the measured[16] transverse and longitudinal sound velocities, respectively, and are given in Table 2. The Debye temperatures were also calculated using Lindemann's melting point rule[17]

$$\theta_D = 120 \, T_M^{1/2} \, A^{-5/6} \, \rho^{1/3} \, , \tag{7}$$

where T_M and A are the melting points and average atomic weights, respectively. The third method for calculating θ_D involved the use of experimental specific heat data from Smith et al.[18] The three values for each compound agree well with each other.

The thermal expansion and Grüneisen constant of SrF_2 have been determined,[19,20] but the authors are unaware of any results on $SrCl_2$. Some initial measurements on $SrCl_2$ indicate that α should be about $29 \times 10^{-6} \, K^{-1}$. We have assumed that γ of $SrCl_2$ is the same as that of SrF_2* and, following Klemens and Ecsedy,[14] that $\gamma \approx \gamma^{'}$ for both materials.

Agreement between calculated and experimental thermal resistivities for both SrF_2 and $SrCl_2$ are excellent as shown in Figure 3. At 100 K, the results for SrF_2 agree to within about 30%, but this improves to 2% above 300 K. The calculated results for $SrCl_2$ are lower than experimental ones by 17% at 100 K and 30% at 400 K. The positive deviation of the experimental curve from linearity is also predicted by the calculated curve for $SrCl_2$.

There have been few studies of the thermal conductivities of other alkaline earth chlorides with the fluorite structure, but based on the large differences in their melting temperatures. Other chlorides and fluorides with a specific cation should have a difference in their thermal conductivities similar to that observed between $SrCl_2$ and SrF_2. Slack[21] has shown that the thermal conductivities of the fluorides such as CoF_2, MnF_2, and ZnF_2 are essentially identical at temperatures above any magnetic ordering temperature. The present data show that SrF_2 has nearly the same λ as do these other three fluorides.

ACKNOWLEDGMENTS

We are grateful to several individuals for various contributions. These include C. S. Morgan for fabricating the $SrCl_2$ specimen, W. A. Simpson for measuring the velocity of sound in both $SrCl_2$ and SrF_2, C. E. Devore for measuring the expansion coefficient of $SrCl_2$ and R. K. Williams for some helpful comments.

---

*Based on a cursory examination of the γ of other alkaline earth compounds, this is probably true to within ±10%.

Table 2. Parameters Used in Eq. (3)

Material	M (Kg/atom) ($\times 10^{29}$)	δ (m) ($\times 10^{10}$)	θ_D^a (K)	θ_D^b (K)	θ_D^c (K)	γ^d	ν_T (m/s) ($\times 10^{-3}$)	ν_L (m/s) ($\times 10^{-3}$)
$SrCl_2$	8.77	3.068	277	216	224	N.A.	2.116	4.122
SrF_2	6.95	2.54	374	353	346	1.62	2.92	5.435

[a]Calculated using Eq. (5).
[b]Calculated using Eq. (7).
[c]Calculated from the low temperature specific heat.[17]
[d]The authors have recently learned that R. K. Williams of the Oak Ridge National Laboratory has calculated a value of 1.7 for γ at 373 K.

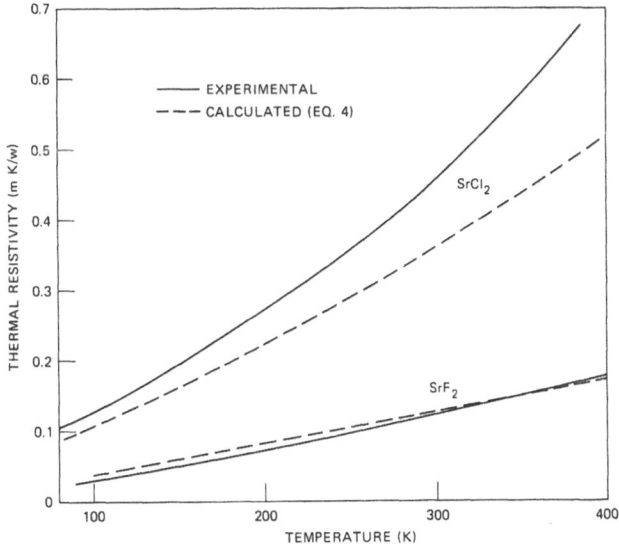

Figure 3: Comparison of the experimental thermal resistivities of SrF_2 and $SrCl_2$ (as corrected for impurity scattering to calculated values using Eq. (4).

REFERENCES

1. L. S. Parfenreva, I. A. Smirnov, and V. V. Tikhonov,
    Soviet Phys. Sol. State 13(5), 1267-1268 (1971).
2. B. M. Mogilevskii and V. F. Tumpurova, Fiz. Tverd. Tela. 17,
    1786–89 (1974).
3. A. S. Dworkin and M. A. Bredig, J. Phys. Chem. 67, 697–98
    (1963).
4. A. S. Dworkin and M. A. Bredig, J. Phys. Chem. 8, 416–17
    (1963).
5. A. S. Dworkin and M. A. Bredig, J. Phys. Chem. 72, 1277–81
    (1968).
6. M. J. Laubitz and D. L. McElroy, Metrologia 7(1), 1–15 (1971).
7. J. P. Moore, R. K. Williams and R. S. Graves, Phys. Rev. B
    11(8), 3107-15 (1975).
8. J. P. Moore, R. K. Williams and R. S. Graves, ORNL/TM-4797,
    Oak Ridge National Laboratory, Oak Ridge, TN., April 1975.
9. A. Eucken, Forsch. Gebiete Ingenieur. B3, Forschungshaft 353
    (1932); Ceram. Abs. 11(11), 576 (1932); Ceram. Abs. 12(6),
    231 (1933).
10. F. R. Charvat and W. D. Kingery, J. Am. Ceram. Soc. 40, 305–15
    (1954).
11. T. L. Inyushkina, I. N. Narenkova and L. N. Dzyubo,
    The Russian Journal of Physics and Chemistry 49, 2458–2459
    (1975).
12. B. Abeles, Phys. Rev. 131, 1906 (1963).
13. G. Slack, Sol. State Phys. 34, 1–71 (1979).
14. K. G. Klemens and D. J. Ecsedy, "Thermal Resistance of
    Dielectric Crystals at High Temperatures" in Phonon
    Scattering in Solids, L. C. Challis, V. W. Rampton and
    A. F. G. Wyatt, eds., Plenum Press, New York (1976).
15. O. L. Andersen, J. Phys. Chem. Sol. 24, 909–17 (1963).
16. Private communication from W. A. Simpson, Metals and Ceramics
    Division, Oak Ridge National Laboratory, P. O. Box X,
    Oak Ridge, TN.
17. J. Ziman, p. 57 in Electrons and Phonons, Oxford University
    Press (1963).
18. D. F. Smith, T. E. Gardner, B. B. Letson, and
    A. R. Taylor, Jr., "Thermodynamic Properties of Strontium
    Chloride and Strontium Fluoride from 0° to 300°K," Rep.
    Investig. 6316, Bureau of Mines, U. S. Department of
    Interior, (1963).
19. D. B. Sirdeshmukh and B. K. Roa, J. Chem. Phys. 57(1), 577–78
    (1972).
20. A. C. Bailey and B. Yates, J. Phys. Soc. 91, 390–98 (1967).
21. G. Slack, Phys. Rev. 122(5), 1451-64 (1961).

THERMAL CONDUCTIVITY OF TELLURIUM

DIOXIDE SINGLE CRYSTALS

K. A. McCarthy and H. H. Sample

Department of Physics
Tufts University
Medford, Massachusetts

ABSTRACT

Tellurium dioxide is a material of interest for acousto-optic devices. Measurements of the thermal conductivity of single crystals of tellurium dioxide have been made in the 1.5-100 K temperature range. The crystals on which these measurements were made have tetragonal (paratellurite) structure; its elastic behavior is characterized by an unusually slow shear wave in the ⟨110⟩ direction.

Measurements were made with the heat flow direction parallel to the c-axis and with heat flow along a ⟨110⟩ direction. For heat flow parallel to the c-axis, K_{PEAK} equals 16.4 watt/cm K at 5 K; for heat flow perpendicular to the c-axis, K_{PEAK} equals 10.8 watt/cm K at 5 K. Although the samples came from two different sources, we believe that there is an orientation dependence of K in this anisotropic material.

These results will be compared with our recent measurements on single crystals of thallium arsenic selenide (TAS or Tl_3AsSe_3), also a material with interesting acousto-optic properties. This material has a trigonal structure, and is one member of the ternary chalcogenide salt group that does not have a very slow shear mode velocity. Measurements made with heat flow both parallel and perpendicular to the c-axis are only slightly different.

CORRELATION BETWEEN THE STRUCTURAL PARAMETERS AND THE THERMAL

CONDUCTIVITY OF CHALCOPYRITE-TYPE TERNARY COMPOUNDS

P. Kistaiah, K. Satyanarayana Murthy
and Leela Iyengar

Department of Physics
University College of Science
Osmania University
Hyderabad-500 007 India

ABSTRACT

The room-temperature thermal conductivity and the tetragonal distortion were calculated for a number of covalently bonded I-III-VI$_2$ (ABC$_2$) compound semiconductors. Comparison of thermal properties and peculiarities of chemical bonding in these compounds lead to the conclusion that there is a close relationship between these parameters which is essentially described by the anharmonicity of the lattice. On the basis of free energy considerations with anharmonic contributions included, a physical model has been suggested to explain the relation between the thermal conductivity and the tetragonal distortion of I-III-VI$_2$ chalcopyrites.

INTRODUCTION

The chalcopyrite compounds of the type I-III-VI$_2$ (ABC$_2$) are the ternary analogues of the II-VI compounds with zinc-blende structure [1]. These compounds crystallize in the tetragonal chalcopyrite structure (space group $I\bar{4}2d$, D_{2d}^{12}, No. 122) in which the ordering in the cation sublattice results in a tetragonal distortion of the lattice defined as $\delta = 2-c/a$, 'c' and 'a' being the lattice parameters [2]. The parameter 'δ' in the chalcopyrite structure affects the structural anisotropy. Some of the physical properties of chalcopyrites depend on the value of 'δ' and the position of the 'C' atom [3]. Recently, these semiconducting compounds have drawn great attention because they show promise for applications in such areas as non-linear optics, solar cells, optical detectors and light emitting diodes [4-9]. However, their physical

127

properties have not been studied in detail. The electrical and optical properties of some of these compounds have been studied at room temperature [2]. Very little work has been reported on the thermal properties of these compounds. The possibility of correlating fundamental physical properties with structural parameters prompted us to take up the present study. The purpose of this study is to calculate the thermal conductivity of all the known I-III-VI$_2$ compounds using the Leibfried and Schlömann theory [10] and correlate it with the tetragonal distortion of the chalcopyrite lattice, both of which depend intrinsically on the anharmonicity of the lattice.

THEORY AND CALCULATION OF PARAMETERS

Tetragonal Distortion

The chalcopyrite structure is closely related to the zincblende structure. In Fig. 1, the chalcopyrite unit cell is shown together with two zincblende unit cells. In the unit cell of the chalcopyrite ABC$_2$, the A and B cations occupy the cation positions of the zincblende structure in an ordered way, so arranged that they alternate along the lines parallel to the c-axis. This reduces the cubic symmetry of the zincblende structure to the tetragonal symmetry of the chalcopyrite structure. Due to the difference in radius of the two cations, the lattice is typically distorted with the unit cell height being slightly less than twice the base dimension. This distortion along the optic or c-axis is known as 'tetragonal distortion' In Table 1, all the known I-III-VI$_2$ chalcopyrites including the mineral chalcopyrite, CuFeS$_2$ are listed together with their lattice parameters, axial ratio and the tetragonal distortion.

Several authors [11-17] have analysed the tetragonal distortion of chalcopyrites in terms of the interactions present in the crystal lattice. Neumann [16] and Binsma et al [17] have shown that the tetragonal distortion of the I-III-VI$_2$ chalcopyrite compounds can be described by free energy considerations. In this approach, the free energy F of the crystal is expanded in terms of the tetragonal distortion 'δ':

$$F = F_o + \sum_{n=1}^{\infty} A_n \delta^n \tag{1}$$

where F_o is the free energy of the ideal chalcopyrite lattice ($\delta = 0$). The expansion coefficients A_n depend on the interatomic forces in the crystal and can be considered more or less complicated functions of the elastic constants [18]. The equilibrium value of 'δ' corresponding to the minimum of the free energy can be evaluated by solving the equation:

$$\frac{dF}{d\delta} = \sum_{n=1}^{\infty} nA_n \delta^{n-1} = 0 \tag{2}$$

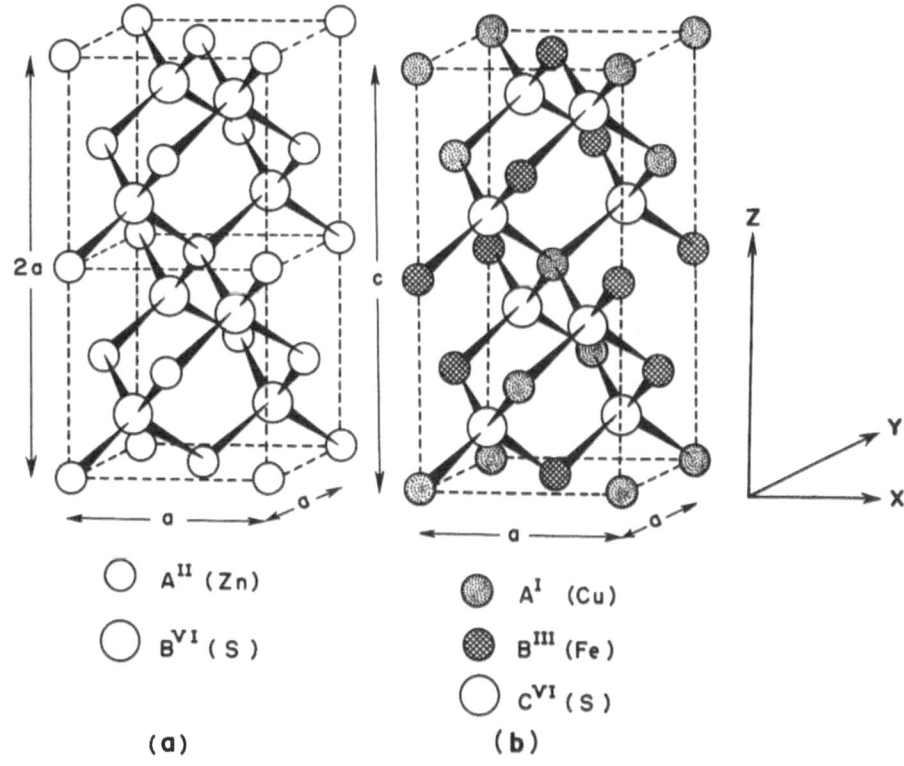

$\bigcirc$ A^{II} (Zn)

$\bigcirc$ B^{VI} (S)

$\quad$ (a)

$\bullet$ A^{I} (Cu)

$\bullet$ B^{III} (Fe)

$\bigcirc$ C^{VI} (S)

$\quad$ (b)

Fig. 1. The tetragonal chalcopyrite structure (b)
compared to two unit cells of the cubic
zincblende structure (a).

From (2) it follows that the change of 'δ' with temperature T
should follow the relation:

$$\frac{d\delta}{dT} = \frac{\sum\limits_{n=1}^{\infty} n\delta^{n-1}\, dA_n/dT}{\sum\limits_{n=2}^{\infty} n(n-1)A_n\,\delta^{n-2}} \tag{3}$$

In previous studies of the tetragonal distortion and its variation with temperature [15-17], terms with n >/ 3 have been neglected
in (1) to (3) which results in a linear relation between dδ/dT and
δ:

$$\frac{d\delta}{dT} = -\frac{1}{2A_2}\frac{dA_1}{dT} - \frac{1}{A_2}\frac{dA_2}{dT}\,\delta \tag{4}$$

129

TABLE 1

Lattice Parameters and Tetragonal Distortion for I-III-VI$_2$ Compounds

Sl. No.	Compound	a (nm)	c (nm)	c/a	δ
1	CuAlS$_2$	0.5334	1.0440	1.9573	0.0427
2	CuAlSe$_2$	0.5604	1.0977	1.9588	0.0412
3	CuAlTe$_2$	0.5967	1.1800	1.9775	0.0225
4	CuGaS$_2$	0.5348	1.0483	1.9602	0.0398
5	CuGaSe$_2$	0.5590	1.0974	1.9632	0.0369
6	CuGaTe$_2$	0.6016	1.1927	1.9825	0.0175
7	CuInS$_2$	0.5523	1.1132	2.0156	−0.0156
8	CuInSe$_2$	0.5783	1.1621	2.0095	−0.0095
9	CuInTe$_2$	0.6205	1.2363	1.9924	0.0076
10	CuTlSe$_2$	0.5844	1.1650	1.9935	0.0065
11	AgAlS$_2$	0.5707	1.0280	1.8013	0.1987
12	AgAlSe$_2$	0.5968	1.0770	1.8046	0.1954
13	AgAlTe$_2$	0.6309	1.1850	1.8783	0.1217
14	AgGaS$_2$	0.5756	1.0299	1.7893	0.2107
15	AgGaSe$_2$	0.5991	1.0885	1.8169	0.1831
16	AgGaTe$_2$	0.6320	1.1983	1.8960	0.1040
17	AgInS$_2$	0.5836	1.1179	1.9155	0.0845
18	AgInSe$_2$	0.6101	1.1691	1.9162	0.0838
19	AgInTe$_2$	0.6419	1.2580	1.9598	0.0402
20	CuFeS$_2$	0.5291	1.0422	1.9698	0.0302

On the other hand, the temperature coefficient $d\delta/dT$ of the tetragonal distortion δ can be evaluated from the linear thermal expansion coefficients α_a and α_c via the relation:

$$\frac{d\delta}{dT} = c/a \ (\alpha_a - \alpha_c) \qquad (5)$$

Recently, we have reported the anisotropic thermal expansion coefficients for many of the I-III-VI$_2$ compounds [19-24] and it was found that the relation (4) contradicts the experimental data. This result is not so surprising taking into account that the equilibrium value of the tetragonal distortion and its temperature dependence are obtained from (2) and (4) on the basis of the harmonic approximation, valid only for very small values of δ. Thus, the failure of relation (4) to describe the 'δ' and its temperature dependence in I-III-VI$_2$ compounds suggests that the harmonic approximation is insufficient to describe the transition from the nonequilibrium ideal chalcopyrite lattice with $\delta = 0$ to the real crystal with nonvanishing δ.

When anharmonic forces are included it follows from general considerations that in the free energy expansion (1) terms up to at least δ^4 must be retained in order to stabilize the lattice [25]. In this anharmonic approximation the equilibrium value of 'δ' is determined from the solution of the third order equation:

$$4A_4\delta^3 + 3A_3\delta^2 + 2A_2\delta + A_1 = 0 \qquad (6)$$

For the temperature variation of 'δ' we get the relation:

$$\frac{d\delta}{dT} = \frac{a_1 + a_2\delta + a_3\delta^2}{1 + a_4\delta + a_5\delta^2} \qquad (7)$$

where the coefficients a_i are functions of the quantities A_n and their temperature derivatives. The detailed results of the temperature dependence of the tetragonal distortion in I-III-VI$_2$ ternary compounds have been given elsewhere [26,27].

Attempts to get a reasonable fit of relation (7) to the data of all I-III-VI$_2$ compounds were not successful. As mentioned in our earlier work [27], this is due to the order-disorder behaviour of these compounds. It is a well known fact that I-III-VI$_2$ compounds with tetragonal distortion $\delta < 0.05$ show a polymorphous phase transition from the chalcopyrite structure to the zincblende structure characterized by disordering in the cation sublattice whilst compounds with $\delta > 0.05$ remain in the chalcopyrite structure up to their melting point [17]. Indeed, considering the Cu-III-VI$_2$ compounds ($\delta < 0.05$) and Ag-III-VI compounds ($\delta > 0.05$) separately we found that the relation (7) gives results in agreement with the experimental data [24]. Thus the physical model based on the free energy considerations including the anharmonic contributions, for the tetragonal distortion and its temperature dependence I-III-VI$_2$ compounds, accounts for all the experimental results reported so far.

Thermal Conductivity

Heat transport by lattice waves in solids is governed by the an-harmonicities of the lattice forces and by the various imperfections of the crystal lattice. Therefore, the study of thermal conductivity offers great diversity and provides an interesting field of study, both from a fundamental point of view, in which one desires to attain agreement between observation and the theoretical concepts, and from the more applied view point in which one desires to use the thermal conductivity as a tool in the study of lattice imperfections.

Leibfried and Schlömann [10] developed a treatment of the thermal conductivity due to the anharmonic terms of the lattice forces. They have calculated the thermal conductivity of a perfect crystal on the basis of the Debye-Callaway model [28]. According to this model, the lattice thermal conductivity is given by:

$$\lambda = \frac{K_B}{2\pi^2 v} \; \frac{K_B T}{h}^3 \; \int_0^{\theta/T} \tau_c \; x^4 e^x (e^x-1)^{-2} \; dx \qquad (8)$$

In this equation $x = h\omega/K_B T$, ω being the phonon frequency, v the average sound velocity, K_B the Boltzmann constant, h the Planck's constant, T the absolute temperature, θ the Debye temperature and τ_c the combined relaxation time for the phonon scattering process. This result was modified by Steigmeier [29] in order to apply it to the tetrahedral semiconductors and is given by:

$$\lambda = \frac{12}{5} \; (4)^{1/3} \; (\frac{K_B}{h})^3 \; \frac{\bar{M} \; V^{1/3} \; \theta^3}{(\gamma + \frac{1}{2})^2 T} \qquad (9)$$

where $\bar{M}$ is the average atomic mass in grams, V is the average volume occupied by one atom of the crystal and γ is the Grüneisen constant.

Steigmeier used this formula to compare the thermal conductivity of the III-V binary semiconductors measured at room temperature, and Slack [30] has extended the work for the II-VI compounds. The I-III-VI$_2$ compounds are the ternary analogues of the II-VI binary semiconductors and exhibit also tetrahedral coordination. Therefore, it is not unreasonable to suppose that the relation (9) can be used to calculate the thermal conductivity of the ternary I-III-VI$_2$ compounds, too. The data employed in investigating the validity of (9) in the case of the I-III-VI$_2$ compounds are compiled in Table 2. The Debye temperatures of these compounds are calculated from their melting temperatures using the relation given by Oshcherin [31].

$$\theta = 125 \; (T_m)^{1/2} \; (d)^{1/3} \; (\bar{M})^{-5/6} \qquad (10)$$

TABLE 2

Material Parameters of the I-III-VI$_2$ Compounds used in Relations (9) and (10)

Sl. No.	Compound	T_m (K)	d (g/cm^3)	M(g)	$(V)^{1/3}$ (10^{-8} cm)	θ(K)	$\bar{M}$ $(V)^{1/3}\theta^3$ (g cm K^3)	λ Calc (W/cm K)	λ exptl. (W/cm K)
1	CuAlS$_2$	1575	3.45	38.66	2.65	386	58.92	0.231	
2	CuAlSe$_2$	1270	4.70	62.11	2.80	272	34.99	0.137	
3	CuAlTe$_2$	1160	5.50	87.18	2.97	303	72.03	0.082	0.032 Ref. [35]
4	CuGaS$_2$	1513	4.29	50.89	2.70	320	45.02	0.176	
5	CuGaSe$_2$	1343	5.45	72.79	2.81	239	27.93	0.109	
6	CuGaTe$_2$	1143	5.87	97.11	3.02	177	16.26	0.064	0.064 Ref. [35]
7	CuInS$_2$	1300	4.71	60.61	2.77	264	30.89	0.121	0.037 Ref. [35]
8	CuInSe$_2$	1263	5.65	84.07	2.91	207	21.70	0.085	0.09 Ref. [36]
9	CuInTe$_2$	1053	6.00	108.99	3.11	156	12.87	0.05	0.063 Ref. [35]
10	CuTlSe$_2$	680	7.08	106.45	2.92	135	7.65	0.03	0.049, 0.028 Ref. [37]
11	AgAlS$_2$	1323	3.94	49.74	2.76	286	32.12	0.126	
12	AgAlSe$_2$	1220	5.07	73.19	2.88	210	19.52	0.076	
13	AgAlTe$_2$	1000	6.18	97.51	2.97	158	11.42	0.045	
14	AgGaS$_2$	1313	4.58	60.42	2.80	259	29.39	0.115	
15	AgGaSe$_2$	1123	5.71	104.52	3.12	156	12.38	0.048	
16	AgGaTe$_2$	993	5.96	130.47	3.31	122	7.84	0.031	0.01 Ref. [35]
17	AgInS$_2$	1120	4.97	71.70	2.88	201	16.77	0.066	
18	AgInSe$_2$	1046	5.80	117.42	3.23	138	9.97	0.039	
19	AgInTe$_2$	953	6.08	141.74	3.38	113	6.91	0.027	0.063, 0.015 Ref. [35]
20	CuFeS$_2$	1150	4.28	46.60	2.62	294	31.03	0.122	0.065 Ref. [35]

where d is the density of the crystal and T_m is the melting point in Kelvins. The Grüneisen constant γ in equation (9) has been used as the adjustable parameter to obtain the best fit with the available experimental data. The last column of the table shows the reported values of λ for different compounds. It was observed that for all the I-III-VI$_2$ compounds a good fit can be obtained by using $\gamma = 1.68$. Therefore it can be concluded that the Steigmeier's equation based on the theory of Leibfried and Schlömann can be extended to the ternary I-III-VI$_2$ chalcopyrites.

CORRELATION BETWEEN TETRAGONAL DISTORTION AND THERMAL CONDUCTIVITY

It is known that a finite thermal conductivity depends essentially on the lattice anharmonicity of the interatomic potential. A perusal of the literature shows that attempts have been made in the past to correlate the thermal conductivity of the semiconducting compounds with the type of chemical bonding in them [32-34]. It was shown that the substances with covalent bonds possess higher thermal conductivity than those with ionic bonds. Since the bonding in the I-III-VI chalcopyrites is predominantly covalent and the tetragonal distortion reflects the nature of chemical bonding in these compounds, it is expected that the compounds with larger tetragonal distortion possess high thermal conductivity.

The above conclusion follows from equations (6) and (9) because the equilibrium value of the tetragonal distortion δ is related to the quantities A_n which are functions of the elastic constants [18]. Since the Debye temperature θ_D of a crystal is closely related to its elastic constants, one would expect a systematic relationship between the tetragonal distortion δ and the Debye temperature θ in equation (9). As is evident from Tables 1 and 2, the Debye temperature θ increases with increasing tetragonal distortion, when the Cu-III-VI$_2$ and Ag-III-VI$_2$ compounds are considered separately. As already stated above, Steigmeier's relation can be extended to the I-III-VI$_2$ ternary compounds and therefore one would expect a close systematic relationship between the thermal conductivity and the tetragonal distortion. In Fig. 2, we have plotted the thermal conductivity (λ) against the tetragonal distortion (δ). As is evident from the figure, the variation of λ with δ of all the I-III-VI$_2$ chalcopyrites cannot be given in terms of a single relation. This may be again due to the order-disorder behaviour of the I-III-VI$_2$ ternary compounds. However, considering the Cu-III-VI$_2$ compounds ($\delta < 0.05$) and the Ag-III-VI$_2$ compounds ($\delta > 0.05$) separately we found rather a close correlation between the thermal conductivity and the tetragonal distortion. As expected, in both the copper and silver chalcopyrites the thermal conductivity increases with increasing tetragonal distortion. Since the tetragonal distortion δ is a measure of the lattice anharmonicity, it can be concluded that the thermal conductivity λ of the I-III-VI$_2$ compounds increases with increasing lattice anharmonicity.

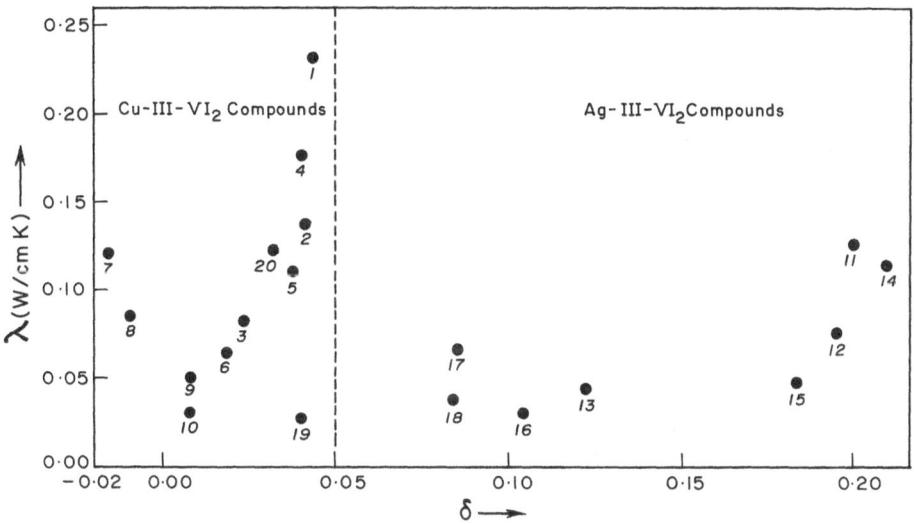

Fig. 2. Room-temperature thermal conductivity of ternary I-III-VI$_2$ chalcopyrites as a function of tetragonal distortion. The numbering of the compounds is as given in tables (1) and (2).

CONCLUSIONS

The tetragonal distortion δ of all the I-III-VI$_2$ chalcopyrites can be analysed in terms of a physical model based on the lattice free energy with anharmonic contributions included. Steigmeier's relation based on the Leibfried and Schlömann theory can be extended to the I-III-VI$_2$ ternary compound semiconductors to estimate their room temperature thermal conductivity.

There is a systematic relation between θ and δ which follows from the theory of tetragonal distortion of these compounds. Since Steigmeier's relation is applicable for these ternary semiconductors, one would expect a relation between the thermal conductivity and the tetragonal distortion, both of which depend on the anharmonicity of the lattice potential. In general, the thermal conductivity λ increases with increasing tetragonal distortion δ. However, the experimental data as well as the results of our theoretical considerations suggest that there seems to be a singularity in the behaviour of the thermal conductivity and the tetragonal distortion of these compounds in the range $\delta = 0.045$ to 0.05. This result can be explained by the structural order-disorder behaviour of these compounds.

ACKNOWLEDGMENT

The authors sincerely thank Prof. K. V. Krishna Rao for his constant attention to the work and for many stimulating discussions. They also cordially thank the referee for his critical comments and useful suggestions. They wish to acknowledge the financial support received for this project from the Department of Science and Technology, Govt. of India, New Delhi.

REFERENCES

[1] Berger, L. I., and Prochukhan, V. D., "Ternary Diamond-Like Semiconductors", Consultants Bureau Enterprises, Inc. New York, 47 (1969).

[2] Shay, J. L., and Wernick, J. H., "Ternary Chalcopyrite Semiconductors:Growth, Electronic Properties and Applications", Pergamon Press, Oxford (1975).

[3] Abrahams, S. C., and Bernstein, J. L., J. Chem. Phys. $\underline{59}$, 5415 (1973).

[4] Smith, R. C., J. Phys. Suppl. $\underline{36}$, C3-89 (1975).

[5] Iseler, G. W., Kildal, H. and Menyuk, N., Inst. Phys. Conf. Ser. $\underline{35}$, 73 (1977).

[6] Wagner, S., Inst. Phys. Conf. Ser. $\underline{35}$, 205 (1977).

[7] Kazmerski, L. L., Inst. Phys. Conf. Ser. $\underline{35}$, 217 (1977).

[8] Bucher, E., Appl. Phys. $\underline{17}$, 1 (1978).

[9] Pamplin, B. R., Kiyosawa, T., and Masumoto, K., Prog. Crystal Growth Charact. $\underline{1}$, 331 (1979).

[10] Leibfried, G., and Schlömann, E., Nachr. Akad. Wiss. Gottingen, Math. - Physik. kl. $\underline{IIa}$, No. 4, 71 (1954).

[11] Noolandi, J., Phys. Rev. B $\underline{10}$, 2490 (1974).

[12] Phillips, J. C., Festkorperprobleme $\underline{XVII}$, 109 (1977).

[13] Simons, C., and Bloch, A. N., Phys. Rev. $\underline{B7}$, 2754 (1973).

[14] Shaukat, A., and Singh, R. D., J. Phys. Chem. Solids, $\underline{39}$, 1269 (1978).

[15] Weaire, D., and Noolandi, J., J. Phys. Suppl. $\underline{36}$, C2-27 (1975).

[16] Neumann, H., Kristall and Technik, $\underline{15}$, 849 (1980).

[17] Binsma, J. J. M., Giling, L. J., and Bloem, J., Phys. Sat. Sol. (a) $\underline{63}$, 595 (1981).

[18] Voigt, W., Lehrbuch der Kristallphysik, Stuttgart (1966).

[19] Kistaiah, P., Venudhar, Y. C., Murthy, K. S., Iyengar, L., and Rao, K. V. K., Pramana, $\underline{16}$, 281 (1981).

[20] Kistaiah, P., Venudhar, Y. C., Murthy, K. S., Iyengar, L., and Rao, K. V. K., J. Phys. D $\underline{14}$, 1311 (1981).

[21] Kistaiah, P., Venudhar, Y. C., Murthy, K. S., Iyengar, L., and Rao, K. V. K., J. Mats. Sci. $\underline{16}$, 1417 (1981).

[22] Kistaiah, P., Venudhar, Y. C., Murthy, K. S., Iyengar, L., and Rao, K. V. K., J. Appl. Cryst. $\underline{14}$, 281 (1981).

[23] Kistaiah, P., Venudhar, Y. C., Murthy, K. S., Iyengar, L., and Rao, K. V. K., J. Less. Common Metals, $\underline{77}$, P 9 (1981).

[24] Kistaiah, P., Ph.D. Thesis, Osmania University (India) (1982).

[25] Ashcroft, N. W., and Mermin, N. D., Solid State Physics, New York (1976).

[26] Neumann, H., Crystal Res. and Technol. $\underline{18}$, 659 (1983).

[27] Kistaiah, P., and Satyanarayana Murthy, K., J. Less. Common Metals, (1984)(In Press).

[28] Callaway, J., Phys. Rev. $\underline{113}$, 1946 (1959).

[29] Steigmeier, E., "Thermal Conductivity", Academic Press, New York, $\underline{2}$, 212 (1969).

[30] Slack, G. A., Phys. Rev. $\underline{B6}$, 3791 (1972).

[31] Oshcherin, B. N., Phys. Stat. Sol. (a) $\underline{35}$, K35 (1976).

[32] Ioffe, A. F., "Semiconductor Physics" (in Russian) [Moscow-Leningrad, Acad. Sci. USSR Press, (1957)].

[33] Krebs, H., Acta Cryst. $\underline{9}$, 95 (1956).

[34] Petrov, A. V., and Shtrum, E. L., Sov. Phys. - Solid State, $\underline{4}$, 1061 (1962).

[35] Spitzer, D. P., J. Phys. Chem Solids, $\underline{31}$, 19 (1970).

[36] Sanchez Porras, G. P., and Wasim, S. W., Phys. Stat. Sol. (a) $\underline{59}$, K175 (1980).

[37] Gasanov, S. A., and Magomedov, Y. B., Sov. Phys. - Semicond. $\underline{4}$, 8 (1971).

THERMAL CONDUCTIVITY BEHAVIOR OF BORON CARBIDES

Charles Wood* and Andrew Zoltan*

Jet Propulsion Laboratory
California Institute of Technology
Pasadena, CA 91109

David Emin**

Sandia National Laboratories
Albuquerque, NM 87185

Paul E. Gray

GA Technologies
San Diego, CA 92138

ABSTRACT

   Knowledge of the thermal conductivity of boron carbide is necessary to evaluate its potential for high temperature thermoelectric energy conversion applications. We have measured the thermal diffusivity of hot-pressed boron carbide $B_{1-x}C_x$ samples as a function of composition ($0.1 \leq x \leq 0.2$), temperature (300 K to 1700 K) and temperature cycling. These data in concert with density and specific heat data yield the thermal conductivities of these materials. We discuss these results in terms of a structural model that has been previously advanced by two of us (D.E. and C.W.) to explain the electrical transport data. Some novel mechanisms for thermal conduction are briefly discussed.

INTRODUCTION

   Boron carbides have been utilized extensively in reactor technology for the absorption of thermal neutrons. More recently

the high-temperature stability and striking electronic transport
properties of these and similar semiconducting material (e.g., the
boron-rich borides) have sparked interest in their use for high-
temperature thermoelectric energy conversion. We have reported
elsewhere [1, 2] on the measurement and analysis of the steady-
state electrical transport properties of a variety of boron
carbides. Here we report on the thermal transport properties of
$B_{1-x}C_x$, in the single-phase compositional range $0.1 \leq x \leq 0.2$ from
300 K to 1700 K. Previous reports [3-7] on the thermal transport
properties of boron carbides have concentrated solely on the
composition $B_{0.8}C_{0.2}(B_4C)$.

Over the compositional range $0.088 \leq x \leq 0.200$ boron carbides
are usually viewed as crystallizing in a rhombohedral structure
$(D_{3d}^5-R\bar{3}m)$, the α-boron structure, with deformed icosahedra shared
at the corners of the unit cell. A structural unit links
icosahedra along a cell diagonal. These units are either three-
atom chains [8, 9] or bridged four-boron atom linkage [10].
Carbon atoms are distributed among the icosahedra and the three-
atom chains [8-10]. With increasing carbon content the unit cell
shrinks slightly. The melting points of compositions within this
single-phase region are about 2600 K.

EXPERIMENTAL PROCEDURE

In this study most boron carbide specimens, $B_{1-x}C_x$, in the
single-phase regime, $0.088 \leq x \leq 0.200$, were prepared having
differing boron to carbon ratios by vacuum (10^{-6} to 10^{-5} torr)
hot-pressing (~2400 K) at 600 psi for two hours using high-purity
(B ~30 ppm; C ~10 ppm maximum impurities) powders (~352 mesh) in
boron-nitride lined graphite dies. Similar samples, examined by
X-ray diffraction, were found to be single-phase with lattice
constants varying with composition essentially in the manner
reported by Bouchacourt and Thevenot [11].

A heat pulse method [12, 13] was employed to determine the
thermal diffusivities. Here a He-Ne laser or Xenon flash lamp
applies a heat pulse to one side of a sample. An InSb infrared
detector or thermocouple measures the temperature rise of the
rear-surface of the sample. Where necessary corrections were
applied for heat losses and finite pulse times [13]. The
apparatus was satisfactorily calibrated against the results of an
International program [14] on thermal diffusivity measurements of
ASM-5Q graphite.

EXPERIMENTAL RESULTS

The thermal diffusivities of various boron carbides, $B_{1-x}C_x$,

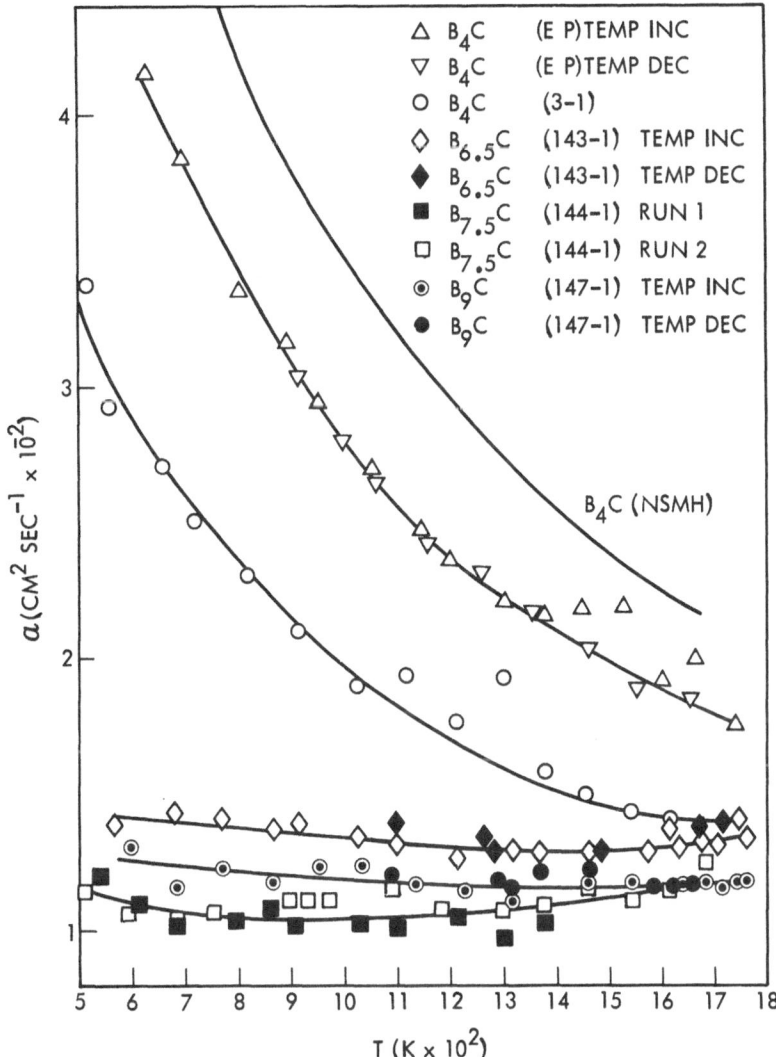

Fig. 1. The Thermal Diffusivity (α) of Boron Carbides as a Function of Temperature (T) ([EP] Refers to Eagle Picher Commercial Grade Boron Carbide and [NSMH] to Nuclear Systems Materials Handbook Data [6]).

141

with compositions ($0.1 \leq x \leq 0.2$) lying within the single-phase region are plotted against temperature in Fig. 1. The general consistency of measurements upon progressive heating of the sample (Temp Inc) with those with subsequent successive cooling (Temp Dec) indicates the insensitivity of these measurements to thermal cycling. Measurements of differently prepared samples of $B_{0.8}C_{0.2}$ (B_4C) are also compared in Fig. 1. Here significant differences in the magnitudes of the thermal diffusivities are found. Nonetheless, the temperature dependences of these thermal diffusivities are similar to one another.

The thermal conductivities (λ) were obtained from the thermal diffusivities (α) using the relationship $\lambda = c\alpha\rho$, where the specific heat, c, was obtained from the Elwell [15] curves in Fig. 2 with c being treated as constant above 900 K. The sample densities (ρ) listed in Table 1 were determined by Archimedes principle and are compared to the theoretical values determined by Bouchacourt and Thevenot [11]. These thermal conductivity curves, referred to the theoretical densities, are plotted in Fig. 3. Two further samples of $B_{0.9}C_{0.1}$, i.e., B_9C, (not shown in Fig. 1 for reasons of space) prepared by hot-pressing at higher pressures (~8000 to 9000 psi) for shorter periods (15 min) are also shown for comparison [16] in Fig. 3. As illustrated both by the $B_{0.8}C_{0.2}$ and $B_{0.9}C_{0.1}$ samples, the magnitudes of the thermal conductivities depend upon the method of sample preparation. In addition, the thermal conductivity manifests a significant compositional dependence. Here both the magnitude and temperature dependence of the thermal conductivities vary significantly with composition.

Table 1. Boron Carbide Densities

Specimen	Experimental Density (gm/cm^3)	Percent Theoretical Density
B_4C EP	2.44[a]	97
B_4C 3-1	2.38	98
$B_{6.5}C(143-1)$	2.38	96
$B_{7.5}C(144-1)$	2.48	100
$B_9C(147-1)$	2.43	98
$B_9C(20-1)$	2.33	94

[a]Average of three samples.

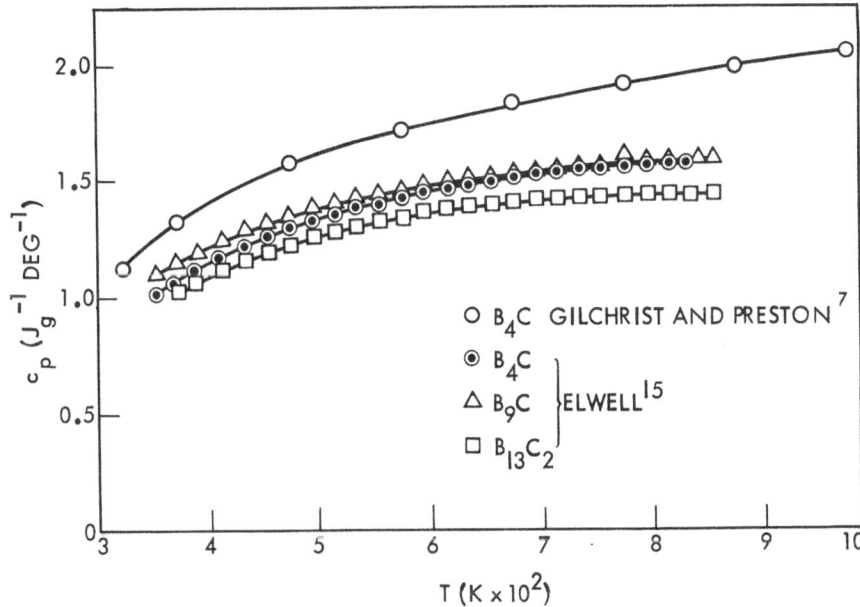

Fig. 2. The Specific Heat (c_p) of Boron Carbides vs Temperature (T)

It should be noted that we have utilized Elwell's specific heat data since his samples, while not ours, were prepared in a manner similar to ours. However, even if the specific heat were to differ substantially, (i.e., be such as those of Gilchrist and Preston of Fig. 2) the fall of the thermal conductivity of the $B_{0.8}C_{0.2}$ with increasing temperature would only be somewhat milder than that shown in Fig. 3. In fact, as shown in Fig. 3, the thermal conductivity of our $B_{0.8}C_{0.2}$ falls with increasing temperature in a manner similar to that found by other researchers.

DISCUSSION

The electrical transport properties of these boron carbides have been studied in detail [2]. It is concluded that the boron carbides are nearly degenerate semiconductors in which the predominant charge carriers (holes) form small polarons and hop between inequivalent positions in the solid. The dependence of the Seebeck coefficient on carbon concentration suggests that the holes hop between sites associated with carbon atoms. In this

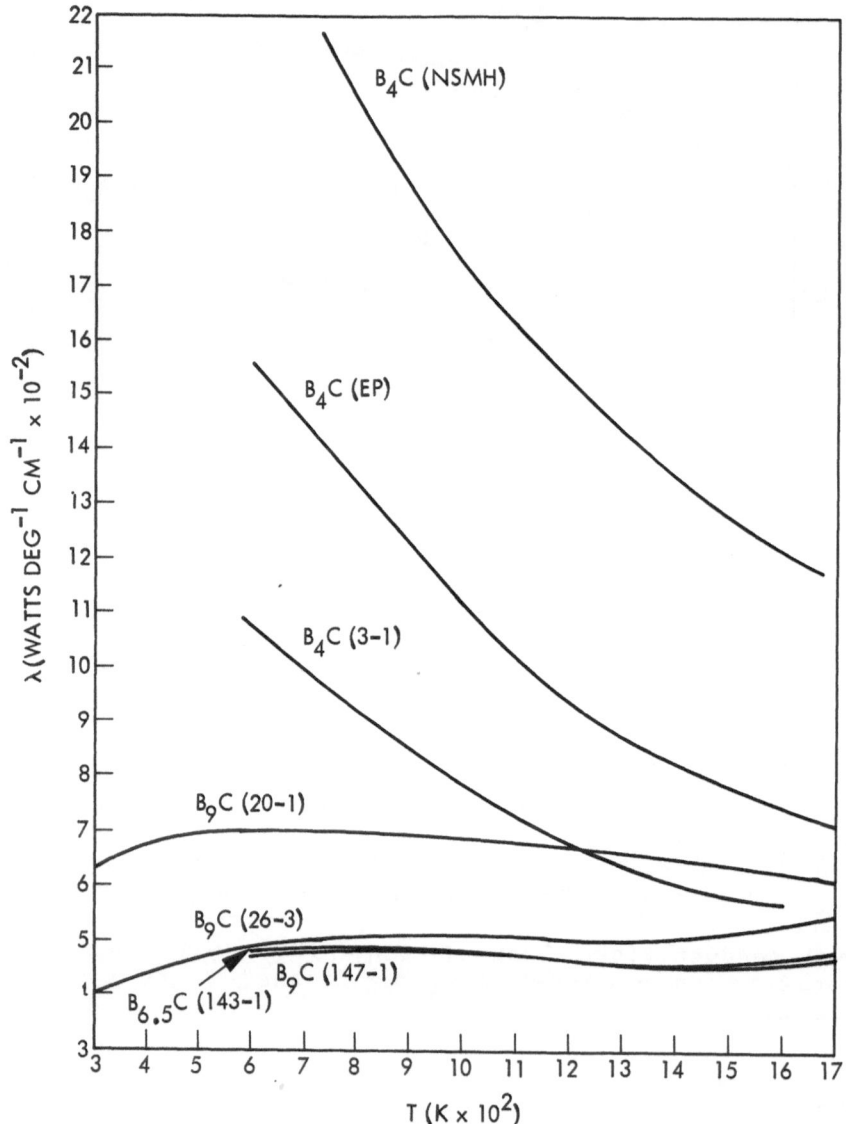

Fig. 3.   The Thermal Conductivity (λ) of Boron Carbides as a
Function of Temperature (T)

case the inequivalence of hopping sites corresponds to jumps between chains and icosahedra. Thus carbon atoms occupy these two types of sites at all carbon compositions within the single-phase region. This contradicts the idea of some that x = 0.133 ($B_{13}C_2$) represents an "ideal" composition in which two carbon atoms reside on each intericosahedral chain, with the icosahedra remaining devoid of carbon atoms. The present thermal conductivity studies also support the view that carbon atoms generally occupy both kinds of sites.

As shown in Fig. 3, the high temperature thermal conductivities of the boron carbides are generally moderately low (~0.1 to 0.01 W/K-cm at 1000 K). At the highest carbon concentrations, x = 0.200 (B_4C) the thermal conductivity is not only at its largest but also is a decreasing function of increasing temperature. At lower carbon concentrations, x = 0.100, 0.133 and 0.176, the thermal conductivities are much smaller with much weaker temperature dependences. The smaller weakly temperature-dependent thermal conductivities of these materials with lower carbon concentrations are similar to those of glasses (e.g., silica). However the thermal conductivity of the high-carbon concentration material (x = 0.200, B_4C) is characteristic of the lattice thermal conductivity of a crystal. This suggests that at the high carbon concentration carbon atoms occupy all available sites. In this situation the thermal transport should be that of a well ordered crystal. At lower carbon concentrations the carbon atoms occupy only a fraction of the available locations. The thermal transport through such a defect structure then resembles that through a disordered material. The fact that the thermal transport through the x = 0.133 composition ($B_{13}C_2$) resembles that through an amorphous structure belies this being an "ideal" composition with only chains being occupied by carbon atoms. Thus both the electronic and thermal transport suggest a consistent picture of the carbon occupancy of the boron carbides.

The thermal conductivity of these solids arises from two sources. First, even in the absence of charge carriers, heat is transferred directly between atoms of the solid. Second, the motion of a charge carrier through a solid can also be associated with a heat flow. In the boron carbides the electronic transport represents a very distinctive type of small-polaron hopping [1, 2]. The associated thermal conductivity is distinctive as well. In particular, vibrational energy is transported with a charge carrier as it hops between inequivalent sites. The average energy carried with a hopping carrier, E_T, is CkT^2, where the constant C is defined by $C \equiv zJ^2/16E_b^3$, z is the number of nearest neighbors, J is the intersite transfer energy and E_b is the small-polaron binding energy characterizing an average site in the material. Thus the energy carried increases with temperature. Preliminary considerations indicate that the electronic contribution to the thermal conductivity is a product of this energy, λE_T, the electronic diffusion constant, D, and the rate of change of the carrier density with temperature (dn/dT), i.e., $_T = E_T D(dn/dT)$. This yields an electronic contribution to the thermal conductivity which increases with temperature and is ~15 mW/K-cm at 1600 K. This is about one third of the lowest value we measure.

Finally we note that the disorder in the masses and spring constants associated with carbon atoms being distributed among distinct sites can greatly alter the nature of the lattice contribution to the thermal conductivity. In particular, in the boron-carbides, reducing their carbon content from its maximum, B_4C, eliminates relatively stiff springs (C-B-C chains) which link relatively soft icosahedra. In this situation a significant fraction of the vibrational modes become localized. In a purely harmonic system these modes will not participate directly in the passage of vibrational energy. However, anharmonic forces provide a means of transferring energy between local and nonlocal vibrational modes. This facilitates the thermal transport. Hence, contrary to customary (phonon-scattering) treatments of thermal conductivity, the thermal conductivity increases as the anharmonic coupling strength increases. Such behavior has been shown in theoretical models [17], when successive phonon scattering events are not sufficiently rare so as to be treated as to be independent. However, it has not, heretofore, been reported in actual solids. The boron carbides, $B_{1-x}C_x$, of low carbon concentrations (e.g., x = 0.100, 0.133 and 0.176) may be examples of this distinctive type of thermal transport. In particular, the usual theory of strong point-defect scattering [18] has the thermal conductivity falling with temperature as $T^{-1/2}$. However, while B_4C behaves in approximately this manner, samples of lower carbon concentrations obviously do not. We are currently investigating whether this behavior is attributable to the previously described anomalous transport.

ACKNOWLEDGEMENTS

We wish to acknowledge the work of N. G. Elsner and G. H. Reynolds of GA Technologies for the materials preparation and A. Lockwood and J. Parker for density measurements.

REFERENCES

*The research described in this paper was carried out by the Jet Propulsion Laboratory, California Institute of Technology, under contract with the National Aeronautics and Space Administration.

**Supported by U.S. Department of Energy Contract No. DE-ACO4-76CDP00789.

[1] Wood, C. and Emin, D., "Conduction Mechanism in Boron Carbide," Phys. Rev. B 29, No. 8 (1984).

[2] Emin, D. and Wood, C., "Small-Polaron Electronic Transport in Boron Carbides," Proceedings of the 18th IECEC, 1, 222-225, Orlando, FL, Aug. 22-26 (1983).

[3] Deem, H. W. and Lucks, C. F., "Thermal Conductivity of Boron Carbide from 100 C to 800 C," Battelle Memorial Institute Report No. BMI-731, 1951 (unpublished); Nucl. Sci. Abstr. 6, 915 (1952).

[4] Hedge, J. C.; Kostenko, C.; and Lang, J. I., Illinois Institute of Technology Research Institute, Technical Documentary Report Nos. ASD-TDR-63-597 and AD-424375, 1963 (unpublished; see Chem. Abstr. 61, 10433d, 1964).

[5] Boiko, N. V. and Shpil'rain, E. E., "Some Questions on the Method of Experimentally Studying the Heat Conductivity of Materials at High Temperatures," Sov. Phys. High Temp. 2, 493-500 (1964)

[6] Nuclear Systems Materials Handbook, Vol. 1 Design Data. Part III Group 1 Section 1. p 1.0,1.1.

[7] Gilchrist, K. E. and Preston, S. D., "Thermophysical Property Measurements on Some Neutron Absorbing Materials," High Temp.-High Pres. 11, 643-651 (1979).

[8] Clark, H. K. and Hoard, J. L., "The Crystal Structure of B_4C," J. Amer. Chem. Soc. 65, 2115 (1943).

[9] Will, G. and Kossobutzki, K. H., "An X-ray Diffraction Analysis of Boron Carbide, $B_{13}C_2$," J. Less-Comm. Met. 47, 43-48 (1976).

[10] Yakel, H. L., "The Crystal Structure of a Boron-Rich Boron Carbide," Acta Cryst. B31, 1797-1806 (1975).

[11] Bouchacourt, M. and Thevenot, F., "The Properties and Structure of the Boron Carbide Phase," J. Less-Comm. Met. 82, 227-235 (1981).

[12] Parker, W. J.; Jenkins, R. J.; Butler, C. P.; and Abbott, G. I., "Flash Method of Determining Thermal Diffusivity, Heat Capacity, and Thermal Conductivity," J. Appl. Phys. 32, No. 9, 1679-1684 (1961).

[13] Taylor, R. E., "Heat-Pulse Thermal Diffusivity Measurements," High-Temp.-High Press. $\underline{11}$, 43-58 (1979).

[14] Minges, M. L., "Evaluation of Selected Refractories as High Temperature Thermophysical Property Calibration Materials," Int. J. Heat Mass Transfer, $\underline{17}$, 1365-1382 (1974).

[15] Data supplied by D. Elwell, Stanford University, using a Dupont Differential Scanning Calorimeter.

[16] From thermal diffusivity data supplied by R. E. Taylor, Purdue University, and confirmed by JPL, using the flash method. An absolute determination of thermal conductivity by G. Slack, G. E. Research, yielded a fairly temperature independent value of ~0.068 $\pm$ 0.002 W/K-cm for $B_{0.9}C_{0.1}$ in the vicinity of room temperature in essential agreement with $B_{0.9}C_{0.1}$ (20-1).

[17] Payton, D. N.; Rich, M.; and Visscher, W. M., "Lattice Thermal Conductivity in Disordered Harmonic and Anharmonic Crystal Models," Phys. Rev. $\underline{160}$, 706 (1967).

[18] Klemens, P. G., "Thermal Resistance Due to Point Defects at High Temperatures," Phys. Rev. $\underline{119}$, 507 (1960).

SESSION F

Data Acquisition Systems

SESSION CHAIRMAN

J. G. Hust
NBS
Boulder, CO.

THE TEMP MICROCOMPUTER AS AN INSTRUMENT

CONTROLLER AND DATA LOGGER

Michael J. Batchelder and A.L. Riemenschneider

South Dakota School of Mines and Technology
Rapid City, South Dakota 57701

ABSTRACT

The microprocessor is becoming an indispensible laboratory tool. To adapt it to one's needs requires some basic understanding of its operation. With this understanding and some inexpensive integrated circuits, one can create custom interfaces for one's own unique requirements.

A microcomputer based on the 8085 microprocessor has been designed at this school around the objective of the teaching of skills of interfacing microprocessors to laboratory instrumentation for data acquisition and control in real time. This microcomputer is simple, powerful and inexpensive.

The microprocessor chip communicates through the system bus consisting of the address bus, the data bus, and the control bus. An address decorder selects which device the microprocessor chip communicates with over the data bus. Every device is connected to the system bus through an interface which performs any required conversion and shifting of logic levels.

Some devices are connected through standard interfaces such as the RS232 interface and the IEEE 488 interface; however other devices, such as A/D and D/A converters, relays, sensors, actuators, and clocks, may have their interfaces designed as needed. These can in many cases be designed by the user.

Useful examples are given of some commonly occurring situations whereby data may be read in or a control signal put out by a microcomputer to interact with a laboratory instrument.

151

A DATA ACQUISITION SYSTEM FOR LOW

LEVEL SIGNALS

Steven C. Maher and T. Ashworth

Physics Department
South Dakota School of Mines and Technology
Rapid City, South Dakota 57701

ABSTRACT

One of the problems encountered in the measurement of thermo-physical properties is the switching and measurement of signals of very low amplitudes. This paper describes a bus system and inter-faces capable of switching such low-level signals and interrogating a high-quality manual potentiometer. A timing system with several interrupt channels is also provided. The circuitry detailed herein has proven to be reliable and flexible in its application. No commercial system has been found which provides the attributes of this system at a reasonable cost. Also detailed is a system, in-cluding the switching circuitry, which can function as a total data acquisition system in conjunction with a host microcomputer.

INTRODUCTION

In the last ten years there has been a revolution in the field of data acquisition, due mainly to the advent of the microprocessor. The price of the microprocessor has become low enough that it can be put to effective use in many fields.

Previously it was not possible to take data at high rates such as those which the microprocessor makes possible. It is now possible to take data at rates of several tens or hundreds of thousands of samples per second. It is also possible to take data without the intensive use of human labor, and therefore to reduce the cost in such instances. Microprocessors can also be used to take data remotely in hazardous or unpleasant environments where it would be difficult or impossible otherwise.

153

A further advantage of a system based on the microprocessor is its ability to accurately monitor and control equipment and therefore take data devoid of human errors. Microprocessors are not difficult to use and require only a limited technical knowledge for many applications, and they are cost-effective. Detailed in this article is a system developed to fit the needs of a particular application in which the taking of accurate measurements over long periods of time was involved.

System Requirements

The instrumentation to be interfaced with the data acquisition system and with control systems consisted of a manual potentiometer and an autoranging nanovoltmeter. These instruments monitor signals related to the operation of equipment such as an adiabatic calorimeter and a guarded hot plate apparatus. Measurement requirements vary. Thermocouples generate signals in the microvolt range which need to be determined to $\pm 0.01\mu V$, and larger voltages across platinum resistance thermometers and associated standard resistors need to be determined to nearly 7 significant figures with precise associated timing to provide the required resolution of temperature changes. The potentiometer system used consists of a Leeds and Northrup K-6 potentiometer, a multiple Leeds and Northrup (type 31) switch, and a sensitive null detector to which we have added a zero-crossing switch to detect a null condition. Despite recent developments of $6\frac{1}{2}$ digit multimeters, the K-6 still gives superior resolution thermometry, hence its continued use. Also included in this system is a timebase permitting the time of measurement to be registered along with the data. Before the advent of the data acquisition system presented here, the data had to be taken by hand over long periods of time for any given experimental run.

Additional multiple selection switches are used ahead of a Keithley 181 nanovoltmeter to allow monitoring of signals from an array of thermocouples. Most commercial relay-based multiplexing systems generate thermal e.m.f.s. of at least $1\mu V$; a multiplexer was needed having the superior electrical and thermal qualities of the Leeds and Northrup switch. One piece of commercial equipment which has a low thermal noise figure is the MS1Ø6N relay multiplexer by Julie Research Labs. This recently available device has thermal emfs of about $0.02\mu V$, but the cost of a single decade of such a system is about $2500. This compares with a cost of less than $500 for the system described here.

The system was designed to have the following characteristics:

I. the ability to obtain readings from the potentiometer and record them automatically without user intervention.

II. the ability to automatically switch between various sense

channels without user intervention and without compromising
the ability to detect voltage changes as small as 0.01μV;

III.   the ability to record the time at which data is taken
without user involvement.

Hardware

The host computer receiving data from the devices is an
APPLE II in the standard configuration with 48K of dynamic RAM. A
parallel interface card was added to the apple for the purpose of
communicating with the control devices (Fig. 1). Two eight-bit
parallel ports plus the interrupt capabilities of the interface card
are dedicated to the control devices.

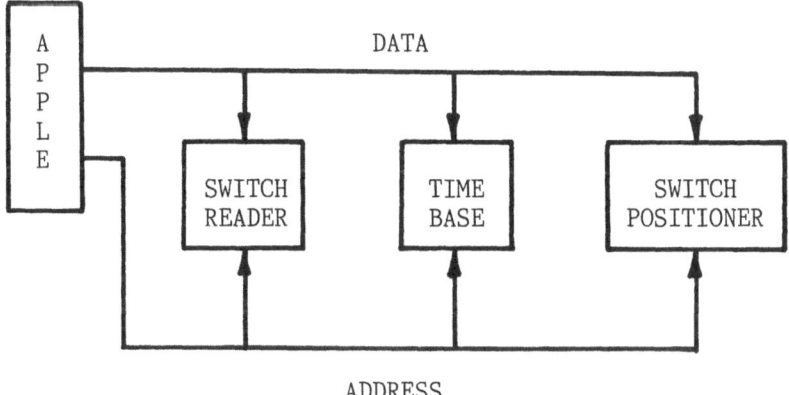

Figure 1. Block and Functional Diagram.

The host communicates with the other devices in the laboratory
over a parallel bus designed by the authors (Table 1). This bus is
buffered with open-collector TTL gates and pull-up resistors on the
terminating end to yield a high noise immunity and good drive
characteristics. Twenty-five lines constitute the bus, the designa-
tions of which are shown in Table 1. The bus has the capabilities
of controlling up to sixteen devices attached to it and allocates
each device sixteen unique addresses. Also allowed for is the
transfer of data from the device back to the Apple in eight-bit
parallel form. Not all of the capabilities of the bus are being
used at this time; this will allow future expansion of the system.

Table 1. Structure of Parallel Communications Bus

Pin	Name	Function	Comments
1	address 7	device select 3	(selects one of 16
2	address 6	device select 2	devices on the bus)
3	address 5	device select 1	
4	address 4	device select 0	
5	address 3		(internal select
6	address 2		addresses for device)
7	address 1		
8	address 0		
9	data 7		(data from device to
10	data 6		host)
11	data 5		
12	data 4		
13	data 3		
14	data 2		
15	data 1		
16	data 0		
17	busy 7		(busy lines from
18	busy 6		devices)
19	busy 5		
20	busy 4		
21	busy 3		
22	busy 2		
23	busy 1		
24	busy 0		
25	GND		(common ground)

The first device attached to the bus is a decade-switch reader
(Fig. 2). This reader will sense the position of multi-position
rotary or linear switches by direct connection to an extra decade
of the switch. It may interface up to sixteen switches of up to
eighteen contacts each. The primary use for this device is the
determination of the position of the various switches on the Leeds
and Northrup potentiometer. Not only does the device detect
numerical setting of the potentiometer, but it also detects the
polarity of the setting, and the range of the potentiometer. All
of this information is transmitted to the host computer over the
parallel bus. The transfer is made by sending the device-select
code to the switch reader along with a code representing the
position which it should read. The switch reader responds by
reading the desired position and returning its value on the data
bus. By performing a series of these data transfers the host may
ascertain the full reading of the potentiometer.

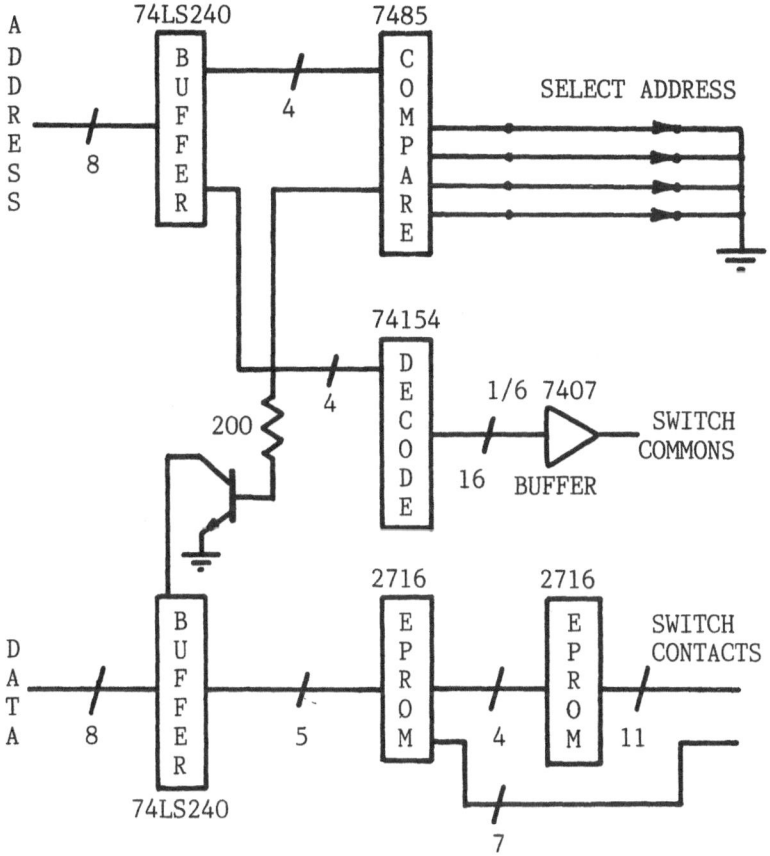

Figure 2. Switch Reader Schematic.

The time base for the system is by far the most complex component of the system. Communication with the host computer is also over the parallel data bus (Fig. 3). It was decided that the real-time reference could be most easily implemented using a separate microprocessor.

The microprocessor selected for this purpose was the TEMP (Tech Education Microprocessor) 8085-based system designed and constructed at the South Dakota School of Mines and Technology campus. The TEMP system is ideally suited in that it provides extensive parallel communication and has a LED display support chip on board.

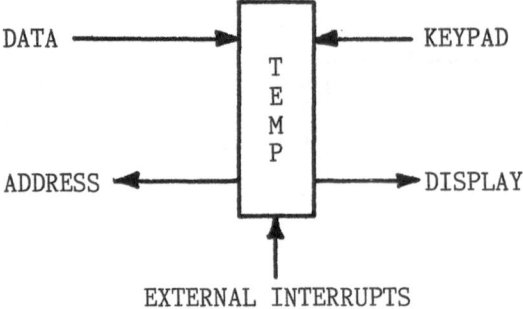

Figure 3. Time Base Block and Functional Diagram.

Real-time reference is done in software on the microprocessor
and is interrupt-driven by a crystal oscillator at one thousand
hertz, yielding an effective resolution of one millisecond for its
time readings. Also provided are four user interrupt channels which
may be connected to any outside equipment which requires that its
activities be time-logged. These channels are enabled/disabled by
front panel switches on the reference. The host has complete control
of the reference via the parallel bus. This includes the ability to
read the time from the clock, to read the stored time of the last
interrupt, or to read the source of the interrupt which stopped the
clock. The host may also reset the clock to zero and may cause the
recover-from-interrupt process to cycle. An interrupt occurs every
time the reference receives an external interrupt on one of its four
channels.

The switch controller (schematic shown in Fig. 4) is the last
device to be attached to the bus. This device provides automatic
selection of sense channels to the potentiometer. A major problem
in the design of this portion of the system was obtaining either
switches or relays with sufficiently low thermal noise to allow
measurement of very low level signals.

The primary reason for using the Leeds and Northrup switch
was the unavailability (at comparable cost) of any system with com-
parably low thermal noise ratings.

The problem then became that of designing an automated positioning system for this rotary switch. As presently implemented, the
design uses a gearing arrangement and a stepper motor to do the
switch positioning. Control of the stepper motor is via a custom
interface card designed by one of us (SCM). This card implements
all of the functions necessary to sense and control the position of
the switch, as well as the capability of communicating with the host
computer over the parallel bus. The controller senses the position

of the switch via an optical shaft encoder attached to the shaft of
the switch. This encoder provides a binary code to the controller
which represents the position of the switch. The controller compares
this position with the desired position as sent to it by the host,
and steps the stepper motor to rotate the switch until the two numbers
are in agreement. Once the switch is in position, the controller
signals the host via a status line to indicate that the switch is
ready for use. Incorporated in this system is a failsafe that will
not allow the host to use the switch until such time as the switch
has stabilized.

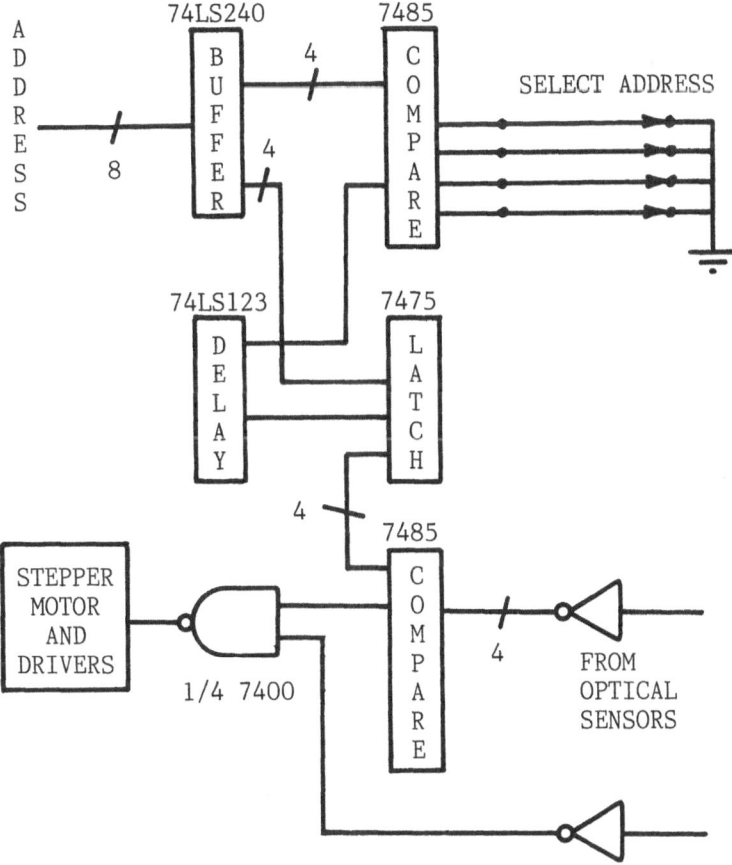

Figure 4. Switch Positioner.

The stepper motors also provide the added benefit of absolute positional accuracy in that they have no cumulative angular error and positive detents at all desired positions. In this way the user may be assured that the switch is indeed in position. The controller will control one switch of up to sixteen positions, and the host may cause the switch to be positioned to any one of these positions with a single command over the bus.

System schematics are shown for the various devices, along with a block diagram of the system, and a description of the bus structure (Table 1 and Figs. 1,2,3,4).

Conclusion

The system developed has proven to be versatile. One of the design criteria of the system was expandability. There are at present thirteen unused device addresses on the parallel bus. In the future these addresses will be dedicated to serve other switch controllers and readers and possibly other devices. One possibility is the construction of an AC power controller for the various pieces of equipment around the laboratory which must be tuned on and off at various times. In the future, the stepper motors may also come under direct microprocessor control. A dedicated microprocessor controller for the stepper motors could make the system much more efficient and allow for the control of multiple stepper motors by a single controller.

The system detailed here is easily constructed with minimal facilities. It could readily be adapted to other systems and other pieces of data acquisition equipment. One very significant advantage of this bus over other (such as Hewlett-Packard's HP-IB) is the ease with which it may be interfaced. The device attached to the bus need only recognize its own unique address, and need not conform to any timing specifications. In interfacing to other busses there are generally more integrated circuits in the interface than in the controller itself.

SESSION G

Metals and Alloys

SESSION CHAIRMAN

G. K. White
CSIRO
Lindfield, NSW
Australia

THERMOELECTRIC POWER OF SELECTED BINARY ALLOY SYSTEMS

PART II - COPPER-PALLADIUM, GOLD-PALLADIUM, AND SILVER-PALLADIUM*

C.Y. Ho, R.H. Bogaard, H.M. James, T.C. Chi,
and T.N. Havill

Center for Information and Numerical Data Analysis
  and Synthesis, Purdue University
2595 Yeager Road, West Lafayette, Indiana 47906

ABSTRACT

Because the scale of the absolute thermoelectric power commonly used throughout the world in the past 45 years or so had become known to be much in error, a revised scale of the absolute thermoelectric power from 0 to 1300 K had been developed to replace the old scale, which was reported in Part I of this work [1] along with the recommended values for the thermoelectric power of aluminum-copper, copper-nickel, and iron-nickel alloy systems.

Reported in this Part II are the recommended values for the thermoelectric power of copper-palladium, gold-palladium, and silver-palladium alloy systems, which were resulted from exhaustive compilation, careful conversion to the revised scale, and critical evaluation, analysis, and synthesis of the available experimental raw data. In the available data, serious gaps exist for both composition and temperature dependences.

In generating the recommended values which cover a full range of composition and temperature, experimental data for both temperature and composition dependences were utilized. Confusion in the thermoelectric power data for Cu-Pd was clarified with the help of the phase diagram. Data below 700 K for alloys with roughly 10-47% Pd which were not completely disordered were therefore

---

*This work has been supported by the Office of Standard Reference Data, National Bureau of Standards, Washington, DC.

justifiably set aside. With no phase-diagram complications arising for Au-Pd and Ag-Pd, the critical evaluation was considerably simplified and recommended values were generated in a relatively straight forward manner.

In addition to the recommended values which are presented in this Conference, experimental data and information have been exhaustively compiled and details of data evaluation, analysis, and synthesis documented [2].

REFERENCES

[1] Ho, C.Y., Chi, T.C., Bogaard, R.H., Havill, T.N., and James, H.M., "Thermoelectric Power of Selected Binary Alloy Systems", in Thermal Conductivity 17 (Hust, J.G., Editor), Plenum Press, New York, NY, 195-205 (1982).
[2] Ho, C.Y., Bogaard, R.H., James, H.M., Chi, T.C., and Havill, T.N., "Thermoelectric Power of Ten Selected Binary Alloy Systems", to be published in Journal of Physical and Chemical Reference Data.

TRANSPORT PROPERTIES OF POLYCRYSTALLINE Ni₃Al*

R. K. Williams, F. J. Weaver, and R. S. Graves

Metals and Ceramics Division
Oak Ridge National Laboratory
Oak Ridge, Tennessee 37831

Recent advances in ductilizing the intermetallic compound Ni₃Al may lead to practical applications. These applications, which are based on the outstanding strength and oxidation resistance of the compound, also require physical property data. In this paper, the room-temperature electrical and thermal conductivities of annealed high purity specimens containing 74 to 76 at. % Ni are presented and compared to theoretical predictions. Residual (4.2 K) electrical resistivity data are also employed in the analysis and these results show a pronounced minimum at the stoichiometric composition. The data show that the thermal conductivity of this compound is quite sensitive to stoichiometry and, at room temperature, has a maximum value at about 74.8 at.% Ni. Calculated and experimentally derived phonon thermal conductivity values agree well, and indicate that this carrier is responsible for about 25% of the room temperature thermal conductivity. The electronic Lorenz function is essentially equal to the Sommerfeld value.

INTRODUCTION

In nickel-base superalloys, the intermetallic compound Ni₃Al (γ´) provides high temperature strength [1]. Experiments on single crystals have shown that the compound is also ductile [2], but polycrystalline samples of Ni₃Al always fracture intergranularly and are not useful as engineering materials. Recently [3]

*Research sponsored by the Division of Materials Science, U.S. Department of Energy, under contract DE-AC05-840R21400 with Martin Marietta Energy Systems, Inc.

it has been found that microalloying can be used to strengthen the grain boundaries of this material, and advanced intermetallic alloys with exceptionally good high temperature mechanical properties are being developed [4]. The experiments described in this paper were conducted to determine the factors which control the temperature and composition dependence of the thermal conductivity of Ni_3Al. The composition dependence is of particular interest because the compound exists over a fairly broad range (72.5—77 at. % Ni), and the ductile alloys are Ni rich.

SPECIMEN PREPARATION AND CHARACTERISTICS

All of the specimens were machined from annealed arc-cast ingots. High purity charges were repeatedly melted to promote homogeneity and drop-cast into a cylindrical mold. The castings were annealed at 1200°C in an inert atmosphere to remove short-range chemical inhomogeneities and step-annealed at lower temperatures to promote ordering. The annealing schedules are shown in Table 1. Radiographs were used to locate casting voids and

Table 1. Chemical Analyses of Ni_3Al of Samples
(ppm)

Element	Group 1[a] Average	Group 2[b] Average	Group 3[c] Average
C	36	242	56
O	55	94	42
H	9	51	3
N	7	N.D.	5
B	0.1	0.2	<1
Co	N.D.	7	10
Cr	0.5	2	13
Cu	7	13	20
Fe	6	24	10
W	12	10	N.D.
Zr	N.D.	4	N.D.

[a]Mond Nickel. Annealed 11 hours at 1200°C, 1 hour at 1100°C, 2 hours at 1000°C, 3 hours at 900°C and furnace cooled.
[b]Commercial Zone-Refined Nickel. Annealed 11 hours at 1200°C, 1 hour at 1100°C, 2 hours at 1000°C, 3 hours at 900°C and furnace cooled.
[c]Mond Nickel. Annealed 20 hours at 1200°C, 1 hour at 1000°C, 2 hours at 900°C, 4 hours at 800°C, 20 hours at 700°C and furnace cooled.

specimens were electrically discharge machined from pore-free regions and ground to final size. Average chemical analyses of the three groups of specimens are shown in Table 1, and it is interesting to note that the Mond Ni obtained by ORNL 30 years ago is superior to currently available zone-refined stock.

Metallographic examination did not show evidence of a residual dendritic cast structure, and the samples all had grain sizes in excess of 100 μm. A few annealing twins were present. The X-ray diffraction patterns showed only Ni_3Al lines, and hardness values did not vary with composition between the 74.7 and 75.2 at. % Ni. The average Diamond Pyramid Hardness (DPH) (1 Kg) was 172 Kg/mm^2.

Room-temperature electrical resistivity checks on the first two groups of specimens revealed that the resistivity varied by several percent along a 75 mm long specimen, indicating either segregation or microcracking. Fluitman et al. [5] described a similar difficulty with their specimens, and microprobe analysis was used in an attempt to determine its cause. The microprobe results did not show significant Ni/Al differences between "good" and "bad" specimens so it was decided that microcracking was a more plausible explanation. In preparing the third group (Table 1) of specimens, extra precautions were taken to minimize machining stresses, and the resistivities of this group of specimens showed much smaller variations. This group included two castings of the stoichiometric composition.

EXPERIMENTAL PROCEDURES

The thermal conductivity, λ, values for the 6.3 mm high, 6.3 mm diameter specimens were determined in a comparative heat flow apparatus which has been described previously [6]. In this method the uninstrumented specimen is compressed between Armco Iron heat meter bars in a vacuum environment. Indium foils are used to reduce the interfacial temperature drops between the specimen and meter bars, and the temperature-distance information obtained on the meter bars above and below the specimen is used to calculate both the heat flux and specimen temperature drop. Measurements on standard specimens indicate an uncertainty of about ±2%.

Room temperature and residual (4.2 K) electrical resistivity, ρ, values were obtained on 75 mm long rod specimens using the standard 4-probe DC method. The previously described variation of these results with position along the specimens led us to make resistivity measurements on the much smaller thermal conductivity specimens. These measurements were made by two methods, an eddy current technique [7] and the 4-probe DC method. In the DC measurements, centered point current contacts were employed and a 2.5 mm knife edge was used to obtain the voltage drops. End effects [8] were significant (~20%), and measurements on three standard specimens were used to convert the measured voltage/current ratios to resistivity values. The scatter of results on the three standard specimens indicates that the DC measurements have an uncertainty of about ±1.5%, and this is comparable to the uncertainty of the eddy current results.

EXPERIMENTAL RESULTS

The experimental data are summarized in Table 2. Three thermal conductivity values, covering the range 305—360 K, were obtained for each specimen. A small temperature-dependent correction of less than 2% was applied to each value because previous comparisons [9] on standard specimens have shown that the λ values are too large in this λ range. The values shown in Table 2 were obtained by fitting these three λ values to a linear equation.

The room-temperature electrical resistivity values are the averages of the eddy current and DC results on the λ specimens, and the range quoted refers to the difference between the two results. Presumably most of this difference arises because the two measurements sample somewhat different portions of the specimen. For the third group of specimens the differences are about what might be expected from the experimental uncertainties. For this group, the ρ values for the 75 mm long specimens also differed from the Table 2 values by an average of 3%. Results on two of these longer specimens indicate that the 300 K resistivity of ideally pure Ni_3Al is $31.1 \pm 0.05 \times 10^{-8}$ Ωm.

DISCUSSION

Residual and ideal electrical resistivity values for compositions close to Ni_3Al are shown in Figs. 1 and 2. The residual resistivity curve is nearly symmetrical and has a minimum at the stoichiometric composition. Our values are in reasonably good agreement with the results of Fluitman et al. [5] and show that both Ni and Al excesses produce unusually strong electron scattering. Other material classes which exhibit strong scattering include Kondo alloys, the Ni-Cr type short-range ordering alloys, and some noble metal semi-metal solutions [10].

The ideal resistivity values shown in Fig. 2 are for the 75 mm long specimens. These values and the Seebeck coefficients increase rapidly with Ni content and the variation appears to be faster on the Ni-rich side of Ni_3Al. The ideal resistivity values for our samples also show a somewhat more regular composition

168

Table 2. Thermal Conductivity and Electrical Resistivity Values for Ni₃Al Samples

Nominal Composition (at. % Ni)	Thermal Conductivity at 300 K (W/m·K)	Temperature Coefficient, dλ/dT (W/m·K²) (300—360 K)	Residual Resistivity (4.2 K) (10⁻⁸ Ωm)	Resistivity at 300 K[a] (10⁻⁸ Ωm)	Resistivity Range (±%)	Absolute Seebeck Coefficient at 300 K (μV/K)
			Sample Group 1[b]			
74.7	30.96	+.0280	3.49	29.7	6.5	+3.8
74.8	31.31	.0292	3.21	28.9	7.4	+4.3
74.88	31.31	.0265	2.92	28.1	10.0	+5.0
75.0	29.08	.0256	2.44	33.1	2.2	+5.0
75.1	27.93	.0277	2.83	33.9	4.2	+6.4
75.2	27.90	.0294	2.86	32.7	7.4	+7.8
			Sample Group 2[c]			
74.88	30.85	.0260	4.75	29.9	8.0	+4.8
75.0	27.70	.0323	4.20	33.6	4.0	+3.5
			Sample Group 3[d]			
74.0	26.60	.0286	10.31	36.5	1.9	+ .8
75.0	30.13	.0254	1.64	32.7	0.8	+5.7
75.0	28.85	.0276	1.54	34.2	0.6	+5.0
76.0	21.37	.0222	9.05	51.3	0.5	+9.9

[a] Average temperature coefficient 0.0707 x 10⁻⁸ Ωm/K.
[b] Mond Nickel. Annealed 11 hours at 1200°C, 1 hour at 1100°C, 2 hours at 1000°C, 3 hours at 900°C and furnace cooled.
[c] Commercial Zone-Refined Nickel. Annealed 11 hours at 1200°C, 1 hour at 1100°C, 2 hours at 1000°C, 3 hours at 900°C and furnace cooled.
[d] Mond Nickel. Annealed 20 hours at 1200°C, 1 hour at 1000°C, 2 hours at 900°C, 4 hours at 800°C, 20 hours at 700°C and furnace cooled.

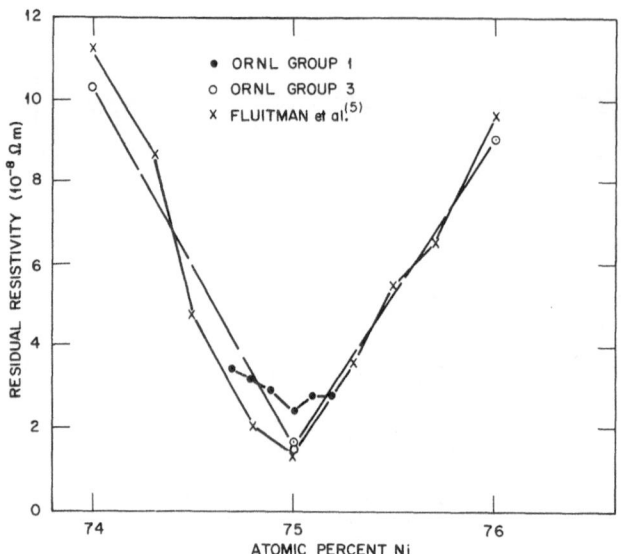

Fig. 1.   Residual resistivity values for compositions close to
Ni₃Al. The groups of specimens are described in Table 1.

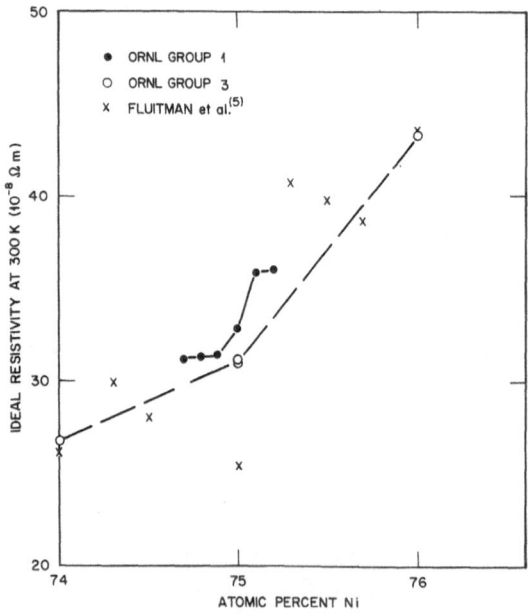

Fig. 2.   Ideal ($\rho_{300\,K} - \rho_{4.2\,K}$) electrical resistivity values for
compositions close to Ni₃Al.

dependence than the results of Fluitman et al. [5]. The specific heat data [11] show that the Debye temperature, Θ_D, does not vary significantly in this composition range, so this increase must reflect either a change in the strength of the electron-phonon interaction or spin scattering. Ni-rich compositions are ferromagnetic at low temperatures [12].

The thermal conductivity values are plotted against nominal composition in Fig. 3. These data show a peak on the Al-rich side of Ni$_3$Al and a scatter of about 5%. The peak in λ corresponds to a minimum in ρ, and the Lorenz functions ($\lambda\rho/T$) exceed the Sommerfeld value, L_0, by as much as 50%. Phonon conduction presumably causes this, and an estimate of the phonon conductivity, λ_p, can be obtained by analyzing data for pairs of specimens which show substantially different ρ and λ values [9]. Data for the third group of specimens are preferred for this purpose because these specimens were found to be more homogeneous, the stoichiometric specimens have lower residual resistivities, and the group includes the highest ρ, lowest λ composition, 76 at. % Ni. Assuming that the electronic L ($\lambda_e\rho/T$) and phonon conductivity do not change with composition [13] yields:

$$\overline{L} = 1.04 L_0$$

$$\overline{\lambda_p} = 6.74 \text{ W/m}\cdot\text{K}$$

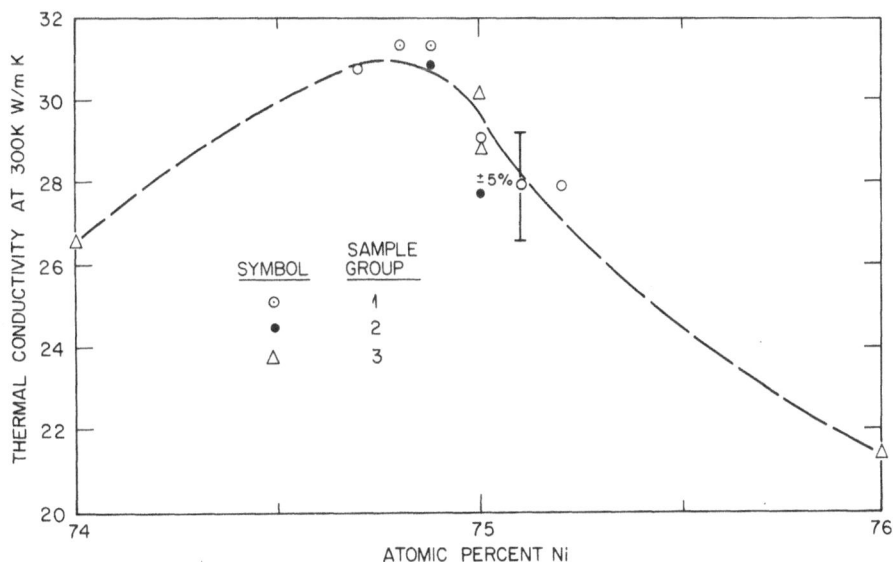

Fig. 3. Room temperature thermal conductivity values for compositions close to Ni$_3$Al.

The average L and λ_p values were obtained by making calculations using the sets of data for both stoichiometric specimens from Group 3. The results differed by about 2%.

Experience has shown [13,14] that estimates of L and λ_p can be improved somewhat by modifying the calculations to include the effects of alloying on λ_p and the intrinsic electronic thermal resistance, W_{ei}. The physical basis for this is that point defect scattering would tend to reduce λ_p in the 76 at. % Ni sample and Fig. 2 shows that the ideal resistivity, and presumably W_{ei}, of this composition is considerably higher than the value for Ni_3Al. The results obtained from this calculation are:

$$\overline{L} = 0.99\ L_o$$

$$\overline{\lambda_p} = 7.50\ W/m \cdot K$$

These values represent our best experimental estimates of the values for well annealed, stoichiometric Ni_3Al at room temperature. Figure 4 shows the composition dependence of λ_e and λ_p which is obtained by using $\overline{L}$ with the λ and ρ measurements (Table 2) for individual samples.

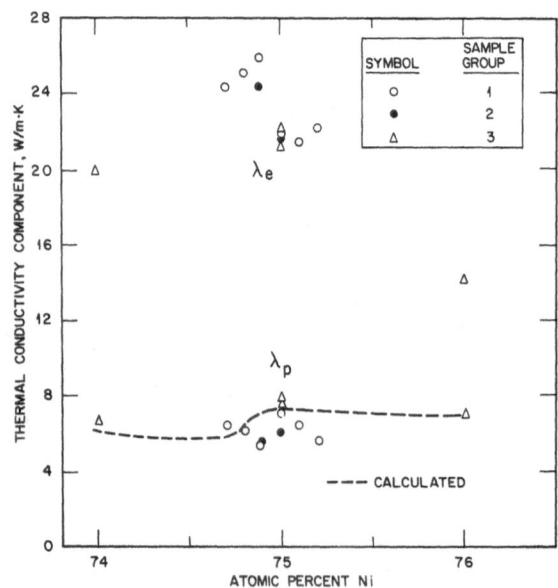

Fig. 4. Derived electronic (λ_e) and phonon (λ_p) values for compositions close to Ni_3Al. The calculated phonon conductivity curve is described in the text.

The plausibility of these results can be tested by making a theoretical estimate of λ_p. The result is also shown in Fig. 4. This calculation employs Callaway's formula [15] with estimates of the phonon-phonon, phonon-electron and phonon-point defect relaxation times. The Leibfried-Schlömann formula [16] was used to obtain the phonon-phonon relaxation time and N-processes were assumed to be three times [17] as frequent as the resistive U-processes. The specific heat data [11] yield a Debye temperature of 391 K, which is in good agreement with values calculated from Lindemann's formula [18]. The Grüneisen constant, 2.05, was estimated from values for Ni and Al [18]. This calculation presumes that the phonon-phonon relaxation time does not depend on composition. The phonon-electron relaxation time was calculated from electronic specific heat, γ_e, data [11] by using a correlation [17] which describes the interelement variation of the high temperature limit of this resistance, W_{ep}:

$$W_{ep} = 3.7 \times 10^{-3} \ \gamma_e^{1.5} \ (m \cdot K/W) \qquad (T \gg \theta_D) \ .$$

The point defect contribution was obtained from a formula due to Ho et al. [19], and an anharmonicity factor of 0.6 was used [20] to describe the contribution from local strain effects. The prediction is that phonon-point defect scattering should be relatively weak because the mass difference and lattice strain terms tend to cancel.

The calculated and derived phonon conductivities are in remarkably good agreement. For the third (best) group of specimens the absolute difference averages only 5%, and part of the scatter in the other values is probably due to the larger uncertainties associated with the ρ values for these specimens. These results indicate that, at room temperature, phonons contribute about 25% of the thermal conductivity of stoichiometric Ni_3Al. Figure 4 shows that the importance of phonon conduction increases rapidly for Ni-rich materials, and this is significant because the ductile intermetallic alloys are based on a 76 at. % Ni composition.

ACKNOWLEDGEMENT

L. Queener melted the specimens for this study, and L. D. Chitwood made eddy current resistivity measurements. The manuscript was prepared by B. B. Hickey, C. P. Whitus, and Judy Young and reviewed by W. H. Butler and D. W. Yarbrough.

REFERENCES

[1] Bieber, G. C. and Randebaugh, R. J., "Some Age Hardening Characteristics of Nickel-Chromium Alloys (Nickel-Rich) Containing Aluminum and Titanium," p. 417 in Precipitation from Solid Solution, American Society for Metals, 1959.

[2] Copley, S. M., and Kerr, B. H., Trans. AIME 239, 977 (1967).

[3] Aoki, K., and Izumi, O., Nippon Kinzoku Gakkaishi 43, 1190 (1979).

[4] Liu, C. T., and Koch, C. C., "Development of Ductile Polycrystalline Ni$_3$Al for High Temperature Applications," p. 42 in Technical Aspects of Critical Materials Use by the Steel Industry, Vol. IIB, NBSIR 83-2679-2, Center for Materials Science, National Bureau of Standards, 1983.

[5] Fluitman, J.H.J., Boom, R., de Chatel, P. F., Schinkel, C. J., Tilanns, J.L.L., and DeVries, B. R., J. Phys F: Metal. Phys. 3, 109 (1973).

[6] Williams, R. K., Graves, R. S., and Moore, J. P., ORNL-5313 (1978).

[7] Dodd, C. V., Materials Evaluation XXVI, 33 (1968).

[8] Jury, S. H., Arnurius, D., Godfrey, T. G., McElroy, D. L., and Moore, J. P., J. Franklin Inst. 298, 151 (1974).

[9] Williams, R. K., Yarbrough, D. W., Masey, J. W., Holder, T. K., and Graves, R. S., J. Appl. Phys. 52, 5167 (1981).

[10] Blatt, F. J., Physics of Electronic Conduction in Solids, McGraw-Hill, New York, NY (1968).

[11] de Dood, W., and de Chatel, P. F., J. Phys. F: Metal. Phys. 3, 1039 (1973).

[12] Kortekaas, T.F.M., and Franse, J., Phys. State. Sal. 40, 479 (1977).

[13] Williams, R. K., Graves, R. S., Hebble, T. L., McElroy, D. L., and Moore, J. P., Phys. Rev. B 26, 2932 (1982).

[14] Williams, R. K., Butler, W. H., Graves, R. S., and Moore, J. P., submitted to Physical Review B.

[15] Callaway, J., Phys. Rev. 113, 1046 (1959).

[16] Roufosse, M., and Klemens, P. G., Phys. Rev. B 7, 5379 (1973).

[17] Yarbrough, D. W., and Williams, R. K., ORNL-5434 (1978).

[18] Gschneidner, K. A., "Solid State Physics 16," p. 275, F. Seitz and D. Turnbull, eds., Academic Press, Inc., New York, NY (1969).

[19] Ho, C. Y., Ackerman, M. W., Wu, K. Y., Oh, S. G., and Havill, N. T., J. Phys. Chem. Ref. Data 7, 959 (1978).

[20] Williams, R. K., and Yarbrough, D. W., unpublished research.

By acceptance of this article, the publisher or recipient acknowledges the U.S. Government's right to retain a nonexclusive, royalty-free license in and to any copyright covering the article.

THERMAL CONDUCTIVITY OF SELECTED STAINLESS STEELS

R. H. Bogaard

Center for Information and
Numerical Data Analysis and Synthesis
Purdue University
2595 Yeager Road
West Lafayette, Indiana   47906

ABSTRACT

The available experimental data and information for the thermal conductivity of stainless steels have been exhaustively searched, comprehensively compiled, and critically evaluated. The analysis took into account factors such as amount and nature of alloying element, presence of cold-work, and thermal history. Also, a significant number of stainless steels produced in other nations were included as equivalent stainless steels in the thermal conductivity data base. Subsequently, recommended values were generated for selected stainless steels with austenitic, ferritic, martensitic, and precipitation-hardening types being included. Results for AISI 302, 304, 310, 316, 321, 410, 430, 446, and 631 stainless steels are presented in survey fashion with AISI 304 steel being discussed in greater detail.

INTRODUCTION

This work on thermal conductivity of stainless steels is part of a larger, ongoing effort at CINDAS aimed at understanding the thermophysical and electrical properties of metals and alloys.

The objectives of the thermal conductivity effort are as follows: review literature data for the thermal conductivity of stainless steels, identify effects due to composition, metallurgical structure, and thermal conditioning treatment, and generate recommended thermal conductivity values consistent with limitations of method and data. A comprehensive and critical literature

175

review has been underway for a number of years at CINDAS. This has resulted in a great deal of information on the thermal conductivity of stainless steels being compiled. Specifically, some 70 odd stainless steels have been identified, 36 of which carry an AISI designation. In addition are a number of stainless steels produced in Czechoslovakia, France, W. Germany, Sweden, U.S.S.R., and the United Kingdom. Roughly one-third of these steels were included as equivalent stainless steels in assembling the thermal conductivity data base. A stainless steel is, of course, identified in terms of its chromium composition (C $\geq$ 12%). In addition, however, a number of other factors have been identified by the literature which affect properties of stainless steels in general, and thermal conductivity in particular. They are metallurgical structure, heat-treatment history, and of course, composition. Ultimately, if we wish to say something about the thermal conductivity for a particular stainless steel, for instance to generate a set of thermal conductivity values carrying an uncertainty at the $\pm$3% to 5% level, then we must be careful about stating the condition of the steel and take seriously an author's comments about his own measurements.

In the following, the data base for one particular stainless steel, AISI 304, will be discussed in greater detail. Then three classes of stainless steels will be surveyed (austenitic, ferritic and martensitic, and precipitation-hardenable) with results shown for selected members from each class.

AISI 304 STAINLESS STEEL

The starting point is the available literature data. AISI 304 has been a popular stainless steel, with experimental information spanning some 30 years and including the fifteen references [1-15] with 20 sets of data shown in Fig. 1. The materials represented are all broadly of the AISI 304 type. This includes AISI 304 [1-5,9-13], the equivalent X 5 CrNi 18 9 [15], a generic 18Cr-11Ni-0.08C steel [14], the low-carbon modification AISI 304L [7], and NS 22S [8] and 3R 12 HV [6] which are low-carbon equivalents. These steels are solution-annealed. Even though material variability is not an issue here it is worthwhile mentioning that reduced carbon reportedly does enhance the electrical resistivity by perhaps 2% or more below room temperature [16]. An effect of this magnitude is likely too small to be observed in the thermal conductivity.

Recommended values for the thermal conductivity were based on an average over all the experimental data. At temperatures above 1000 K care was taken to assure a slope consistent with the comparatively fewer data available for those temperatures. The steepening slope at the lowest temperatures shown in Fig. 1 is due

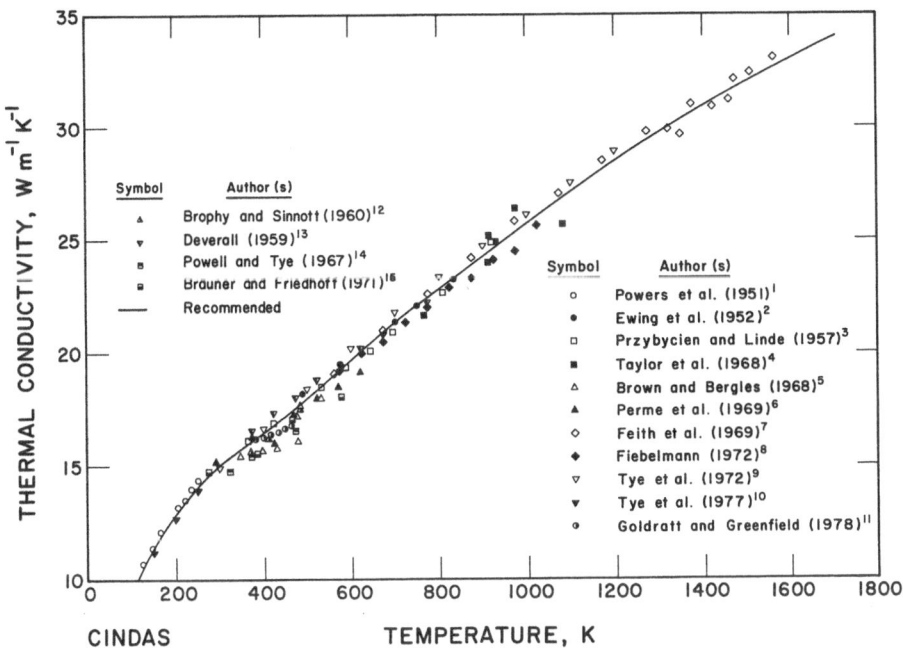

Fig. 1. Thermal Conductivity of AISI 304 Stainless Steel.

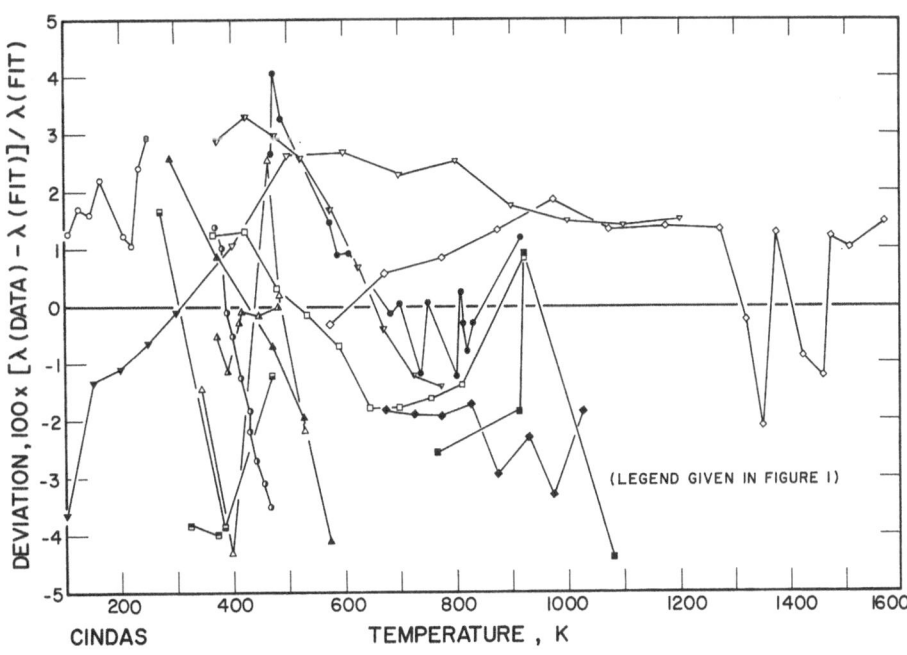

Fig. 2. Deviation Plot for AISI 304 Stainless Steel.

Table 1. Thermal Conductivity of AISI 304 Stainless Steel[a]

[Temperature, T, K; Thermal Conductivity, λ, W $m^{-1}K^{-1}$]

T	λ	T	λ	T	λ
100	9.34	400	16.43	1000	25.71
150	11.35	450	17.13	1100	27.12
200	12.84	500	17.93	1200	28.47
250	13.99	550	18.81	1300	29.74
273	14.44	600	19.67	1400	30.92
293	14.79	700	21.31	1500	32.00
300	14.92	800	22.81	1600	33.00
350	15.71	900	24.28	1707[b]	34.01

[a] Values are for the solution-annealed condition. The uncertainty is estimated to be within ±4%.
[b] Solidus temperature.

in part to the lattice thermal conductivity which, as will be discussed below, has a maximum at roughly 200 K and amounts to not quite half the measured value at that temperature. Above 300 K the existing data [3,5,6,9-15] are divided according to slope with two data sets [5,11] indicating a quite flat slope from 300-500 K, and also according to magnitude with several data sets [5,11,14, 15] indicating rather low values between 300 and 500 K. Since at present no physical mechanism has been identified that would reduce the thermal conductivity over this span of temperatures, the recommended curve essentially bypassed the lowest lying data. Generally, the recommended curve was drawn such that the deviations of the literature data scattered symmetrically as shown in Fig. 2. The recommended values are given in Table 1. They carry an uncertainty estimated to be within ±4% based upon the deviations shown in Fig. 2.

AISI 302, 304, 310, 316 AND 321 STAINLESS STEELS

The thermal conductivity for a selected group of austenitic stainless steels is shown in Fig. 3. The overall behavior of these steels is consistent with the thermal conductivity of fcc γ-Fe which is, above 1200 K, only about 3 to 5% higher than the curve for AISI 304 steel. In terms of composition these steels are variations on the basic '18-8' (18%Cr-8%Ni) theme as indicated in Table 2. AISI 302 contains a few percent less Cr and Ni which probably accounts for its higher level of thermal conductivity. The AISI 316 and 310 stainless steels are more heavily alloyed;

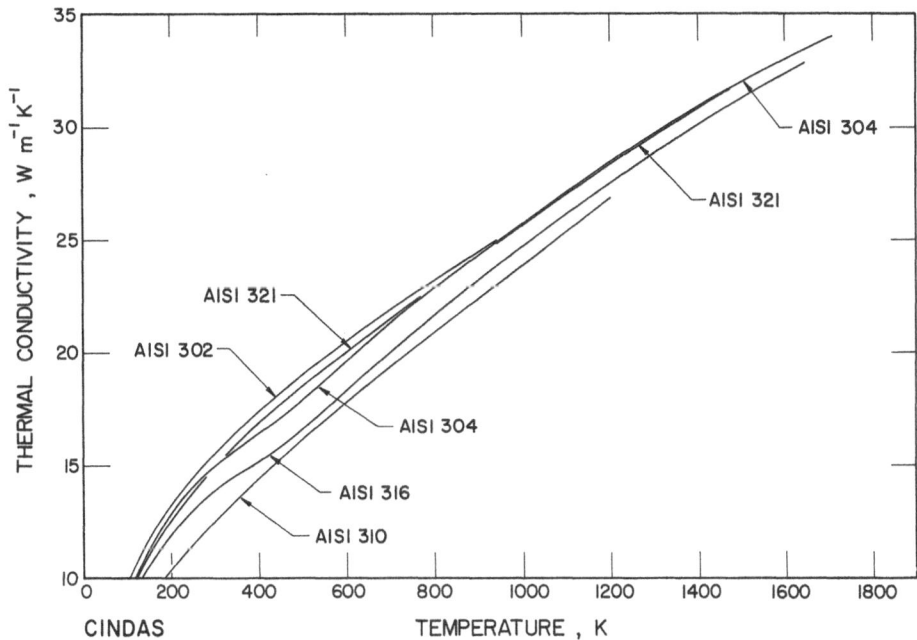

Fig. 3. Thermal Conductivity of AISI 300-Series Stainless Steels.

Table 2. Chemical Compositions of AISI Stainless Steels

AISI Type	C max.	Mn max.	Si max.	Cr Range	Ni Range	Other Elements
302	0.15	2.00	1.00	17.0–19.0	8.0–10.0	
304	0.08	2.00	1.00	18.0–20.0	8.0–12.0	
310	0.25	2.00	1.50	24.0–26.0	19.0–22.0	
316	0.08	2.00	1.00	16.0–18.0	10.0–14.0	Mo,2.0–3.0
321	0.08	2.00	1.00	17.0–19.0	9.0–12.0	Ti,5xC min.
410	0.15	1.00	1.00	11.5–13.5	–	
430	0.12	1.00	1.00	14.0–18.0	–	
446	0.20	1.50	1.00	23.0–27.0	–	Ni,0.25 max.
631	0.09	1.00	1.00	16.0–18.0	6.50–7.75	Al,0.75–1.50

AISI 316 with 2-3% Mo and AISI 310 with an additional 6% Cr and 10% Ni. The effect of composition is observed to be considerably reduced at high temperatures. It is noted that the procedure for each stainless steel was to generate recommended values of thermal conductivity by averaging together the existing data. Exception was made only in regions having too few data such as at the highest temperatures where an attempt was made to maintain consistent behavior from steel-to-steel. Generally, however, the resulting curve for each stainless steel is just that indicated by the data.

The lattice thermal conductivity can be obtained from the total thermal conductivity by estimating the electronic component from the electrical resistivity. One commonly employs the Wiedemann–Franz–Lorenz Rule, $\lambda_e/\rho = LT$, along with the Sommerfeld value of the Lorenz constant, L_0, which is valid for high residual-resistivity alloys, and the absolute temperature, T, to give the electronic thermal conductivity, $\lambda_e = L_0 T/\rho$. An estimate of the lattice thermal conductivity, λ_g, may then be obtained from

$$\lambda_g = \lambda - L_0 T/\rho \tag{1}$$

Values of the lattice thermal conductivity estimated from Eq. (1) for the AISI 300-Series austenitic stainless steels are shown in Fig. 4. The maximum in the conductivity curves occurs roughly in the 150 to 250 K region. It is noted that the maximum lattice-component values are about half of the total thermal conductivity. Below the maximum the lattice thermal conductivity varies monotonically with temperature. Above the maximum the estimated values are somewhat scattered and, consequently, the curves shown represent only smoothed values at higher temperatures. The

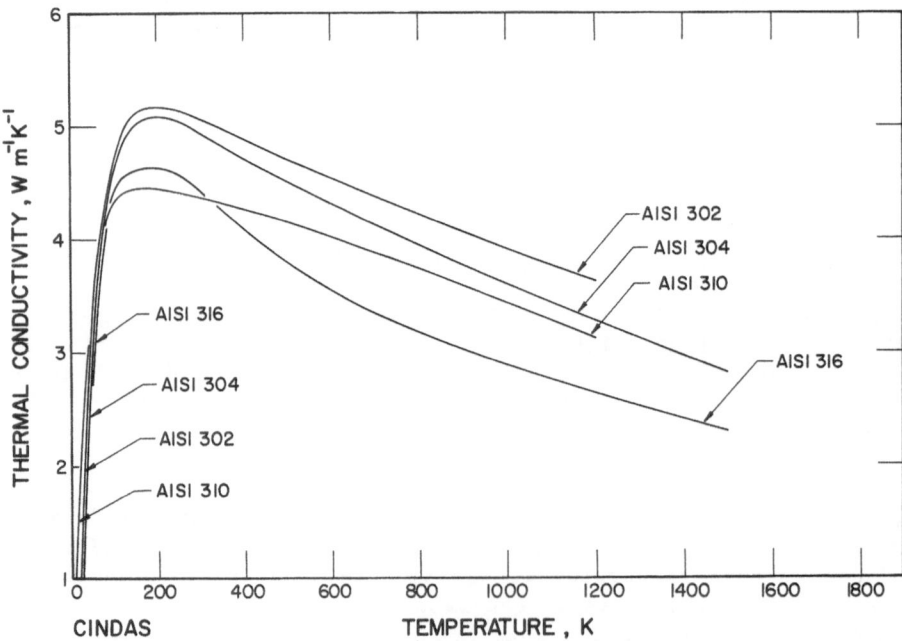

Fig. 4. Lattice Thermal Conductivity of AISI 300-Series Steels.

180

temperature dependence, except for the AISI 310 stainless steel curve, is roughly consistent with $\lambda_g \propto T^{-1}$.

AISI 410, 430, AND 446 STAINLESS STEELS

For martensitic and ferritic stainless steels the information base is smaller and the selection of steels is more limited. AISI 410 stainless steel when solution-treated has an austenitic structure which upon quenching is mostly transformed to martensite and becomes quite hard. As shown in Fig. 5, the data indicate that the thermal conductivity is enhanced by annealing and slow-cooling, perhaps by ~20% at room temperature. AISI 430 and 446 stainless steels have sufficiently high chromium content, 16% and 25%, respectively, to remain ferritic. The enhancement of the thermal conductivity due to the essentially bcc structure is evident for temperatures over much of the region shown (see also Fig. 3). Also evident is the reduction in the thermal conductivity due to increased chromium content.

In Fig. 6 is shown the lattice thermal conductivity for AISI 410, 430, and 446 stainless steels. It is noted that the thermal conductivity maxima occur at roughly 100-200 K. Even though the

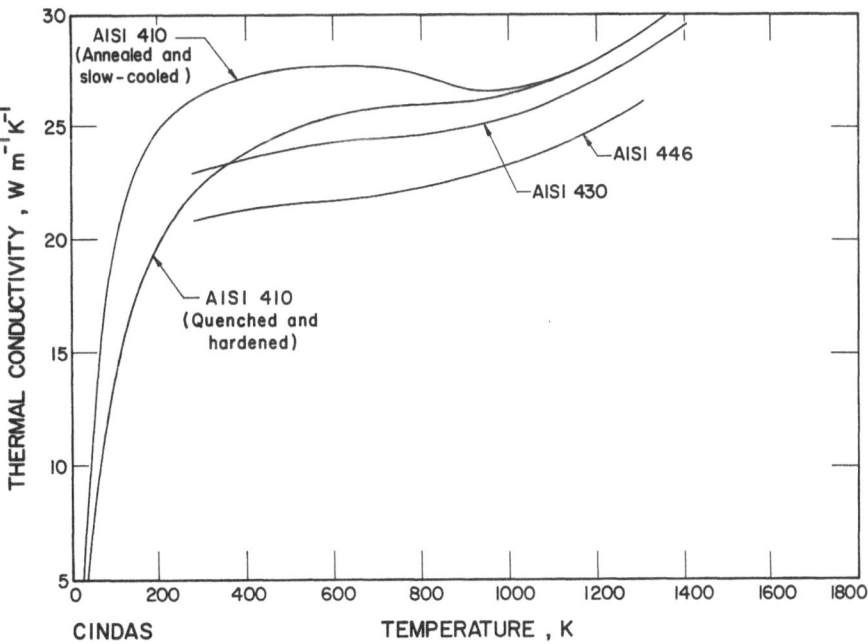

Fig. 5. Thermal Conductivity of AISI 400-Series Stainless Steels.

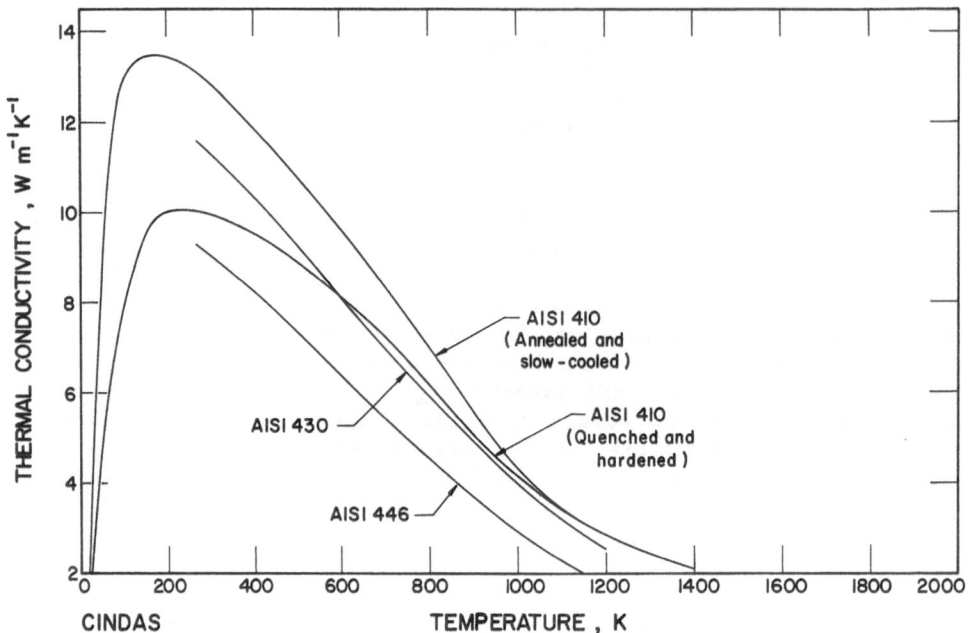

Fig. 6. Lattice Thermal Conductivity of AISI 400-Series Steels.

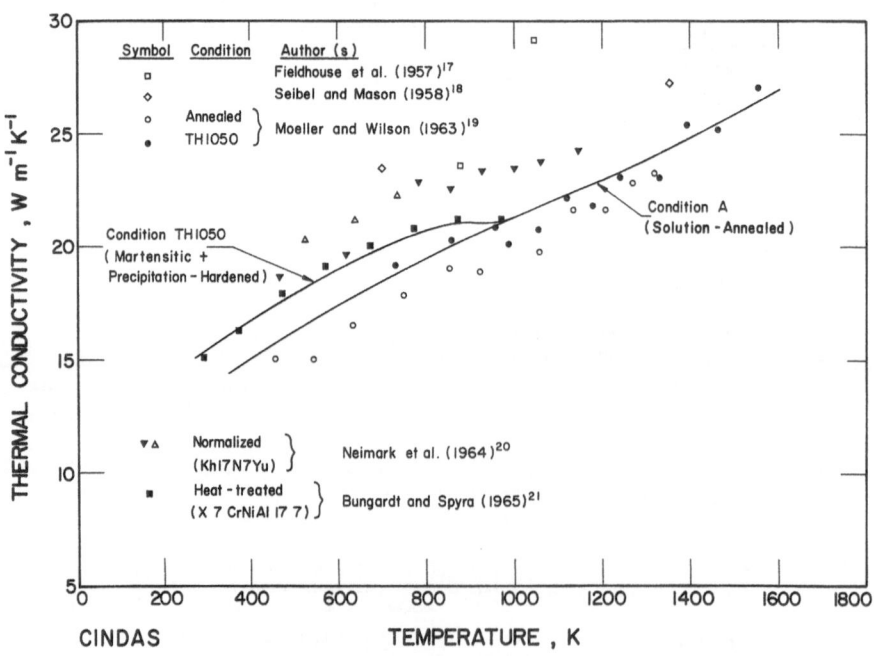

Fig. 7. Thermal Conductivity of AISI 631 Stainless Steel.

maximum value in the lattice component is a factor of roughly x2-3 larger than for the austenitic stainless steels, it remains roughly half of the total thermal conductivity.

AISI 631 STAINLESS STEEL [17-7PH]

The available literature data for the thermal conductivity of AISI 631 stainless steel are shown in Fig. 7. Data are reported for both annealed and precipitation-hardened conditions [19], while the early data sets [17,18] are uncharacterized. Also shown are data reported for equivalent stainless steels, X 7 CrNiAl 17 7 [21] and Kh17N7Yu [20]. In both instances the condition of the steels was reported as indicated in Fig. 7. It is observed that the precipitation-hardening treatment enhances the thermal conductivity above that for the solution-annealed condition [19,21]. It turns out that this behavior is paralleled in the electrical resistivity of these materials [16]. The recommended curve is drawn so as to take into account these very complex phase changes.

ACKNOWLEDGMENTS

The author would like to express his thanks to all those CINDAS staff members, past and present, who have participated in the critical evaluation and generation of recommended values for thermal conductivity. Notable among these are T.K. Chu and M.W. Ackerman.

The extensive literature search, acquisition, and documentary activities essential to this work were supported by the Defense Logistics Agency (DLA), U.S. Department of Defense, Arlington, VA.

REFERENCES

[1]   Powers, R.W., Ziegler, J.B., and Johnston, H.L., 'The Thermal Conductivity of Metals and Alloys at Low Temperatures. I. Apparatus for Measurements Between 25 and 300 K. Data on Pure Aluminum, OFHC Copper, and 'L' Nickel', U.S. Air Force Rept. TR-264-6, 14 pp. (1951). [ATI-105 923]

[2]   Ewing, C.T., Grand, J.A., and Miller, R.R., 'Thermal Conductivity of Liquid Sodium and Potassium', J. Am. Chem. Soc., 74, 11-14 (1952).

[3]   Przybycien, W.M. and Linde, D.W., 'Thermal Conductivities of Gases, Metals, and Liquid Metals', USAEC Rept. KAPL-MWMP 1, 24 pp. (1957).

[4]   Taylor, R.E., Powell, R.W., Nalbantyan, M., and Davis, F.E., 'Evaluation of Direct Electrical Heating Methods for the Determination of Thermal Conductivity at Elevated

Temperatures', Thermophysical Properties Research Center, Purdue Univ., U.S. Air Force Rept. AFML-TR-68-227, 74 pp. (1968). [AD 679 639]

[5] Brown, W.T., Jr. and Bergles, A.E., 'Measurement of Thermal Conductivity for Electrically Heated Heat-Transfer Test Sections', in Proceedings of the Symposium on Thermophysical Properties, 4th Ed., 184-188 (1968).

[6] Perme, T., Krstic, D.J., and Vodopivec, F., 'Measurements of the Thermal Conductivity and Electrical Resistivity of High-Alloy Ferritic Steels', Nukl. Inst. Jozef Stefan, Nijs Porocilo, p-245, 1-31 (1969).

[7] Feith, A.D., Hein, R.A., Johnstone, C.P., and Flagella, P.N., 'Thermophysical Properties of Low Carbon 304 Stainless Steel to 1350°C', in Thermal Conductivity, Proceedings of the Eighth Conference [Ho, C.Y. and Taylor, R.E., Editors], Plenum Press, New York, NY, 1051-1065 (1969).

[8] Fiebelmann, P., 'Determination of the Radial Thermal Conductivity of Multilayer Tubes for Thermionic Converters', Forsch. Ingenieurw., 38(5), 133-138 (1972).

[9] Tye, R.P., Hayden, R.W., and Spinney, S.C., 'The Thermal Conductivity of a Number of Alloys at Elevated Temperatures', High Temp.-High Pressures, 4(5), 503-511 (1972).

[10] Tye, R.P., Hayden, R.W., and Spinney, S.C., 'Thermal Conductivity of Selected Alloys at Low Temperatures', Adv. Cryog. Eng., 22, 136-144 (1977).

[11] Goldratt, E. and Greenfield, A.J., 'New Method for Measuring the Thermal Conductivity', Rev. Sci. Instrum., 49(11), 1531-1536 (1978).

[12] Brophy, J.H. and Sinnott, M.J., 'The Thermal and Electrical Conductivities of Ductile Cast Iron and Several Gray Cast Irons', Trans. Am. Soc. Met., 52, 567-581 (1960).

[13] Deverall, J.E., 'The Thermal Conductivity of a Molten Pu-Fe Eutectic (9.5 Atomic % Fe)', USAEC Rept. LA-2269, 62 pp. (1959).

[14] Powell, R.W. and Tye, R.P., 'New Measurements on Thermal Conductivity Reference Materials', Int. J. Heat Mass Transfer, 10(5), 581-596, 1967.

[15] Brauner, J. and Friedhoff, P., 'Simple Procedure for the Determination of the Thermal Conductivity of Steels', Materialpruefung, 13(7), 230-233 (1971).

[16] Bogaard, R.H., Desai, P.D., Li, H.H., Chu, T.K., Chi, T.C., Matula, R.A., and Ho, C.Y., Properties of Stainless Steels, Vol. IV-1 of McGraw-Hill/CINDAS Data Series on Material Properties (Touloukian, Y.S. and Ho, C.Y., Editors), McGraw-Hill Book Co., New York, NY, 450 pp., in course of publication.

[17] Fieldhouse, I.B., Hedge, J.C., Lang, J.I., and Waterman, T.E., 'Thermal Properties of High Temperature Materials', U.S. Air Force Rept. WADC-TR-57-487, 78 pp. (1957). [PB-131 718]

184

[18] Seibel, R.D. and Mason, G.L., 'Thermal Properties of High
      Temperature Materials', U.S. Air Force Rept. WADC-TR-57-468,
      58 pp. (1958). [AD 155 605]
[19] Moeller, C.E. and Wilson, D.R., 'Thermal Conductivities of
      Several Metals and Non-Metals from 200 to 1300°C by the
      Radial Heat-Flow Technique', in Proceedings of the 3rd
      Conference on Thermal Conductivity (McElroy, D.R., Editor)
      [Proceedings of Conference at Gatlinburg, TN on October
      16-18, 1963], 224-251 (1964).
[20] Neimark, B.E., Lyusternik, V.E., and Korytina, S.F.,
      'Physical Properties of Kh17N7Yu Steel', Teplofiz. Vys.
      Temp., 2(5), 725-729 (1964); Engl. Transl.: High Temp.,
      2(5), 652-655 (1964).
[21] Bungardt, K. and Spyra, W., 'Thermal Conductivity of Alloyed
      and Plain Steels and Alloys at Temperatures Between 20 and
      700°', Arch. Eisenhuettenwes., 36(4), 257-267 (1965).

THERMAL CONDUCTANCE MEASUREMENTS OF PRESSED OFHC COPPER

CONTACTS AT LIQUID HELIUM TEMPERATURES

[1]Louis J. Salerno, [1]P. Kittel and [2]A.L. Spivak

[1]NASA Ames Research Center
[2]Moffett Field, Calif. 94035
Trans-Bay Electronics
Richmond, Calif.

ABSTRACT

The thermal conductance of pairs of oxygen-free high conductivity (OFHC) copper specimens with surface finishes ranging from 0.1 to 1.6-µm rms roughness has been investigated over the range of 1.6 to 6.0-K under applied contact forces up to 670 N. The thermal conductance increases with increasing contact force; however, no correlation can be drawn with respect to surface finish.

INTRODUCTION

To optimize performance of cryogenic instruments, a knowledge of the thermal conductance of pressed contacts is necessary. This is especially true for instruments whose performance is temperature-dependent, as is the case with many infrared astronomical instruments. Facilities such as the Infrared Astronomical Satellite (IRAS), Shuttle Infrared Telescope Facility (SIRTF), and the Large Deployable Reflector (LDR) depend on accurate knowledge of the behavior of pressed contacts at liquid helium temperatures.

Whereas estimates of the thermal conductance can be derived from measurements of the electrical conductance from the Wiedemann-Franz law, it has been shown [1] that such estimates may deviate from the actual values by a factor of 10^5.

Several theoretical models have been developed to account for the thermal resistance of pressed contact pairs [2-8]; however, most usable data in the field are empirical. Previous work has shown that the thermal conductance is independent of contact area, and is dependent on applied contact force [1,9]. At liquid helium temperatures, conductance follows a T^2-temperature dependence. Whereas surface finish effects have been studied, most data available deal with specific contact geometries, such as cup and cone, copper rods, etc., and often correspond to particular applications. A need exists for more general data covering a variety of specimen pairs over a wide range of temperatures and applied contact forces.

Method

The present work examines the thermal conductance of pressed contact OFHC copper specimen pairs as a function of temperature from 1.6 to 6.0 K with applied force up to 670 N as a parameter. An apparatus has been fabricated and tested and is pictured in Figs. 1 and 2. Several specimen pairs have been prepared, with surface finishes ranging from 0.1-µm surface roughness to 1.6-µm surface roughness.

The general form of the relation involving thermal conductance is given as

$$\dot{Q} = \int k \ dT.$$

Although the method and theory have been covered in a previous paper [10], the equation of condition which is applicable in the present case

$$\dot{Q} = \int_{T_L}^{T_u} \alpha T^n \ dT$$

is employed where:

$\dot{Q}$ = the applied heater power

T_L = the lower sample temperature

T_u = the upper sample temperature

α = a constant of proportionality

n = an exponent

In this case, the thermal conductance is assumed to follow a power-law function of temperature, where $k = \alpha T^n$. The values of α and n are obtained by using a computer program and by linearizing the equation of condition. A Gauss-Jordan elimination is

188

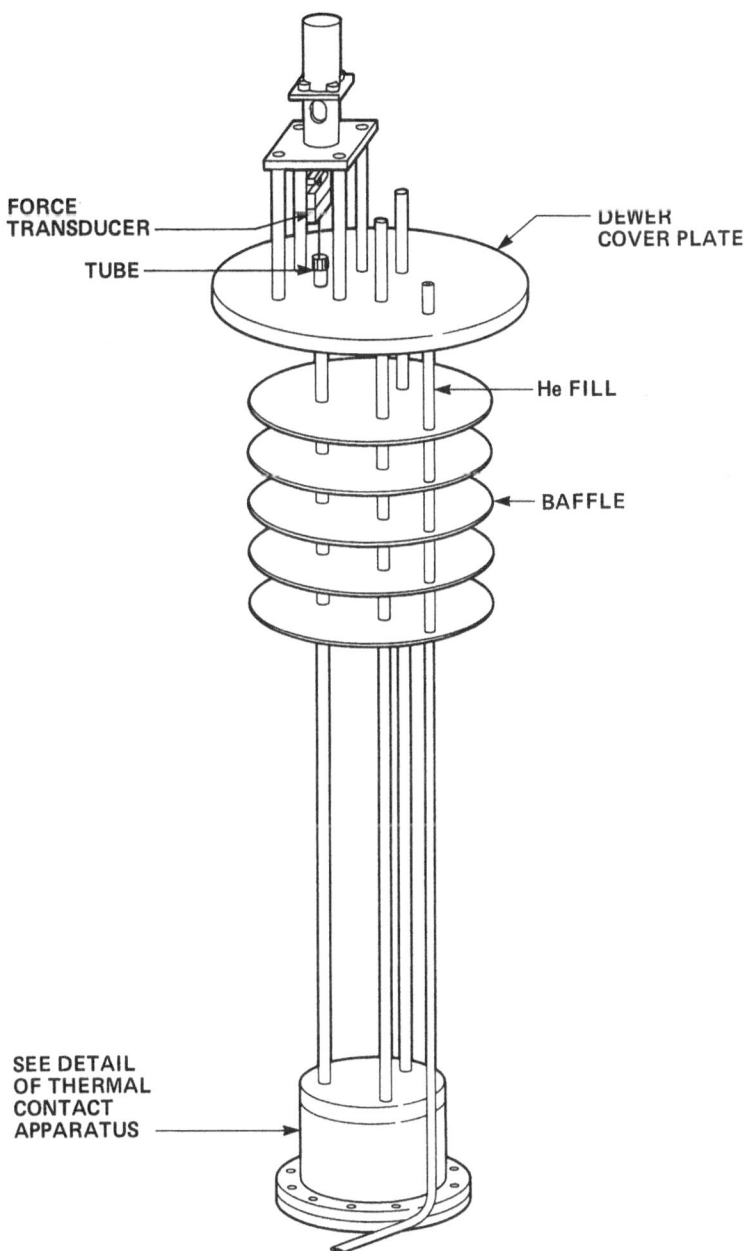

FORCE TRANSDUCER

TUBE

DEWER COVER PLATE

He FILL

BAFFLE

SEE DETAIL OF THERMAL CONTACT APPARATUS

Fig. 1. Overall view of thermal contact apparatus.

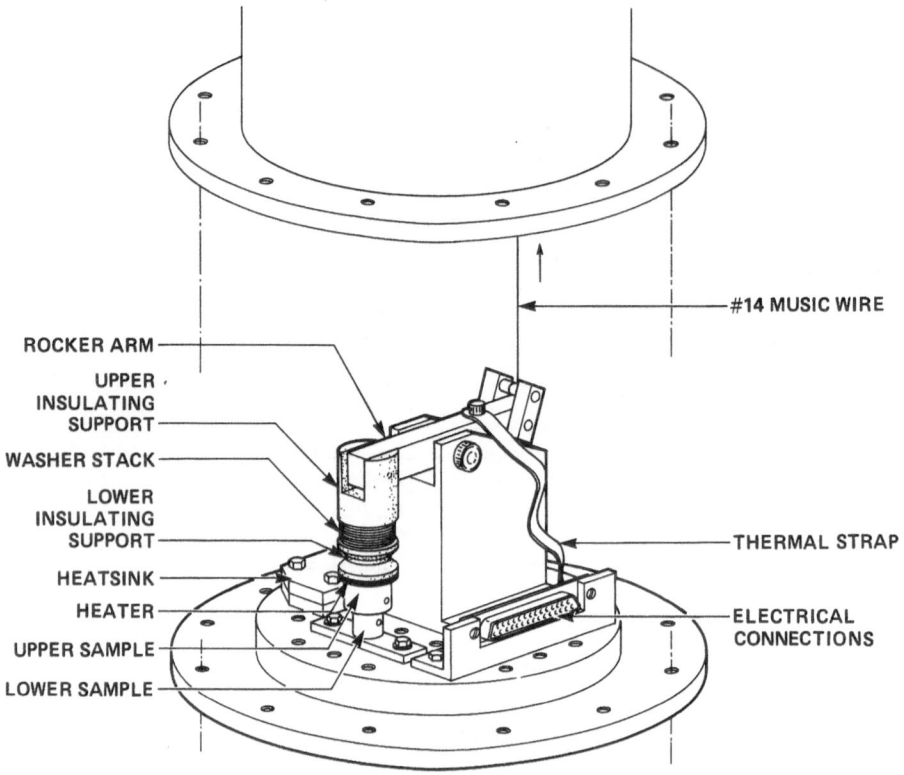

Fig. 2.   Cold plate detail.

performed to solve a set of linear equations representing the sum
of the squares of the deviations between the measured and the computed values and n normal equations, so that the deviation can be
minimized. The computer program also performs a perturbation of
the input parameters according to the standard deviation of the
known measurement tolerances including instrument accuracies,
roundoff, and truncation errors. By means of a random number generator, 99 replications are performed with the result that the
output values of α and n are averages of the replications representing the error as a result of the input uncertainties.

Results

Figures 3-7 plot thermal contact conductance vs temperature
with applied force as a parameter for each of the surface finishes. Curves were obtained by calculating αT^n from the program
ouput parameters for a given temperature over the measured range
from the lowest specimen temperature to the highest. The errors

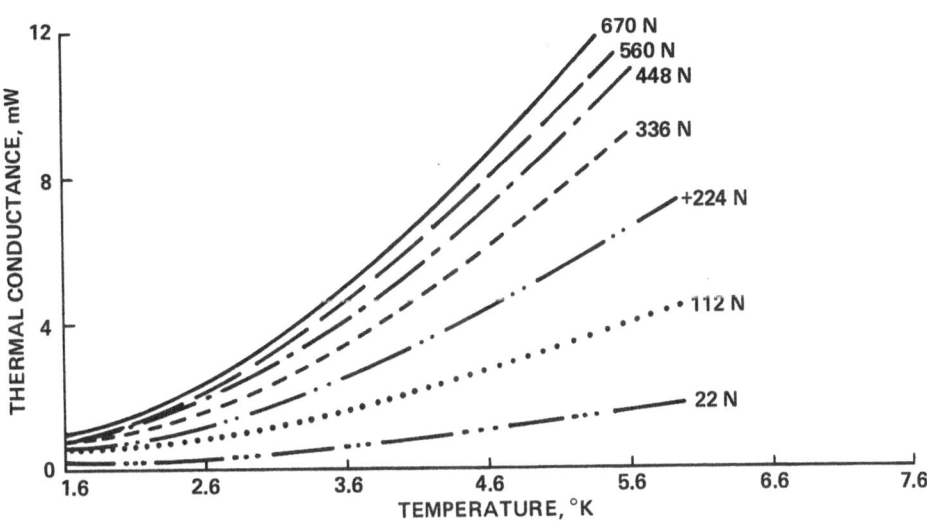

Fig. 3. Results for a pair of specimens having 0.1 μm surface
roughness.

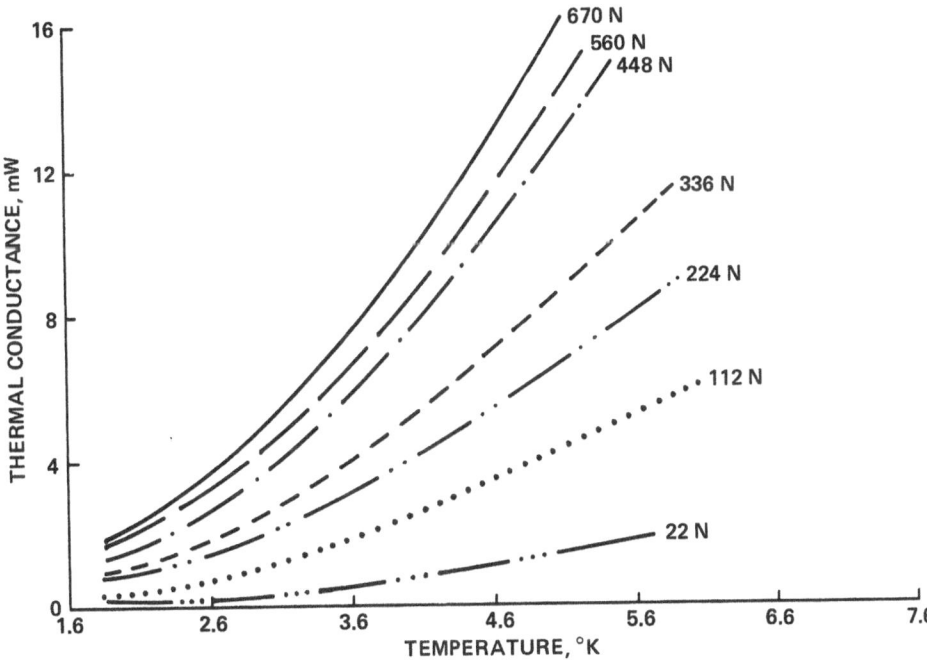

Fig. 4. Results for a pair of specimens having 0.2 μm surface
roughness.

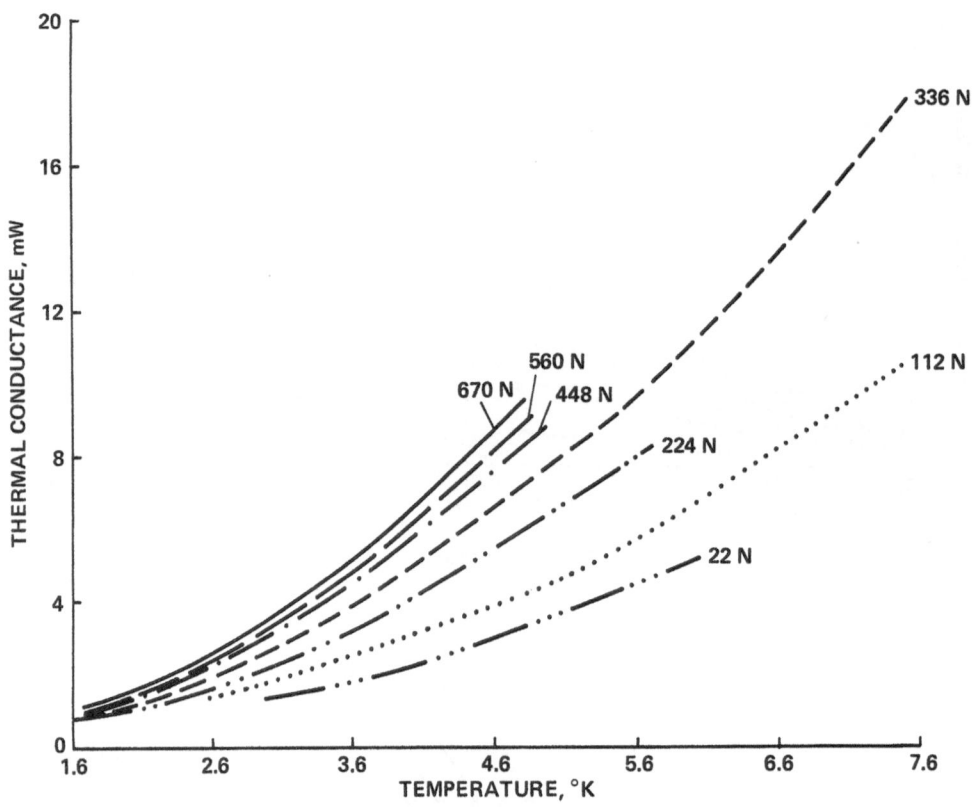

Fig. 5.   Results for a pair of specimens having 0.4 μm surface
roughness.

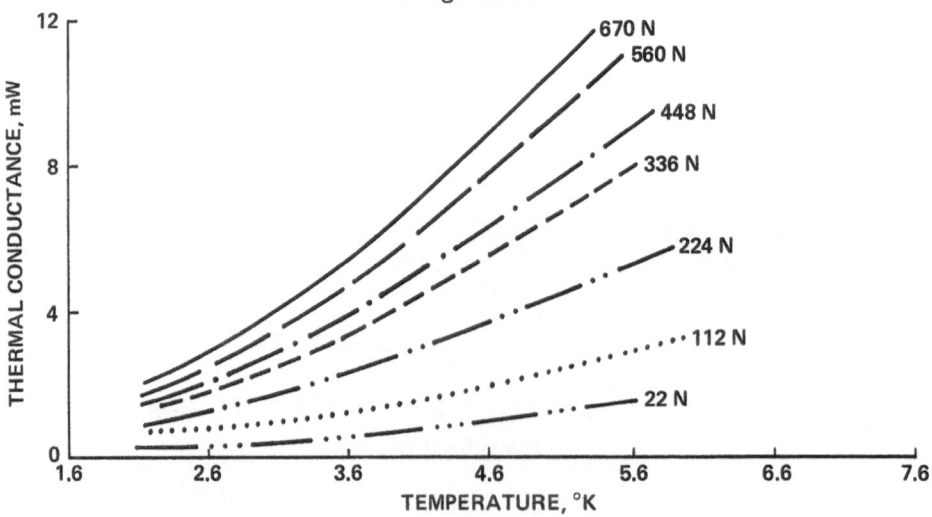

Fig. 6.   Results for a pair of specimens having 0.8 μm surface
roughness.

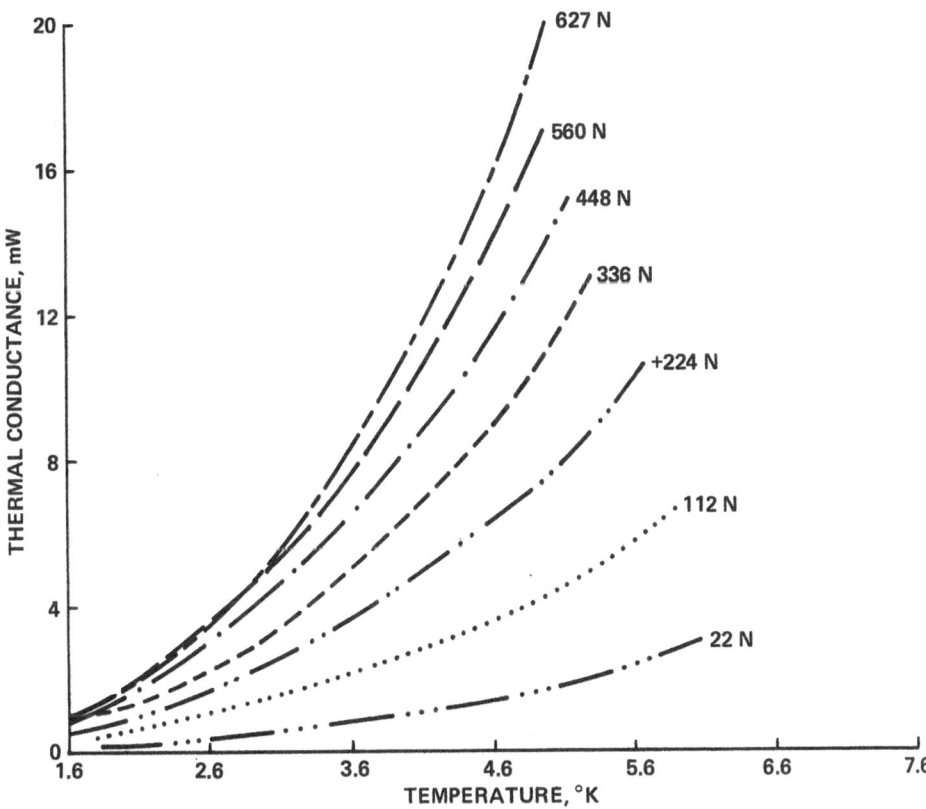

Fig. 7. Results for a pair of specimens having 1.6 μm surface
roughness.

obtained in α and in n from the program are on the order of
10^{-3}, which indicates little effect on the curves, since the values of α and n are on the order of 0.1 and 2.0, respectively.
Figure 8 compares the overall performance of the surface finishes
with respect to thermal contact conductance.

DISCUSSION

    In examining Figs. 3-7, it can readily be seen that thermal
conductance very definitely increases with the increasing contact
force, thus confirming earlier work. To ensure repeatability, the
0.2-μm sample pair was tested twice over a 90-day period. The
two sets of results agreed within experimental error. The thermal
conductance obtained appears to be a function of temperature to

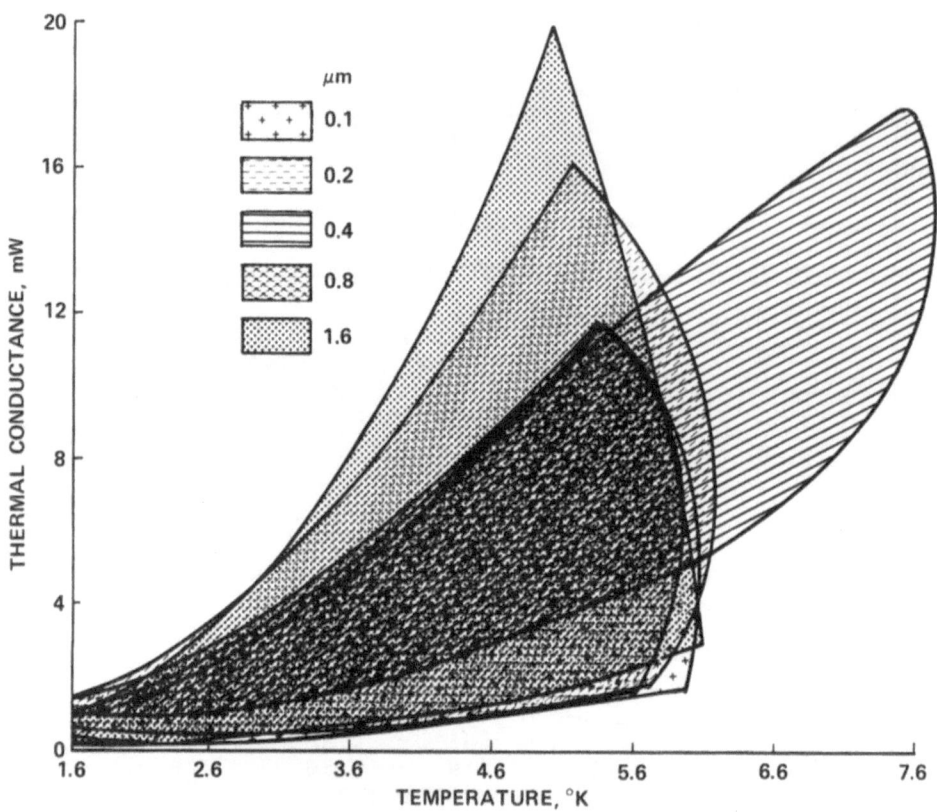

Fig. 8. Surface finish comparison.

the second power, again as found in previous work. Although no precise quantitative correlations can be drawn from Fig. 8, it appears that the surface finishes tested are essentially equivalent in terms of thermal conductance, with the exception of the 1.6-μm and 0.2-μm surfaces. If indeed the thermal conductance is dependent only upon the applied contact force and independent of area as earlier work would suggest, the observed results are not surprising. The reason is that the energy transfer is thought to occur only at a few discrete points which represent the asperities of the surfaces. In this case, the higher roughness of the 1.6-μm surface would explain the increased conductance shown in Fig. 8, since the points would provide elastically deformable contact areas. It would be expected that conductance would increase because of the effect of cold welding of the surfaces as the surfaces became very fine (in particular, highly polished surfaces).

CONCLUSION

At the present time it appears that the thermal contact conductance of OFHC copper sample pairs increases with increasing applied force, which is supported by earlier work; however, no correlation can be drawn with respect to surface finishes. Future work examining the contact conductance of brass, stainless steel, and aluminum may provide further insight into the effect of surface finish.

REFERENCES

[1] Berman, R., "Some Experiments on Thermal Contact at Low Temperatures," J. Applied Physics 27, No. 4 (1956).
[2] Bobeth, M. and Diener, G., "Variational Bounds for the Effective Thermal Contact Resistance Between Bodies with Rough Surfaces," Int. J. Heat Mass Trnsf. 25, No. 1, 111-117 (1982).
[3] Jeng, D. R., "Thermal Contact Resistance in Vacuum," J. Heat Transfer, Trans. ASME, 275-276 (1967).
[4] Bobeth, M. and Diener, G., "Upper Bounds for the Effective Thermal Contact Resistance Between Bodies with Rough Surfaces," Int. J. Heat Mass Trnsf. 25, No. 8, 1231-1238 (1982).
[5] Cooper, M. G., Mikiv, B. B., and Yovanovich, M. M., "Thermal Contact Conductance," Int. J. Heat Mass Trnsf. 12, (1969).
[6] Mian, M. N., Al-Astrabadi, F. R., O'Callaghan, P. W., and Probert, S. D., "Thermal Resistance of Pressed Contacts Between Steel Surfaces: Influence of Oxide Films, J. Mech. Eng. Sci. 21, No. 3 (1979).
[7] Mikesell, R. P. and Scott, R. B., "Heat Conduction Through Insulating Supports in Very Low Temperature Equipment," J. Research of Nat'l Bureau Standards 57, No. 6 (1956).
[8] Thomas, T. R. and Probert, S. D., "Thermal Contact Resistance — the Directional Effect and Other Problems," Int. J. Heat Mass Trnsf. 13, (1970).
[9] Berman, R. and Mate, C. F., "Thermal Contact at Low Temperatures," Nature, Dec. 13, 1958.
[10] Salerno, L. J., Kittel, P., and Spivak, A. L., "Thermal Conductance of Pressed Contacts at Liquid Helium Temperatures," AIAA Paper 83-1436 presented at AIAA 18th Thermophysics Conference, June 1-3, 1983, Montreal, Canada.

SESSION H

Miscellaneous Materials

SESSION CHAIRMAN

V. V. Mirkovich
Energy, Mines & Resources Canada
Ottawa, Canada

THERMOPHYSICAL PROPERTIES OF PROPELLANTS

R. L. Shoemaker, J. A. Stark, L. G. Koshigoe[*] and
R. E. Taylor

Thermophysical Properties Research Laboratory
Perdue University, School of Mechanical Engineering
West Lafayette, Indiana

ABSTRACT

The determination of the thermophysical properties of some
solid propellants have been made. The two thermal properties
measured were specific heat and thermal diffusivity. The specific
heat was measured by using a differential scanning calorimeter (DSC)
and thermal diffusivity was measured by the flash diffusivity
method. The samples used included both single crystal and pressed
powders of ammonium perchlorate (AP), HMX and RDX.

The measurements of the propellants' specific heat required
the modification of both the DSC equipment and the technique to
accommodate the decomposition products from the samples. A flow-
through cover was installed for this purpose. The computerized
control and analysis for the DSC was also changed. The specific
heat of AP, HMX and RDX was measured using both small single crystals and small amounts of powder. The specific heat of the HMX
in both its normal phase (beta) and the delta phase were obtained.
There proved to be little difference in the specific heat between
these two phases of the HMX.

The thermal diffusivity of HMX in pressed powder form was
determined. It was necessary to develop a method of holding powdered samples so that propellants would not see the laser pulse
directly in order to avoid excessive temperature rises or radiant
energy transmission. A cup was designed to allow the powder to be
pressed into it and to not allow an alternate heat path when the
sample was under test.

---

*Presently at Naval Weapons Center, China Lake, CA

INTRODUCTION

An investigation was undertaken to determine the thermophysical properties of selected propellant materials. Specific heat, thermal diffusivity and thermal conductivity were the properties of interest. Although much work has been done on specific heat by other researchers it was desirable to verify and extend specific heat values for these materials. The materials were ammonium perchlorate (AP), RDX and HMX which are commonly used in the aerospace and defense industry. In actual use these oxidizers are combined with binders to give the propellants better burning and mechanical characteristics. In the case of AP the binder may actually be the fuel for the propellant. HMX and RDX are self-burning and may need no additional fuel. There is much interest in modeling these fuels as an aid to improving their burning characteristics and to improving rocket engine designs. As part of this modeling effort, characteristics of the pure components are needed. Propellants are somewhat difficult to work with due to the possibility of detonation or at least combustion, and as a result a small sample mass is desired. In addition, such materials are necessarily unstable in their useful temperature range and rapid measurement techniques are required. Thus, traditional methods for measuring thermal properties such as drop calorimetry and guarded hot plate were not considered. Specific heat values were determined by using a differential scanning calorimeter and thermal conductivity values were determined by calculation from measured thermal diffusivity results.

PROPELLANTS

Ammonium perchlorate, NH_4ClO_4, has three phases; a low-temperature phase, a second phase which is orthorhombic and stable from 83 to 513K and a third cubic phase stable from 513 to 723K. Densities of the crystals in the last two phases are 1957 kg/m^3 and 1756 kg/m^3, respectively. AP melts at 723K. AP has been used as the common oxidizer in chemical propellants for quite some time. The material is easily obtained from most chemical suppliers. The major handling difficulty is that AP is very hydroscopic and precautions must be taken to protect the material from water. Both single crystal and powdered AP were used for this work. The AP was obtained from Fisher Scientific Company as a certified grade material.

RDX, 1,3,5-trinitro-s-triazine, has a stable solid phase which is orthorhombic. At higher temperatures there is an unstable beta phase. Pure RDX melts at 477K. RDX is mostly used as an experimental propellant and is commonly used as a munition. The material decomposes at the lowest temperature of the three propellants investigated. Military grade RDX powder was used in this work. This grade of RDX has approximately 5% HMX as an impurity, which depressed the melting point about 14K.

200

HMX, Octahydro-1,3,5,7-tetranitro-1,3,5,7-tetrazocine, has four phases; the alpha phase which is the lowest temperature phase, the beta phase which has a monoclinic structure and exists from room temperature to 388K, the gamma phase which is unstable and the delta phase, which has a hexagonal structure, is stable from 420 to 552K, the melting point. Density values for the beta phase and the delta phase are 1900 kg/m^3 and 1700 kg/m^3 respectively. Samples used in this work included single crystals of high purity and powdered samples composed of three powder sizes.

Crystalline samples of HMX were prepared by cleaving a large crystal until a single crystal was obtained small enough to fit into the DSC sample pan. Crystal samples were typically 3mm square by 1mm thick and weighed about 17mg. Highly pure, large crystals were grown and furnished by Boggs et. al. [1]. Electrochemical analysis of the crystals showed the concentration of RDX, the major impurity, to be less than 0.1%.

Powder samples consisted of a blend of various particle sizes and were 99.7% pure HMX. Sample masses of about 19mg were used. The powders' descriptions are in Table 1. The thermal diffusivity samples were prepared by pressing the HMX powder into a stainless steel and teflon cup to yield a pellet-shaped sample. The delta-phase samples were prepared by transforming the beta-phase samples. Table 2 contains source and purity information for the propellants.

Table 1. Propellant Powder Composition

MATERIAL	HOLSTON LOT NUMBER	PARTICLE SIZE (μm)	PERCENT OCCUPYING BLEND
RDX	5775	16.0	50
	7017	6.0	30
	5789	2.6	20
HMX	5486	18.9	50
	7016	6.0	30
	7016	3.3	20

Table 2. Source and Purity of Materials

MATERIAL	SOURCE	PURITY
AP	Fisher Scientific Company	99%
RDX	U.S. Government	95%
HMX	T. L. Boggs [1] China Lake	99.9%

SPECIFIC HEAT

The specific heat of the propellants was measured by using small samples in a Perkin-Elmer Differential Scanning Calorimeter II (DSC) with a flow-through cover. The flow-through cover was installed to decrease the amount of contamination to the enclosure due to decomposition and vaporization of the sample. Nitrogen gas was continuously allowed to pass through the enclosure and cover, thus purging out most of the evolving gases from the sample. The DSC measures the differential power required to keep a sample's temperature increasing at a given rate when compared to that of a standard reference material. Figure 1 is a diagram of the basic components of the DSC. The equipment has been computerized to improve experimental accuracy. Samples were prepared by placing them in an aluminum sample holder and crimping the holder shut to seal in the sample. The mass of the sample was recorded before and after each measurement to determine if there was any sample loss during testing. Several tests were made to determine if results for the specific heat measurements were sensitive to heating rate. This involved using heating rates of 2.5 K/min, 5 K/min and 10 K/min as shown in Figure 2. There was found to be no appreciable variation in the specific heat results at these heating rates. This implied that the samples were making good thermal contact with the sample holder and that there were no significant thermal gradients in the sample while under test.

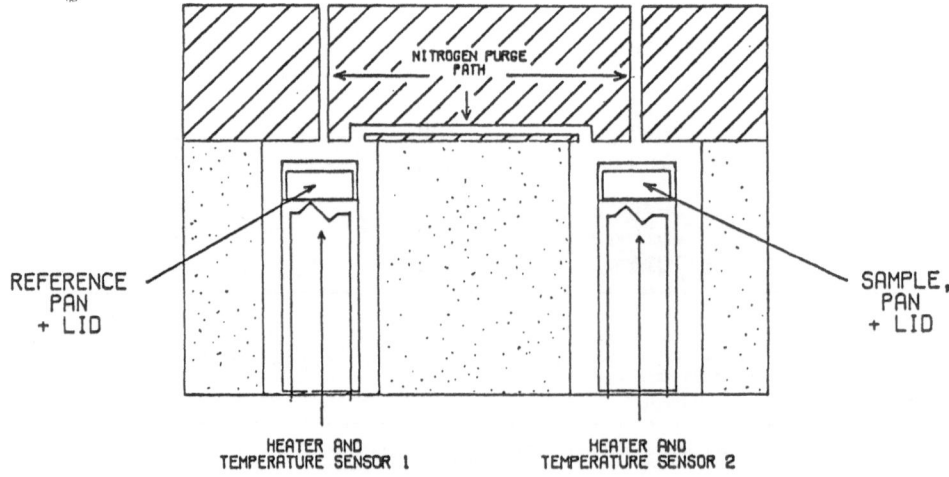

Figure 1. Differential Scanning Calorimeter (DSC) Head

Specific heat results from 310 to 510K for powdered samples of AP are given in Figure 2. Above 510K the AP sample began decomposing. Also plotted in the figure are results from Westrum and Justice [2] and data from the JANNAF tables [3] for purposes of comparison.

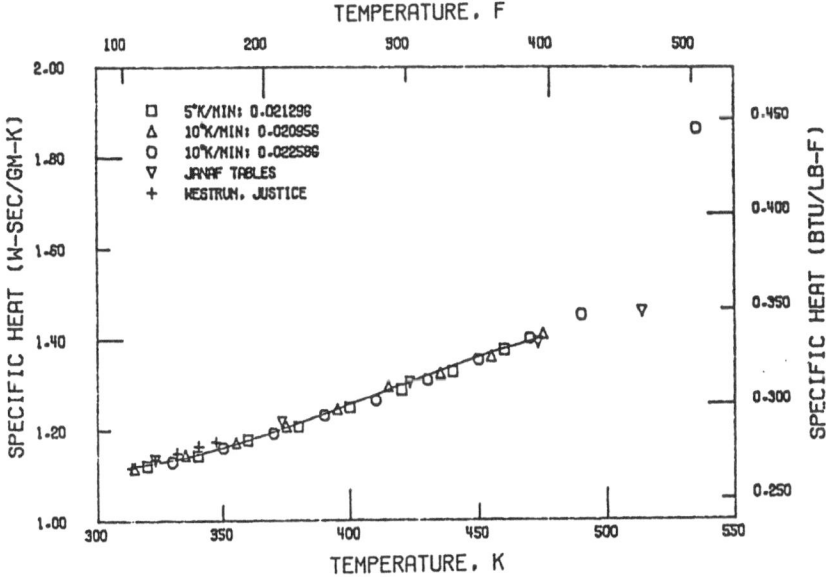

Figure 2. AP Specific Heat

Specific heat results for the temperature range from 310 to 370K for RDX powder are given in Figure 3. Above 370K decomposition of the sample became significant. Heating rate variations were checked and found to be insignificant. The melting point for the military grade RDX that was used was found to be 463K. This was due to the melting point depression caused by the HMX impurity. RDX specific heat values measured by three other authors; Velicky, et al. [4], Wilcox [5] and Baytos [6], are also presented in the figure. Pure RDX was used in the work of these authors. In spite of several attempts, pure RDX has thus far been unobtainable.

The HMX specific heat results for the beta phase are shown in Figure 4 for the single crystal and the powdered blend HMX samples heated at a rate of 5 K/min. The samples were measured over a temperature range of 310 to 445K. The difference between the specific heat of the single crystal and powdered samples was at the experimental error limit of 3%. Figure 4 also contains the results of Velicky et. al. [4] and Wilcox [5]. Velicky et. al. used an ice calorimeter while Wilcox used a DSC. The differences are at the 6% level.

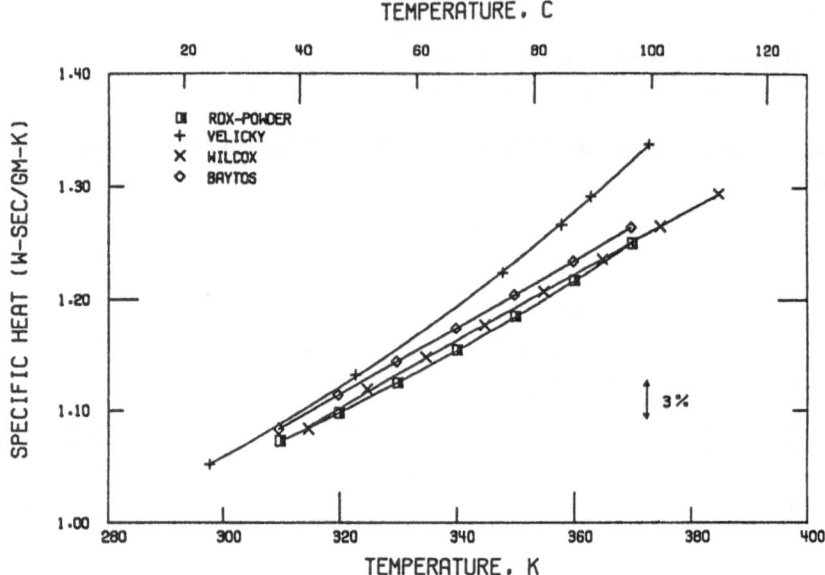

Figure 3.   Specific Heat of RDX

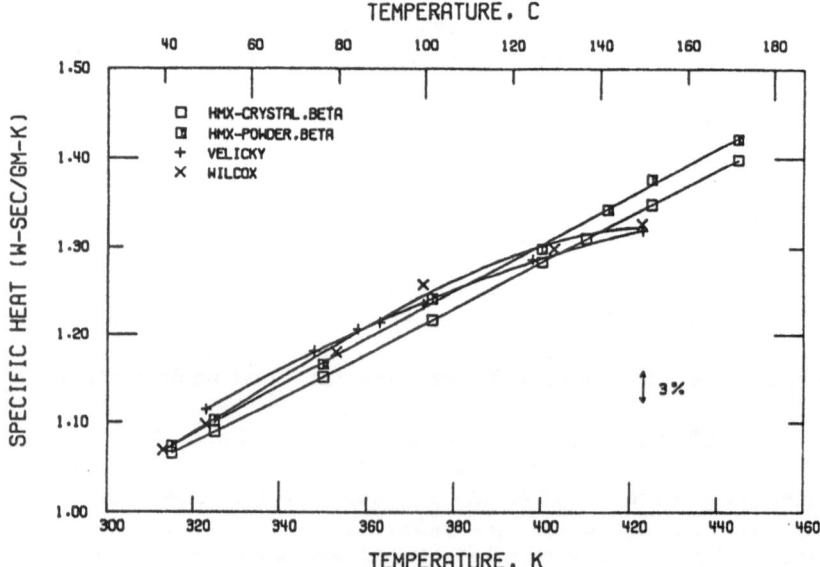

Figure 4.   Specific Heat of Beta HMX

The beta HMX was transformed to the delta phase by using the DSC. The transformation curve from the DSC output is shown in Figure 5. This was done by heating the sample in the DSC until the sample underwent transformation, as shown by an exothermal peak. The sample was then cooled and reheated to determine if any of the sample had returned to the beta phase. This demonstrated that the transformation had occurred and, upon repeated runs, that the transformation was complete and that the HMX did not immediately transform back to the beta phase. The transformation started near the temperatures reported in the literature [4], [7]. The sample could not be cooled below 400K without some of the HMX transforming back to beta phase. The upper limit for the specific heat measurement of the delta phase was 490K because of appreciable sample decomposition above that temperature. The delta phase results for both single crystal and powdered HMX are shown in Figure 6. The specific heat values of the powdered HMX were again approximately 3% higher than the results for a single crystal.

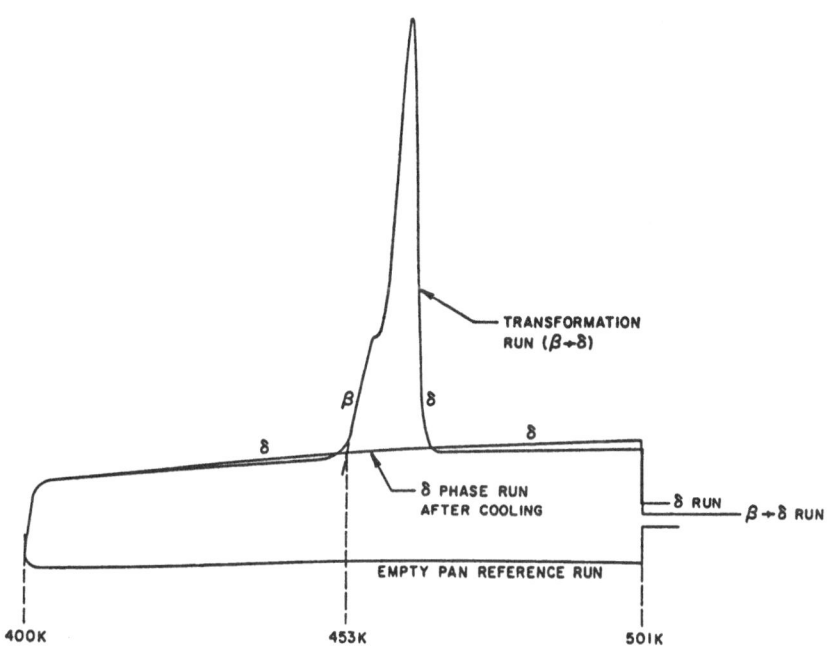

Figure 5. DSC Output for Beta to Delta Transformation

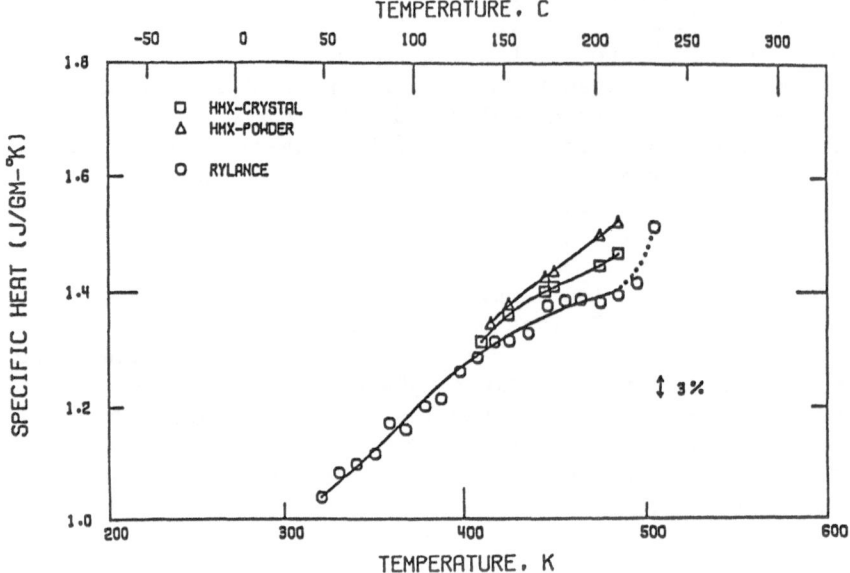

Figure 6. Specific Heat of Delta HMX

    Samples that had undergone some decomposition were measured
to determine how sensitive the specific heat is to decomposition
products in the HMX lattice. The samples were prepared by heating
them in a furnace at 510K until they had lost approximately one
eighth of their original mass. There was no attempt to determine
the amount of decomposition products in the samples. The specific
heat was then determined over the temperature range of 315 to 445K,
which was below the temperature where significant decomposition
occurred. The results are plotted in Figure 7. Data for samples
that had undergone no decomposition as well as for partially
decomposed samples are shown for comparison in the figure. The
partially decomposed samples had specific heats approximately 6%
higher than those of the virgin samples for both the single
crystal and the powdered HMX. No transformation from beta to
delta was observed during the specific heat measurement, which
indicates that the HMX was in the delta phase. The samples were
weighed before and after the DSC run to check for additional decomposition; however, the mass loss was insignificant. After
letting the samples age at room temperature for several days and
then remeasuring them, the transformation from beta to delta phase
was again observed.

    Figure 8 summarizes the results of the specific heat work for
all three propellants.

206

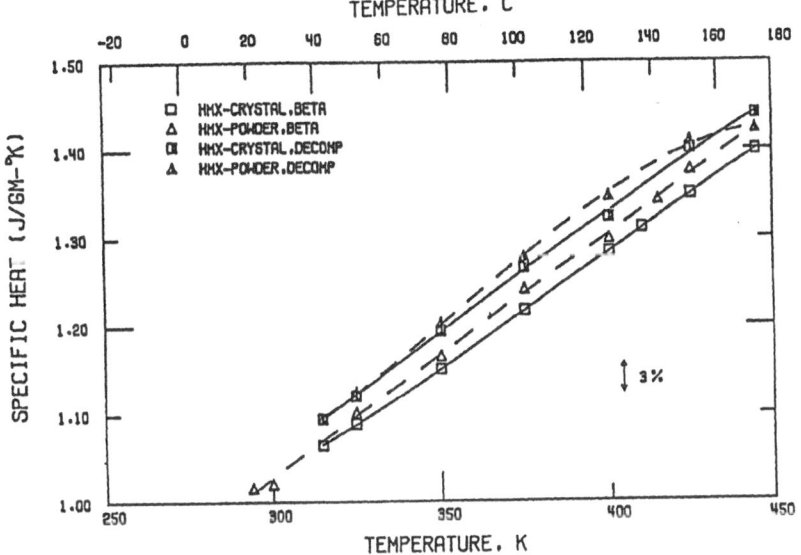

Figure 7. Specific Heat of Partially Decomposed HMX

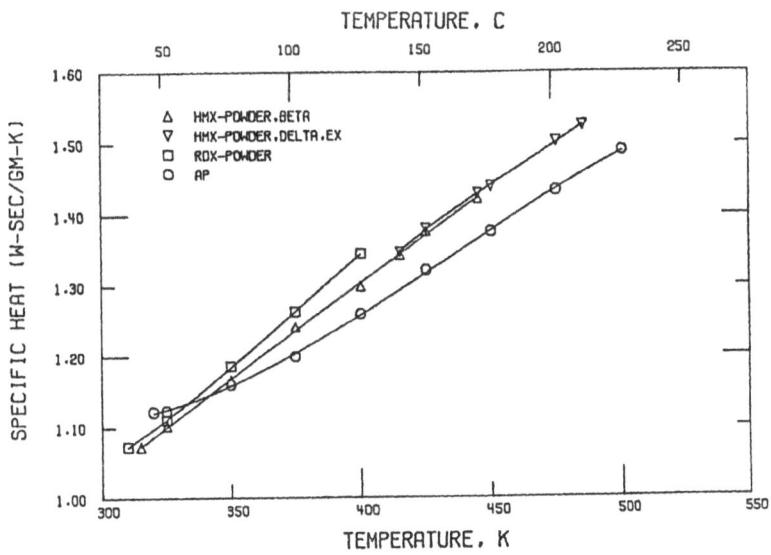

Figure 8. Specific Heat of AP, HMX and RDX

THERMAL DIFFUSIVITY

The thermal diffusivity of HMX was measured by using the flash diffusivity method. Figure 9 is a block diagram of the system. Containing the sample posed the most significant difficulty. A sample normally consists of a 1/2 inch disc that is about 1/8 inch thick. The powders could not easily be made into freestanding discs. Also, it was desirable to avoid exposing the propellant directly to the laser beam. Therefore, it was necessary to develop a sample holder that would protect the propellant from direct laser irradiation and which would still yield accurate diffusivity results. The first approach was to press the powder into a stainless steel cup; however, the resulting temperature rise curves did not follow the theoretical prediction. Finite element modeling was then used to examine the arrangement of sample and sample holder. The results showed significant heat flow around the sample to the back of the sample holder. Figure 10 shows two diagrams of the sample configuration, one immediately after the laser pulse and one at a long time after the pulse. The heat pulse was applied to the bottom face of the stainless steel cup. The lines on the cup are the isotherm lines and the arrows show the direction of heat flow. The sample holder was redesigned and checked with the finite element model, Figure 11. The diagram follows the same format as the previous figure. Measurements showed that this design permitted little heat loss to the sample holder and hence better agreement with theory. An infrared detector was used to measure the temperature rise of the back face during the diffusivity experiment. Test were made to assure that the detector was not seeing into the sample.

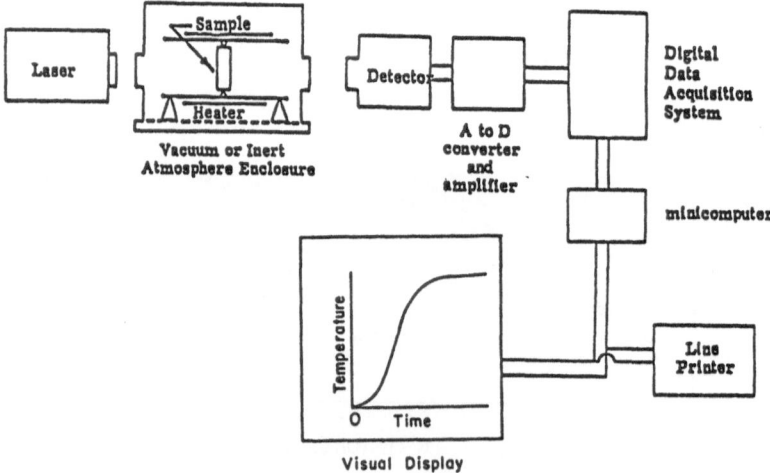

Figure 9. Flash Diffusivity System

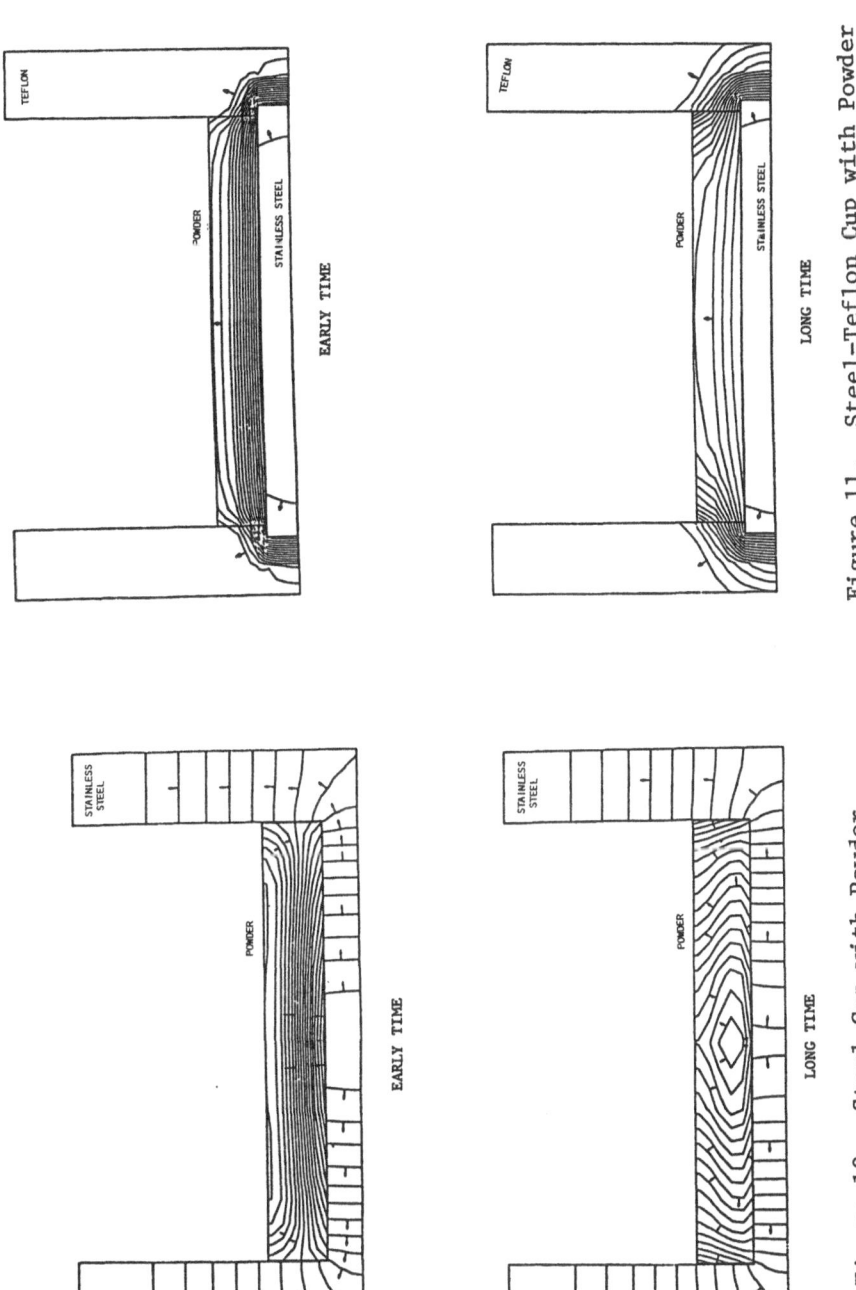

Figure 10. Steel Cup with Powder

Figure 11. Steel-Teflon Cup with Powder

209

Thermal diffusivity is a structurally sensitive property
and can vary considerably for the same material as it goes from a
loosely packed powder to a single crystal. Since suitable single
crystal samples of HMX have not yet been obtained, the powder was
pressed into the sample holders to obtain as high a density as
possible. The thermal diffusivity for powdered HMX is presented
in Figure 12. The heating curve followed the normal 1/T depend-
ency; however, the cooling curve was temperature independent.
Since the sample was heated past the temperature of transition for
beta to delta phase, it is assumed that the HMX was in the delta
phase upon cooling. The temperature independence could be due
to microcracking in the powder sample when the transition from
monoclinic to hexagonal occurred. The work in progress includes
measuring HMX powders of different densities and in single crystal
form as well as measuring the thermal diffusivity of the other
two propellants; AP and RDX in both single crystal and powder
forms.

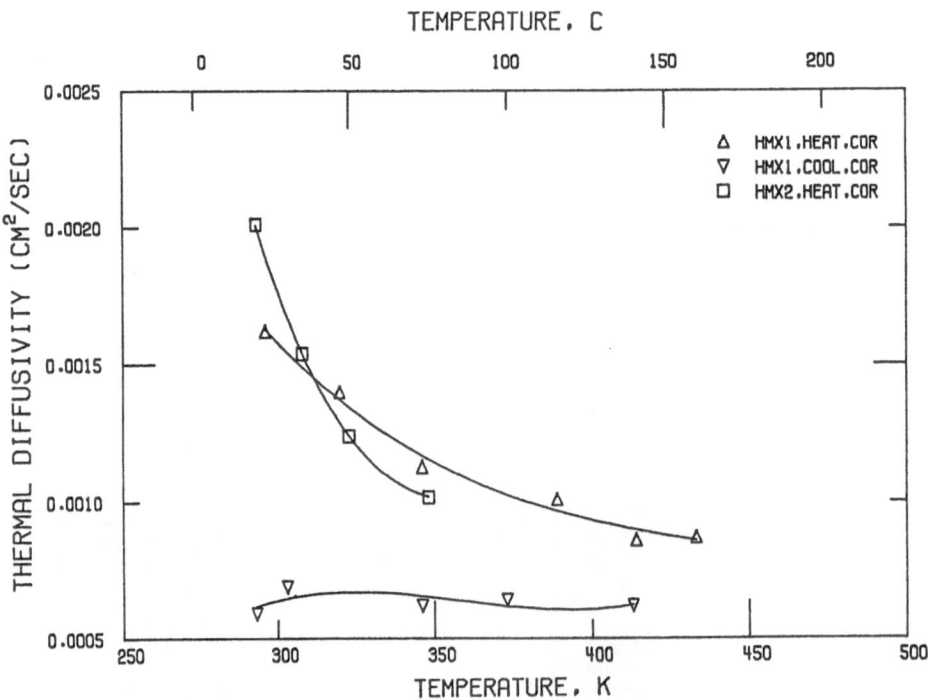

Figure 12.   Thermal Diffusivity of HMX Powder

ACKNOWLEDGEMENT

This work was performed under contract from the Air Force Office of Scientific Research with Dr. Len Caveny as program monitor.

REFERENCES

[1] Boggs, T.L., Price, C.F., Zurn, D.E., Derr, R.L. and Dibble, E.J., "The Self-Deflagration of Cyclotetramethylenetranitramine (HMX)," AIAA/SAE 13th Propulsion Conference, Orlando, Florida, July 11 - 13, 1977.

[2] Westrum, E.F. and Justice, B.H., "Molecular Freedom of the Ammonium Ion. Heat Capacity and Thermodynamic Properties of Ammonium Perchlorate from 5-300K," J. Chem. Phys., Vol 50, No. 12, June 1969.

[3] JANNAF Thermochemical Tables, Second Edition, NSRDS-NBS 37, D.R. Stull and H. Prophet, June 1971.

[4] Velicky, R., Lenchitz, C. and Beach, W., "Enthalpy Change Heat of Fusion and Specific Heat of Basic Explosives," Picatinny Arsenal, Dover, N.J., Technical Report 2504, Jan. 1959.

[5] Wilcox, J.D., "Differential Scanning Calorimeter Methods in the Determination of Thermal Properties of Explosives," Air Force Institute of Technology, Air University, Wright-Patterson Air Force Base, OH, Masters of Science Thesis, GAW/ME/67B-3, June 1976.

[6] Baytos, J.F., "Specific Heat and Thermal Conductivity of Explosives, Mixtures and Plastic-bonded Explosives Determined Experimentally," LA-8034-MS, Los Alamos Scientific Laboratory, Los Alamos, N.M.

[7] Brill, T.B. and Karpowicz, R.J., "Solid Phase Transition Kinetics: The Role of Intermolecular Forces in the Condensed Phase Decomposition of Octahydro-1,3,5,7-Tetranitro-1,3,5,7-Tetrazocine (HMX)," J. Phys. Chem., Vol 86, 1982, pp 4260-4265.

THERMAL CONDUCTIVITY OF MIXTURES

Marcos Ortiz[1], Natalia Gonzales[2], Carlos Guzman[2], Ernesto Paiva[2], and Enrique Bello[2]

[1]Department of CEME, The Univesity of Vermont
Burlington, Vermont 05405-0156
[2]Independents

ABSTRACT

An experimental study was conducted to determine the effect of composition on the thermal conductivity of three kinds of mixtures: Gas-Gas, Liquid-Liquid and Gel-Solid. An apparatus, based on the "hot wire transient technique," measured the conductivity of each type of mixture. For each mixture type, the apparatus was calibrated with substances of known thermal conductivity. Gas mixtures yielded the largest errors (up to 10% when compared to the theory), while liquid mixtures yielded small errors (as low as 0.1%). The measurements were compared to currently available theoretical models, and comparison indicates that some of the models accurately describe the measured properties of the mixture.

INTRODUCTION

The world's increasing demand for energy, and the depletion of enery sources of easy access, call for new ways and improved techniques to recover oil resources, of which heavy crude oil constitutes a large percentage. One such technique is the mixing of heavy viscous crude with a much lighter crude to facilitate the crude oil flow through extraction, and transportation systems. The proper design and operation of these systems requires knowledge of the thermophysical properties of the crude oil mixtures, particularly their thermal conductivity. This property is often estimated from empirical correlations, and no previous studies were found in the available literature.

The purpose of this project was to design and implement an experimental apparatus for the routine measurement of the thermal conductivity of mixtures of crude oils, and to determine the effect that the composition of the mixture has upon this property. The method chosen was the well-known [1,2,3] "Hot Wire Transient Technique". It was selected because of its ease of implementation and accuracy, and because background data are available from previous studies based on this method [4,5,6].

Samples of gas mixtures and liquid mixtures were first chosen to adjust the equipment and to test it against existing theoretical models and empirical correlations, as reported in references [7,8,9,10]. High purity Helium, Nitrogen, Argon, Oxygen, Hydrogen, Acetone, water, and Toluene were substances used for calibration, and experimentation. Then, measurements were performed in a mixture of two Venezuelan crude oils, "Cabimas", and "Cerro Negro". Experiments were then extended to suspensions of solids in a gel to examine the validity of the experimental technique in non-homogeneous materials. The results were satisfactory and a selected group of the experimental observations is reported and discussed in this paper.

EXPERIMENTAL APPARATUS AND PROCEDURE

A glass cell was used for liquid mixtures as well as gel-solid suspensions. A metallic frame held the thin wire that served as heat source, and measuring element. A perforated rubber plug sealed the tube and supported the frame. The plug also contained all connections between the wire and the rest of the equipment. The sensing wire was made of copper (0.06mm). The wire was cleaned or changed when necessary by disassembling the cell. For gas mixtures the container was sealed, and had a valve and a manometer attached. The wire in these tests was made of platinum, and fixed inside the cell. No cleaning or changing of the wire was found to be necessary.

A plotter was used to register the variation of electrical resistance of the wire, and hence its temperature, using a constant current. The current typically was low (1ma) and the test time short (6s) to avoid the onset of convection around the wire. The presence of convection was easily detected as a sudden drop of the temperature of the wire. This usually occurred within the first 20 seconds of the experiment. Fig. 1 shows a schematic description of the experimental setup.

A step input of current was generated by closing the switch, and the voltage variation in time was recorded by the plotter. The thermal conductivity of the medium was determined with Eq.1

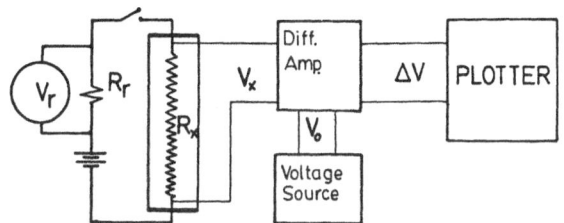

Fig. 1. Measuring Circuit

$$\lambda = \frac{(Vr/Rr)^3 . Rx . S . Ln(t1/t2)}{4.\eta .L.(\Delta V(t2) - \Delta V(t1))} \quad (1)$$

More conventional forms of this equation use dRx/dT instead of S [2.11]. However, the value of dRx/dT for a particular wire is not readily available, and it is impractical to determine dRx/dT for every time the wire is replaced. Using S, a calibration factor described below, is much more practical.

Consider the situation shown in Fig.2: Zone I represents the wire of finite diameter 2a and conductivity $\lambda 1$; zone II represents the specimen in which the heat has propagated some distance δ in the time t. The specimen conductivity is $\lambda 2$.

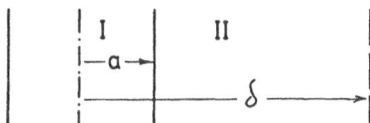

Fig. 2. Schematic representation of the 'real' wire-specimen
system. At some time t after the experiment begins,
the heat wave has diffused into the medium.

The 'effective' thermal conductivity of the system, as measured in our apparatus, will then be:

$$\lambda = \lambda 2 \ \frac{a + \delta}{a. \ \lambda 1/ \ \lambda 2 + \delta} \quad (2)$$

This expression involves the conductivity and dimensions of the wire, as well as the conductivity of the medium and the time duration of the experiment ($\delta \equiv$ t). From this relation (Eq.2), one obtains an approximate expression for the thermal conductivity of the specimen, $\lambda 2$, in terms of the measured conductivity, λ, which in turn defines S.

$$S = \frac{a. \ \lambda 2/\lambda 1 + \delta}{a + \delta} \ .dRx/dT \quad (3)$$

S, and therefore dRx/dT, is determined by measuring a reference substance.

The value of S, Eq.3, is affected by the properties of the test substance. Hence, calibration must be made with a reference substance whose thermal conductivity is rather close to that of the test substance. If the properties of the substance under study are completely unknown, an iterative procedure is required. Similarly, some of the mixtures examined exhibited great variation of their thermal conductivity with composition. It was then necessary to use different calibration factors S for the same wire for different ranges of thermal conducitivity. A variety of substances were used as reference materials for this purpose. Water, toluene, ethyl alcohol, helium and nitrogen, covering a wide spectrum of thermal conductivities, were among those used.

EXERIMENTS CONDUCTED AND RESULTS

Three distinct types of experiment were conducted, depending on the nature of the mixture: gas, liquid or gel-solid suspensions. In most cases, a theoretical prediction was obtained for its comparison with experiments. The results are described as follows:

Gas Mixtures

Measurements in gases yielded the largest departures from theoretical values. Convection appeared very early in the experiments, therefore S (Eq.3) had to be carefully checked for each composition of the gas mixture examined.

Figure 3 shows how the error in the absolute measurement changes as the thermal conductivity of the specimen (Argon, Nitrogen, Oxygen, and Hydrogen) varies from the reference substance (ambient air in this case).

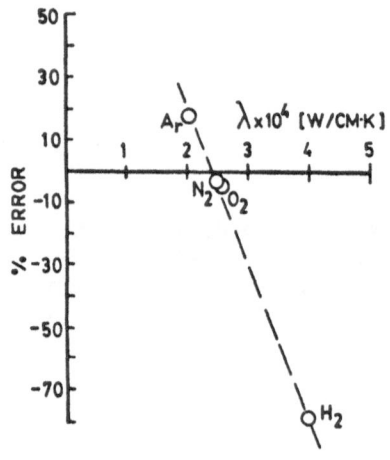

Fig. 3 Variation of the error of the measurement with the difference between sample gas conductivity and pattern gas conductivity. The pattern gas for this case was air.

The results of Fig. 3 were used to correct the measurements obtained in binary mixtures of nitrogen and helium. These corrected measurements were compared to the values predicted by the relation derived by Wassiljewa, and reported by Tsederberg [10], from kinetic theory of gases. The theory was used to predict the thermal conductivity of the mixture, and that of the individual gases. This comparison is shown in Fig. 4.

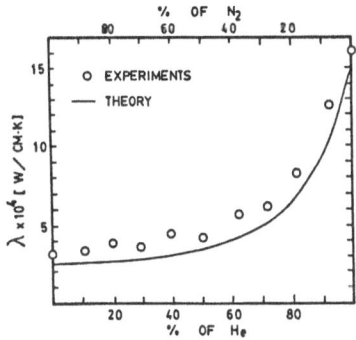

Fig. 4 Thermal conductivity of Helium – Nitrogen binary mixtures, Experimental values and theoretical predictions according to Wassiljewa's relation.

Helium-Nitrogen was the combination of gases chosen for this comparison, because they were expected to show large variations of thermal conductivity with composition. Argon-Helium mixtures could have shown greater variations, but the availability of Argon, at the time of the experiments, was not sufficient.

As indicated by these results (Fig 4), theoretical predictions and experimental values follow parallel trends. Experimental values are consistently above the theoretical predictions by what seems to be a fixed small quantity. This may be an effect of convection not detected in the experiment. One can conclude from this, that this technique using S gives reasonable results for gas mixtures, where it yields a maximum error of about 10%.

Liquid Mixtures

Three binary liquid mixtures were examined: tap water-ethyl alcohol (reagent grade), tap water-acetone (reagent grade) and a mixture of Cabimas, a light crude oil, and Cerro Negro, a heavy crude oil. The first two mixtures were measured for the purpose of adjusting the apparatus and checking the technique. The third mixture was tested to determine the nature of the mixture and its constituents. Samples of a fourth mixture (crude oil and water) were also measured to evaluate the effect of moisture present in crude oil on its thermal conductivity.

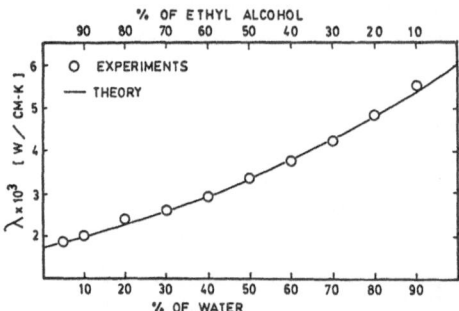

Fig. 5 Thermal conductivity of the binary mixture of water and ethyl alcohol. The solid line represents the predicted values according to the so-called Predvoditelev-Vargaftik equation [10].

Figures 5 and 6 show the results obtained for the first two mixtures. The so-called theoretical prediction corresponds to an empirical correlation suggested by Predvoditelev and Vargaftik, and reported by Tsederberg [10].

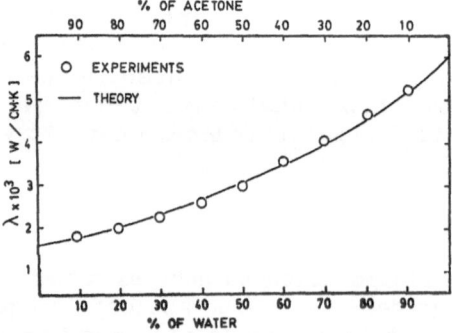

Fig. 6 Thermal Conductivity of binary mixtures of acetone and water. The solid line represents theoretical predictions according to the so-called Predvoditelev-Vargaftik equation [10]

Other predictions, such as the additive rule as well as more complicated computer models (TRAPP [7] and NRTL [12]), were examined and rejected for the purposes of this study, because they were either inaccurate or required extensive knowledge of the substances involved.

Figures 5 and 6 show that the experimental results can be described accurately by the relation proposed by Predvoditelev and Vargaftik. The repeatibility of each data point in those figures was good, falling always within 3% of the computed value.

One concludes from these results that the apparatus can be used to determine the thermal conductivity of polar fluids and that it can be used with precision for measurements in electrically conducting fluids (i.e., tap water, Fig's 5 & 6).

A mixture that is of interest to the oil industry is the combination of a heavy crude oil with a lighter crude oil. This is done for the purpose of recuperating heavy oils. The mixture of heavy oil with light oil has lower viscosity than the heavy oil alone. Therefore, it can be pumped and transported more efficiently as a mixture. The properties of both constituents in this mixtures are in many cases unknown. The properties of their mixture, although important, are even less known. An experiment was conducted to examine the behavior of one of such mixtures and the results are presented in Fig. 7.

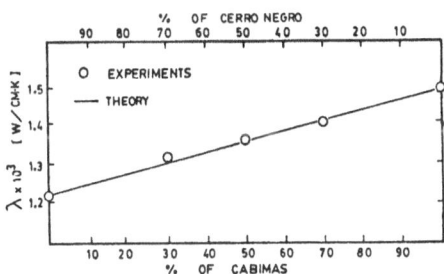

Fig. 7. Thermal conductivity of a mixture of heavy crude oil, Cerro Negro, and light crude oil, Cabimas.

The thermal conductivity of each crude oil was measured separately and a set of mixtures was prepared to examine the effect of mixture composition. The most important observation was that the thermal conductivity of these mixtures follows the additive rule.

Tests were also performed adding small amounts of water to another crude oil. Results are shown in Table I. They indicate a strong increase in the thermal conductivity of the crude with increased water content.

Mixtures of Lagotreco crude oil with more than four percent

of added water were not measured, because the oil did not dissolve any more water and the excess water formed a deposit at the bottom of the cell. Close observation of other samples of crude oil, as delivered out of the well, indicated that they are not a homogeneous substance but rather a suspension of particles of sand and very small bubbles of water. A third set of experiments was then designed to study this type of non-homogeneous suspensions.

TABLE I
Binary system "Lagotreco oil" - water

% of Lagotreco Crude oil	λ W/cm-k x 10^3
100	1.528
98	1.683
96	1.688

Gel-Solid Mixtures

Three gel-solid mixtures were chosen to examine the effect of composition in non-homogeneous substances. This selection was made on the basis of availability of materials and the properties of polystyrene, PVC and iron.

Oil-metal dust suspensions have been examined [13] in a qualitative fashion. The results indicated that the particles of metal tend to increase the thermal conductivity of the mixture. Quantitative studies for large particle concentration were not possible because the dust precipitated to the bottom of the cell at a slow but appreciable rate.

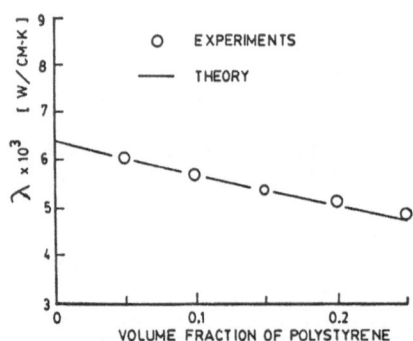

Fig. 8 Thermal Conductivity of gel-solid suspensions of particles of polystyrene.

Figures 8, 9 and 10 show the experimental results obtained with three mixtures: gel-polystyrene, gel-PVC, and gel-iron particles. Theoretical computations obtained using a relation proposed by Hamilton [14], are also shown in the figures for comparison. The figures indicate that experiments and theory lead to equivalent results.

Special care was taken in these experiments to assure that the measurement corresponded to a representative sample of the mixture. That is, the time duration of the experiment was long enough to include within δ(Fig. 2) a representative part of the total mixture. If the time is too short, there might be insufficient solid particles within δ to equal the average concentration present in that particular specimen. For this reason, it was not possible to measure at very low concentrations of suspended solids. Such mixtures required a time exceeding that at which convection appeared in the tests.

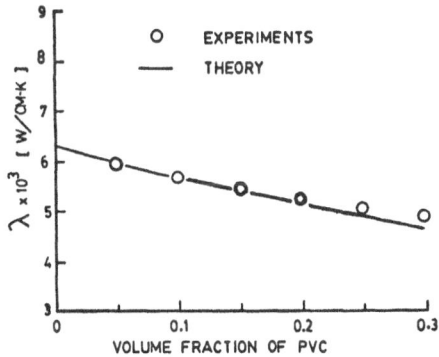

Fig. 9 Thermal conductivity of gel-solid suspensions of particles of PVC.

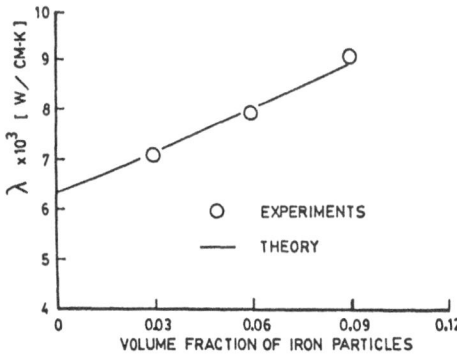

Fig. 10 Thermal conductivity of gel-solid suspensions of particles of iron.

221

CONCLUSIONS

The most important result from this work is that the calibration factor S, as defined by Eq. 3, allows measurements of thermal conductivity of fluids with an accuracy that otherwise requires highly sophisticated peripheral equipment. Other conclusions are: (1) Wires thinner than that utilized here exist, but they break too easily; (2) One could reduce the input heat to delay the onset of convection, and therefore increase accuracy by increasing the duration of the experiment. But the signals would be too small and require better amplification since ambient noise would become a problem.

Another important observation is that crude oil mixtures seem to follow the additive rule. The two experiences with crude-water mixtures (Table I) seem to follow a different trend. More experiments are necessary to arrive at a general conclusion. The authors feel at this point that the thermal conductivity of crudes, although apparently confined to a small range [4,5] is unpredictable. Effort is currently directed toward developing means of in-situ measurement of these properties.

From the results on non-homogeneous mixtures, one may say that the method and the appartus can still be used for more complex substances.

REFERENCES

1. Schleiermacher, A. Annalen Phys. Chem. vol. 34,n6,623. (1888)
2. Tye, R. P., "Thermal Conductivity", Vol 2, Ch. 2 Academic Press London (1969).
3. Clifford, A. A., Kestin, J., Wakeham, W. A. A Further Contribution to the Theory of Transient Hot Wire Technique for Thermal Conductivity Measurements, Physica, 100A, 370-374. (1980).
4. Ortiz, M., Arroyo, D., Osorio, J., Gonzalez, N., Guzman, C. An Experimental Investigation of the Effect of Temperature on the Thermal Conductivity of Heavy Crudes. High Temperatures-High Pressures, Vol 15 pp. 413-418. (1983)
5. Ortiz, M., Osorio, J., Arroyo, D. Medicion de la Conductividad Termica de Crudos Pesados Venezolanos y su Variacion con la Temperatura, Proceedings of VII Congreso Venezolano de Ingenieria Electrica y Mecanica, Caraballeda. (1982)
6. Ortiz, M., Calarotti, R., Quintero, P., Smicht, P. Medicion de Conductividad Termica de Hidrocarburos Liquidos Mediante Tecnicas de Exitacion Transitoria, XXV Convencion Nacional de AsoVAC, Maracaibo. (1974)

7. Ely, J. F., Hanley, H.J.M. Prediction of the Viscosity and Thermal Conductivity in Hydrocarbon Mixtures - Computer Program TRAPP, Proceedings of the Sixtieth Gas Processors Association Annual Convention. (1981)

8. Ogiwara, K., Arai, Y., Saito, S. Thermal Conductivity of Liquid Hydrocarbons and Their Binary Mixtures, Ind. Eng. Chem. Fundam., 19, 295-300. (1980).

9. Teja, A. S., Rice, P. A Generalized Corresponding States Method for the Prediction of the Thermal Conductivity of Liquids and Liquid Mixtures, Chemical Engineering Science, 36, 417-422, (1981)

10. Tsederberg, N.V., "Thermal Conductivity of Gases and Liquids". Chapters V & VII. MIT press, Cambridge (1965).

11. Carslaw and Jaeger, "Conduction of Heat in Solids", second edition, Oxford University Press, London (1959).

12. INTEVEP S.A., Prediction of Liquid Mixtures Thermal Conductivities. Unpublished report, Los Teques (1981).

13. Ortiz, M. "Diseno y Construccion de un Aparato para Medir la Conductividad Termica de Hidrocarburos Liquidos". Mech. Eng. Thesis, University Central de Venezuela, Caracas (1974).

14. Hamilton, R., L., Ph.D. Thesis, University of Oklahoma (1960)

15. Souto, M.I., Gil, N., Cabrera, O., "Medicion de Conductividades Termicas en Sustancias Granulares", Chem. Eng. Report, University Simon Bolivar, Caracas (1983).

16. Hernandez, D., Corsi, A., "Modelaje y Simulacion del Metodo Transitorio del Hilo Caliente para la Determinacion de Conductividad Termica de Medios Compuestos", Mech. Eng. Thesis, University Simon Bolivar, Caracas (1983).

PHONON CONDUCTION IN HEXAGONAL CRYSTALS

EXHIBITING EXTREME ELASTIC ANISOTROPHY

A. K. McCurdy

Worcester Polytechnic Institute
Worcester, Massachusetts 01609

ABSTRACT

Striking differences (more than several orders of magnitude) are predicted in the intensity of phonons propagating ballistically along different directions to the c axis in hexagonal crystals exhibiting extreme elastic anisotropy. Calculations of the boundary-scattered phonon conductivity for long rods of square cross section predict anisotropy ratios $\lambda_{11}/\lambda_{\perp}$ due to phonon focusing of more than 70 for samples having the elastic constants of CAP graphite. Furthermore, the anisotropy ratio $\lambda_{11}/\lambda_{\perp}$ increased from less than unity to more than 100 as the sample dimension ratio $d_{11}/d_{\perp}$ was varied from 0.1 to 10. Corrections to Casimir's theory generalized to include phonon focusing have been derived for samples in which a fraction of the phonons are specularly reflected from the side and end faces, and for samples in which a fraction of the phonons are specularly transmitted through domain walls. Calculations performed for CAP graphite near $1^{\circ}K$ show little significant change in the anisotropy ratio $\lambda_{11}/\lambda_{\perp}$ if transmission of ballistic phonons occurs through grain boundaries, but reductions in $\lambda_{11}/\lambda_{\perp}$ by more than a factor of 20 can be realized if reflection rather than transmission occurs at grain boundaries perpendicular to the c axis.

INTRODUCTION

Phonon focusing in hexagonal crystals depends upon the second-order elastic constants C_{11}, C_{33}, C_{44}, C_{66}, and C_{13}. The ratios between these constants determine the anisotropy, the shape of the phase and group-velocity surfaces, and the phonon focusing properties for each of the 3 modes of propagation [1]. The boundary-scattered phonon conductivity, λ, in turn depends upon the phonon

focusing properties of the crystal. Calculations of λ in the boundary - scattering regime depend only upon the elastic constants, the
density, and the linear dimensions of the sample [2]. Along strongly focused directions λ is significantly larger than the end-correct-
ed Casimir value [3]. Predictions for silicon and calcium fluoride
have been in good agreement with experiments [2].

Striking effects of phonon focusing upon phonon conduction
arising from the extreme elastic anisotropy of graphites are predicted in the following sections. Expected results are presented
and discussed for graphites of both near-ideal and pyrolytic structure.

RESULTS

The directions of intense phonon focusing for two contrasting
cases of extreme elastic anisotropy are illustrated schematically in
Figs. 1a and b, respectively. Graphite (Fig. 1a) has pronounced
cuspidal features about a collinear direction, $<\Theta_s>$, with respect to
the c axis. Angle Θ_s is given in terms of the second-order elastic
constants as [1]:

$$\tan \Theta_s = [(C_{33}+C_{13})/(C_{11}+C_{13})]^{\frac{1}{2}}. \tag{1}$$

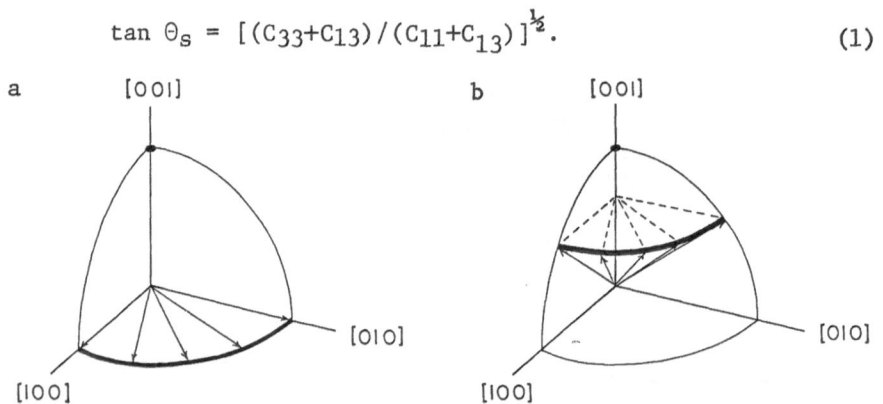

Fig. 1a,b. Simplified schematic diagrams illustrating phonon focus-
            ing for all 3 modes of 2 hexagonals crystals with con-
            trasting elastic anisotropies: (a) CAP graphite (b) a
            contrasting solid. Directions of high phonon intensity
            are given by constructing lines from the origin to the
            heavy dark areas.

The contrasting solid (Fig. 1b) was chosen to have cuspidal extrema
about the c axis which extend to within several degrees of the cuspidal extrema perpendicular to the c axis. In addition, there is
strong focusing of longitudinal waves about a collinear direction
45 degrees to the c axis. As a result there is a circular band of
intense longitudinal phonons sandwiched between 2 narrow bands of
transverse phonons 45 degrees to the c axis. The group-velocity

Table 1. Elastic constants of compressed annealed pyrolytic (CAP) graphite at liquid helium temperatures obtained from the curves of Gauster and Fritz [4]. The elastic constants are in units of 10^9 n/m^2; the density is 2267 Kg/m^3. These constants are believed to be nearly equal to those in single crystals.

C_{11}	C_{33}	C_{44}	C_{66}	C_{13}
1127.0	40.8	0.289	463.5	40.0

Table 2. Calculated phonon intensities in CAP graphite parallel and perpendicular to the c axis. For elastically isotropic solids the corresponding phonon intensities are unity. Because the cusp in the T2 mode extends to within 0.5 deg. of both the c and $c_\perp$ axes, calculations for this mode were performed for detectors subtending 0.5 and 1°. Calculations for the other 2 modes apply for an infinitesimal detector.

Axis	C_{44}	L	T1	T2 (0.5°)	T2 (1°)
$<c>$	normal	1.022	3.89×10^{-7}	2078	3316
$<c>$	doubled	0.966	1.56×10^{-6}	....	1355
$<c>$	halved	1.011	9.72×10^{-8}	12044	4564
$<c_\perp>$	normal	651.6	1604	0.889	0.606
$<c_\perp>$	doubled	552.5	801.9	0.550	0.563
$<c_\perp>$	halved	715.7	3208	0.983	0.621

surfaces for graphite using the elastic constants in Table 1 are given in Figs. 2a and b, and the angular deviation between the directions of the wave vector and the corresponding group-velocity direction for each of the 3 modes is given in Fig. 3. Phonon intensities parallel and perpendicular to the c axis are given for each mode in Table 2. Intensity values are also given for values of C_{44} twice and half the values quoted by Gauster and Fritz [4]. Values of λ for phonon conduction in near-ideal graphite samples are given in Table 3 and Fig. 4. Similar calculations have been described in detail elsewhere [2,3]. The thermal length, L, of the sample is defined as the distance between the centers of the heat source and heat sink (i.e., length of the thermal gradient

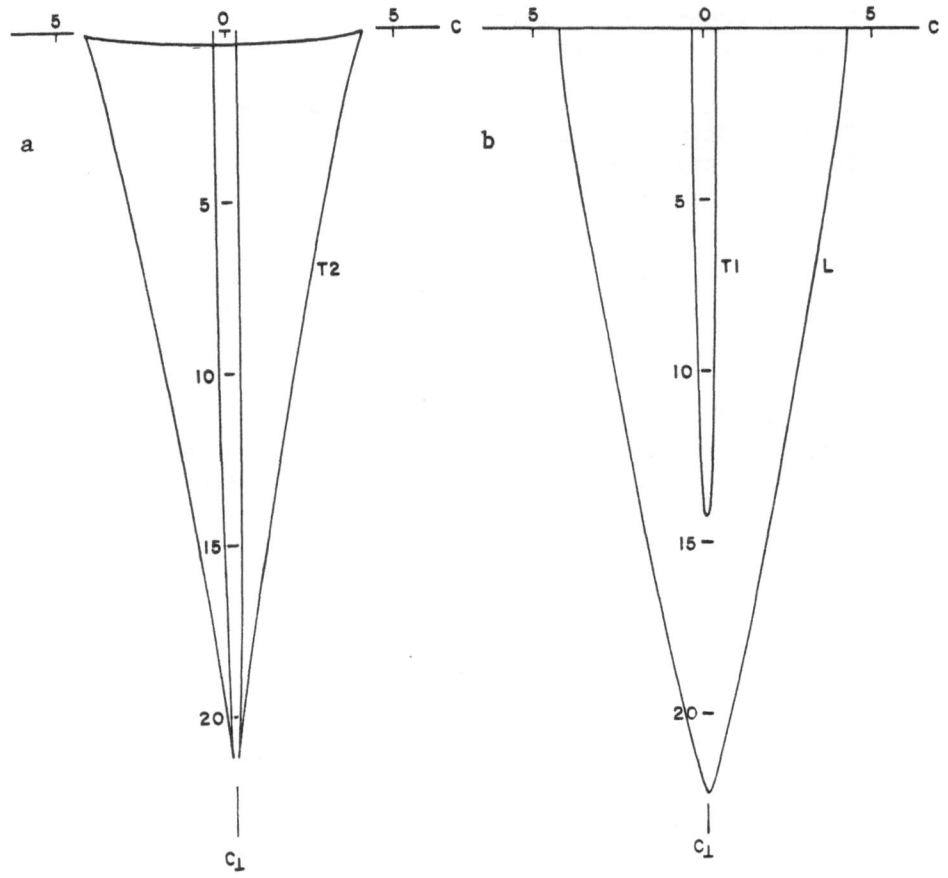

Fig. 2a,b. Polar plots of the group-velocity surfaces (km/s) in
CAP graphite for (a) the cuspidal transverse (T2)
(b) the longitudinal (L), and the cusp-free transverse
(T1) mode.

along the heat-flow axis) in a conventional thermal conductivity
measurement, and thus is usually less than the overall length of
the sample. These calculations apply for samples in which the phonon
mean-free path is restricted only by the perfectly absorbing outside
surfaces of the sample. Results are expressed in dimensionless units
as $\lambda/<\lambda_c>$ where $<\lambda_c>$ is the Casimir phonon conductivity for infinitely long, elastically anisotropic crystals, defined elsewhere [3].
Note, however, that for hexagonal crystals the averaging of the phase
velocities is different; averages of the inverse square and inverse
cube phase velocity, respectively, are taken over the [001], [100],
and [010] directions. The value of $<\lambda_c>/T^3$ for graphite is 142.45 Λ_c
kw m^{-1}K^{-4} where Λ_c is the Casimir length for very long samples [3].

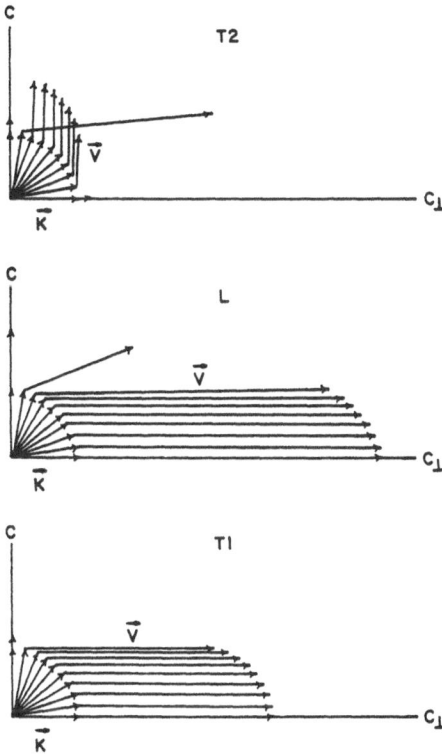

Fig. 3. Schematic diagrams showing phonon focusing in graphite for
the cuspidal transverse (T2), the longitudinal (L), and the
cusp-free transverse (T1) mode. A uniform angular distribution of wave vectors, $\vec{k}$ is shown, but the relative magnitudes of the group velocities, $\vec{v}$, are not to scale. The
deviation in direction of the group-velocity vectors, $\vec{v}$,
relative to their corresponding wave vectors, $\vec{k}$, is such
that nearly all phonons in the T1 and L modes travel
parallel to $c_{\perp}$, whereas in the T2 mode most phonons travel
parallel to the c axes.

The phonon-conductivity anisotropy ratio, $\lambda_{11}/\lambda_{\perp}$, for a heat-
flow axis parallel and perpendicular, respectively, to the c axis is
given in Fig. 5 for rods of CAP graphite. Samples are assumed to
have periodic phonon-absorbing internal boundaries with c-axis ordering of the crystallites. Results are given as a function of $d_{11}/d_{\perp}$,
where d_{11} and $d_{\perp}$ are the dimensions of these crystallites parallel
and perpendicular, respectively, to the c axis (see inset). Curves
1 through 4 apply for samples with perfectly absorbing boundaries
(i.e., the absorption coefficient is unity for incident phonons).
Curves 5 through 8 apply for samples in which a fraction, f, of the
incident phonons are absorbed and then diffusively reradiated at

Table 3. Dimensionless boundary-scattered phonon conductivity, $\lambda_{11}/\langle\lambda_c\rangle$, parallel to the c axis for near-ideal graphite rods of square cross section as a function of the dimension ratio $(d_1d_2)^{\frac{1}{2}}/L$.

$(d_1d_2)^{\frac{1}{2}}/L$	0	0.05	0.10	0.20	0.40
$\lambda_{11}/\langle\lambda_c\rangle$	8.03	5.18	3.48	2.07	1.14

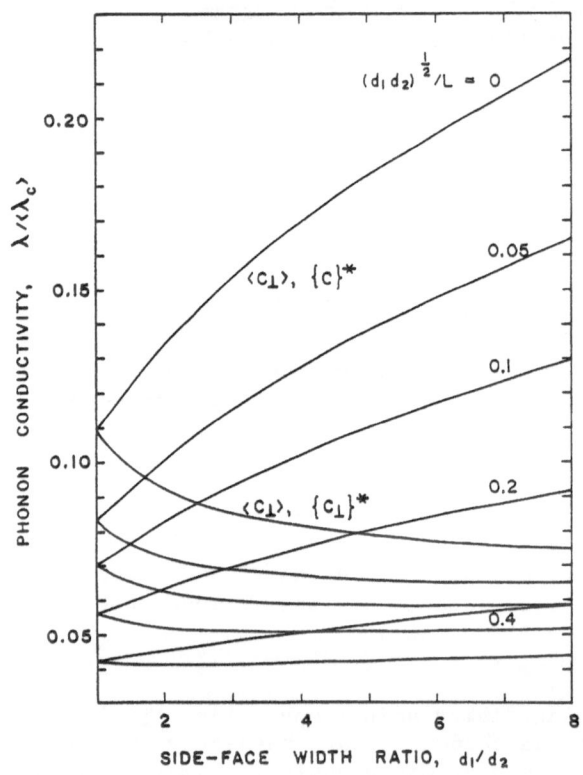

Fig. 4. Dimensionless boundary-scattered phonon conductivity, $\lambda/\langle\lambda_c\rangle$, for near-ideal graphite rods of rectangular cross section and varying side-face width ratio, d_1/d_2, with heat-flow axis perpendicular to the c axis. The wider side face is indicated by an asterisk. The incident phonons are assumed to be completely absorbed and diffusively reradiated by the outside surfaces.

230

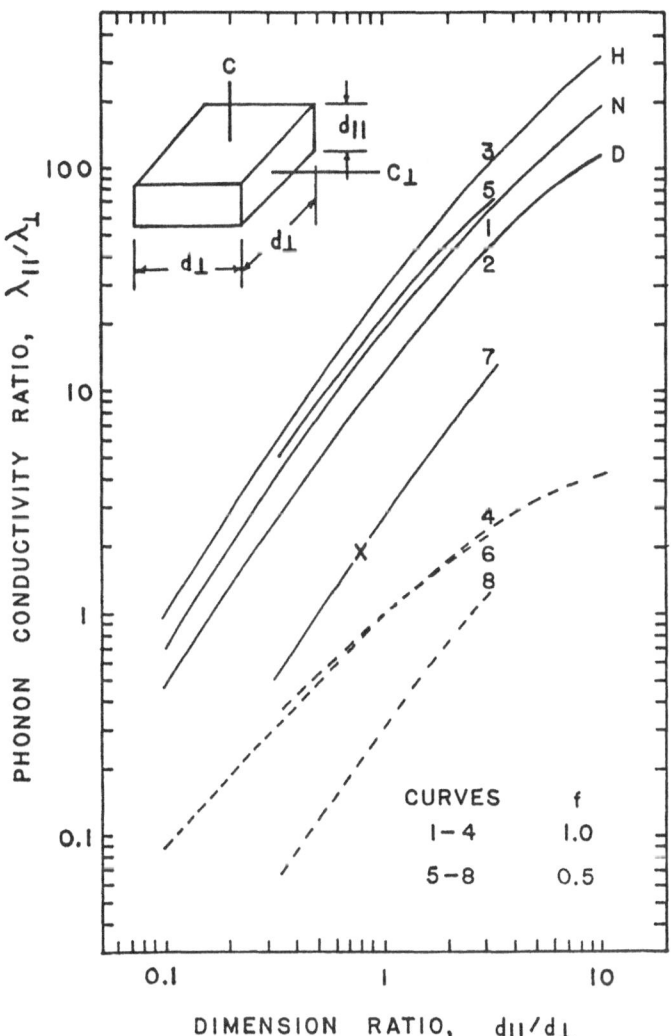

Fig. 5. The phonon-conductivity anisotropy ratio, $\lambda_{11}/\lambda_{\perp}$, as a function of the crystallite dimension ratio, $d_{11}/d_{\perp}$, for CAP graphite and for elastic isotropy (dotted lines) for different values of the phonon absorption coefficient, f. Boundaries perpendicular to the c axis are assumed to specularly transmit (curves 5 and 6), or specularly reflect (curves 7 and 8) (1-f) of the incident phonons, whereas boundaries parallel to the c axis specularly transmit (1-f) of the incident phonons. Distances between the phonon-scattering boundaries are defined in the inset.

these boundaries. Boundaries parallel to the c axis (i.e., {$c_\perp$} faces) are assumed to transmit (1-f) of the incident phonons, whereas boundaries perpendicular to the c axis (i.e., {c} faces) transmit (1-f) of the incident phonons (curves 5 and 6) or reflect (1-f) of the incident phonons (curves 7 and 8). Curves 1-3, 5, and 7 apply to CAP graphite, whereas the dotted curves 4, 6, and 8 apply to an elastically isotropic solid. The letter x marks the approximate coordinates at liquid helium for the graphite samples reported by Klein and Holland [5].

DISCUSSION

In graphite strong focusing of longitudinal and T1 transverse waves occurs perpendicular to the c axis. In addition, the cusp about $<\Theta_s>$ (see Fig. 1a) in the T2 transverse wave gives rise to a narrow beam of high-intensity phonons approximately 0.5 degree to the c axis. As a result, there are narrow beams of phonons parallel and perpendicular to the c axis having rotational symmetry about the c axis (Fig. 2a). Phonon intensities in Table 2 show the remarkable variation in the values parallel and perpendicular to the c axis for each of the 3 modes. In the other solid, however, there is strong focusing of cusp-free transverse waves parallel to the c axis and strong focusing of longitudinal waves about a collinear direction 45 degrees to the c axis. Furthermore, large cusps in the other transverse mode about the c and $c_\perp$ axes nearly meet so that strong focusing in this mode is confined to a narrow beam several degrees wide at an angle 45 degrees to the c axis (see Fig. 2b).

Graphite Rods of Near-ideal Structure

Rods with a heat-flow axis aligned along a strongly focused direction will exhibit an enhanced phonon conductivity. Furthermore, rods of rectangular cross section aligned along the same strongly focused direction will exhibit a higher phonon conductivity when the wider side face is oriented so that the sample cross section coincides with the high-intensity phonon beam cross section. This can best be realized in a hexagonal crystal of rectangular cross section when the narrow beam of phonons lies in a plane perpendicular to the c axis (Fig. 1a), the heat-flow axis is parallel to $c_\perp$, and the wider side face is in the plane perpendicular to the c axis.

Phonon conductivity is approximately inversely proportional to the square of the phonon velocity and thus the slower velocity modes make the major contribution to the flow of heat. Furthermore, because of phonon focusing about the c axis, 99.99% of the phonon conduction along this axis is carried by the T2 mode so that along the c axis graphite is essentially a one-mode solid. Along the $c_\perp$ directions, however, 53.1% of the phonon conduction in long rods of square cross section is carried by the T1 mode, and 43.4% by the T2 mode so that along these directions graphite is nearly a 2-mode solid. Since phonon velocities along $c_\perp$ are much larger (see Fig. 2a, b) the phonon conductivities along this axis in samples of square cross section is lower by a factor of up to 74 than values parallel to the c axis (compare Table 4 and Fig. 4).

In samples of rectangular cross section, however, the phonon conductivity along $c_\perp$ is strongly dependent upon the orientation of the wider side face. A similar effect was observed in silicon [2] and discussed in detail [3], but in graphite the effect is more intense. Because phonon intensities in hexagonal crystals have rotational symmetry about the c axis, the phonon conductivity along this direction depends little upon the side-face width ratio.

Corrections for finite length depend upon the focusing properties along the heat-flow axis. Longer samples have longer phonon flight paths to reach the end surfaces and thus have longer mean-free paths. Strong focusing along the heat-flow axis therefore increases the phonon mean-free path and thus increases the dependence of the phonon conductivity upon sample length.

Rods of CAP Graphite

The extreme phonon-focusing properties of graphite should be evident even in pyrolytic samples where phonon scattering occurs along the internal crystallite boundaries. The high ordering of the c axis in pyrolytic graphites and the intense focusing of T2 waves along the c axis should give a strong enhancement of the phonon conductivity, λ_{11}. Preliminary calculations using tetragonally-shaped crystallites (see inset, Fig. 5) as a crude approximation illustrate the dependence of $\lambda_{11}/\lambda_\perp$ upon the crystallite dimension ratio $d_{11}/d_\perp$ compared to an elastically isotropic solid. Results for hexagonally-shaped crystallites are not expected to be significantly different. As expected the results for CAP graphite strongly depend upon the value of C_{44} (curves 1-3); a smaller C_{44} increases the degree of c axis focusing and increases the $\lambda_{11}/\lambda_\perp$ ratio. Increasing $d_{11}/d_\perp$ ($d_{11}/d_\perp > 1$) gives long rod-like crystallites aligned along the c axis so that λ_{11} approaches the values for long rods of square cross section. The conductivity $\lambda_\perp$, however, is the value for short rods of length $d_\perp$ and rectangular cross section with the wider side face oriented parallel to the c axis. The resulting anisotropy ratio, $\lambda_{11}/\lambda_\perp$, thus exceeds unity even for elastically isotropic samples, but for all 3 CAP graphite samples exceeds the isotropic values by more than a decade. Conversely, decreasing $d_{11}/d_\perp$ ($d_{11}/d_\perp < 1$) gives very thin plate-like crystallites aligned along the c axis, decreasing λ_{11}. The conductivity, $\lambda_\perp$, however, is the value for short rods of length $d_\perp$ and rectangular cross section but with the wider side face oriented perpendicular to the c axis. The predicted $\lambda_{11}/\lambda_\perp$, as expected, is less than unity for isotropic samples, but doesn't drop below unity in all CAP graphite samples until $d_{11}/d_\perp$ falls below 0.15.

When only a fraction, f, of the incident phonons are absorbed on the crystallite boundaries one must consider specular reflection and transmission through crystallite boundaries. If each {c} face transmits (1-f) phonons, the effective phonon mean-free path Λ_{11} will be increased, giving an increase in $\lambda_{11}/\lambda_\perp$. Conversely, if each {c} face reflects (1-f) phonons then Λ_{11} will be decreased, giving a decrease in $\lambda_{11}/\lambda_\perp$. Similarly, if all {$c_\perp$} faces transmit (1-f) phonons the effective phonon mean-free path $\Lambda_\perp$ will be increased, increasing $\lambda_\perp$. Reflection from {c} faces and transmission through {$c_\perp$} faces thus gives a significant reduction in the ratio $\lambda_{11}/\lambda_\perp$ for f = 0.5* (see curves 5 and 6, Fig. 5).

If specular reflection/transmission as discussed does indeed occur for ballistic phonons in CAP graphite then a lowering of the temperature should decrease the phonon absorption coefficient, f, as the crystallite boundaries appear smoother to the lower frequency phonons. This in turn causes a decrease in $\lambda_{11}/\lambda_\perp$ with decrease in temperature if reflection rather than transmission occurs at {c} faces. Indeed a significant reduction in $\lambda_{11}/\lambda_\perp$ was observed in the measurements of Klein and Holland [5] as the temperature was lowered. However, the sensitivity of C_{44} to dislocations, and the sensitivity of the phonon focusing property to the value of C_{44} makes further comparison of these results with reported experiments difficult. It is clear, however, that effects of phonon focusing are very strong, and must be considered in any detailed analysis of the low-temperature thermal conductivity of graphites.

———————————

*Because phonon velocities along $c_\perp$ are much larger, the {$c_\perp$} faces should appear smoother than the {c} faces, with a resulting smaller absorption coefficient.

REFERENCES

[1] McCurdy, A.K., "Phonon Focusing and Phonon Conduction in Hexagonal Crystals in the Boundary-Scattering Regime", Phys. Rev. B9, 466-480 (1974).
[2] McCurdy, A.K., Maris, H.J., and Elbaum, C., "Anisotropic Heat Conduction in Cubic Crystals in the Boundary-Scattering Regime", Phys. Rev. B2, 4077-4083 (1970).
[3] McCurdy, A.K., "Phonon Conduction in Elastically Anisotropic Cubic Crystals", Phys. Rev. B26, 6971-6986 (1982).
[4] Gauster, W.D., and Fritz, I.J., "Pressure and Temperature Dependences of the Elastic Constants of Compression-Annealed Graphite", J.A.P. 45 3309-3314 (1974).
[5] Klein, G.A., and Holland, M.G., "Thermal Conductivity of Pryolytic Graphite at Low Temperatures. I. Turbostatic Structures", Phys. Rev. 136, A575-A590 (1964).

SESSION J

Measurement Apparatus

SESSION CHAIRMAN

R. E. Taylor
TPRC/Purdue
W. Lafayette, IN.

A GUARDED HOT PLATE APPARATUS FOR EFFECTIVE THERMAL

CONDUCTIVITY OF INSULATIONS AT 80-360 K

David R. Smith

Department of Physics
South Dakota School of Mines and Technology
Rapid City, South Dakota 57701

Lambert J. Van Poolen

Department of Engineering
Calvin College
Grand Rapids, Michigan 49501

ABSTRACT

This report describes modifications made to a guarded hot plate apparatus located at NBS/Boulder and used to measure materials having very low thermal conductivity, such as glass-fiber insulations. Measurements can be performed at temperatures from 80 K to 360 K, and from atmospheric pressure to a vacuum of 10^{-4} Pa. Various fill gasses (air, nitrogen, argon or helium) can be used. Overall uncertainties in thermal conductivity at atmospheric pressure are estimated to be 1% at the higher temperatures and 5% at the lower temperatures.

INTRODUCTION

Standard Reference Materials (SRM's) are needed to measure thermal conductivity and an accurate apparatus is required to establish such SRM's. The guarded hot plate (GHP) apparatus is the accepted standard apparatus. This paper describes an effort made to establish such an apparatus of high accuracy.

Modifications were made to a commercial guarded hot plate apparatus to obtain the accuracy and precision needed for the establishment of SRM certification data on thermal insulation specimens. The goal of the measurement program was to establish thermal transmission data within $\pm$ 1% for insulations over the temperature range 80 K to 360 K (-190°C to 90°C). This uncertainty is considered to be approximately state of the art.

BRIEF DESCRIPTION OF APPARATUS

The apparatus is of standard design [1], with a central stack of specimens and heaters 20 cm (8 in) in diameter and of circular cross-section (Fig. 1). The stack consists of a guarded-main-heater assembly sandwiched between two specimens, two auxiliary heater (AH) plates and two cooling plates. The AH's provide for additional flexibility in controlling the temperatures of the cooler faces of the specimens. The annular portion of the main-heater/inner guard (MH/IG) heater combination contains the inner guard heater element, which surrounds the 10 cm main heater (MH). An electronic temperature controller powers the inner guard. The plate surfaces are anodized black to increase the emittance, thereby promoting radiative heat transfer to the specimens to simulate common in-use conditions. The stack is isolated from external influences by loose-fill insulation, an outer guard, a fluid-cooled shroud and an environmental chamber (Fig. 2). The purpose of the stack and associated guarding is to create and measure steady-state one-dimensional heat flow under a set of defined conditions (temperature differences and thicknesses of the specimens). From these measurements an effective thermal conductivity may be calculated using Fourier's Law.

The MH section of the MH/IG plate provides the measured quantity of heat which flows axially through the specimens to the lower temperature surfaces provided by the AH plates. The temperatures of the AH plates and guard heaters are electronically controlled to maintain stable temperatures for the specimen surfaces. The inner guard, outer guard, shroud and insulating fill minimize the radial heat flow from the metered area. The cold plates intercept the heat from the heaters and are maintained at lower temperatures by means of water, chilled alcohol or liquid nitrogen (LN). In connection with the use of the AH plates this allows a wide range of choices for the median temperatures and temperature gradients in the specimens.

MODIFICATIONS

Stack Elements

The thermal unbalance signal from the inner guard heater to its controller was originally produced by a Type-K 10-pair differential thermocouple (DTC). This thermopile was constructed of wires

BP(BP') = baseplate
CI(CI') = cold plate offset insulation
MS(MS') = composite spacers
F(F') = coolant fluid lines
BCP = bottom cold plate
HI(HI') = heater offset insulation
BAH = bottom auxiliary heater
S(S') = specimen
SS(SS') = specimen thickness spacers
MH/IG = main heater/inner guard
G = gap between MH and IG
SR = strain relief
TAH = top auxiliary heater
TCP = top cold plate
OG = outer guard
OGH = outer guard heater
OGS = outer guard supports

Approximate Overall Dimensions:
BP to BP' = 30cm
Diameter of OG = 24cm

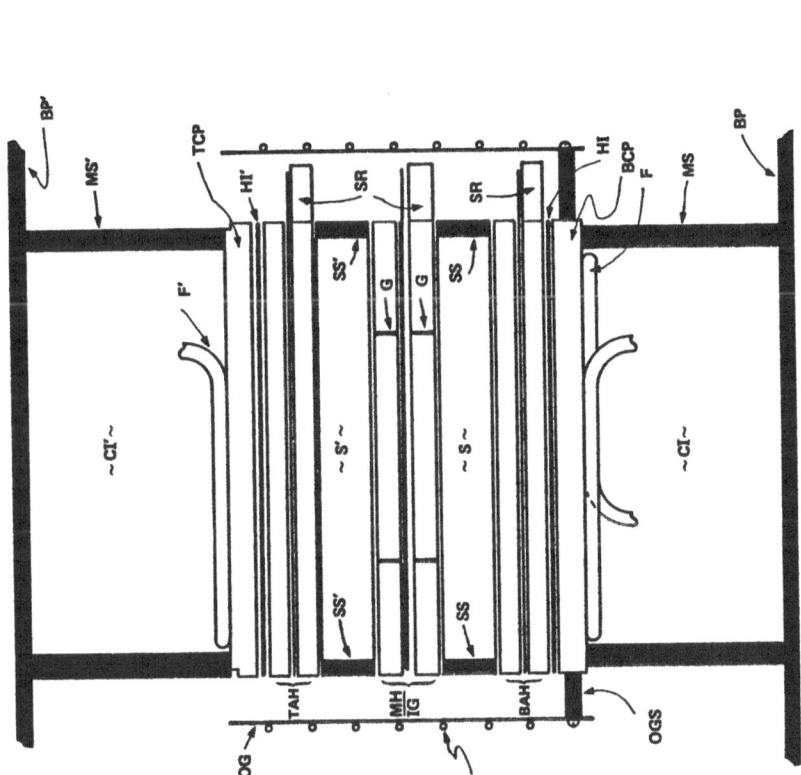

Figure 1. Guarded Hot Plate Stack

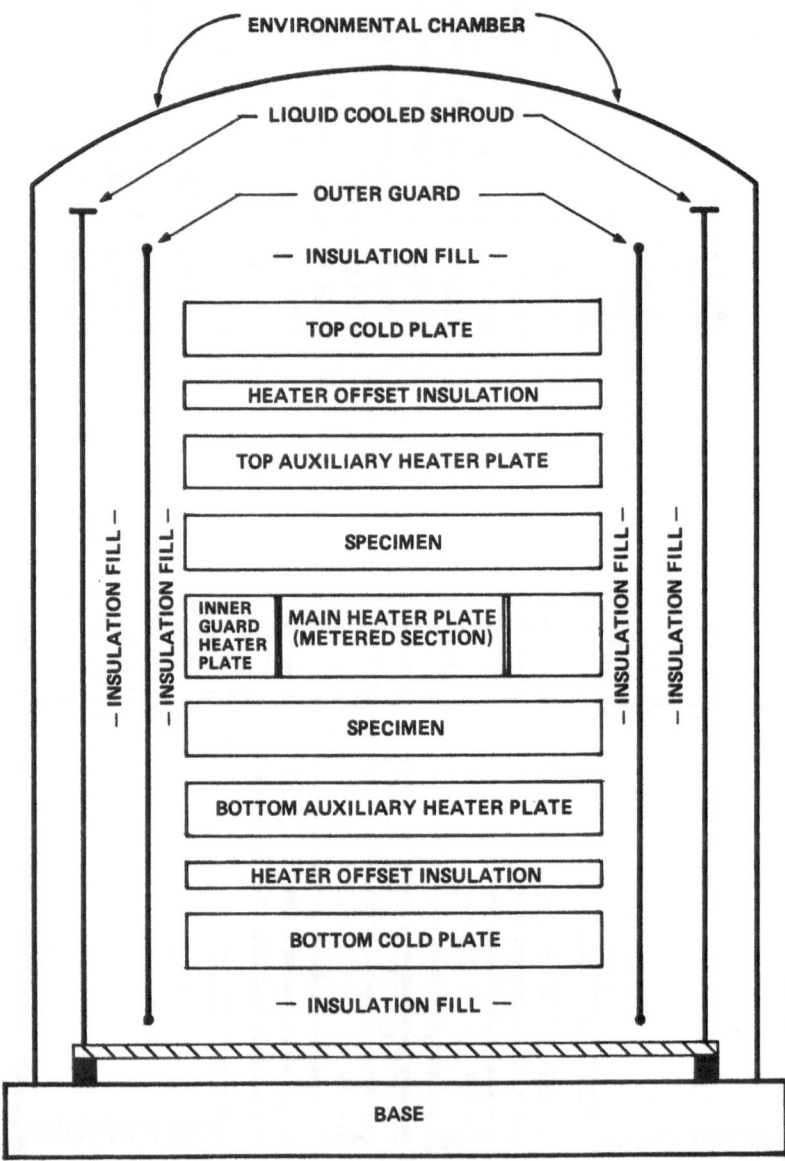

Figure 2.   Guarded Hot Plate System

of small diameter to minimize heat leaks across the gap between the inner guard and MH. Increasing the number of pairs of junctions in the DTC to 20 and changing to Type E wire increased the sensitivity of the unbalance signal by approximately a factor of four.

The DTC junctions are cemented to, but electrically isolated from, the MH/IG plates. One half of the junctions are on the upper MH/IG plates above the heater element and the other half are on the lower plates. This placement is chosen to average the unbalance signal of the top and bottom plates to zero. To minimize induced signals this thermopile is noninductively configured. The sensitivity of this thermopile is about 1200 μV/K at a temperature of 80 K. An average thermopile unbalance of less than about 5 μV output is readily maintained, corresponding to a 4 mK MH/IG unbalance at room temperature.

The commercial version of the apparatus employed thermocouples (TC) enclosed in twin-tube ceramic insulators laid in grooves machined in the surfaces of the heater plates. This practice tends to give poor thermal contact between the TC and its associated heater plate because the TC is not bonded to its insulator tube.

TC's were cemented with a refractory cement of high thermal conductivity into the grooves of the heater plates. The wires near the TC beads were raised to the surface plane of the plate.

To provide further thermal tempering of the TC leads to the MH/IG surface, cylindrical copper lugs were bolted to the edge of the MH/IG plate. The exiting TC leads were wrapped around and cemented to the lugs. Heat conduction along the leads external to the stack is thus diverted to the inner guard, reducing the influence of this heat on the metered area.

Early measurements suggested the presence of a significant heat leak to the stack along the heavy heater leads. These leads were replaced by coiled, lighter leads wrapped on copper tempering lugs mounted on the outer guard. This minimizes the heat leak to the stack by reducing the temperature gradient along the adjacent leads and diverts most of the remaining heat leak to the outer guard.

Thermocouples are mounted on the top and bottom AH's in a fashion similar to that used in the MH/IG assembly. For temperature control two resistive temperature sensors were made of high-resistance alloy wire, wound noninductively. The sensor coils were cemented into a diagonal groove machined into the inner surface of each AH. The sensors have a sensitivity of about 3.8 Ω/K over the entire temperature range. Each resistive sensor forms one resistance element in a Wheatstone Bridge measuring circuit. These sensors are independent of ice reference baths and are not subject to drift of more than 1 mK/h.

For efficient operation at temperatures significantly cooler than ambient, each cold plate is mounted on three composite spacer rods with a length of 9 cm. Fibrous glass insulation was packed into the space between the cold plates and baseplates.

Secondary Guarding

To improve the guarding of the stack, the light-weight outer guard was made of sheet copper half-shells bolted together and weakly heat-sinked to the bottom cold plate. The outer guard bears copper tempering lugs for the leads to the stack. A temperature controller maintains the temperature of this outer guard at the mean temperature of the specimens, using a Type E DTC to compare the outer guard's temperature to that of the stack.

Thermocouple Terminal Strips

The thermocouple terminal strips were moved from their original location to the lower baseplate that support the environmental chamber; this baseplate is directly connected to a massive support ring at or near room temperature. This position for the terminal strips minimizes gradients between TC junctions and along the un-calibrated TC extension wires.

Fluid Transfer

The shroud fluid lines were connected in series with the cold plate coolant lines. The coolant leaving the stack is then efficiently used to cool the shroud, which is the largest cooled element intercepting inflowing heat.

The stainless-steel shroud was mounted on a band of epoxy-fiberglass composite to provide thermal insulation between the shroud and its baseplate. The resulting elevation of the shroud permitted the use of a larger volume of insulation fill (expanded mica) within the shroud, thereby reducing usage of liquid nitrogen during operation. Sliding doors were constructed in the composite band under the shroud to permit easier removal of the insulation during disassembly.

The glass bell jar that was originally the environmental chamber was replaced by a large, metal vacuum-insulated dewar. It provided additional guarding from ambient temperatures and conserved cryogenic coolant when very low temperatures were being maintained at the cold plates.

The cold-plate entrance line is a double-walled vacuum-insulated stainless-steel tube assembly with a built-in pump-out valve.

Double-walled vacuum-insulated construction was continued in the fluid line from the bottom cold plate to the top cold plate to minimize heat leak to the coolant. A vacuum-insulated flexible line with internal teflon spacers was used to bring coolant to the top cold plate that is suspended from the movable top support plate.

A bayonet-type interconnection was constructed at the entrance to the external coolant line to permit rapid change of the LN dewar and minimize the heat leak into the transferred LN at the interconnection. The transfer line from the LN storage dewar also has a built-in pump-out valve. All three vacuum-insulated lines are pumped out to an insulating vacuum before beginning LN transfer for operation of the apparatus at cryogenic temperatures. The bayonet-type connection is sized to accept ordinary (3/8" OD) soft copper tubing into the O-ring seal when water or chilled alcohol cooling is desired.

Temperature Control

After a considerable amount of diagnostic testing, drift problems with the temperature controllers were traced to: a) electronic ice-point thermocouple reference devices, b) unstable and temperature-sensitive operational amplifiers in the low-signal-level circuitry of the temperature controllers, c) excessive spurious emf's generated at TC junction blocks and switches, and d) an excessively high instantaneous power delivery to each of the heaters, precluding the continuous low-power operation required for optimum control-point stability.

Extensive changes were made. Four on/off power switches provide independent operation of each controller. The need for a constant-temperature reference point for the thermocouples was eliminated by use of either differential thermocouples or resistive temperature sensors in a Wheatstone Bridge circuit as described earlier. A power-control circuit was added (using rheostats) to allow regulation of the instantaneous power delivered to the heaters, and a monitor circuit was installed using neon lamps in parallel with the control heaters so the operator can qualitatively gauge from the flash rate the average power being delivered.

A commercial nanovoltmeter of high stability was used in the input circuit between each electronic controller and its control thermocouple. The nanovoltmeter output was connected directly to the printed circuit card of the controller at a high-signal-level point bypassing the controller's temperature-sensitive operational amplifiers that were subject to drift.

After the above modifications were made, tests showed marked improvement in the temperature stability of the system. The long-term (24 h) stability of the current system as measured by the primary thermocouples is 0.1 K. The short-term (1 h) stability is 0.02 K. These instabilities include not only the effect of the controller instabilities but also the main heater power supply fluctuations, spurious thermocouple and switch emf's, and read-out instrumentation variation.

Selector Switches

To read TC emf's and MH voltage or current values, two double-pole rotary switches specifically designed for microvolt-level measurements are used. Experience has shown that these switches, when mounted in an isothermal environment, are noise-free to well below 1 μV.

Measurement

The d-c power supply for the main heater was replaced by a commercial high-accuracy voltage-current calibrator with a specified stability of 0.0025% of setting or 0.005% of range (whichever is greater) per hour after 2 minutes of constant current output, when operated in the constant-current mode.

To reduce the uncertainty in measurement of power to the main heater the circuit contains a 1-ohm temperature-controlled standard resistor in series with the heater for determining the heater current. The heater voltage is measured directly with a digital voltmeter (DVM) having a high imput impedance and a resolution of 0.01 μV. With this arrangement both the heater voltage and the voltage across the standard resistor are high enough that spurious thermal emf's are negligible, eliminating the need for forward and reverse readings.

Analysis indicated that the measured current supplied to the main heater is accurate to within 0.01%, including power supply variations and all measurement uncertainties. This is equivalent to 0.02% uncertainty in power to the main heater.

Type K TC's are used to measure all critical stack temperatures and temperature differences. One spool of Type KP and one of Type KN thermocouple wire were reserved for use exclusively in this system and a Type K thermocouple made of wire from them was calibrated using liquid He, LN, dry-ice/alcohol, water/ice, room-temperature water and boiling-water baths.

One cause of uncertainty in temperatures measured by the thermocouples is spurious thermal emf's generated by nonisothermal conditions at terminal blocks within the environmental chamber. The terminal strips were mounted on heavy aluminum angle stock to provide an isothermal region. Additional aluminum blocks were used in conjunction with the TC terminal strips to temper the TC leads entering and leaving the terminal strips. The TC wire was laid in grooves milled into the opposing surfaces of two blocks. The wires are compressed between two layers of electrically insulating film bonded to the blocks with rubber cement. Silicone grease is used to hold the leads in place during assembly as well as for thermal contact between the wires and blocks.

In the control console the TC wires were brought to a terminal strip mounted on a heavy metal plate thermally insulated inside a heavy metal enclosure. The TC circuits then continue to an ice reference bath, where they are joined to thermocouple-grade copper wires. The copper wires connect to the TC emf selection switches. Within the ice bath the wires are wrapped around copper tubes over a length of 20 cm for thermal tempering. The wires and copper tubes are divided into three groups and placed in glass tubes filled with pure mineral oil. Rubber stoppers at the tops of the tubes keep the wires from being pulled out and prevent ice from entering the tubes.

Two additional thermocouples are used to gain information on conditions outside the stack. One is mounted on the fluid line between the TCP and the shroud to indicate the coolant temperature at this point. The other is mounted where it detects the cabinet temperature; a drift in this temperature may affect the performance of the enclosed electronic control units.

Further details on this apparatus are found in an NBS Internal Report [2].

RESULTS

Preliminary tests on the original apparatus produced results which varied by as much as ± 15% due to thermal effects at various regions within the measurement system. The improvements described here reduced the scatter and uncertainty to about one-fourth their original values and also provided the operator with an increased capability to detect improper functioning of the system.

Measurements on glass fiber board and blanket specimens using this apparatus are reported elsewhere [3,4].

SUMMARY

A commercial guarded-hot-plate apparatus was extensively modified: (a) the sensitivity of the differential thermocouple controlling the inner guard to the main heater was increased; (b) much care was taken to thermally temper thermocouples to the temperatures they are intended to measure and to reduce spurious heat leaks along wires: (c) thermocouple circuits were made less susceptible to effects of thermal gradients; (d) fluid lines were vacuum-insulated to allow efficient use of liquid nitrogen in measurements at low temperatures; (e) electronic temperature controllers were made more sensitive to temperature unbalances at controlled regions and less susceptible to control-point drift due to thermal effects on electronics; (f) metered power was made more stable and more accurately measurable.

This apparatus is very versatile, being usable over wide ranges of temperature and fill-gas pressure. It has been used to assist in establishing insulation SRM's and in studying the importance of various heat transfer mechanisms (radiation, gas conduction and solid-solid conduction) in fibrous insulation [5].

REFERENCES

[1] "Standard Test Method for Steady-State Thermal Transmission Properties by Means of the Guarded Hot Plate", C177-76, 1981 Annual Book of ASTM Standards Part 18, Thermal Insulation, etc. pp. 20.

[2] "A Guarded-Hot-Plate Apparatus for Measuring Effective Thermal Conductivity of Insulations between 80 K and 360 K", NBS1R 81-1657, National Bureau of Standards, U. S. Dept. of Commerce, Boulder, CO. (1982).

[3] Smith, D. R., and Hust, J. G., "Measurement of Effective Thermal Conductivity of a Glass Fiberboard Standard Reference Material", Cryogenics 21/7, 408-410 (July 1981).

[4] Smith, D. R., Hust, J. G., and Van Poolen, L. J.,"Measurement of Effective Thermal Conductivity of a Glass Fiber Blanket Standard Reference Material", Cryogenics 21/8, 460-462 (Aug. 1981).

[5] Van Poolen, L. J., Hust, J. G., and Smith, D. R., "A Model of Apparent Thermal Conductivity for Glass-Fiber Insulations", Thermal Conductivity 17, J. G. Hust, ed., (Plenum, NY 1983), pp. 777-788.

AN AUTOMATED THERMAL CONDUCTIVITY PROBE AND APPLICATIONS TO POWDERS*

William D. Drotning

Sandia National Laboratories
P. O. Box 5800
Albuquerque, NM   87185

ABSTRACT

A thermal conductivity probe has been developed for measurements of powders and porous media from ambient temperature to 1300 K in vacuum or in an inert gas atmosphere. Automated data acquisition and graphical analysis programs have been developed in FORTRAN for use with a laboratory minicomputer. Nonlinear data analysis techniques are available which can account for the probe response due to the effects of contact resistance and the thermal mass of the probe. Programs have also been developed with terminal BASIC for use with a small, portable desktop computer system. Examples of the use of the probe and data acquisition and analysis system are shown for measurement of the effective thermal conductivity of ceramic and metal powders.

INTRODUCTION

The transient probe technique generally allows one to determine more rapidly the thermal conductivity of materials than do steady state methods. In addition, the probe method usually requires a smaller sampling volume and less measurement equipment and is more suitable for nondestructive in situ measurements. As a result, the probe technique is often employed for use in a portable thermal conductivity measurement device for remote applications. However, there is generally a lack of portable automated instrumentation for the acquisition and immediate analysis of probe data.

---

*This work performed at Sandia National Laboratories supported by the U.S. Department of Energy under Contract DE-AC04-76-DP00789.

247

In this report, we describe a probe system for the measurement of thermal conductivity which has been used for powdered materials. The probe designed for this work has been used from ambient temperature to 1300 K in both vacuum and inert gas atmospheres. The data acquisition and analysis system uses a desktop microcomputer, which allows the entire system to operate as a portable, stand-alone instrument. Applications of the probe system are provided which show measurements of the thermal conductivity of ceramic and metal powders.

ANALYTICAL BACKGROUND

The reader may consult additional references for analytical background in the measurement technique [1-3]. Basically, the thermal conductivity probe is based on the transient line source technique, where heat is supplied to the test material at a constant rate along the axis in a cylindrical geometry. The thermal response of the probe as a function of time is used to determine the effective thermal conductivity of the test material.

We have used two analytical methods to analyze probe data in this laboratory. An integral solution, due to Jaeger [4], describes the probe temperature rise as a function of time through the entire time period of the probe response. This solution includes the effects of the thermal mass of the probe and sample material, thermal contact resistance at the probe-sample interface, and the finite dimensions of the probe; an infinite probe thermal conductivity is assumed. Our implementation of this solution follows the work of Koski and McVey [2], which is useful when only short time data are available.

The second solution which has been employed provides a simpler, but approximate, analytical method. It has been shown [4] that after an initial time period during which probe thermal mass and contact resistance effects are important, the probe temperature rise is linear with the natural logarithm of time. The slope of the temperature rise is then proportional to the probe heater input power and inversely proportional to the effective thermal conductivity of the material. When the linear temperature rise is fully developed, this so-called straight line method is readily employed as the basis for probe data analysis techniques.

EXPERIMENTAL DETAILS

Thermal conductivity probes were built incorporating constantan heater wire for use to temperatures approaching 1300 K with minimum electrical resistance change with temperature. Such a probe is depicted in Fig. 1. The constantan heater comprises a dual

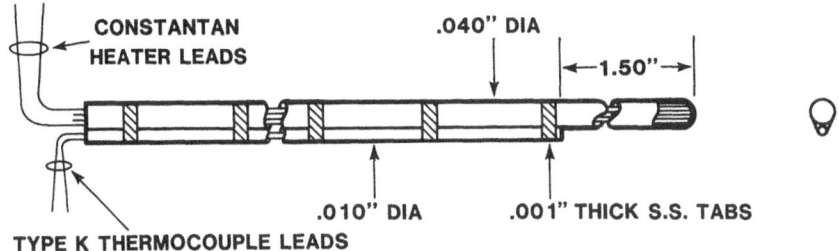

Fig. 1.   Construction detail of high temperature constantan probe.

chromel-constantan (type E) replacement thermocouple with the two
junctions grounded to the stainless steel external sheath [5]
(e.g., Omega Engineering, Inc., part number Dual SCXSS-040G RP).
The constantan legs of the two thermocouples are used to provide a
continuous constantan heater element. The heater leads are surrounded by a refractory insulation inside the 1.0 mm (0.040 in.)
O.D. 304 stainless steel sheath. Depending on the application,
probe heaters of lengths from 30 to 66 cm (12 to 26 in.) have been
constructed in our laboratory.

The temperature sensor consists of a type K (chromel-alumel)
thermocouple in a 0.25 mm (0.010 in.) stainless steel sheath (e.g.
Omega P/N SCASS-010G), welded to the probe heater using stainless
steel tabs, as shown in the figure. To a first approximation, the
attachment of the temperature sensor modifies the probe radius, but
the radius is not employed in the calculation of the conductivity.
As shown later, this probe gives results in good agreement with results from a different measurement technique. These probes typically
exhibit resistance of the order of 0.4 ohms per cm of length. The
figure shows the temperature sensor junction attached 3.8 cm (1.5
in.) from the heater end, for use on the axis of a 75 mm (3 in.)
cylindrical crucible. In this application, the length to diameter
ratio of the probe is approximately 65 for minimal errors due to
axial heat loss.

We have successfully used the high temperature constantan
probes in inert gas environments to 1300 K. The probe is typically
placed on the axis of ceramic cylindrical crucibles, 75 mm long x
35 mm inside diameter. Once filled, the crucible and probe are
placed in a high temperature tube furnace, typically 40 cm in length
and 7 cm I.D. The constantan probe extends beyond the sample material and crucible to regions outside of the furnace. A small correction (typically less than 5%) is applied which accounts for the
variation in heater resistance with temperature due to the temperature gradient existing along the probe [6].

A diagram of the portable thermal conductivity probe system is shown in Fig. 2. The hardware components comprise the Hewlett-Packard HP-85 Personal Computer [5], the HP 3497A Data Acquisition/Control Unit (referred to as a DVM/scanner), an HP 6266B Constant Voltage DC Power Supply, a probe switching system of our own design, and the thermal conductivity probe. The entire system requires not more than 56 cm (22 in.) of vertical space in a standard electronics rack.

The constant voltage DC power supply provides power to the resistive load selected by the probe switching system. The probe switching system provides a means of switching DC power to the probe, either manually or by an external contact closure, in this latter case provided by computer control from the 3497A. When the probe is off, power is supplied to a dummy load whose resistance is matched to the probe; this allows power supply stabilizaton at the appropriate load.

The HP 3497A DVM/Scanner is used for analog input, analog-to-digital conversion, and digital output. Hardware thermocouple compensation is provided by an external electronic ice point reference. The probe thermocouple voltage, probe heater voltage, and probe heater current are measured on three scanner channels for analog input. Analog-to-digital conversion in the 3497A is accomplished by the 5 1/2 digit DVM , with 1 microvolt resolution. The DVM/scanner and HP-85 are connected by the HPIB interface bus. The contact closure necessary for operation of the probe switching system is provided by a digital actuator card in the 3497A.

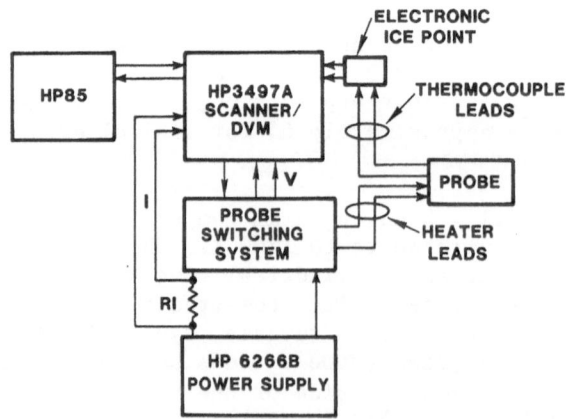

Fig. 2. Block diagram of thermal conductivity probe measurement system.

DATA ACQUISITION AND ANALYSIS SYSTEMS

We have developed two independent data acquisition and analysis systems for use with the thermal conductivity probe. In our laboratory, a Hewlett-Packard HP 1000 minicomputer system is used for data acquisition, graphical data analysis, and nonlinear data fitting of thermal conductivity probe data. For this computer, software was developed in FORTRAN. This minicomputer has sufficient computational size and speed to allow the use of the Jaeger integral solution for data simulation and nonlinear parameter estimation techniques. Software documentation and examples are provided in Ref. [7].

For a portable, stand-alone probe system, data acquisition and analysis software was developed in BASIC for use with the system depicted in Fig. 2, incorporating the straight line analysis method. The HP-85 Personal Computer is used for data acquisition, device control, data storage and retrieval, graphical data analysis, and data presentation. The HP-85 microcomputer consists of the HP-85A Personal Computer, with keyboard, CRT display, cassette magnetic tape, an integral thermal printer, 32K memory, and an HPIB interface (IEEE-488). Additional hardware and software documentation may be found in Ref. [6].

The software developed for the HP-85 for data acquisition and analysis is indicated in Fig. 3. The program uses the special function keys on the HP-85 to direct the operator to the various functional sections of the program. These functional sections include data acquisition (ACQUIRE), graphical data analysis (ANALYZE), and visual readout of the instrumentation parameters (VIEW).

Once loaded and initialized, the program waits for the operator to select one of the desired functions. During this period, the system periodically samples the probe temperature in order to determine a linear approximation to the baseline temperature drift of the probe. This information is available to the operator with the VIEW function once sufficient data are obtained for a least squares baseline determination. VIEW also provides the operator with system parameter information such as time, probe temperature, voltage, current, and resistance of the load (either the dummy load or the resistive probe heater).

The ACQUIRE section is used to acquire probe data. The operator may optionally specify a file name for data storage on cassette tape following data collection. Once acquisition has begun, the system switches power from the dummy load to the probe heater, and readings of the time and probe temperature are recorded. The power (current and voltage) applied to the probe is also recorded. Acquisition of data may be terminated either by operator intervention or by automatic control after a specified run duration. The applied

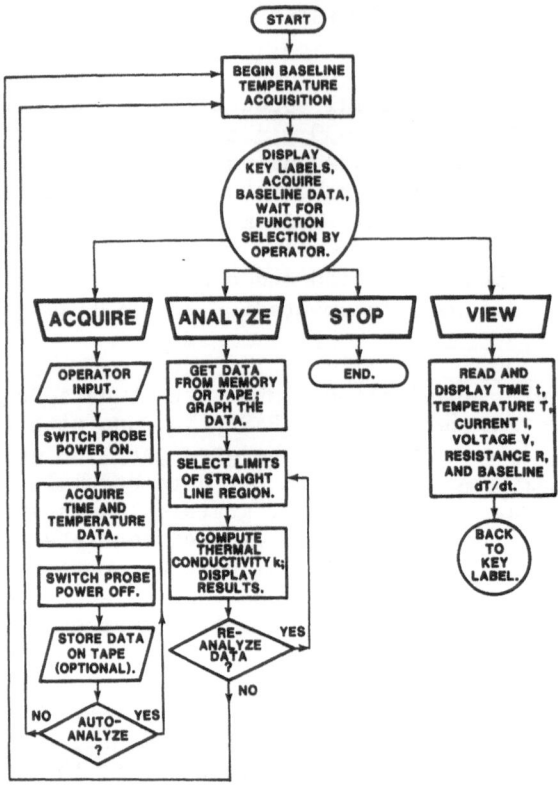

Fig. 3.   Block diagram of probe data acquisition and analysis
software.

power is then switched back to the dummy load. The program either
returns to the function selection menu or proceeds to the analysis
section, depending on the software configuration.

In ANALYZE, acquired data are obtained from storage on tape or
in memory for subsequent graphical data analysis using the straight
line analysis method. The data are plotted on the CRT screen as
temperature vs the natural logarithm of time. The operator selects
the linear region using a movable cursor, a least squares fit is
made to the data, and the thermal conductivity is computed.

APPLICATIONS

Ceramic Microspheres

A series of measurements were made on the thermal conductivity
of alumina microspheres in nitrogen and argon from 25 to 650 C. The

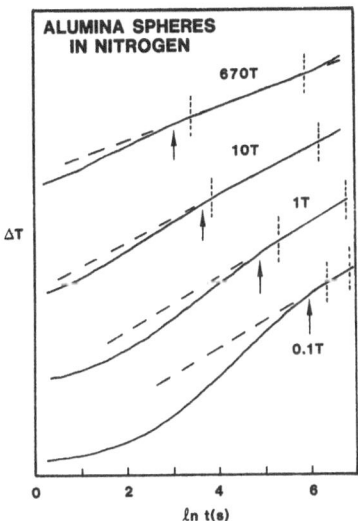

Fig. 4. Probe response curves for alumina microspheres at ambient temperature at various nitrogen pressures. The linear region used in the data analysis is bounded by vertical dashed lines. The arrows mark the inflection region, where the slope of the response curve changes.

microspheres were originally supplied by General Atomic Company to Oak Ridge National Laboratory (ORNL), where the conductivity was measured in nitrogen at 1 atm. using the steady state radial heat flow technique [8]. Subsequently, samples of the microspheres were measured at AERE Harwell by Hall [9], again at 1 atm. nitrogen. McElroy at ORNL supplied our laboratory with samples as well; our results are described in this section.

The alumina (98.9% pure) microspheres ranged in diameter from 495 to 505 micrometers. Measurements were made in argon and nitrogen, for the most part at 670 Torr (1 psig, Albuquerque). The tap density of the microspheres in these measurements was approximately 2.35 g/cm^3, which is about 60% of theoretical density.

The probe response (temperature rise vs ln time) for the microspheres in nitrogen at room temperature is shown in Fig. 4 as a function of gas pressure. The linear regions used in the analysis are bounded by the vertical dashed lines. As the gas pressure is reduced, the effective conductivity of the material decreases, and the contact resistance between probe and specimen increases. The conductivity values are as follows: 0.325 W/m-K at 670 Torr nitrogen, 0.169 at 10 T, 0.066 at 1 T, and 0.031 at 0.1 T. As the contact resistance increases, the time at which the linear region begins (the inflection in the curve) is delayed, as may be observed by inspection of the arrows in the figure. The observed reduction in effective conductivity with decreasing gas pressure is typical of measurements on powder-gas systems [10], and is due to effects of changing mean free path. Prior to the inflection point, a "false" straight line region is evident, particularly at the lower pressures.

Fig. 5 shows the results of measurements on the microspheres at this and other laboratories. The curves were determined from quadratic polynomial fits to the data points using a least squares technique. For the ORNL and Harwell fits, a limited amount of data was available (4 and 8 points, respectively). Throughout the reported range of measurement, the Harwell data are 6-7 % higher than the ORNL values. For an MgO powder, Moore, et al. [10] reported results from the ORNL apparatus to be 20% lower than previous measurements. Our data are in reasonable agreement with the Harwell data to about 400 C. At 600 C, the difference between the results is about 9 %, which suggests some systematic error at the higher temperatures between the different measurement devices. The fit to the conductivity data from the probe technique is described by:

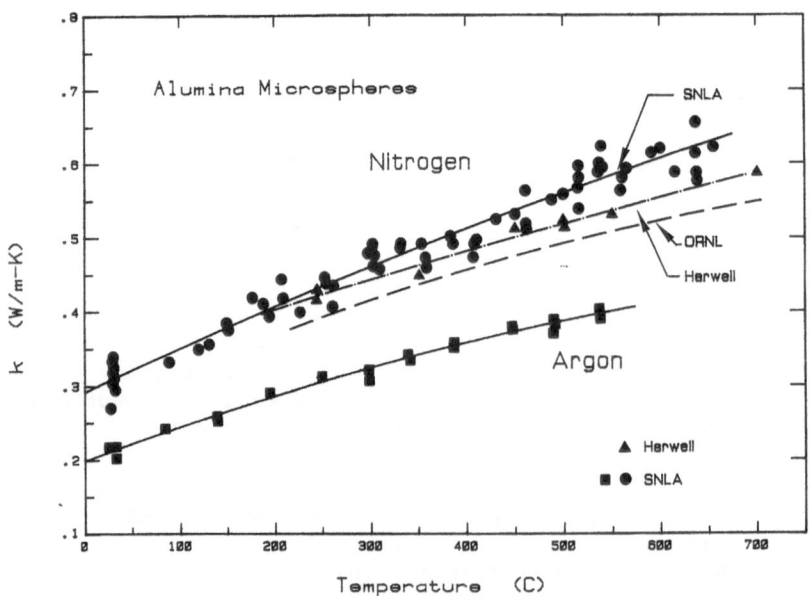

Fig. 5. Thermal conductivity of ORNL alumina microspheres vs temperature in nitrogen and argon at 670 Torr. Curves represent quadratic fits to the data (this work, at SNLA, probe method, as well as ORNL and Harwell data by the radial heat flow method).

$$k \text{ (W/m-K)} = 0.297 + (5.739 \times 10^{-4}) \times T - (1.0714 \times 10^{-7}) \times T^2.$$

for T in degrees C (rms deviation = 0.020 W/m-K). This expression yields a conductivity at ambient temperature (T=25 C) of 0.311 W/m-K in 1 atm. nitrogen. For argon at 670 Torr, the fitted conductivity values give

$$k \text{ (W/m-K)} = 0.200 + (4.80 \times 10^{-4}) \times T - (2.073 \times 10^{-7}) \times T^2.$$

This expression yields a conductivity at room temperature of 0.211 W/m-K (rms deviation = 0.007 W/m-K).

Metal Powders

An example of the probe response in metal powders is shown in Fig. 6. These measurements on Te powders were done at room temperature in air. The solid curve shows the temperature rise vs ln time (in seconds), and the characteristic inflection in the curve (change in slope) near ln t = 4.

Curve S represents a powder of 5 - 20 micrometer spherical Te particles; the packing fraction was 59% (41% void); the measured conductivity was 0.135 W/m-K. Curve A represents a Te powder of 1-100 micrometer irregularly-shaped particles, 50% void fraction, and a conductivity of 0.115 W/m-K. Generally, we would expect the higher density powder to have a higher conductivity, as was measured.

We have also made measurements on a powder of HP-9-4-20 stainless steel spherical particles, 10 - 150 micrometer diameter, in air at ambient temperature. The void fraction was 0.34; the thermal conductivity was 0.390 W/m-K. A 50 w/o mixture of the spherical Te in the steel gave a powder with a density of 4.63 g/cm^3 and a conductivity of 0.265 W/m-K. These results are summarized in the table.

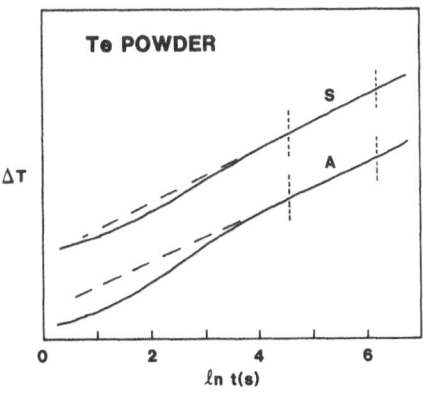

Fig. 6. Thermal conductivity probe response for two Te metal powders.

Table 1. Thermal Conductivity of Metal Powders by the Probe Method,
Ambient Temperature, Air Atmosphere.

Material	Particle Diameter (micrometer)	Powder Density (g/cm^3)	Void Fraction	Thermal Conductivity (W/m-K)
Te, spherical	5 - 20	3.67	0.41	0.135
Stainless steel, HP-9-4-20, spherical	10 - 150	5.20	0.34	0.390
Mixture, steel + Te, 50 w/o	5 - 150	4.63	-----	0.265

SUMMARY

A thermal conductivity probe measurement system was described
which uses a probe capable of operation to 1300 K and a portable,
stand-alone data acquisition and analysis system. Device software
was developed to allow automated acquisition and conductivity
analysis of thermal probe data by the straight line method. Measurements on ceramic and metal powders were presented which demonstrate the use of the system.

ACKNOWLEDGEMENT

We are grateful to D. L. McElroy, Oak Ridge National Laboratory, for supplying the alumina microspheres used in this study.

REFERENCES

1. W. D. Drotning, J. A. Koski, and P. E. Havey, "Development of
a Technique for Measurement of the Thermal Conductivity of
Powders at High Temperatures," SAND81-0631, Sandia National
Laboratories, Albuquerque, NM, August, 1981.*
2. J. A. Koski and D. F. McVey, "Application of Parameter Estimation Techniques to Thermal Conductivity Probe Data Reduction,"
in Thermal Conductivity 17, edit by J. G. Hust, Proc. of the
17th Inter. Thermal Conductivity Conference, June 15-18, 1981,
Gaithersburg, MD (Plenum Press, NY, 1983), p. 587.

3. A. W. Pratt, "Heat Transmission in Low Conductivity Materials," in R. P. Tye, <u>Thermal Conductivity</u>, Vol. 1, Academic Press, NY, 1969.

4. J. C. Jaeger, Aust. J. Phys. <u>9</u>, 167 (1956).

5. Reference to a particular product or company implies neither a recommendation nor an endorsement by Sandia National Laboratories or the U. S. Department of Energy, nor a lack of suitable substitutes.

6. W. Drotning and T. Tormey, "Thermal Conductivity Probe Measurement System," SAND83-2057, Sandia National Laboratories, Albuquerque, NM (to be published).*

7. J. A. Koski, "CONFIT: A Computer Code for Thermal Conductivity Probe Data Reduction With the Use of Parameter Estimation Techniques," SAND82-0741, Sandia National Laboratories, Albuquerque, NM, May, 1982.*

8. D. L. McElroy, Oak Ridge National Laboratory, personal communication.

9. R. O. A. Hall, AERE Harwell, personal communication.

10. J. P. Moore, R. J. Dippenaar, R. O. A. Hall, and D. L. McElroy, "Thermal Conductivity of Powders with UO_2 or ThO_2 Microspheres in Various Gases from 300 to 1300 K," ORNL/TM-8196, Oak Ridge National Laboratory, Oak Ridge, TN, June, 1982.

*Available from: National Technical Information Service (NTIS)
                  U. S. Department of Commerce
                  5285 Port Royal Road
                  Springfield, VA   22161

ANALYSIS OF UNCERTAINTIES IN CALIBRATION

OF A HEAT FLOW METER APPARATUS

M. Bomberg — National Research Council of Canada,
Ottawa, Ontario
C.M. Pelanne — Manville Corporation of U.S., Denver,
Colorado
Wendy S. Newton — Manville Canada Inc., Innisfail, Alta.

ABSTRACT

This paper reviews the main sources of uncertainty in laboratory measurement of thermal resistance with a heat flow meter apparatus over a wide range of testing conditions. It deals with calibration uncertainties in general terms, indicates techniques for controlling instrument precision, analyses the sensitivity of results to some changes in testing conditions, and describes two procedures for establishing transfer standards that permit comparison of absolute and relative measurements.

INTRODUCTION

Errors in measurements with the heat flow meter (HFM) apparatus arise out of uncertainties in electrical and thickness measurements, as well as out of deviations from one-dimensional heat flow and uncertainty in the HFM calibration factor. Bomberg and Solvason [1] have compared various calibration techniques and concluded that the HFM apparatus calibration using transfer standards is the preferred technique. The use of transfer standards (or calibrated specimens) requires, however, that guarded hot plate (GHP) tests be performed on a variety of materials. If the HFM apparatus is to be used for quality control in manufacturing of thermal insulation, the transfer standard should be made of the same material [2], and this may introduce a significant increase in the uncertainties of both HFM and GHP test results.

Errors associated with GHP equipment have already been discussed [3], but the effect of the characteristics of the material under test have not. Precision of the GHP apparatus was shown to be strongly dependent on uniformity of heat flow across the gap between the metering and guard ring areas; and only homogeneous, uniform, "ideal" specimens could be used for experimental determination of the uncertainties in GHP testing.

Ideally, calibration of the heat flow sensor (HFS) should not vary with specimen thickness. Thin, uniform specimens could then be used to utilize, as the calibration base, the high level of precision and accuracy possible with the GHP apparatus. Changes in the design of the heat flow meter apparatus to reduce the effects of lateral heat flow on the calibration of the heat flow transducer have been described [1]. The new design makes the calibration coefficient less dependent on the characteristics of the material being tested and produces the same calibration curve for both thick, low-density specimens and thin, high-density ones. The range of conditions under which materials may be tested accurately is thus enlarged. These aspects will be discussed in the present paper and in another [4] that describes the verification of heat flow meters used for quality control in the production of low-density mineral fiber insulation. Both papers are based on a cooperative study carried out by the Division of Building Research, National Research Council of Canada (NRCC), the Research and Development Division of Manville Corporation, USA, and the Quality Control Group in the Innisfail Plant of Manville Canada Inc.

This paper reviews a selection of materials and test conditions used for calibration and certain aspects of HFM apparatus performance (i.e., measurements of temperature and specimen thickness), and discusses three techniques suitable for examining uncertainties due to testing procedure. All utilize air layers, layered low-density glass fiber specimens, and adjustment of ambient temperature to study the lateral heat flow effect. The importance of material structure in the development of a transfer standard is stressed.

Test Conditions for Calibration of HFM Apparatus

As the precision and accuracy of an HFM test depends on the construction of the apparatus as well as the testing procedure, both factors must be considered in selecting a range of test conditions. At NRCC a 600- × 600-mm HFM apparatus is used to test low-density thermal insulations up to 150 mm thick (or up to 200 mm with somewhat lower test precision). The working range is between 6 and 180 mm, corresponding to a thermal resistance range of 0.2 to 7 m^2K/W.

Material acceptance standards usually require thermal testing at a mean temperature of 24 ±1°C, with a temperature difference of 22 deg (Canada), or 27 deg (USA). To establish the dependence of thermal resistance on mean temperature, tests are normally conducted over a range of mean temperatures from 0 to 50 deg C.

Transfer Standard and Reference Materials

A change in testing conditions, even one so small that it may not be observed by the operator, may produce significant deviations in measured results, depending on the nature of the tested material. Although quality control can select a reference specimen of the same material as that being controlled, an independent testing laboratory must have several reference specimens. At NRCC the following materials are used to establish transfer standards because of their uniformity, time stability, and handling qualities: expanded polystyrene (preferably high density) or, alternatively, air-filled extruded polystyrene; medium or high-density glass fiber, e.g., standard reference materials (SRM) 1450 from the National Bureau of Standards [5]. In addition, NRCC has a set of layered, low-density glass fiber transfer standards to verify testing of materials with a significant fraction of radiative heat transport.

Apparatus Construction

The accuracy of an HFM depends on several factors related to the construction of the apparatus: for example, flatness and parallelism of the hot and cold plates (these factors are described in ASTM C518 [2]).

Temperature Measurement

The following characteristics related to temperature measurement must be checked:

(1) temperature uniformity - uniformity or differences in temperatures measured over the surface of the plates,
(2) temperature stability with time, and
(3) repeatability of temperature measurements.

Temperature uniformity and stability can be determined by adjusting both HFM plates to the same temperature and placing them in contact with each other. The standard deviation for each thermocouple recording will indicate stability, while the range between minimum and maximum measured surface temperatures will indicate temperature uniformity.

The precision of measurement of a temperature difference depends on the temperature difference used in the test. For example, 0.1 C deg means an uncertainty of 1% for a 10-deg temperature difference but an uncertainty of only 0.5% for the usual 22- to 27-deg difference. A good HFM apparatus has, for example, temperature uniformity better than ±0.15°C, temperature stability better than ±0.02°C, and repeatability precision (with readings on DVM) of 1.0%.

Thickness Measurement

If the HFM apparatus is provided with a thickness indicator, the zero position and linearity of the thickness gauge should be checked as well as the parallelism of the two plates enclosing the specimen [4]. At NRCC, specimen thickness is calculated as the average of the distance between plates measured on each side of the apparatus. This procedure includes a check of parallelism as well as an averaging of a number of measurements.

Additional consideration is given to all high precision measurements (such as calibration), namely, comparison of specimen thickness measured inside and outside the HFM apparatus. The two thicknesses are permitted to differ by as much as ±0.2%. If the thickness measured in the apparatus differs by more than this, the test is excluded from the calibration series; if the difference is less than 0.1%, no correction is made; if the difference is between 0.1 and 0.2%, an appropriate correction is applied.

UNCERTAINTIES RELATING TO TESTING PROCEDURE

A laboratory routine for testing must be established, including methods of preparing specimens (drying or conditioning): measure density and thickness of specimen outside and inside the apparatus, estimate time to reach the steady-state condition; record, calculate and report temperatures and heat flux. Then select a material known for uniformity and time stability to establish "a standard repeatability precision," i.e. repeatability obtained in standard testing conditions on the reference material. This basic measure of repeatability precision (two standard deviations on minimum nine test results) may then be compared to the repeatability precision obtained for different materials or changing test conditions, for example, when testing in the range of mean temperature. These concepts may be illustrated by the following: standard repeatability precision related to 90% probability level, i.e., two standard deviations, is approximately 0.2% for high-density glass fiber specimens 344-6 (Table 1) on the 600-mm GHP and for specimens 290-48 on the 450-mm GHP.

Repeatability precision determined on low-density glass fiber (LDGF) specimens, code 354-79, is in the same range as that for the HDGF specimens; for LDGF specimens, code 354-68, repeatability

Table 1. Repeatability Precision of 450- and 600-mm GHP on 26-mm thick Glass Fiber Specimens

| | Specimen | | | | |
Code	Density kg/m^3	Equipment Size, mm	λ Mean Value* W/m K	Standard Deviation %
344 - 6	91 ± 2	600	0.03174	0.10
290 - 48	80 ± 1	450	0.03136	0.10
354 - 79	8.7 ± 0.1	450	0.04817	0.09
354 - 68	10.6 ± 0.1	450	0.04416	0.26

* Last digit shown for calculation purposes only

precision is almost three times poorer. It appears that repeatability precision of the GHP apparatus depends on the specimens selected for testing. From tests on the 600- and 450-mm GHP's shown in Table 1 repeatability precision of the instruments on a 95% probability level appears to be 0.2%. The lower precision apparent in some of the tests must be attributed to the effects of specimen preparation and conditioning, and perhaps to some other material variable.

Effect of Lateral Heat Flow

In testing to determine the significance of lateral heat flow, the same temperature is maintained on both plates and ambient air temperature is constant but much lower. Heat loss from the edge of the specimen can be assumed symmetrical, half from the "hot" plate and half from the "cold" plate. Output from the heat flow transducer, corrected for any small temperature difference across the specimen, will give the edge loss associated with the temperature difference between the two plates and ambient air.

As an example, Table 2 shows the results of a series of lateral heat flow tests conducted at NRCC on a particular apparatus* using 95- and 190-mm thick polystyrene specimens. It is evident that the temperature difference between the temperature-controlled enclosure surrounding the specimen (the sleeve) and the HFM plates causes a measurable output from the transducer (error). It is also evident that the use of edge insulation reduces this error.

* Manufactured by Foundation Electronic Instruments, Ottawa, Canada.

Table 2. Lateral Heat Flow

EPS Spec. Thickness, mm	25-mm Edge Insulation	Temperature, degree C			Output, mV	
		Sleeve	Spec. Edge	Plates*	Meas.	Corrected for ΔT of Plates
190.35	No	–	29.9	36.0	34	37
190.52	Yes	26.9	30.1	36.1	21	16
95.12	Yes	28.0	32.9	36.5	-14	-11

* Both plates maintained at the same temperature

If the lateral heat flow component is less than 0.5% of the average output of the heat flow meter (i.e., 3-5 μV in Table 2), edge insulation should not be required when testing thin specimens. A difference of up to 2 C deg between sleeve temperature and mean specimen temperature may be accepted if the specimen thickness does not exceed 40 mm. Specimens thicker than 60 mm should, however, be tested with edge insulation. For those thicker than 90 mm, the difference between sleeve and mean specimen temperatures should be limited to 1 C deg.

Use of Air Layers

Air layers can be used as test specimens to verify the accuracy of the HFM apparatus when convection of air is eliminated and the thermal conductivity of still air is either known or determined from test results for at least two levels of plate emittance and two or three air layer thicknesses. The latter approach was proposed by Klarsfeld [6] as an extrapolative procedure in which thermal conductivity of still air (λ_g) can be determined. This procedure consists of extrapolation to zero thickness for the following:

(1) apparent thermal conductivity* of an air layer at a given mean air temperature for each value of boundary emittance,
(2) apparent thermal conductivity* of an air layer and known boundary emittance for a given value of mean air temperature.

Figure 1 shows NRCC data for apparent thermal conductivities of three air layers tested with and without aluminum foils (on both

* For brevity, "apparent" is dropped in subsequent discussions. Thermal conductivity of an air layer or of low-density thermal insulations means, however, apparent thermal conductivity.

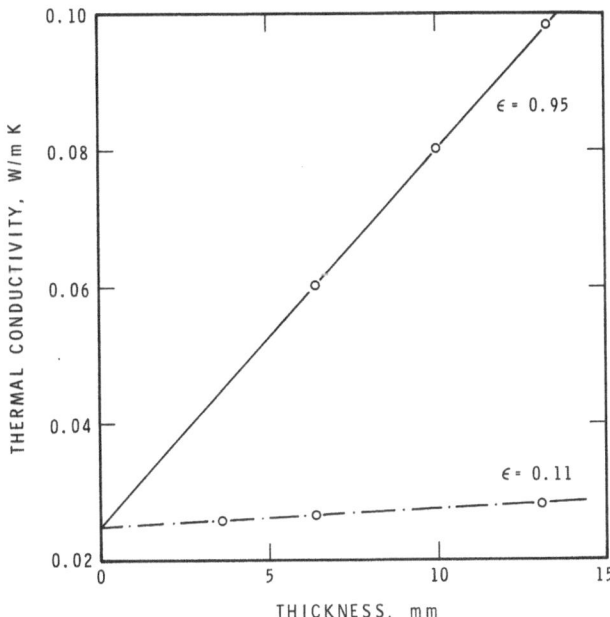

Fig. 1. Thermal conductivities of an air layer, with and without
aluminum foil, extrapolated to determine still-air
conductivity. Measured in 600-mm GHP apparatus

sides of the air layer), at the same mean temperature, and with a
temperature difference of 28 C deg across the air gap [7].
Extrapolating the data to zero thickness, one obtains a still-air
conductivity of 0.0247 W/m K at a mean temperature of 24°C. It is
somewhat lower than the 0.0254 W/m K shown for air at 297 K by
Raznjevic [8] or the 0.0252 W/m K quoted in the Smithsonian
Physical Tables [9]. The level still-air conductivity can then be
used to calculate plate emittance using techniques described by
Hager [10] and Pelanne [11].

In addition to the data shown in Fig. 1, eight other tests on
a 10-mm air gap over a mean temperature range of 0 to 45°C yielded
the following equation:

$$\lambda_a = 0.0652 + 0.0061 \ T_m$$

Data generated from it can be compared with values in Refs. 8
and 9, making this method useful for verifying the temperature
coefficient of the calibration curve.

Use of Layered, Low-density, Glass Fiber Specimens

A technique used at NRCC to fabricate low-density transfer
standards ranging in thickness from 50 to 250 mm is slightly
different from that described in Ref. 4. Using a light table [12],
thirty 12-mm thick, low-density glass fiber specimens were selected
as most uniform among one hundred and thirty. Their densities
varied between 8.2 and 10.0 kg/m^3. To reduce the number of tests
to be performed, pairs of specimens, each separated by a paper
septum to intercept radiative heat transfer, were tested at one
time. The results of the thermal conductivity tests are shown in
Fig. 2. Some of the specimens were recombined in new pairs and
nine pairs were finally selected. The results of thermal
conductivity tests on a glass fiber stack, with and without paper
septa, are shown in Fig. 3.

Layered specimens are particularly useful in examining the
effect of specimen thickness [13,14] on apparent thermal
conductivity measured for a given set of test conditions. The
instrument error has been estimated (another approximative model
has also been presented [15]) by assuming that all test results
with radiation-absorbing barriers (paper septa) placed 25 mm apart
should be equal to the results obtained on 25-mm thick specimens
alone [4]. Although such an approximation appears to be acceptable
for mineral fiber insulations, it has not been proved for low-
density cellular plastics.

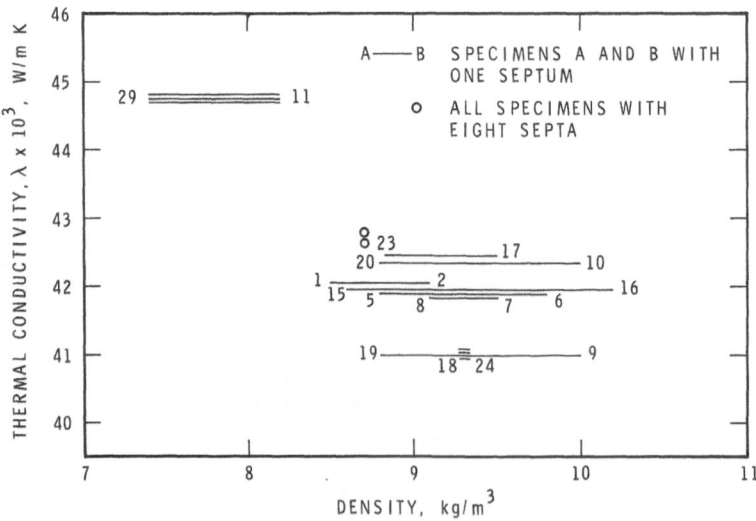

Fig. 2. Thermal conductivities of two specimens separated by paper
septum. Result of preliminary series used for specimen
selection

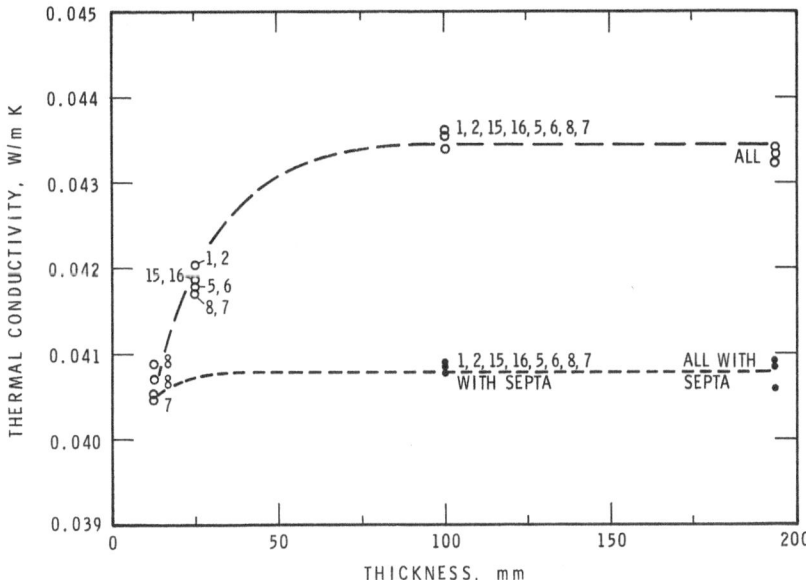

Fig. 3. Thermal conductivity of glass fiber stack with and without
paper septa. Eighteen layers, each 12.5 mm, were tested

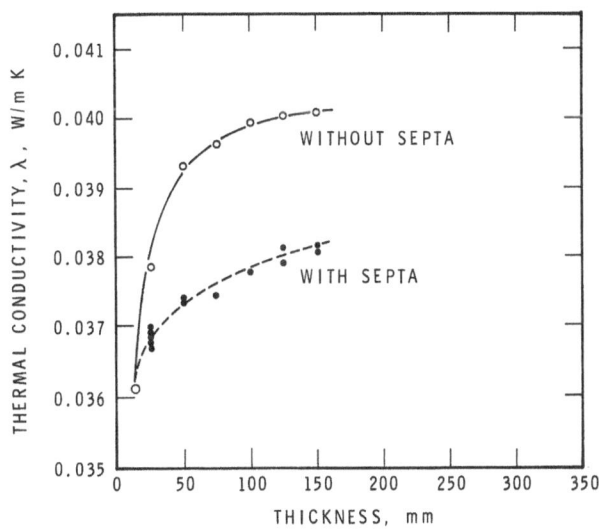

Fig. 4. Thermal conductivity of polystyrene stack with and without
paper septa. Twelve layers, each 12.5 mm, were tested

For comparison, Fig. 4 shows the results of two series of tests on the same batch of expanded polystyrene, density 15–16 kg/m^3, with and without paper septa. The curve of thermal conductivity versus specimen thickness shows a different character from that of Fig. 3, implying a somewhat different mechanism of radiative heat transport. Layered low-density glass fiber insulation appears, therefore, to be more suitable for verification of the HFM apparatus [16,17] than low-density expanded polystyrene insulation, although radiative heat transport is a significant fraction of the total heat transfer for both types of insulation.

Importance of Material Structure

The foregoing discussion of repeatability (reproducibility) precision illustrates the importance of test specimens of uniform material structure. Some low-density glass fiber specimens (e.g., 354-79, Table 1) were shown to have the same precision of repeatability as the high-density glass fiber standard reference material (SRM) produced by the National Bureau of Standards [5], but much lower precision was achieved with other specimens. Use of thick specimens of low-density thermal insulation for calibration purposes usually led to lower precision than could be achieved with high-density, standard reference material [18]. As the precision of thermal conductivity determination is inversely proportional to specimen thickness, an alternative approach would be to fabricate thick specimens from thin layers already tested separately. This can only be done, however, if the material has a high degree of structural uniformity.

Specimens of almost identical density and thermal conductivity were selected at the Innisfail plant [4]. Specimens with thermal conductivity very close to that calculated as a function of density from the model [15] were used for the NRCC tests shown in Fig. 3. In both cases, however, the specimens were chosen with a light table to ensure structure uniformity.

Figure 5 illustrates the importance of structure uniformity on precision, showing the range of measured thermal conductivities as a function of specimen thickness for two sets of specimens of different uniformity:

Set 1 contains six 25-mm thick low-density glass fiber specimens of good uniformity;

Set 2 contains six 25-mm thick specimens of poor uniformity selected as the extremes from the same sample as Set 1.

The variability in thermal conductivity was about 1% for a density variation of 9% in Set 1; a range of 5% in thermal conductivity was observed for a 15% range in density in Set 2. Figure 5 also shows

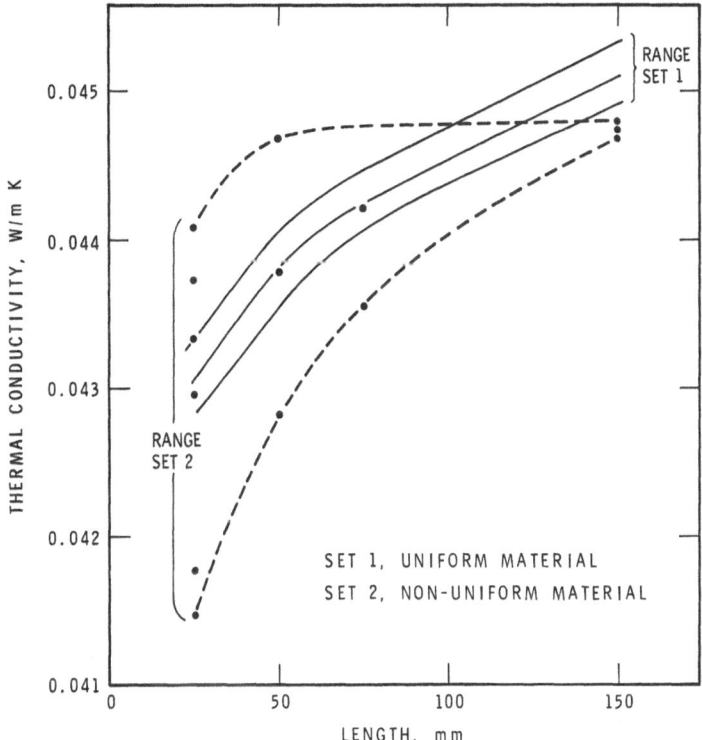

Fig. 5. Range of thermal conductivities as a function of specimen
thickness for low-density glass fiber
Set 1 - six 25-mm thick, uniform specimens
Set 2 - six 25-mm thick, non-uniform specimens

that a different curve of thermal conductivity as a function of
thickness is obtained for each of the sets, although a large and
statistically significant number of tests were performed.

One must remember that properties of a material vary in all
three directions and that the manner in which specimens are
selected for comparative tests may have a decisive effect on the
studied dependence. In previous work [4] specimens with the same
thermal conductivity and density but different thickness were
selected. The relation between thermal conductivity and density of
specimens was also maintained in the final selection of the matched
pairs in the present research (Fig. 2). It was not, however,
observed in selecting Sets 1 and 2 of Fig. 5.

These results are not surprising at all if one also remembers
that for a given fiber diameter the relation between thermal

conductivity and glass fiber specimen density* describes material structure [15]. Comparison of Figs. 3 and 5 shows that in studying functional dependence one must use specimens with identical structure.

CONCLUSIONS

Calibration of a heat flow meter apparatus used for quality control of manufactured products can be extended to a wide range of testing conditions. Air layers and layered, low-density glass fiber transfer standards have been very useful in calibrating and verifying HFM apparatus precision and accuracy.

Lateral heat flow tests performed at the extreme thickness to be tested and determination of the mean temperature dependence of the calibration coefficient are also helpful in establishing the range of uncertainties caused by changes in test conditions.

This and previous NRCC studies [1,3,4] suggest that with proper calibration and careful verification of its errors, the HFM apparatus can replace the GHP apparatus in many industrial and commercial testing programs.

ACKNOWLEDGEMENT

The authors wish to thank J.G. Theriault, R.G. Marchand and Nicole Normandin, of the Thermal Insulation Laboratory, DBR/NRCC, for their contribution to the development of testing facilities and transfer standards and for making the measurements. They also wish to express their gratitude to S. Klarsfeld, C.R.I.R., Isover-St. Gobain, Rantigny, France, to K.R. Solvason, DBR/NRCC, for detailed discussion of the development of transfer standards for calibration and verification of the HFM apparatus, and to J.R. Sasaki, DBR/NRCC, for careful review of the manuscript. Particular thanks are offered to R. Boisvert, Manager of the Manville Plant in Montreal, for preparation of the 130 specimens used for selection and development of NRCC low-density glass fiber transfer standards.

This paper is a contribution of the Division of Building Research, National Research Council of Canada, and is published with the approval of the Director of the Division.

---

* Only if a material is uniform will the specimen density be identical to that of the metering section.

REFERENCES

1. Bomberg, M., and Solvason, K.R., Comments on calibration and design of a heat flow meter; ASTM, STP 789, 277-292 (1983).

2. ASTM C518-1976, Part 18, 1982 Book of ASTM Standards.

3. Bomberg, M., and Solvason, K.R., Precision and accuracy of guarded hot plate method; Proc., 17th International Thermal Conductivity Conf., Plenum Press, 393-410 (1983).

4. Newton, Wendy S., Pelanne, C.M., and Bomberg, M., Calibration of heat flow meter apparatus used for quality control of low-density mineral fiber insulations; Proc., 18th Internatinal Conductivity Conference, Rapid City, SD. (1983).

5. Siu, M.C.I., Fibrous glass board as a standard reference material for thermal resistance measurement systems; ASTM, STP 718, 343-360 (1980).

6. Klarsfeld, S., Letter to the ISO International Thermal Conductance Round-Robin proposing foil-faced frames for air-gap layer study, 20 February 1981.

7. Cammerer, W.F., Experimental determination of the equivalent thermal conductivity of air space at low temperatures; Presented to Committee BI, Int. Institute of Refrigeration, Washington, D.C. (Sept. 1976).

8. Raznjevic, K., Thermal tables and diagrams based on Käl temaschinen Regeln (in Polish), 5th ed., Karlsruhe, Müller Press, 1958.

9. Smithsonian Physical Tables, 9th Ed. (ed. by W.E. Forsythe), Smithsonian Institution, Washington, D.C., Publ. 4169, Vol. 120, 1954.

10. Hager, N.E., Method for measuring total hemispheric emissivity of plane surfaces with conventional thermal conductivity apparatus; 7th Conference on Thermal Conductivity, Gaithersburg, MD, National Bureau of Standards Spec. Publ. 302, 1968.

11. Pelanne, C.M., Experiments on the separation of heat transfer mechanisms in low-density fibrous insulation; 8th Conference on Thermal Conductivity, Plenum Press, 897-911, 1969.

12. Pelanne, C.M., Light transmission measurements through glass fiber insulations, ASTM, STP 660, 263-280, 1978.

13. Albers, M.A. and Pelanne, C.M., An experimental and mathematical study of effect of thickness in low density glass fiber insulation; Thermal Conductivity 17, Proc., 17th International Thermal Conductivity Conference, Plenum Press, 471-482, 1983.

14. Pelanne, C.M., Discussion on experiments to separate the 'effect of thickness' from systematic equipment errors in thermal transmission measurements; ASTM, STP 718, 322-334, 1980.

15. Bomberg, M., and Klarsfeld, S., Semi-empirical model of heat transfer in dry mineral fiber insulations; J. Thermal Insulation, Vol. 6, 156-173 (1983).

16. Pelanne, C.M., The development of low density glass fiber insulation as thermal transmission reference standards; Thermal Conductivity 17, Proc., International Thermal Conductivity Conference, Plennum Press, 763-775, 1983.

17. Pelanne, C.M., Development of a company wide heat flow meter calibration program based on the N.B.S. certified transfer specimens; Forum on the Guarded Hot Plate and the Heat Flow Meter, Quebec City, Canada, 1982. (To be published by ASTM)

18. Rennex, B., Low-density thermal insulation calibrated transfer sample - a description of a discussion of the material variability; National Bureau of Standards, NBSIR82-2538, 1982.

A NEW TRANSIENT HOT WIRE THERMAL CONDUCTIVITY INSTRUMENT FOR USE

WITH BOTH STEP POWER AND RAMP POWER FORCING

R.A. Perkins, R. McAllister, E.D. Sloan
and M.S. Graboski

Colorado School of Mines
Golden, Colorado 80401

ABSTRACT

A new transient-hot-wire thermal conductivity instrument is
described for use with non-electrolytic liquids. The initial
state of the fluid of interest can be varied from ambient to 500 C
and 200 atmospheres. This range of experimental conditions allows
most hydrocarbon liquids to be studied up to either their critical
point or their limit of thermal stability. The hot wire is 12.7
micrometers in diameter and 12 centimeters long. The hot wire is
constructed of platinum, with all electrical connections silver-
soldered, and has been proven reliable in a wide range of liquids
including several unstable coal liquid fractions.

The transient hot wire experiment is controlled and monitored
by a Rockwell AIM 65 microcomputer. The drive voltage to the hot
wire is directly controlled by the computer through a 12-bit digital-to-analog converter. The fact that the drive voltage is
directly under software control provides great flexibility in both
the shape as well as the magnitude and timing of the forcing
function. This has been used to advantage in this instrument to
provide a choice of either a step-power or ramp-power forcing
function. The resistance of the platinum hot wire is monitored
with a 14-bit analog-to-digital converter connected to an amplified Wheatstone bridge system. The typical thermal conductivity
experiment is one second in duration, and consists of 1000
measurements of temperature rise versus time. The data points are
transferred to a mainframe computer over phone lines for detailed
data analysis and permanent storage on magnetic tape.

The performance and accuracy of the instrument have been verified through a study of toluene with both ramp-power and step-power forcing. This provides a check of internal consistency through comparison of step and ramp results, as well as a check of accuracy through a comparison with the transient step-forced, and radiation-free steady-state parallel-plate data of other experimenters. This testing has shown that both the step and ramp forcing functions give comparable results, with an accuracy on the order of one percent. The thermal conductivity data obtained are believed to be nearly free of the effects of radiation due to use of a back extrapolation procedure to obtain the apparent thermal conductivity at zero time. The instrument has been utilized to study meta-xylene, methylcyclohexane, decahydronaphthalene, tetrahydronaphthalene, and 1-methylnaphthalene. In addition, the instrument has been used to study two well-characterized coal liquid materials, an SRC-I naphtha, and a Utah COED fraction. All data are from ambient to the limit of thermal stability, or to the critical point. The effect of liquid-phase compressibility on thermal conductivity is clearly resolvable by this instrument.

BODY

This paper describes work with a new transient-hot-wire thermal conductivity instrument performed over a wide range of temperature and pressure. The current instrument is capable of measuring the thermal conductivity of non-electrolytic fluids from ambient conditions to 500 C and 200 atmospheres. The transient hot wire technique has become popular and reliable in recent years due to the advances of solid state electronics. The only forcing function which has been utilized in these instruments is the step power forcing function. The work described here represents the first attempt to explore the use of other forcing functions.

The use of alternative power forcing functions has been explored theoretically by Mohammadi, et al. (1981). They showed that the line-source heat conduction model has analytic long-time solutions for Dirac delta, step, and ramp power forcing functions. They further demonstrated that the Dirac delta forcing function has very poor sensitivity due to a rapid decay to the unperturbed initial conditions. Both the step power and ramp power forcing functions are shown to have some relative advantages. The line-source solution for the step power forcing function is given by:

$$T(r_w, t) - T_o = \frac{q}{4\pi\lambda} \ln \frac{4\alpha t}{r_w^2 C} \tag{1}$$

274

The line-source solution for the ramp power forcing function is given by:

$$\frac{T-r_w,t) - T_o}{t} = \frac{\hat{q}}{4\pi\lambda} \ln \frac{4\alpha t}{r_w^2 D} \tag{2}$$

In these equations $C=\exp(\gamma)$, $D=\exp(1+\gamma)$, q is the magnitude of the step power change, and $\hat{q}$ is the slope of the ramp power forcing function with respect to time. Additional details on the mathematics of the ramp power and step power forcing function solutions are given by Mohammadi et al. (1981).

The use of the step-power forcing function in an actual transient hot wire has been demonstrated experimentally to provide a fast reliable alternative to tedious steady-state methods for fluids. A description of the first transient hot wire instrument based on a ramp power forcing function is provided by Perkins, et al. (1981). The ramp power forcing function has been shown to provide a viable alternative to the step power forcing function. The instrument described by Perkins, et al. (1981) has been modified to extend its upper temperature limit from 200 C to 500 C, and to enable both step-forced and ramp-forced experiments to be run with the same hot wire and data acquisition system. The fact that either forcing function can be selected at any desired temperature and pressure allows direct comparison of the relative advantages of each forcing function at identical fluid conditions.

The transient hot wire cell consists of a 12.7 micrometer platinum wire which is surrounded by the fluid of interest, and is contained in a stainless steel pressure vessel. The hot wire is 12 centimeters long and is held taut in a vertical position by a pyrex support frame and a platinum spring. The ratio of length to diameter is such that non-ideal effects due finite wire heat capacity and axial conduction are insignificant with liquids with both ramp-and step-power forcing. Two electrical leads are taken from each end of the hot wire to allow four-terminal resistance measurements. The temperature of the pressure vessel is controlled to 0.001 C with an isothermal bronze block. Electrical connections in the pressure vessel are silver soldered for reliability.

A detailed description of the data acquisition system hardware is provided by Perkins, et al. (1981). A detailed description of all aspects of the present instrument is provided by Perkins (1983). The drive power is generated under software control by a 12-bit digital-to-analog converter under the supervision of a Rockwell AIM 65 microcomputer. The fact that the forcing function is under software control allows any desired forcing function to be created as a piecewise continuous set of steps, at any desired power level and timing. It is this versatility in forcing function generation which enables the step versus ramp comparison. The resistance of the platinum hot wire is monitored by the AIM 65 with a 14-bit analog-to-digital converter connected to an amplified Wheatstone bridge which contains the wire as one leg.

A typical data set consists of 1000 measurements of temperature rise obtained for a single one-second experiment. The quantity of data obtained is too large to allow processing with a microcomputer, so the data are transferred over phone lines to a mainframe computer for analysis and storage. The fluid thermal conductivity is obtained from the linear line source solution for both the step and ramp cases. For the step case, a plot of temperature rise versus the logarithm of time should be linear. For the ramp, a plot of temperature rise over time versus log of time should be linear. Figures 1 and 2 show uncorrected data plots for the step-forced and ramp-forced cases. These runs were obtained within minutes of one another at identical temperatures and pressures for toluene. As is evident in the plots, the data are linear over a wide range of the experiment and have minimal noise.

The data sets are analyzed by a regression fit of the local thermal conductivity at incremental times over the course of the experiment, with the constraint that the apparent thermal conductivity is not changing at zero time. The measured apparent thermal conductivity depends on all modes of heat transfer, including convection and radiation; however the zero-time apparent conductivity obtained from the regression is expected to be nearly free from the effects of radiation and convection.

Since toluene has been studied by a wide variety of experimental methods to a high degree of accuracy it was selected in order to compare the step and ramp forcing functions in measurements on liquids. The experimental range extends from ambient to the critical temperature and 200 atmospheres. The ramp-forced data and step-forced data are plotted separately, along with 500 psi isobars obtained from curve fits, in Figures 3 and 4. The effect of liquid-phase compressibility is clearly shown in the data. Approximately 100 experiments were used in the comparison. In order to demonstrate that the reported thermal conductivity is not a function of the forcing function selected, a plot of the deviation between the step and ramp data as a function of thermal conductivity is provided as in Figure 5. No trend is noted other than the fact that the experimental uncertainty increases at higher temperatures (lower thermal conductivity) as the critical point is approached.

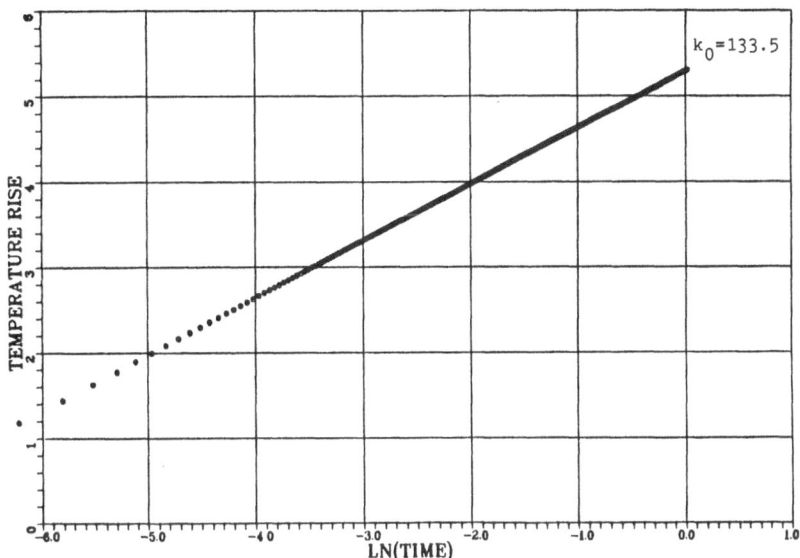

Figure 1.   Raw Data Plot of TL053

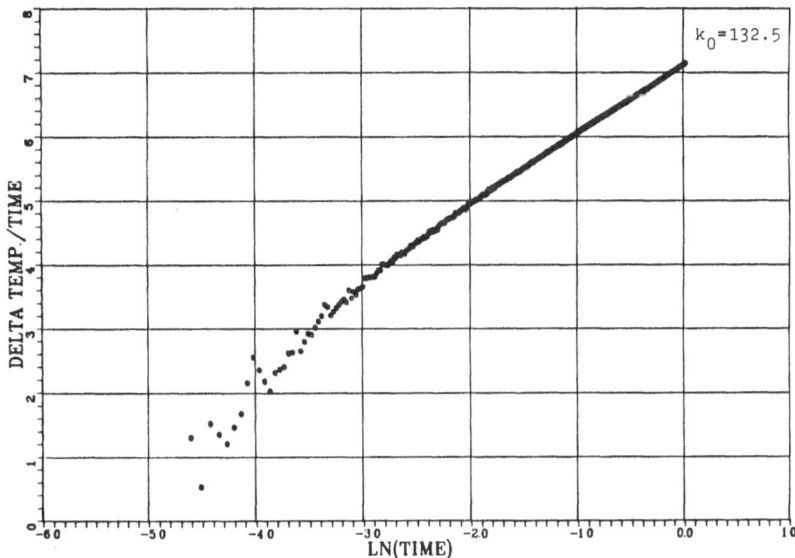

Figure 2.   Raw Data Plot of TL052

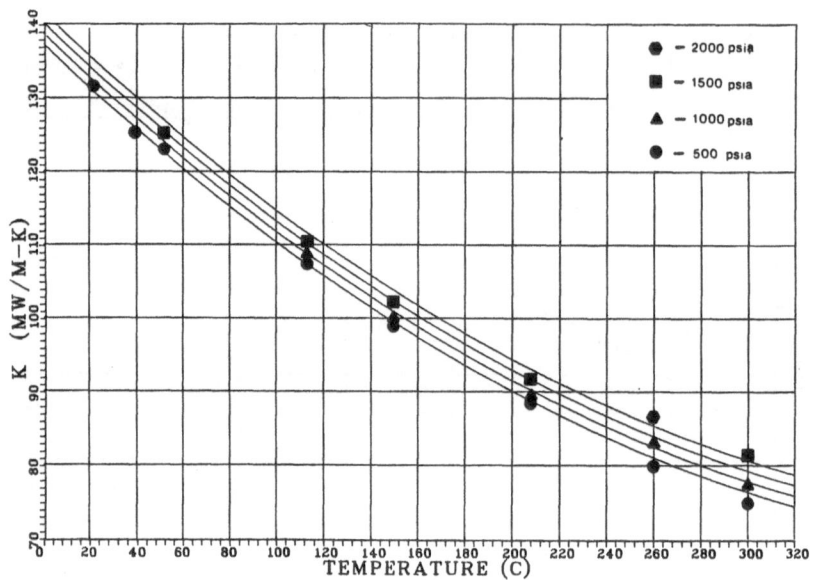

Figure 3.   Ramp Forced Toluene Thermal Conductivity

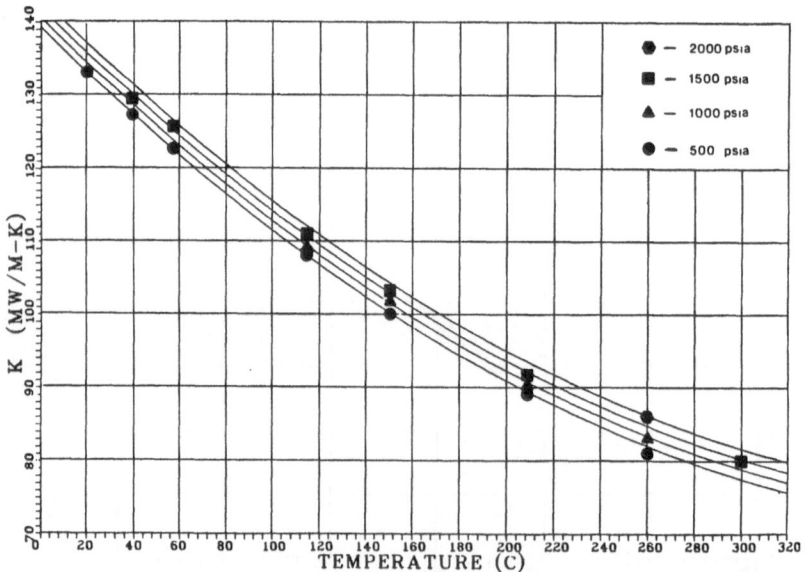

Figure 4.   Step Forced Toluene Thermal Conductivity

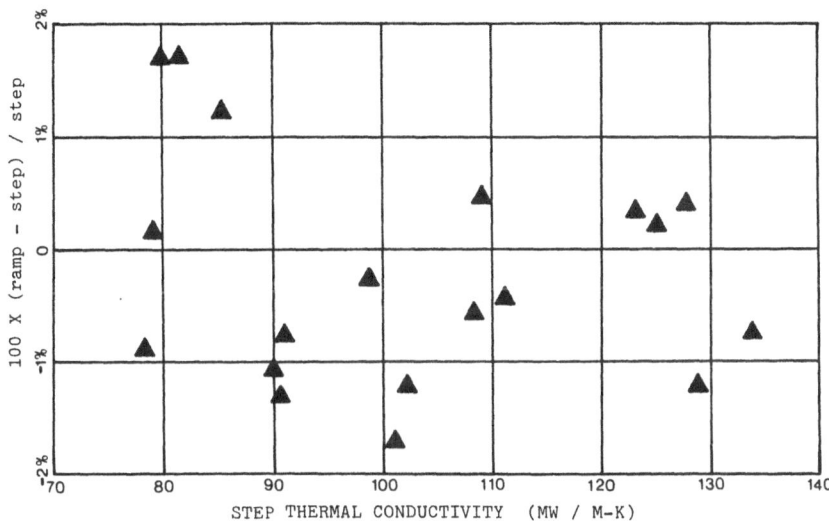

Figure 5. Independence of Toluene Thermal Conductivity From Forcing Function.

The data for toluene are then compared to the data of Mani-
Venart (1973), which represent the only other set of data for
toluene over such a wide range of temperatures. As is evident
in Figure 6, there is excellent agreement between our step and
ramp data and the step data of Mani-Venart (1973). The accuracy
of our toluene data is further indicated by comparing them with
data from other recent experiments using step-force functions, as
well as with the radiation-free, steady-state parallel-plate data
of Poltz and Jugel (1967), in Figure 7. It is noted in this plot
that the transient hot wire data obtained from a back extrapolation
to zero experiment time (i.e., Mani-Venart, and this work) is in
excellent agreement with the steady-state data of Poltz and Jugel
corrected for radiative effects. The step-forced transient hot
wire data obtained from a single fit of the entire data set are
higher by about 1%. It is thought that this trend is due to
radiation effects, which are minimized by the extrapolation to
zero time. A recent mathematical study of combined conduction and
radiation (Menasche and Wakeham, 1982) found the radiative contribution to the effective thermal conductivity should increase
approximately linearly with elapsed time. The radiation contribution should then decay exponentially in the limit of zero time.

279

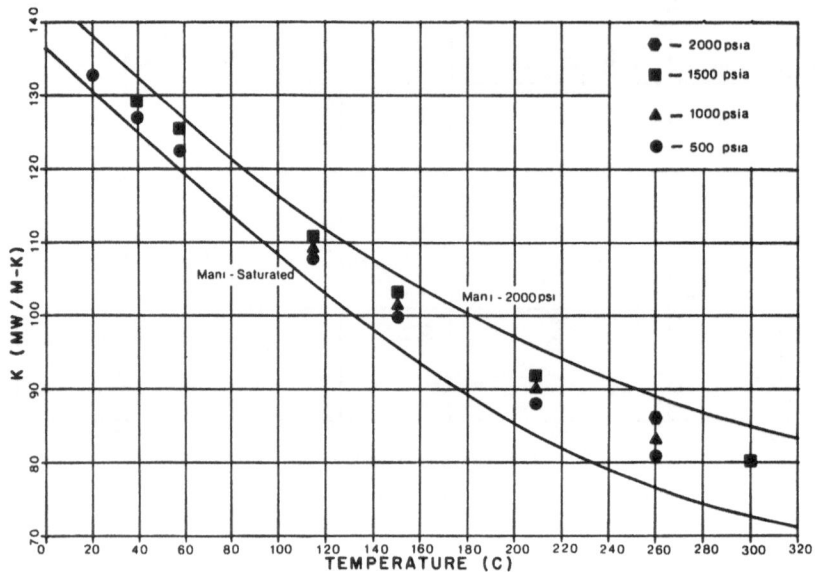

Figure 6. Toluene Data Comapred to Mani-Venart (1973) Data

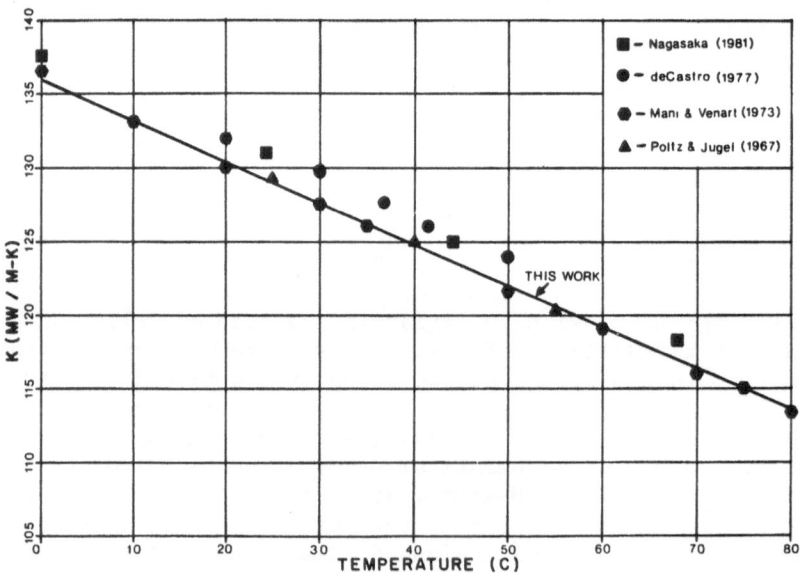

Figure 7. Saturation Pressure Toluene Thermal Conductivity Comparison

Transient hot wire zero-time thermal conductivity data are
therefore analogous to the zero-plate-spacing steady-state data
of Poltz and Jugel.

Both the ramp-forced and step-forced transient hot wire data
appear to be of comparable accuracy for an equivalent data acquisition system. Both the ramp and step data are in excellent agreement with radiation-free measurements at steady state. The
mathematical study of Mohammadi (1981) shows that the ramp power
forcing function should have advantages in delaying the onset of
convection. Since the onset of convection in liquid systems
occurs at times greater than 1 second at any reasonable power
level, the delay in the onset of convection with ramp-power
forcing does not provide any increase in accuracy in these measurements. If the ramp-power forcing function does have significant
advantages over the step function, they will be realized in
measurements on a low-density gas. The ramp-power forcing function
is a more gentle, continuous forcing function which is much more
sensitive to the initial state of the fluid. The very small rises
in temperature which are present at short times in the ramp experiment require data acquisition systems which are more sensitive
than the systems required for good step-forced data. The ramp
power forcing function has the advantage of not decaying to a
steady state value. This is counterbalanced by the fact that less
accurate temperature control systems are required for step forced
instruments because of decreased sensitivity of step forced
function to the fluid initial temperature. The "forgiving" nature
of the step-forcing function relative to the ramp means that for
the liquid phase measurements studied thus far, a step forced
instrument can be constructed at a lower cost than a ramp forced
instrument.

The current instrument will be used to study both pure component and mixture thermal conductivities of materials common in
synthetic fuel materials over the range of ambient to their
critical point or thermal stability limit. Materials studied
thus far include toluene, meta-xylene, methylcyclohexane, tetrahydronaphthalene, decahydronaphthalene, an SRC I fraction, and
a Utah COED fraction. In addition, the ramp versus step comparison
will be extended to low-density gas-phase measurements.

REFERENCES

1) DeCastro, C.A., Calado, J.C.G., Wakeham, W.J., Absolute
Measurements of the Thermal Conductivity of Liquids Using a
Transient Hot Wire Technique, 7th Symposium on Thermophysical
Properties, National Bureau of Standards, Gaithersburg, MD,
(1977).

2) Mani, N., Venart, J.E.S., The Thermal Conductivity of Some Organic Fluids: HB-40, Toluene, Dimethylphthalate, Proceedings 6th Symposium on Thermophysical Properties, ASME, (1973).

3) Menashe, J., Wakeham, W.A., Effect of Absorption of Radiation on Thermal Conductivity by the Transient Hot Wire Technique, Int. J. Heat Mass Trans., 25, 5, (1982).

4) Mohammadi, S.S., Development and Application of a Ramp Forced Transient Technique for Thermal Conductivity Measurement of Non-Electrolytic Liquids and Gases, Ph.D. Thesis, Colorado School of Mines, (1980).

5) Mohammadi, S.S., Graboski, M.S., Sloan, E.D., A Mathematical Model of a Ramp Forced Hot Wire Thermal Conductivity Instrument, Int. J. Heat Mass Trans., 24, 4, (1981).

6) Nagasaka, Y., Nagashima, A., Simultaneous Measurement of the Thermal Conductivity and the Thermal Diffusivity of Liquids by the Transient Hot Wire Method, Rev. Sci. Instr., 52, 2, (1981).

7) Perkins, R.A., Mohammadi, S.S., McAllister, R., Graboski, M.S., Sloan, E.D., A New Transient Vertical Hot Wire Thermal Conductivity Instrument for Fluids Utilizing a Ramp Power Input, J. Phys. E., 14, (1981).

8) Perkins, R.A., Development of a Transient Hot Wire Instrument to Measure the Thermal Conductivity of Fluids at High Temperatures and Pressures, Ph.D. Thesis, Colorado School of Mines, (1983).

9) Poltz, H., Jugel, R., The Thermal Conductivity of Liquids - Temperature Dependence of Thermal Conductivity, Int. J. Heat Mass Trans., 8, 4, (1967).

A THERMAL PROPERTIES MEASURING SYSTEM FOR GEOLOGIC MATERIALS

W. B. Krause

Materials Laboratory
RE/SPEC Inc.
Rapid City, South Dakota   57701

B. G. Everett,
D. K. Schwemle and
D. C. Springhetti

Engineering Staff
Innovative Systems, Inc.
Rapid City, South Dakota   57701

ABSTRACT

This paper discusses the design concepts of a thermal property measuring system for laboratory measurements on geologic materials. Topics considered are the pressure vessel, laser-interferometer dilatometer system, the thermal subsystems, and the computer-related data acquisition and control systems. Also discussed are the measurement techniques for thermal conductivity, thermal diffusivity, and linear thermal expansion.

INTRODUCTION AND BACKGROUND

The thermophysical property measuring system (TPMS) discussed in this paper has the unique capability to measure thermophysical properties under hydrostatic pressure. Because of the depths

---

RSI PUBL. NO. 83-17

involved, the influence of hydrostatic pressure on the thermo-physical properties for geologic materials is required for thermo-mechanical modeling related to fields such as geothermal energy, storage cavern operations, oil and gas recovery and high-level nuclear waste disposal. Emphasis is placed on design concepts related to the measurement of thermal conductivity (by the steady-state radial heat flow method), thermal diffusivity (by the transient-pulse technique), and also linear thermal expansion (by a laser-interferometer dilatometer system). The system discussed is similar in function and design to the thermophysical property measuring system in use at Lawrence Livermore National Laboratory [1].

The TPMS includes elements to control and measure temperature and pressure via computer. The laser-interferometer dilatometer system measures linear expansion of materials while exposed to elevated temperature and hydrostatic pressure. The hydrostatic confining pressure is applied to viton-jacketed, 200-mm-long spec-imens having outer diameters ranging from 50 mm to 100 mm. For heater emplacement in the specimen, a 12.5-mm-diameter heater hole is drilled along the specimen axis. The large-diameter specimens compensate for material inhomogeneity. The system maximum operating temperature is 225°C with a maximum operating pressure of 50 MPa (7250 psi).

SYSTEM COMPONENTS

The major components of the TPMS are:

1. Pressure vessel;

2. Computer, software, peripherals, and data acquisition; and

3. Measurement and control.

A computerized data acquisition and control system are used in conjunction with the pressure vessel and instrumentation to pro-vide the capability to monitor and control test parameters. An overview representation of the TPMS is shown in Figure 1.

THE PRESSURE VESSEL

The pressure vessel is a 230-mm-diameter by 460-mm-high vessel of heat-treated high-strength steel, designed to contain a 200-mm-long specimen. The vessel is surrounded by external heating ele-ments and a thermal insulating blanket to allow heating the vessel to 225°C.

The specimen is mounted vertically inside the vessel between two stainless steel platens, which in turn mount between guard heaters. Thermocouple probes enter the interior of the specimen from the bottom of the vessel. High-pressure feed-throughs are provided in the vessel base to allow wires to enter the vessel for additional thermocouples and interior heaters. A DC-powered central heater is placed in the hole drilled along the axis of the specimen. The vessel is filled with silicone oil during testing and can withstand pressures to 50 MPa.

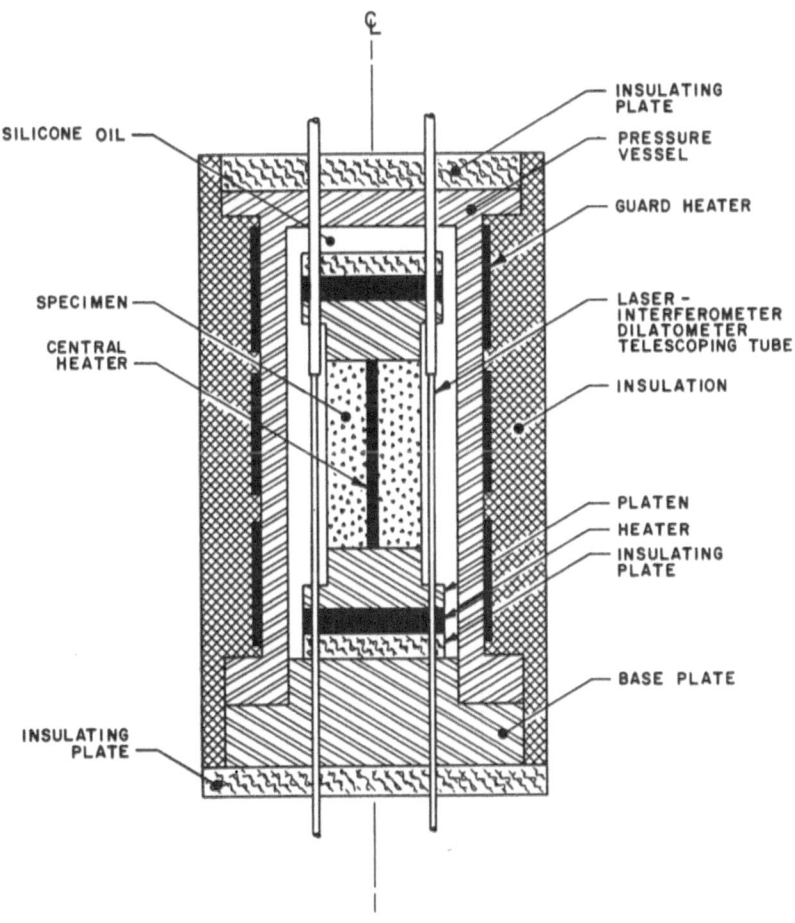

Fig. 1. Overview schematic of TPMS.

COMPUTER, SOFTWARE, PERIPHERALS, AND DATA ACQUISITION

The computer is a Digital Equipment Corporation PDP-11/23 system with RL01 cartridge disk drives, hard copy terminal, and analog and digital I/O subsystems (see block diagram in Figure 2). The RSX11M operating system and Fortran 77 language are used because of their flexible programming features and good real-time control features.

The digital subsystem provides distributed control for the AC heaters, the central heater, the pressure valves, and the interferometer. It also monitors status of the test apparatus. This subsystem communicates with the computer using a serial interface (RS-232). These functions are implemented using a controller designed by Rushmore Micro Systems [2].

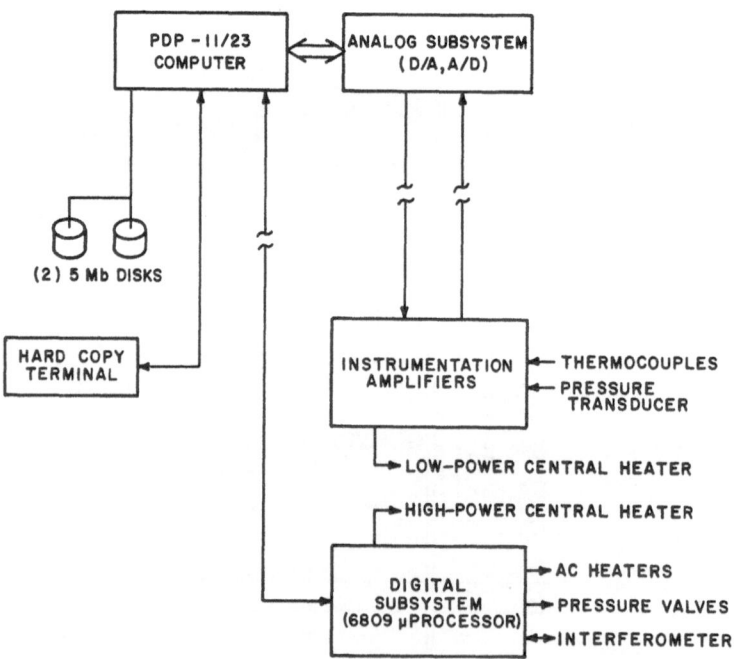

Fig. 2. Block diagram of computer-based measurement and control system.

MEASUREMENT AND CONTROL

The measurement and control subsystem for the TPMS consists of the central heater, components for control and measurement of temperature and pressure, and the laser-interferometer dilatometer.

Central Heater

The purpose of the central heater is to provide power to three heater zones located along the axis of the specimen. The central heater measurement and control subsystem consists of a circuit for temperature control of the three-zone central heater and also a power measurement circuit for each zone. The heaters are temperature controlled to within ± 0.5°C of indicated value. Power is measured to 0.5 percent of the indicated measurement.

Temperature and Pressure

Confining Oil Temperature Control. This is an AC heater system that consists of the outer vessel band heaters, the specimen end guard heaters, and the electrical and electronic equipment used to control the proportional power provided to the heaters.

The power delivered to the AC heaters is computer software controlled. The digital subsystem, through use of timers, provides proportional timing signals to SCR power switches that control the heater power. The oil temperature is controlled to ± 1°C of the indicated value.

Temperature Measurement. Type T thermocouples are used throughout the system. The thermocouple outputs are presented to a connection block that provides cold-junction temperature compensation. The thermocouple signals are amplified and filtered by signal conditioning cards and are then sent to the data acquisition subsystem. Type T thermocouple linearization is provided by software. A digital display meter is provided to display confining oil temperature. Thermocouples are calibrated against resistance temperature detectors with National Bureau of Standards traceablity. Tolerances are ± 0.1°C of true value.

Confining Oil Pressure. The hydraulic system uses a piston-type intensifier in a nitrogen-over-oil configuration. The pressure of the nitrogen side of the intensifier is amplified with a gain of six to provide the required oil pressure. A hydraulic pump and the high-pressure valves are directly controlled by the digital subsystem. The oil pressure is controlled to 1 percent of indicated pressure.

Laser-Interferometer Dilatometer

Previous application of a conventional push-rod dilatometer is documented for the measurement of thermal linear expansion under pressure [3]. Because of the limited resolution of a conventional system, other methods such as laser interferometry have been designed to provide precision greater than that provided by conventional push-rod dilatometers. The design of a high-precision dilatometer, which uses laser interferometry as the basis for length change measurement, has been presented by Drotning [4]. The TPMS uses a spherical Fabry-Perot interferometer. To operate in a varying pressure environment, the TPMS interferometer mirrors are mounted within telescoping stainless steel tubes, as shown in Figure 3. The telescoping tubes are sealed to prevent the silicone oil from contacting the interferometer, and the ends of the tube assemblies extend outside of the pressure vessel at the top and bottom through high-pressure seals. The external surfaces of the telescoping tubes are kept in a hydrostatic condition, minimizing the effects of pressure on the tube ends. The vertical dilatometer consists of two diametrically opposed tube assemblies placed around the specimen.

A stabilized single-frequency He-Ne laser is used to provide the light source for the interferometers. Mirrors are used to direct the laser beam into the interferometers, generating fringe patterns. As the distance between the interferometer mirrors changes, the interference pattern changes correspondingly. Photo-detectors and decoding circuitry are used to count interference fringes caused by movement of the interferometer mirrors. The fringe-counting circuit is interfaced to the digital subsystem, allowing the microprocessor to read and zero the count. The system is designed such that length change is represented by:

$$\Delta L = \frac{\lambda K}{16} \text{ count} \qquad (1)$$

where:

ΔL = length change
K = index of refraction of the media between the interferometer mirrors
λ = the wavelength of the laser beam in vacuum
count = four times number of interference fringes.

The interferometer measures specimen length change with an accuracy of $\pm 5 \times 10^{-4}$ mm. Errors include laser wavelength accuracy, variations in the index of refraction K, and influence of varying temperature on the interferometer mirrors.

288

The laser employed, a Coherent model 200, operates at 473.6128 x 10^{12} Hz (λ = 632.9914 x 10^{-9} m) in vacuum and is stabilized by a temperature feedback system. The fundamental frequency varies at most by 5 MHz/°C ambient temperature. Since the laser temperature is maintained within ±5°C, the total frequency drift is held to under ±25 MHz or 0.05 ppm [5]. Mielenz et al [6] document similar laser accuracies to approximately one part in 10^7.

Ready [7] reports the effects of pressure, temperature, and vapor pressure upon the index of refraction. He found the index of refraction for dry air to be:

K_{dry} = 1.002765 at 15°C and 760 Torr with 0.03 percent CO_2 at the He-Ne wavelength

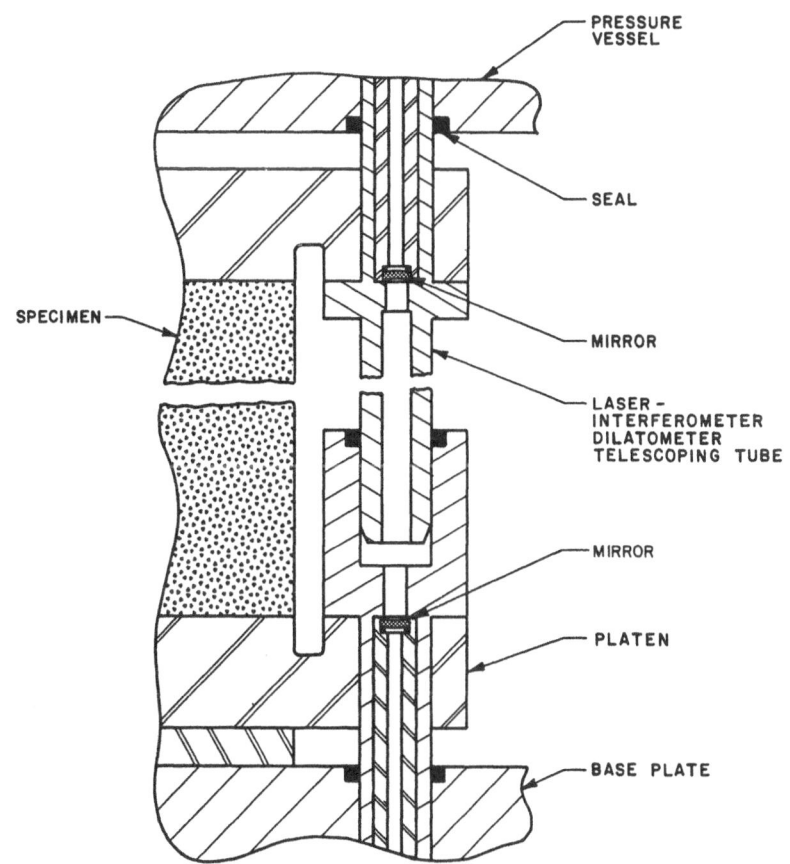

Fig. 3. Schematic diagram of the laser-interferometer dilatometer telescoping tube and mirror arrangement.

289

The expression for correction for temperature, pressure, and vapor pressure is:

$$K - 1 = (K_{dry} - 1) \frac{P}{720.775} \frac{A(T,P)}{(1 + 0.003662T)} - 5.6079 \times 10^{-8} f \qquad (2)$$

where:

P = pressure in Torr
T = temperature in °C
f = partial vapor pressure (Torr) of water vapor at T and P
$A(T,P)$ = $1 + P(0.817 - 0.0133T) \times 10^{-6}$.

ANALYTICAL CONSIDERATIONS

Temperature-time responses of the specimen at fixed radii are the required data for thermal diffusivity computation. Power input to the specimen along with temperature measurement at the inner and outer radii are the required quantities for steady-state thermal conductivity measurement. The laser-interferometer dilatometer measures length change, which is used to compute linear thermal expansion.

Thermal Diffusivity Measurement

By using the central heater, the specimen is heated with application of a pulse (duration of about 100 seconds). The transient temperature response is measured at several radii within the specimen. The temperature history is then used to compute thermal diffusivity.

The differential equation for heat flow in axisymmetric coordinates is:

$$\frac{d^2 T}{dr^2} + \frac{1}{r} \frac{dT}{dr} = \frac{1}{\alpha} \frac{dT}{dt} \qquad (3)$$

where:
T = temperature
r = radius
α = thermal diffusivity
t = time.

This equation and an initial temperature condition can be solved with time-dependent boundary conditions:

$$T(r_i, t) = T_i(t) \tag{4}$$

$$T(r_o, t) = T_o(t). \tag{5}$$

The radii specified as r_i and r_o denote the inner and outer thermocouple locations, respectively. Either the finite element method or an exact solution to the boundary value problem may be used to compute temperature. The solution to the boundary-value problem, along with the measured temperature data and a nonlinear least squares fitting routine, provides a method to determine the thermal diffusivity.

Thermal Conductivity Measurement

The steady-state thermal conductivity measurement is based on the measured temperatures, T_i and T_o, and the amount of power per unit length that is supplied to the specimen by the central heater. The power input for the radial coordinate system may be defined as:

$$\dot{Q}/L = 2\pi k (T_i - T_o)/\ln(r_o/r_i) \tag{6}$$

where:

$$\dot{Q}/L = \text{power input/unit length}$$
$$k = \text{thermal conductivity.}$$

Linear Thermal Expansion Measurement

Linear thermal expansion may be defined as:

$$\alpha_{LIN} = \frac{1}{L} \frac{\partial L}{\partial T_p} \tag{7}$$

where:

L = the specimen original length

$\dfrac{\partial L}{\partial T_p}$ = the rate of change of length with temperature for a constant pressure.

CONCLUSION

This paper presented an overview of the design and development concepts for a thermophysical property measuring system. The paper outlined the specific capabilities and design details related to system components.

REFERENCES

1. Abey, A. E., Durham, W. B., Trimmer, D. A., and Dibley, L. L., Apparatus for Determining the Thermal Properties of Large Geologic Samples at Pressures to 0.2 GPa and Temperatures to 750 K, Rev. Sci. Instrum., 53, 6, pp. 876-879 (1982).
2. Rushmore Micro Systems, "Users Manual 6809 Industrial Controller," Rev. B (1983).
3. Durham, W. B. and Abey, A. E., "Thermal Properties of Avery Island Salt to 573 K and 50 MPa Confining Pressure," Lawrence Livermore National Laboratory, UCRL-53128 (1981).
4. Drotning, W. D., "Design of a High Precision Dilatometer Using Laser Interferometry," Sandia National Laboratories Publication, SAND78-1796 (1978).
5. "Coherent (Laser Products Division) Model 200 Application Notes" (1982).
6. Mielenz, K. D., Nefflen, K. F., Rowley, W. R. C., Wilson, D. C., and Engelhard, E., Reproducibility of Helium-Neon Laser Wavelengths at 633 nm, Applied Optics, 7, 2, pp. 289-292 (1968).
7. Ready, J. F., "Industrial Applications of Lasers," Academic Press, New York, NY (1978).

ACKNOWLEDGEMENT

This project was supported by Battelle Memorial Institute for the United States Department of Energy through Contract Number E512-02300. The authors are grateful to Paul E. Senseny for his thorough review of the manuscript. Judy C. Hey and Fay L. Swenson are acknowledged for their expertise in typing the manuscript. Thanks are extended to Danny P. Nelson for his drafting expertise.

SESSION L

Invited Papers

SPEAKERS

G. K. White
CSIRO
Australia

P. G. Klemens
University of Connecticut
Storrs, CT.

SESSION CHAIRMAN

R. Berman
University of Oxford
England

LATTICE THERMAL CONDUCTIVITY AT NORMAL AND HIGH TEMPERATURES

Guy K. White

CSIRO Division of Applied Physics
Sydney, Australia 2070

ABSTRACT

The heat conductivity λ of a crystal lattice is limited at normal and high temperatures by mutual scattering of the lattice waves. Theoretical models have shown that this scattering depends strongly on the lattice spectrum (or appropriate value of Debye θ_D) and on the anharmonic coupling (or Grüneisen parameter γ). Comparison of measured $\lambda(T)$ for representative crystals with the values calculated from a modified Leibfried-Schlömann equation illustrate difficulties inherent in choosing appropriate values for γ and θ.

INTRODUCTION

At temperatures T above $\theta_D/5$ (θ_D is a Debye characteristic temperature), the lattice thermal conductivity λ of a crystalline solid is determined by the anharmonic interaction between the lattice waves. The forces binding together a solid in three dimensions are necessarily anharmonic so that when a lattice wave or phonon disturbs the local density, this will affect the frequency of a second wave and interaction occurs. There have been numerous efforts to quantify this interaction, expressing λ in terms of mass, volume, T, θ and an anharmonicity factor γ, beginning with Akhieser, Leibfried and Schlömann, Lawson, Kontorova, Roufosse and Klemens, Julian, etc. (see review by Slack [1] and book by Berman [2]). A convenient form at $T \geqslant \theta_\infty$ is

$$\lambda \approx 3.0 \times 10^4 \, n^{1/3} \, \overline{M} \, \delta \, \widetilde{\theta}_\infty^3 \, / \gamma_\infty^2 T \quad \text{W/m·K}$$

where δ^3 is the volume per atom (δ is in meters), $\overline{M}$ is the average atomic mass, n is the number of atoms per unit cell and $\tilde{\theta}_\infty$ is the high temperature value of θ_D appropriate to acoustic modes of vibration.[1] If the acoustic frequency spectrum is known from inelastic neutron scattering then $\tilde{\theta}_\infty$ can be calculated from the second moment of the frequency distribution, or we assume (see Slack [1]) that $\theta_0 = \tilde{\theta}_0 \times n^{1/3}$

$$\text{whence } \lambda \text{ (at } \tilde{\theta}_0) = 3.0 \times 10^4 \, \overline{M} \, \delta \, \theta_0^{\,2}/n^{1/3} \, \gamma^2 \text{ W/m·K.} \qquad (1)$$

The low temperature limiting value θ_0 may be obtained from the T^3-term in the heat capacity, or from the elastic constants. [3,4]

The choice of γ in eq. 1 is more complex than the choice of θ as it depends on a weighted mean of the strain dependences, γ_i, of the individual vibrational frequencies, ω_i, and these γ_i may vary in sign as well as magnitude. Which is the most appropriate average for θ or γ in determining $\lambda(T)$ is not obvious. This paper is concerned with choosing θ and γ for some sample materials and examining the agreement between calculation and experiment as well as the effect of changing temperature and pressure P.

DEBYE TEMPERATURE, θ_D

Figure 1 illustrated the dispersion curves relating frequency ω_i and wave number q_i determined by neutron scattering for one of the sample materials ZnS.[5] The initial slopes of $\omega_i(q)$ curves at q = 0 are the wave velocities for lowest frequency phonons and "averaging" gives the value of θ_0. Alternatively the ultrasonic wave velocities can be integrated over a sufficient number of directions in reciprocal space or C_v may be determined calorimetrically to give θ_0 as $T \to 0$. As T increases high frequency modes are excited and dispersion occurs, i.e., $d\omega/dq$ decreases and causes θ_D to fall from its θ_0 value. Generally θ_0 has a minimum near $T \approx \theta_0/15$ and then increases with T as the influence of the longitudinal acoustic (LA) branch and finally optic modes become more effective (Fig. 2). The optic modes may play only a minor role in conducting heat due to their small group velocity [1,2] and the same may be true of the acoustic modes near to the zone edge. Both may be important, however in the phonon scattering process and determining γ.

The effects of dispersion in the temperature region $T < \theta/15$ are most marked for crystals having the zinc blende structure such as CuCl and HgTe. For these the bonding is more ionic than covalent and results in a very low-lying transverse acoustic mode (TA) in the (110) direction: $\theta_D(T)$ falls to nearly half its limiting value or in other words C_v increases to nearly eight times its Debye continuum value. We may ask whether this should be observed in the dependence of $\lambda(T)$ in the low-temperature (boundary scattering)

296

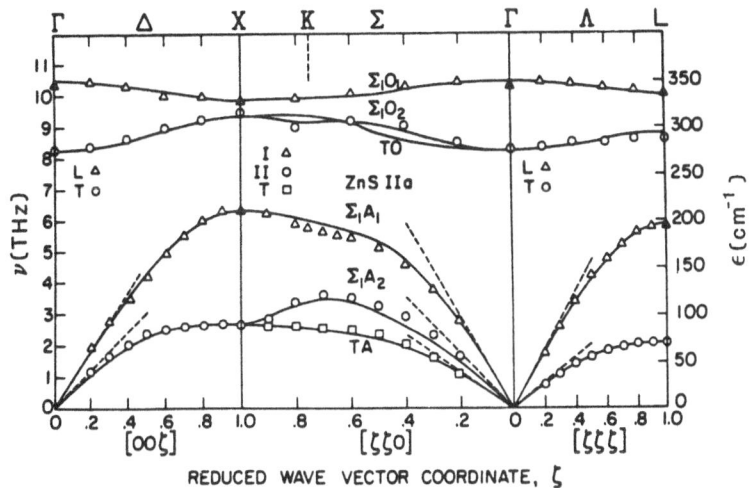

Figure 1.   Frequency versus wave number for ZnS.[5]

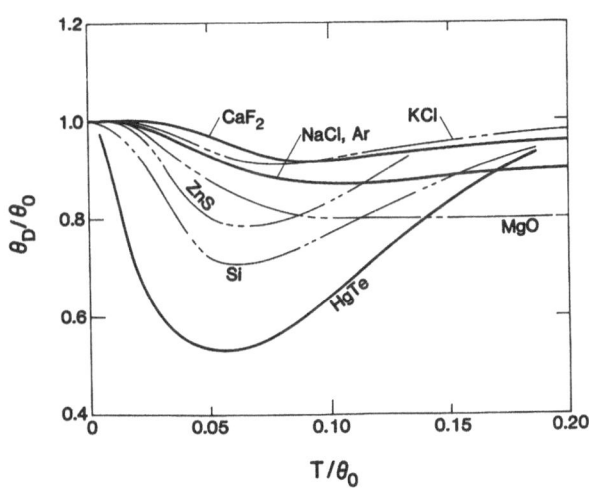

Figure 2. Variations of θ_D/θ_0 with reduced temperature, T/θ_0.

region or is the effect offset by the smaller group velocity of the modes of higher wave number?

GRÜNEISEN FUNCTION, γ

It has been said that there is a θ and γ for every occasion (e.g., see Appendix A [6]). The problem is to pick the right one. The thermodynamic Grüneisen function (we do not call it "constant") is defined by

$$\gamma^{th} = \beta \, B_T \, V/C_V = \beta \, B_S \, V/C_P = -(d \ln T/d \ln V)_S$$

where β is the volume coefficient of thermal expansion, B_T, B_S are isothermal and adiabatic values of the bulk modulus, V is the molar volume. For systems in which all the energy is from lattice vibrations, γ^{th} will be weighted average

$$\gamma = \Sigma \, \gamma_i \, C_i/\Sigma \, C_i$$

where $\gamma_i = -d \ln \omega_i/d \ln V$ and C_i is the contribution of the i-th mode to the heat capacity. γ has a low-temperature limiting value, γ_0 analogous to θ_0 which can be calculated from the pressure derivatives of the principal elastic constants:

$$\gamma_i = -\frac{1}{6} + \frac{B \, d\ln c_i}{2 \, dP} \, .$$

For a cubic crystal, three measurements of dc_{11}/dP, dc_{44}/dP and dc'/dP (where $c' = (c_{11}-c_{12})/2$) suffice to give γ_0 (elastic) using the same procedures of numerical integration as to calculate θ_0 (elastic) (see Appendix C [6]). For anisotropic crystals, five or more elastic constants are needed at two or more pressures in order to calculate θ_0 at different volumes and thence

$$\gamma_0 = - \, d \ln \theta_0/d \ln V.$$

Alternatively, γ_0^{th} can be determined from expansivity data at temperatures $T \to 0$ if sufficiently sensitive techniques are available.

At high temperatures, $\gamma^{th} = \gamma_\infty$ is simply the arithmetic average of the values of γ_i but these values are not available unless the entire phonon spectrum has been measured as a function of pressure.

For metals γ_∞ and γ_0 do not differ usually by more than 20% because the γ_i values do not change greatly with polarization or frequency. The same may be said of the close-packed rare-gas solids.

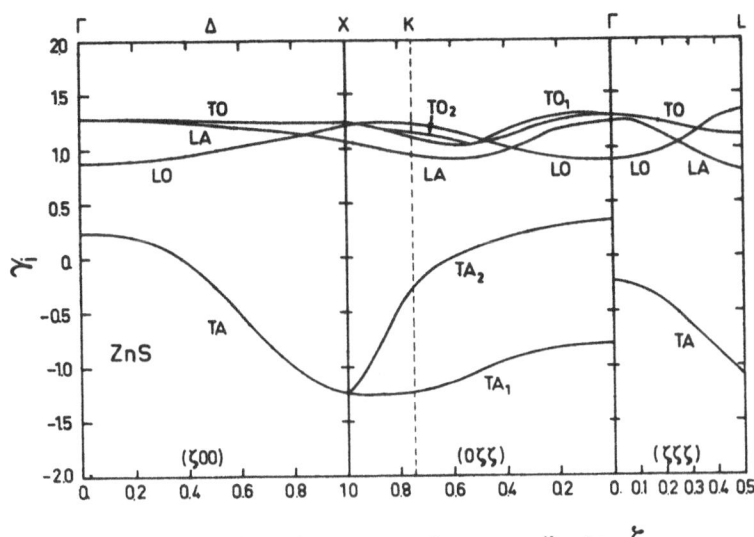

Figure 3. γ_i versus wave number calculated for ZnS.[5]

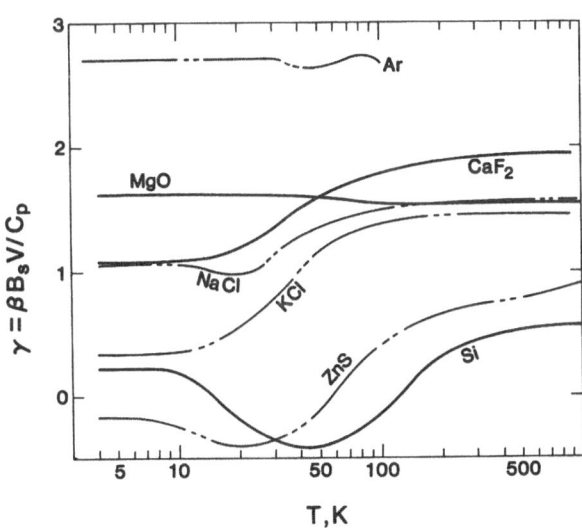

Figure 4. Variation of γ (thermal) with temperature.

But this is not the case with most non-metallic crystals, particularly those of low coordination number for which a low-lying TA branch may have values of γ_j which are very different from those of LA or optic branches (Fig. 3).

Note that other approximations for γ based on pressure derivatives of B have been introduced by Slater, Dugdale and MacDonald, etc., and used by some geophysicists and others but they have no theoretical foundation and depend for their derivation on assuming that all γ_j are equal (see Appendix B [6].

For want of a better alternative, we shall often use γ^{th} (Fig. 4) (from expansivity) in eq. [1] but we should be aware that the γ^2 might be better considered as the average of γ_j^2, i.e., $\overline{\gamma_j^2}$ and not as the square of the average $\overline{\gamma}_j$, otherwise we might be led to $\lambda = \infty$ as $\gamma^{th} \rightarrow 0$.

CALCULATION OF $\lambda(T)$

Table 1 lists values of θ_0, mean atomic mass $\overline{M}$, δ (where δ^3 is volume per atom), limiting values of γth taken from reviews [1,6] λ calculated at $T = \tilde{\theta}_0 = \theta_0/n^{1/3}$ and λ measured.[7] Also included are values of $\gamma_j(q = 0)$ from elastic constants and their pressure derivatives (referenced in [6]) which illustrate the large anisotropy of γ_j in most crystals. Bracketed values for Ar are based on neutron data obtained for Ne under pressure.[8]

The T^{-1} dependence of λ expressed in eq. (1) is derived for a condition of constant volume whereas most measurements are made at constant pressure. Expansion will affect θ, γ, δ and hence λ. We may estimate the extent by differentiating [1] to obtain

$$\frac{d \ln \lambda}{d \ln V} = 3 \frac{d \ln \theta}{d \ln V} - 2 \frac{d \ln \gamma}{d \ln V} + \frac{1}{3} \qquad (2)$$

For example, volume increase of MgO in warming from 293 to 756 K is 1.9% which should cause θ_∞ to decrease by 3% ($\gamma = 1.5$), and γ to increase by ca 3% (see discussion below) with resulting decrease of 15% in λ.

Slack [1] shows that the power law becomes $\lambda \propto T^{-\varepsilon}$ where $\varepsilon = 1 + 3 \alpha g T$ and $g = - d \ln \lambda/d \ln V$. For MgO at 756, $\alpha = 14.3 \times 10^{-6} K^{-1}$ so that if $g \approx 7$ (see below), then $\varepsilon = 1.23$.

λ, CALCULATED AND MEASURED

The measurements quoted for Ar were done at constant volume [9] and agree reasonably with eq. (1). Agreement is remarkably good if we use $\theta_\infty = 84$ K (see Slack [1]). The elastic constants and γ_j are

300

Table 1. Values of parameters used in Equation 1 to calculate lattice conductivity $\lambda(W/m \cdot K)$ at $\tilde{\theta}_0$.

	Ar	MgO	NaCl	KCl	CaF$_2$	Si	ZnS
$\theta_0(K)$	94	953	321	235	510	645	339
$\tilde{\theta}_0 = \theta_0/n^{1/3}$	94	756	255	187	354	512	269
$\overline{M}$	40.0	20.2	29.4	37.3	26.0	28.1	48.75
δ (nm)	0.334	0.210	0.282	0.314	0.238	0.268	0.270
γ_∞	2.75	1.54	1.58	1.48	1.95	0.56	0.95
γ_0	2.7	1.62	1.06	0.34	1.04	0.24	- 0.15
λ (calc)	0.47	38	8	9.6	8.8	240	44
λ (meas)	0.39	~ 20	7	12	8	80	33
$\gamma(c_{11})$	(3.0)	2.22	2.48	2.44	1.34	1.12	1.31
$\gamma(c')$	(2.5)	2.8	2.62	2.39	0.41	- 0.10	- 0.9
$\gamma(c_{44})$	(2.7)	0.4	0.11	- 0.84	1.43	+ 0.33	+ 0.21

rather isotropic for Ar so that we should hope for agreement if eq. (1) is well based.

For MgO agreement is not so satisfactory. The measured values in TPM [7] show scatter of ± 20% and the measured value could be increased from 20 to 23 W/m·K to correct for expansion. Slack obtains closer agreement using θ_∞ = 600 K in his calculation. If we make a crude calculation of $(\overline{\gamma_i^2})$ based just on the three γ_i(q = 0) values in Table 1, we obtain γ = 2.1 which would reduce λ(calc) to about 21 W/m·K at 756 K.

For NaCℓ and KCℓ, agreement is good. Expansion effects should not be significant at 200 K compared with experimental uncertainties. Elastic anisotropy for the rock-salt halides increased markedly as we proceed from Li to Na to K to Rb halides; that is, c_{44} decreases relative to c_{11} (or B) and its pressure derivative (and associated γ_i value) becomes negative. This causes γ_0 to fall from 1.6 (LiF) to 1.0 (Na halides), 0.3 (K halides) and 0 (Rb halides). The γ_∞ from expansivity does not change much due presumably to influence of optic modes. If we calculate $\overline{\gamma_i^2}$ from the three values in Table 1 we obtain a value of 2.0 which reduces λ(calc) well below the measured values.

For CaF$_2$, agreement is good if we use γ_∞ from expansivity data but would be poor if we used an average γ_i^2 based on three elastic values in Table 1.

Finally, Si and ZnS representing the zincblende family present additional problems. One is that $\lambda \propto T^{-\varepsilon}$ where $\varepsilon \sim 1.3 \pm 0.1$ and thermal expansion is too small to account for this. Consider the lattice dynamics for this family. At low temperatures, the properties are dominated by a low-lying highly dispersive TA mode corresponding to a wave in the (110) direction, namely to c'. Progressing from Si to ZnS to CuCℓ, for example, the binding becomes less covalent and more ionic. This is accompanied by a relative weakening of c' and increased softening of the mode under pressure thus γ_0 becomes lower, reaching - 2.3 for CuCℓ. The importance of this mode as T → 0 and increasing importance of optic modes (for which $\gamma_i \sim 1$) at high temperatures makes $\gamma(T)$ vary greatly with T (Fig. 4). How important are these different modes in determining the phonon scattering? For Si and ZnS there are good expansivity data at T ⩾ θ_D which lead to γ^{th} values of 0.56 and 0.95 near 1000 K. Using these in (1) gives λ(calc) which is bad for Si but not so bad for ZnS. Using the three-mode average, $\overline{\gamma_i^2}$ = 0.7 for Si helps a little but not much. It seems likely that optic phonons are improtant in the scattering and for them γ_i values lie between 1.0 and 1.3. Clearly using γ_i^2 = 1 would reduce λ(calc) to the same magnitude as λ (meas). The increasing importance of the optic phonons as T increases would also help to explain the temperature dependence.

PRESSURE EFFECTS AT $T \geqslant \theta_D$

The behaviour of λ for non-metals at elevated temperatures and pressures is particularly important to the geophysicist. Differentiating eq. (1) with respect to volume, we obtain (2) as

$$g = 3 \gamma + 2q - \frac{1}{3} \tag{3}$$

where $g = - d \ln \lambda/d \ln V$, $\gamma = - d \ln \theta/d \ln V$ and $q = d \ln \gamma/d \ln V$.

It is usual to choose $\gamma^{th} = \gamma_\infty$ for (3) although there are the same difficulties of resolving acoustic and optic mode contributions as were discussed above. Most of the measurements of $d\lambda/dP$ have been for alkali halides and yield values of g from 6 to 10.[10] The data in Table 1 and accompanying discussion make it seem unlikely that the appropriate γ for alkali halides lies outside the range 1.5 to 2.0. Therefore values of g as high as 10 would suggest values for q much greater than 1. There are difficulties in measuring the volume dependence of β with any accuracy so that there have been few direct measurements of q.

Swenson [11] discussed the equation of state at moderately elevated temperatures and concludes that γ is linearly dependent on V to a first approximation, i.e., q = 1. This depends on the observations that βB_T is roughly constant for $T \geqslant \theta_D$ and so is C_v.

Roberts and Ruppin [12] calculate q for some alkali halides using a thermodynamic relation derived from $\gamma = \beta V B_S/C_p$, namely

$$q = 1 + \gamma - B_S' + (1 + \beta \gamma T) \delta_S$$

where $B_S' = (d B_S/dP)_T$ and $\delta_S = -(d \ln B_S/d \ln V)_p$ is one of the Anderson-Grüneisen functions (see Appendix B[6]). They obtain q values lying between 1.2 and 2.0 and conclude $q \sim 1.5$ is a reasonable approximation. Demarest [13] obtained a similar value from a shell model calculation for Na and K halides but got $q \approx 2$ for Rb halides.

There is another Anderson-Grüneisen function defined by

$$\delta_T = -(d \ln B_T/d \ln V)_p$$

which may be used in the approximation

$$q \approx \delta_T + 1 - B_S' \quad \text{for} \quad T > \theta_D$$

Barron (see Appendix B of [6]) has discussed the relations between δ_S, δ_T, q and pointed out some of the invalid approximations that have crept into the literature.

If we assume q = 1.5 ± 0.5, then

$$g \approx 3 \; \gamma_\infty + 3 - 1/3 \approx 3 \; \gamma_\infty + 2.7$$

Using γ_∞ from Table 1, this would lead to g = 7.2 (KCℓ), 7.5 (NaCℓ), and 9.0 (CsBr) compared with respective measured values of 5.7, 6.7, and 9.3 [1, 10].

CONCLUSION

Comparison of experimental values for the lattice thermal conductivity at $T \sim \theta_D$ with values calculated from the modified Leibfried-Schlömann equation [1] indicates difficulties in choosing appropriate values for γ as well as for θ. In close-packed crystals where there are no optic modes, choice is clearer and agreement is reasonable. Note that an earlier paper [14] calculating λ_g for Cu, Ag, Au gave values which were a factor of two too large but this discrepancy is removed in the modified version, viz., eq. [1].

For crystals having the zincblende structure, elastic constants and γ_i values are highly anisotropic so that any average of γ_i changes with T. The dispersion relations are well established for many of these crystals but $\gamma_i(q)$ curves are still speculative. Can they be integrated to provide more useful values of γ?

ACKNOWLEDGEMENTS

I thank Mr. Jerome G. Hust and his colleagues at the National Bureau of Standards (Boulder Laboratories) for their hospitality while preparing this paper.

REFERENCES

1. G.A. Slack, The Thermal Conductivity of Non-Metallic Crystals, Solid State Physics 34:1 (1979).

2. R. Berman, Thermal Conduction in Solids, Clarendon Press, Oxford (1976).

3. G.A. Alers, Use of Sound Velocity Measurements in Determining the Debye Temperature of Solids, Physical Acoustics (Ed. W.P. Mason) Vol. III-B, Academic Press, New York (1965).

4. O.L. Anderson, A. Simplified Method for Calculating the Debye Temperature from Elastic Constants, J. Phys. Chem. Solids 24:909 (1963).

5. D.N. Talwar, M. Vandevyer, M. Kunc and M. Zigone, Lattice Dynamics of Zinc Chalcogenides Under Compression: Phonon Dispersion, Mode Grüneisen and Thermal Expansion, Phys. Rev. B24:741 (1981).

6. T.H.K. Barron, J.G. Collins and G.K. White, Thermal Expansion of Solids at Low Temperatures, Adv. Phys. 29:609 (1980).

7. Y.S. Touloukian, R.W. Powell, C.Y. Ho and P.G. Klemens, Thermal Conductivity of Non-Metallic Solids in Thermophysical Properties of Matter, Vol. 2, Plenum Press, New York (1970).

8. J. Eckert, W.B. Daniels, and J.D. Axe, Phonon Dispersion and mode Grüneisen Parameters in Neon at High Density, Phys. Rev. B14:3649 (1976).

9. F. Clayton and D.N. Batchelder, Temperature and Volume Dependence of the Thermal Conductivity of Solid Argon, J. Phys. C. 6:1213 (1973).

10. D. Gerlich and P. Andersson, Temperature and Pressure Affects on the Thermal Conductivity and Heat Capacity of CsCℓ, CsBr and CsI, J. Phys. C. 15:5211 (1982).

11. C.A. Swenson, Equation of State of Cubic Solids: Some Generalizations, J. Phys. Chem. Solids 29:1337 (1968).

12. R.W. Roberts and R. Ruppin, Volume Dependence of the Grüneisen Parameter of Alkali Halides, Phys. Rev. B4:2041 (1971).

13. H.H. Demarest, Lattice Model Calculation of Hugoniot Curves and the Grüneisen Parameter at High Pressure for the Alkali Halides, J. Phys. Chem. Solids 35:1393 (1974).

14. G.K. White, Lattice Conductivity and Lorenz Ratio of Standard Metals in Thermal Conductivity, p. 37 (Ed. C.Y. Ho and R.E. Taylor) Plenum Press, New York (1969).

THEORY OF THE THERMAL CONDUCTIVITY OF

AMORPHOUS SOLIDS

P.G. Klemens

Dept. of Physics and Institute of Materials Science
University of Connecticut
Storrs, Connecticut

Thermal conduction by means of lattice waves in glasses will be discussed. Random arrangements of molecular units can be treated as point defects which scatter phonons, with possible enhancement of the scattering due to correlations over longer distances. However, such a model does not result in sufficient scattering at low frequencies and accounts neither for the thermal resistivity at ordinary temperatures nor at low temperatures, unless correlations over unlikely long distances are assumed. An additional interaction mechanism is provided by phonon-assisted transitions between the states of two-level systems, with a wide distribution of level spacings. These systems, whose presence in amorphous systems are postulated, also give rise to an excess specific heat at low temperatures, linear in temperature. This model also accounts for saturation effects in the absorption of high-frequency ultrasonic waves. Unfortunately we know too little about the basic structure of amorphous solids to derive the density, the level-spacing distribution and the phonon interaction strength of these centers. In a simpler case of crystalline Ti-V alloys, where such centers have also been observed, it is possible to derive the first two of these parameters, and hence the excess specific heat, from the known size difference of the two ions.

INTRODUCTION

The thermal conductivity of amorphous dielectric solids is less than that of the corresponding crystals. This can be ascribed to their disorder, which causes additional phonon scattering. Defining an average phonon mean free path ℓ in terms of the thermal conductivity λ by

$$\lambda = (1/3) \; C \; v \; \ell \qquad\qquad\qquad\qquad (1)$$

where C is the specific heat per unit volume and v the sound velocity, one can deduce ℓ from measurements of λ. Kittel[1] pointed out that in glass ℓ is roughly constant above 50K, but increases at lower temperatures. This seemed reasonable, since the phonons have long wavelengths at low temperatures, so that amorphous solids should then approach the behavior of homogeneous elastic media. Subsequent measurements, starting with those of Berman[2], confirmed that ℓ continues to increase as T decreases.

A detailed explanation of the behavior of ℓ in terms of phonon scattering by structural disorder has proved far from easy. The most obvious model of disorder would consider each molecular unit as a phonon scattering center, because the local phonon velocity fluctuates from point to point. For point-defect scattering, this change occurs over an elementary volume a^3. If δv is the root mean square velocity fluctuation, one can show that the attenuation probability per unit length, or the reciprocal of the mean free path, is given by [3]

$$1/\ell_p(\omega) = c(\delta v/v)^2 (a^3/\pi v^4)\omega^4 \qquad\qquad (2)$$

where c is the concentration of point defects (per atom or molecule) and ω the (angular) phonon frequency. If δv is due to a relative mass fluctuation $\Delta M/M$, then $\delta v/v = -\Delta M/2M$.

In the case of glasses such as vitreous silica, δv arises not from fluctuations in mass or density but from a variation of the orientation of molecular units. We can estimate $\delta v/v$ from the variability with direction of the ultrasonic wave velocity in crystalline quartz; thus we take $\delta v/v \simeq 0.15$ and c=1.

The effects on the thermal conductivity due to atoms differing in mass have been extensively studied. For crystalline Ge-Si of 50-50 composition, c=1 and $|\Delta M|/M=1/3$, so that $\delta v/v=0.16$. Thus one would expect vitreous silica and mixed crystals of 50-50 Ge-Si to have comparable point defect scattering for phonons of the same wavelength, except that a^3 is larger for SiO_2, so that ℓ_p should be correspondingly smaller. At 500K the thermal conductivity of vitreous silica is 1.6 $W\text{-}m^{-1}\text{-}K^{-1}$, while that of 50Ge-50Si is 6 $W\text{-}m^{-1}\text{-}K^{-1}$. It thus appears that the thermal conductivity of vitreous silica is slightly lower than one would expect from the ratio of the point defect mean free paths.

However, the disagreement rapidly worsens as the temperature is lowered, since the thermal conductivity of the mixed crystal increases with decreasing temperature, while that of glass decreases. Therefore in amorphous solids, there must be a further scattering mechanism which becomes relatively more important at low temperatures, hence at low frequencies, and which becomes overwhelmingly important at very low temperatures where the thermal conductivity of glass is orders of magnitude lower than that of the corresponding crystal.

THE LOW-FREQUENCY MECHANISM

One cannot treat the thermal conductivity as mainly due to phonons of one dominant frequency at any given temperature, but must consider that λ is made up of contributions from a wide frequency spectrum, i.e.

$$\lambda = \int \kappa(\omega) \, d\omega \tag{3}$$

where

$$\kappa(\omega) = (1/3) \, C(\omega) \, v \, \ell(\omega) \tag{4}$$

and where $C(\omega)d\omega$ is the contribution to C from modes in the frequency interval ω, $d\omega$. At low frequencies $C(\omega) \propto \omega^2$, so that point defects, for which $l(\omega) \propto \omega^{-4}$, cannot by themselves yield a finite thermal conductivity, for the integral of equation (3) would diverge at low frequencies. One must always treat point defects in combination with some other scattering mechanism which varies more slowly at low frequencies.

In the case of disordered crystals, such as Ge-Si, at ordinary temperatures, the low-frequency divergence is removed by anharmonic processes, with a mean free path $\ell_i \propto \omega^{-2} T^{-1}$. This makes $\kappa(\omega)$ independent of ω but varying as T^{-1} at frequencies well below ω_0, where ω_0 is defined by $\ell_p(\omega_0) = \ell_i(\omega_0)$. Thus $\omega_0 \propto (T/c)$, and one can show[4] that, based on a simple Debye model

$$(\lambda/\lambda_i) = (\omega_0/\omega_D) \arctan(\omega_D/\omega_0) \tag{5}$$

where λ_i is the intrinsic thermal conductivity and ω_D the Debye frequency. At ordinary temperatures for strong point defect scattering ($\omega_0 < \omega_D$), $\lambda \propto T^{-\frac{1}{2}} c^{-\frac{1}{2}}$. This simple theory accounts in a rough manner for the thermal conductivity of mixed crystals, including Ge-Si alloys, but neglects the effect of normal three-phonon processes and tends to underestimate the reduction in λ due to point defects[5].

In the case of amorphous solids too, point defect scattering must be supplemented by another scattering process at low frequencies. However, since λ decreases with decreasing temperature, it seems that anharmonic processes, which depend on T, do not contribute significantly to that unknown process. The unknown scattering seems to be a function of frequency – but varying more slowly than ω^3 – and seems to be practically independent of temperature.

More information about this scattering mechanism is provided by observations of the thermal conductivity at low temperatures, where only low-frequency modes are excited. In general[6] if $\ell(\omega)$ depends on frequency but not on temperature, and if $\ell(\omega) \propto \omega^{-n}$, where $n<3$, $\lambda(T) \propto T^{3-n}$. Berman's observations above 1K suggest n=2, but later measurements below 1K suggest n=1 for the low-frequency scattering process[7].

STRUCTURAL CORRELATIONS

It was pointed out[8] that scattering due to a local velocity fluctuation δv can be enhanced if there are correlations in the values of δv at neighboring locations. If the fluctuations δv vary randomly from one elementary volume a^3 to the next, scattering is like point-defect scattering and varies as $a^3 \delta v^2 \omega^4$. If δv at one location is correlated with δv at a location a distance r away, scattering of waves of wave-number q is enhanced by a factor

$$a^{-3} \langle \delta v \rangle^{-2} \int \langle \delta v(0) \delta v(r) \rangle \; \sin(qr) \; rq^{-1} \; dr \qquad (6)$$

where the pointed brackets denote averages, and the integral is the q'th Fourier transform of a spherically symmetrical correlation function f(r). If $f(r) \propto r^{-m}$, $\ell(\omega) \propto \omega^{-(m+1)}$ and $\lambda(T) \propto T^{2-m}$. In the liquid helium range where $\lambda(T) \propto T$, we require m=1, and quantitatively one requires [9]

$$f(r) \approx 0.01 \; a/r \qquad (7)$$

for r below about 300Å. Since such correlations are not seen in small-angle X-ray scattering, they would have to be orientational correlations, possibly arising from residual crystallinity over distances governed by the grain size of the precursor material.

As one goes to lower temperatures, well below 1K, one finds $\lambda \propto T^2$, so that m=0. No reasonable model of spherically symmetric structural correlations would yield such a dependence; one would have to invoke correlations of cylindrical symmetry. Furthermore, the low temperature values of λ for any single material do not show the variability as function of specimen history that one would expect if long-range structural correlations were the sole factor governing phonon scattering[10]. One must conclude that another mechanism dominates below 1K. At higher temperatures structural correlations may be significant in phonon scattering and they may be responsible for the variability in thermal conductivity around the plateau region (10 to 20K), but they may not be the major mechanism, and they are not important below 1K. There are strong arguments in favor of another mechanism, a distribution of two-level centers, proposed by

Phillips[11] and by Anderson, Halperin and Varma[12]. Perhaps the strongest argument in favor of this mechanism is that the same model also predicts an excess specific heat linear in T and a saturation in the absorption of high-frequency ultrasonic waves, both of which have been observed.

DISTRIBUTION OF TWO-LEVEL CENTERS

Consider a solid which contains a number of defect centers, each of which can be in two distinct quantum-mechanical states, separated by a level difference $E = \hbar\omega_0$, where $\hbar$ is the reduced Planck constant. If the energy of the defect can be perturbed by strain, transitions between the two levels occur with the emission or absorption of a phonon of frequency ω_0. This will reduce the mean free path of phonons of that frequency, and since the process is of first order in the local strain produced by the phonon, this reduction will be stronger than that due to phonon scattering by the same defect, since the latter is second order in strain, by a factor of order $(v/a\omega_0)^2$.

However, if all defects had the same level spacing E, the reduction in thermal conductivity would be minor, since only a small portion of the phonon spectrum would be affected. To account for the reduction in $\ell(\omega)$ over a wide range of frequencies, one must assume a distribution of E-values for different centers, no doubt because of the stochastic nature of amorphous structures.

Let us assume that the number of centers per unit volume with E in the range E, dE is given by a Gaussian

$$n(E)dE = 2ca^{-3} \; \pi^{-\frac{1}{2}}\sigma^{-1} \; \exp(-\omega_0^2/2\sigma^2) \; d\omega_0 \qquad (8)$$

so that $a^{-3}c$ is the total number of such centers per unit volume and σ is the root mean square width of the corresponding ω_0 distribution. If the level population is in thermal equilibrium at temperature T, and if $\hbar\sigma \gg kT$, where k is the Boltzmann constant, there is an excess specific heat per unit volume due to these centers given by

$$C_E = a^{-3}c(\pi^{3/2}/3) \; k^2 T/\hbar\sigma. \qquad (9)$$

Glasses do show an excess specific heat linear in T over the corresponding crystalline phase, and one can treat both c and σ as adjustable parameters. Typical values of C_E/T are 30 erg-cm^{-3}-K^{-2}, whence one can deduce values of $ck/\hbar\sigma$ of the order of 10^{-6} per degree Kelvin. It was assumed, of course, that the two quantum states are distinct, which requires a potential barrier between them of some minimum height[11].

The phonon mean free path is limited by processes in which a phonon is absorbed in the transition between the levels and subsequently re-emitted in a random direction, thus restoring thermal equilibrium. If the two levels are linked by a phenomenological perturbation Hamiltonian

$$H' = Ae \hspace{5cm} (10)$$

where e is the strain at the center due to the phonon and A is a parameter having the dimensions of energy, one obtains the following expression for the reciprocal mean free path of the phonons

$$1/\ell(\omega) = 8\pi^{\frac{1}{2}}cA^2(Mv^2\hbar\sigma)^{-1}(e^x-1)(e^x+1)^{-1}\omega/v \hspace{2cm} (11)$$

where $x=\hbar\omega/kT$. Except at the very bottom of the thermal phonon distribution, equation (11) makes $\ell(\omega)\propto\omega^{-1}$, so that $\lambda(T)\propto T^2$, as observed below 1K. The phonon mean free path, and thus λ, varies inversely as the factor c/σ, which appears as an empirical parameter in the excess specific heat; however it also varies inversely as A^2, and thus depends on a second parameter.

The role of these distributed two-level systems in limiting the thermal conductivity of glasses has been reviewed by A.C. Anderson [13]. The temperature dependence of the thermal conductivity is explained in terms of three mechanisms:-
(a) At lowest frequencies $\ell(\omega)$ is limited by interactions with the two-level systems, with a reciprocal mean free path given by (11).
(b) At intermediate frequencies, ℓ is controlled by point defect scattering with a reciprocal mean free path given by (2). However, $\delta v/v$ is treated as an adjustable parameter, and is larger than the value of 0.15 suggested by the anisotropy of the phonon velocity in the crystal.
(c) There is therefore a relatively wide range of high frequencies such that (2) breaks down because it would make ℓ_p less than the molecular diameter. Over that range, ℓ is chosen to be independent of frequency and of the order of 5Å.

I propose a modification of the model. Point defect scattering is not as strong, but of strength appropriate to $\delta v/v$ 0.15. The frequency where ℓ_p reaches its constant limiting value is reached only at high frequencies ($\omega>\omega_D/2$). To compensate for this increase in λ, one needs an additional scattering mechanism, namely inter-mediate-range correlations in the orientation of molecular units, given by (7) over distances of up to perhaps 100 to 300Å, but rapidly disappearing over greater distances. This enhances point-defect scattering by the factor of expression (6), and effectively adds a scattering mechanism at frequencies intermediate between those of (a) and (b). In this range $1/\ell(\omega)\propto\omega^2$. Until we know more about orientational correlations in amorphous solids, this enhancement factor has to be treated as an empirical function of phonon frequency. It could well vary between specimens of the same glass and account for the modest variability between specimens seen at intermediate temperatures. I regard point defect scattering as weaker, which would allow one to explain the lower thermal conductivity at ordinary temperatures of lead-containing glasses[14] in terms of additional point-defect scattering.

We must still explain the magnitude of the mean free path at lowest frequencies, where phonons interact with two-level systems of distributed level spacing. There is good support for this mechanism from the excess specific heat and the saturation of the absorption of ultrasonic waves, but we need a way of explaining the parameters in equation (11) from basic principles.

A CALCULABLE TWO-LEVEL SYSTEM

The following illustrates how one can calculate parameters of a two-level system in a better-defined situation: a solid solution of two atoms of different size. Consider a Ti-V alloy; the atomic radius of Ti is 15% larger than that of V. The thermal conductivity of Ti-V containing 34% V has been measured by Chandrasekhar et al. [15] below 0.1K. At those low temperatures the superconducting alloy has the thermal conductivity of a dielectric crystal: it is entirely due to lattice waves, and the phonon mean free path is not influenced by interactions with free electrons. These authors found a thermal conductivity proportional to T^2, and a specific heat[16] well above the expected lattice specific heat and proportional to T. It thus appears that the alloy contains a distribution of two-level systems, which the authors attribute to some atoms being in a bistable configuration as a result of the size difference between the atoms.

Let us assume that the only bistable configuration is one in which a Ti atom (surrounded by 8 atoms in that lattice) is surrounded by 6 Ti atoms and 2 V atoms, and that the two V atoms are diagonally opposite each other. The central Ti atom has then two stable positions on either side of its normal position at the center of its atomic cage. If x is the V concentration, the number of such sites per atom site is

$$c = 4 \, x^2 \, (1-x)^7 \tag{12}$$

provided there is no short-range order.

To find σ we use elasticity theory to calculate the fluctuating strain, and hence the strain energy, seen by the Ti atom due to the fluctuations in size of all the atoms beyond the nearest neighbor shell. This depends on the difference in atomic size and also on x. In the case considered here

$$\hbar\sigma/k \cong 8000 \text{ K} \tag{13}$$

while the observed excess specific heat $C_E/T = 1.4 \times 10^{-4} \text{J-mole}^{-1}\text{-K}^{-2}$

would demand a value of 6000K. Furthermore we would expect the specific heat for the complementary 67%V alloy to be less by a factor 32, because σ is the same, but c changes according to (12). Observations[16] show a reduction by at least a factor of 7.

To calculate the phonon scattering we must estimate A of equation (10). The effects of strain due to phonons can be related to the effects of static strains, hence to σ. However, there is an additional factor in A due to the quantum-mechanical overlap integral between the atomic wave-functions centered on the two stable sites. This factor is very model-sensitive, and only crude estimates seem possible at present. Thus we can calculate the excess specific heat with greater confidence than the thermal conductivity. Whether similar considerations can be used in amorphous systems remains to be seen.

REFERENCES

[1] C. Kittel, Phys. Rev. 75, 972 (1949).
[2] R. Berman, Proc. Roy. Soc. (London) A208, 90 (1951).
[3] P.G. Klemens, in "Thermal Conductivity 16" ed. D.C. Larsen, p. 15, Plenum Press, New York, 1983.
[4] P.G. Klemens, Phys. Rev. 119, 507 (1960).
[5] J.E. Parrott, Proc. Phys. Soc. (London) 81, 726 (1963).
[6] P.G. Klemens, in "Solid State Physics" eds. F. Seitz and D. Turnbull, vol.7, p.1, Academic Press, New York, 1958.
[7] R.C. Zeller and R.O. Pohl, Phys. Rev. B4, 2029 (1971); see other citations in ref. [13] below.
[8] P.G. Klemens, in "Non-Crystalline Solids" ed. V.D. Frechette, p. 508, J. Wiley & Sons, New York, 1960.
[9] P.G. Klemens, in "Physics of Non-Crystalline Solids" ed. J.A. Prins, p. 162, North Holland, Amsterdam, 1965.
[10] D.H. Damon, Phys. Rev. B8, 5860 (1973).
[11] W.A. Phillips, J. Low Temp. Phys. 7, 351 (1972).
[12] P.W. Anderson, B.J. Halperin and C.M. Varma, Phil. Mag. 25, 1 (1972).
[13] A.C. Anderson, in "Amorphous Solids" ed. W.A. Phillips, p. 65, Springer Verlag, Berlin (1981).
[14] E.H. Ratcliffe, Glass Technol. 4, 113 (1963).
[15] B.S. Chandrasekhar, H.R. Ott and H. Rudigier, Solid State Communications 42, 419 (1982).
[16] H.R. Ott, private communication.

SESSION M

Insulations I

SESSION CHAIRMEN

F. Cabannes
University - CNRS
Orleans, France

R. P. Tye
Dynatech R/D Co.
Cambridge, MA.

THE THERMAL TRANSMISSION PROPERTIES OF HIGH TEMPERATURE THERMAL INSULATION MATERIALS

R.P. Tye, A.O. Desjarlais, and S.E. Smith

Dynatech R/D Company
99 Erie Street, Cambridge, Massachusetts 02139

ABSTRACT

   The increased need for energy conservation measures for furnace and other high temperature applications has focused attention on the thermal performance of available thermal insulation materials. Thermal transmission properties have been measured on a number of alumino-silicate, alumina, and fibrous yttria stabilized zirconia-blanket, felt and board products in accordance with ASTM C177-76, "Steady-State Thermal Transmission Properties by Means of the Guarded Hot Plate." All measurements were made with the heat flow normal to the fiber direction.

   Thermal transmission property data are presented for the temperature range 100 to 1000C. The results are analyzed and discussed with regard to the following considerations:

- Comparisons with data in manufacturer's literature;
- Variability in recovered thickness and density when compared to nominal values;
- Effect of temperature difference on the thermal performance at the same mean temperature;
- Effects of density of a single manufactured product on the thermal performance;
- Comparison of similar products from different manufacturers;
- Relative performance of different fibers at a similar density of product;

INTRODUCTION

   As energy conservation measures become more important the ther-

mal transmission properties of high temperature refractory fiber
insulation materials become critical. Processes and systems have
to be analyzed in detail to justify the use of such materials which
are often expensive. For such analyses it is desirable to have
available thermal transmission properties over a wide temperature
range. In addition since these fibrous materials are available in
a number of densities the effects of radiation transport are of
direct interest. Radiation is the dominant mechanism of heat transfer at elevated temperatures [1,2] and the performance of an insulation will depend on its effectiveness in attenuating the radiative
heat transfer.

These fibrous materials are heterogeneous products. Their properties, particularly for the lower densities, can be influenced
markedly by variability in both density and thickness. Products are
manufactured and shipped to specifications of nominal density at a
nominal thickness. However, after unpacking it is often found that
the materials do not recover to these nominal values.

Finally, several different experimental methods have been or
are being used to evaluate the thermal transmission properties.
These include the guarded hot plate method [3], the flat plate calorimeter [4], and the hot wire method [5]. The guarded hot plate
can be used with large or small temperature differences across two
close to identical test specimens; the calorimeter, in general, uses
a large temperature difference across one specimen and the hot wire
technique involves only a small difference generated within a test
specimen. Results vary with the method and the variations need to
be quantified.

On behalf of the US Department of Energy, a preliminary study
of some of these factors was undertaken on a selected number of high
temperature fibrous insulations. The overall study encompassed
different densities of each of the major products and involved the
use of three measurement methods. The present paper describes the
major portion of the investigation which utilized the guarded hot
plate method.

The major objectives of the program were:

- provide reliable thermal performance data on available
  ceramic fiber products
- investigate variability in performance of different products
- compare the performance of similar products from different
  manufacturers
- study density effects in this product of one manufacturer
- compare performance of different fibers at similar product
  density
- study effect of temperature difference on thermal performance

at two particular mean temperatures
- provide initial data to assist in development of models of heat transmission in fibrous products

The complete test program consisted of the following items:

- evaluate a total of sixteen products covering five different manufacturers purchased from commercial sources at the appropriate nominal density at a nominal thickness of 25.4 mm
- undertake a preliminary characterization of density and thickness at various positions in the purchased materials and after stabilization for 48 h. at 1000C, followed by initial thermal performance measurements at 24 and 160C using the heat flow meter method [6]
- undertake high temperature thermal performance measurements to 1000C on specimen pairs fabricated to close to the nominal density utilizing either a data set GHP1 of six mean temperatures with a temperature difference, ΔT, of 50C, or a set GHP2 of the same six mean temperatures and ΔT, plus measurements at 300 and 800C with additional ΔT of 150 and 400C. The mean temperatures were 100, 300, 500, 700, 800, and 1000C with one repeat after attaining 1000C

DETAILS OF MATERIALS AND TEST PROCEDURES

Lots of each material to be tested were purchased by and the thicknesses and densities were measured at Oak Ridge National Laboratory. Individual specimens of each lot were placed in an oven at 1000C for 48 hrs. After cooling, individual 300 mm square specimen pairs of each were carefully packed and shipped to Dynatech. On receipt they were unpacked carefully, shaken gently, placed on a flat surface and allowed to recover for at least 16 hrs. The dimensions and masses were measured and densities derived in accordance with ASTM C167 [7] followed by the heat flow meter [6] measurements of apparent thermal conductivity at 24 and 160C.

Table 1 contains relevant details of the materials and the results of the preliminary measurements. The results indicate that there is considerable variability in the thickness and density of the fibrous products. This variability, while not having too significant an effect on the thermal transmission properties at or near room temperature, will have a considerable influence on the high temperature performance.

The variability encountered did have some effect on the final choice of test specimens for the guarded hot plate measurement. The original intention had been evaluation at a thickness of 25.4 mm. Due to the variability, however, it was decided to undertake the measurements on specimens at close to the nominal density, and for thicknesses as close to the nominal as could be attained. Thus a test

Table I

DETAILS OF MATERIALS INVESTIGATED

	MATERIAL	MANUFACTURER[+]	NOMINAL DENSITY* kg/m³	VARIABILITY % THICKNESS	VARIABILITY % DENSITY	λapp. 24C	λapp. 160C	DATA SET
A	Zirconia Board	A	480	+2	+2.5 -3.8	0.061	0.070	GHP2
B	Zirconia Blanket	A	96	-8.5 +2.5	-56 -32	0.047	0.047	GHP2
C	Aluminosilicate Blanket	B	64	-22.5 -18.7	+41.2 +38.4	0.034	0.051	GHP2
D	Aluminosilicate Blanket	B	96	-51.1 -41.9	+60.8 +67.3	0.034	0.046	GHP2
E	Aluminosilicate Blanket	B	160	<1	11.5 11.5	0.052	0.065	GHP2
F	Aluminosilicate Felt	C	64	-9.5	+52.8 +35.8	0.034	0.051	GHP2
G	Aluminosilicate Blanket	C	96	+5.1 +5.9	-20.4 -15.4	0.033	0.050	GHP2
H	Alumina Blanket	D	96	+35.8 +24	-60 -41.8	0.033	0.050	GHP1
I	Alumina Board	A	240	+16 0	-11.4 +3.5	0.056	0.066	GHP1
J	Aluminosilicate Blanket	B	128	-17.6 -16	+24.7 25.8	0.036	0.050	GHP1
K	Aluminosilicate Blanket	C	128	0.3 -12.4	-2.6 +16	0.037	0.058	GHP1
L	Aluminosilicate Blanket	E	64	-25.1	-43.8 -53.4	0.035	0.060	GHP1
M	Aluminosilicate Blanket	E	96	-1.6	-52.3 -29.6	0.035	0.057	GHP1
N	Aluminosilicate Blanket	E	128	+46 36.5	-18.5 -29.8	0.034	0.056	GHP1
O	Aluminosilicate Felt[X]	C	96	-11.4 -10.4	0.5 +3.5	0.033	0.048	GHP1
P	Aluminosilicate Felt	C	96	-5.4	-5 -5.5	0.035	0.055	GHP1

+ Manufacturers were Babcock and Wilcox Co., Carborundum Co., Imperial Chemical Industries, Johns Manville Corporation, and Zircar Products

* At 25.4 mm   X Not heat-treated

specimen 200 mm diameter was cut from each of the square pieces and weighed. It was then either compressed to or fluffed up to a uniform thickness which would provide a test specimen at or close to the nominal density at or close to the nominal thickness. Care was taken to ensure that the thicknesses of each specimen of a pair were the same, but there were small differences in the densities of a pair. Details of the specimens prepared for the guarded hot plate measurements are given in Table II.

All measurements were made in accordance with ASTM C177-6 [3] using a 200 mm diameter horizontal high temperature guarded hot plate which has been described elsewhere [8,9]. In these measurements, every effort was made to ensure that heat losses were kept to a minimum. This was accomplished by ensuring that the edge loss ratio, defined in C177 as the difference in the edge and mean temperatures divided by the temperature difference across the specimen was always maintained at 0.04 or less. This is well below the 0.1 maximum value allowed by the specification.

The total ±1.4% determinate error parameters affecting the accuracy of measurement are:

power	±0.1%
dimensions	±0.2%
temperature difference	±1%
electronic measurement circuits	±0.1%

In all cases the thickness of the test specimen was close to 25 mm. This was well within the limits of the recommended thickness in the specification for measurements to ±5% at elevated temperatures using a guarded hot plate having overall dimensions of 200 mm.

Since no reference materials are available for measurements at elevated temperatures, it is not yet possible to establish the limits of accuracy of the guarded hot plate. However some measurements on the same or similar specimens have been carried out on both MinK-2000 [10] and on an aluminosilicate material [11]. The results of these measurements indicate that there is agreement between different guarded hot plates on the same material of ±4% or better over the range 200 to 900C. Consideration of the above factors results in the conclusion that the overall accuracy of the apparent thermal conductivity measurements in this study is better than ±6%.

RESULTS AND DISCUSSION

The λapp results for the sixteen materials tested are presented in Table II. The values of λapp were derived from smoothed curves drawn through the individual data points which were each obtained within ±5C of the desired mean temperature. Table III contains details of the measurements undertaken to investigate both the effect

Table II

THERMAL PERFORMANCE OF CERAMIC FIBER INSULATION: COLLECTED RESULTS OF HEAT
TRANSMISSION PROPERTIES OF SIXTEEN FIBROUS INSULATIONS

MATERIAL	TEST THICKNESS mm	DENSITY kg/m³		PAIR VARIATION %	λapp WITH ΔT 50C AND MEAN TEMPERATURE OF						
		NOMINAL	TEST AVERAGE		100	300	500	700	800	1000	300C
A	25.8	480	478	0	0.065	0.080	0.092	0.115	0.13	0.16	0.081
B	24.8	96	78.5	8.5	0.040	0.064	0.100	0.155	0.185	0.25	0.064
C	24.8	64	73.7	3.5	0.045	0.085	0.15	0.24	0.28	0.37	0.085
D	21.3	96	127	3.6	0.041	0.066	0.115	0.18	0.23	0.32	0.067
E	25.2	160	180.3	3.0	0.059	0.081	0.11	0.145	0.165	0.205	0.080
F	21.4	64	104	2.8	0.044	0.068	0.11	0.155	0.19	0.265	0.067
G	23.4	96	93	0	0.044	0.075	0.12	0.185	0.205	0.32	0.074
H	24.8	96	94.6	9.6	0.042	0.069	0.11	0.17	0.21	0.31	0.068
I	25.4	240	240	4.1	0.062	0.08	0.105	0.13	0.15	0.19	0.08
J	25.0	128	141	1.6	0.043	0.075	0.115	0.18	0.23	0.30	0.074
K	22.7	128	147.5	5.8	0.046	0.085	0.14	0.21	0.25	0.32	0.085
L	27.9	64	67.3	2.4	0.047	0.088	0.155	0.24	0.29	0.40	0.088
M	21.6	96	83.3	22.7	0.047	0.083	0.135	0.195	0.24	0.33	0.084
N	24.8	128	147.5	19.0	0.045	0.068	0.105	0.155	0.19	0.27	0.068
O	24.8	96	93	0.8	0.042	0.071	0.12	0.18	0.23	0.31	0.072
P	21.3	96	109	0.1	0.042	0.069	0.105	0.15	0.185	0.27	0.069

Table III

THERMAL PERFORMANCE OF CERAMIC FIBER INSULATION:
(A) EFFECT OF TEMPERATURE DIFFERENCE

TEMPERATURE, C MEAN	ΔT	ZIROCONIA		ALUMINOSILICATE				
		$79kg/m^3$	$478kg/m^3$	$73.7kg/m^3$	$93kg/m^3$	$104kg/m^3$	$127kg/m^3$	$180kg/m^3$
300	50	0.064	0.080	0.085	0.0765	0.068	0.066	0.0805
	150	0.066	0.081	0.086	0.079	0.071	0.066	0.0815
	400	0.071	0.081	0.089	0.081	0.072	0.069	0.083
800	50	0.185	0.13	0.28	0.205	0.19	0.23	0.165
	150	0.195	0.135	0.30	0.22	0.205	0.23	0.175
	400	0.21	0.145	0.31	0.23	0.22	0.24	0.185

(B) EFFECT OF FIBER TYPE FOR SIMILAR DENSITY PRODUCTS

TEMPERATURE, C	Zirconia (78.5)	ALUMINA (94.6)	ALUMINOSILICATE (83.3)	ALUMINOSILICATE (93)
100	0.04	0.042	0.047	0.044
500	0.10	0.11	0.135	0.12
1000	0.25	0.31	0.33	0.32

DENSITY, kg/m^3, in parentheses

of increasing temperature difference and of the effect of fiber type for the same order of product density.

All materials exhibit the typical approximate T^3 proportionality performance versus temperature behavior as the mean temperature is increased. The high temperature radiation component becomes increasingly large as the density decreases. As expected the curves for the higher density products A, E, and I are much flatter than those of the lower density materials. In addition, the small but significant increases in λapp at 300 and 800C obtained as the temperature difference is increased successively from 50 to 150 to 400C are a further indication of the extent of radiation transmission in these fibrous materials particularly at the lower densities.

The present results have been compared with those taken from the literature of the appropriate manufacturer for the nominal density product. In general, it is found that there is agreement of 10% or better below 500C increasing to 25% or better at higher temperatures up to 1000C especially when density differences have to be considered. As an example, the data at 1000C is summarized in Table IV.

Table IV

Summary of Data Comparison at 1000C

DIFFERENCE, %	NUMBER	COMMENT
25 > X > 15	7	All manufacturers low
10 > X > 5	6	1 manufacturer high
5 > X > 0	3	All manufacturers low

It is seen that the present λapp results, with one exception, are higher than those of the manufacturer. For the majority of products, any differences in λapp will then increase as the mean temperature is increased above 1000C due to the divergence mentioned earlier. While density differences need to be considered, this factor alone is not seen to be large enough to account for all of the variability between the results of the present study and the manufacturer's information.

The major factor to be considered in the total analysis of the results concerns the impact of the density and thickness variability when combined with possible low values of λ in the literature. Practical thermal performance values may be considerably less than the design figures if the products do not meet any or all of the thickness, density, and thermal conductivity criteria.

The following major conclusions can be drawn:

1.) Fourteen of the materials exhibit significant (>10%) variation
     in thickness or density or both from the nominal manufacturers
     specification values. This is particularly noticeable for the
     lower density products. It was possible to undertake tests on
     samples reasonably close to the nominal density by either com-
     pressing or fluffing the original pieces to thicknesses some-
     what different to the nominal.

2.) The original heat treatment given to these materials appears to
     provide a stabilization since no significant changes in dimen-
     sions or masses were noted after the measurements cycle. Only
     one material type was evaluated without heat-treatment. The
     results for this material, O and for the same heat treated
     material, G, indicate that any difference in property at the
     same mean temperature is less than 6%.

3.) For the same generic products, small differences in absolute
     value of λapp exist between similar density products of differ-
     ent manufacturers. Some of these may be attributed to the
     density variations in the products although there may be some
     differences in the particular fiber of a respective manufacturer.

4.) Comparison of the results for λapp of the pure alumina material
     H and those for the aluminosilicates at approximately the same
     nominal density, 96 kg/m^3, indicates that the thermal performance
     of the pure fiber product is better by 5% at 1000C. Furthermore
     for the lower density pure zirconia product B at 78.5 kg/m^3,
     there is an even larger improvement in performance of over 25%.
     This confirms that the type, size, and form of the fibers does
     have a very significant effect [12] on the thermal performance.

SUMMARY

     The thermal performance of a number of different refractory
fiber insulation products of different densities and fiber types have
been measured up to 1000C. Results are presented for the products
to provide designers and analysts with reliable data for their needs.
Some indications of the extent of radiation transmission at elevated
temperatures are also presented. Further work is required in a
number of areas in order to obtain more complete information on the
behavior of high temperature fiber insulation products of different
fiber types, sizes, and shot and impurity contents. In particular it
is recommended that studies include:

  • a wider selection of materials and test methods
  • anisotropy of thermal performance
  • performance in different working environments
  • development of high temperature reference materials for
    validation of test techniques

- development of a new method(s) suitable for temperatures at 1000C
- development of reliable analytical and empirical models

ACKNOWLEDGEMENT

This study was carried out under Contract 72X038961V for Oak Ridge National Laboratory with Dr. D.L. McElroy, as program manager. The authors wish to thank their colleague, M.J. Saunders, who assisted with some of the experimental work.

REFERENCES

1. Linford, R.M.F., Schmitt, R.J., and Hughes, T.A., "Radiative Contribution to the Thermal Conductivity of Fibrous Insulations", Heat Transmission Measurements in Thermal Insulations, ASTM STP 544, ASTM Philadelphia, PA, pp. 68 - 84, (1974

2. Striepens, A.H., "Heat Transfer in Refractory Fiber Insulations", Thermal Transmission Measurements of Insulations, ASTM STP 660, R.P. Tye, Editor, ASTM Philadelphia, PA, pp. 293 - 309, (1978).

3. ASTM C177 (1945, revised 1976), A.S.T.M. Stand.

4. ASTM C201 (1947, revised 1979), A.S.T.M. Stand.

5. Jackson, A.J., Adams, J., and Millar, R.C., "Thermal Conductivity Measurements on High-Temperature Fibrous Insulations by the Hot-Wire Method", Thermal Transmission Measurements of Insulations, ASTM STP 660, R.P. Tye, Editor, ASTM Philadelphia, PA, pp. 154 - 171, (1978).

6. ASTM C518 (1967, revised 1976), A.S.T.M. Stand.

7. ASTM C167 (1964, revised 1976), A.S.T.M. Stand.

8. Tye, R.P. "Effects of Edge Losses on the Thermal Conductivity of Thermal Insulations at High Temperatures", Rev. Int. Hautes Temper. et Refract., 7, 308, (1970).

9. Brazel, J.P. and Tye, R.P., "Thermal Characterization of Reuseable External Insulation for the Space Shuttle", High Temperatures-High Pressures, 4, 639, (1972).

10. Tye, R.P., "The Thermal Conductivity of MINK-2000 Thermal Insulation in Different Environments to High Temperatures", Proceedings Ninth Thermal Conductivity Conference, USAEC Conference-691002, pp. 341 - 351, (1967).

11. Ober, D.M., "ASTM C8 Round Robin on Thermal Conductivity of a Ceramic Fiber Insulation Material", In Course of Publication.

12. Miller, W.C. and Scripps, T.A., "Relating Apparent Thermal Conductivity to Physical Properties of Refractory Fiber", Ceramic Bulletin, 61 (7), 711, (1982).

STATUS OF THERMAL CONDUCTIVITY STANDARD REFERENCE MATERIALS AT THE

NATIONAL BUREAU OF STANDARDS

J. G. Hust

Chemical Engineering Science Division
Center for Chemical Engineering
National Engineering Laboratory
National Bureau of Standards
Boulder, Colorado 80303

ABSTRACT

This paper describes the present status of NBS thermal conductivity Standard Reference Materials (SRM's) and Calibrated Transfer Specimens (CTS's). Included are the metal SRM's, tungsten, electrolytic iron, and austenitic stainless steel. Also discussed is graphite, a soon-to-be-established SRM, and candidate SRM's, such as black quartz. Finally, a description is given of the insulation SRM's and CTS's.

INTRODUCTION

The National Bureau of Standards (NBS) has supplied calibrated transfer specimens (CTS) to the insulation industry for over a half century. These calibrated reference specimens are distinguished from Standard Reference Materials (SRM's) in that each CTS supplied to the user is measured by NBS. An SRM is defined as a lot of material whose inhomogeneities and instabilities have been characterized, so that each individual specimen distributed for use does not have to be measured. Generally this leads to a somewhat higher degree of uncertainty since it includes the measurement uncertainty as well as the effect of inhomogeneity. More recently, NBS has initiated a program to establish SRM's for thermal and electrical conductivity over a wide range of conductivity.

The earlier efforts by NBS to establish thermal SRM's were concerned with relatively high conductivity solids. A brief review of this earlier work, including similar efforts by other laboratories, is given by Hust [1]. Since that review, the NBS SRM's were extended in temperature range and they were involved in round-robin

327

programs, such as those described by Minges [2] and Berman, Hardy, Sahota, Hust, Tainsh [3]. The results were used to further improve the reliability and range of these SRM's.

During the late 1970's, NBS decided to establish insulation SRM's, i.e., very low conductivity standards. These SRM's are not intended to totally replace the CTS's, supplied for so many years by NBS, but rather to make available the less expensive SRM's for those users who can tolerate a somewhat larger uncertainty.

This paper describes the status and recent progress on the thermal SRM's available from NBS. The following description is divided into four ranges of conductivity: a) high conductivity, b) medium conductivity, c) low conductivity, and d) very low conductivity. The high conductivity range includes two pure metals. The medium conductivity range includes steel and graphite. The low conductivity range is presently not covered by any SRM. Finally, the very low conductivity range includes fibrous insulations, with air as the interstitial gas. The very low conductivity materials frequently exhibit non-conductive heat transfer mechanisms. Because of this, measured thermal conductivity is not a true bulk property and it is referred to as apparent thermal conductivity. The modifier "apparent" signifies that the measured conductivity is dependent, to some degree, on the specimen dimensions, especially the thickness in the direction of heat flow. It is generally dependent also on boundary emittances. The approximate ranges and temperature dependencies of conductivity are illustrated in figure 1.

HIGH CONDUCTIVITY

The first thermal conductivity SRM's established by NBS are the pure metals, electrolytic iron (SRM's 1463 and 1464) and sintered and arc cast tungsten (SRM's 1465, 1466, 1467, 1468, and 1469) are described by Hust and Giarratano [4,5]. Note that the SRM numbers have been changed since the original issue in 1975. Since that time, these SRM's have been measured by other laboratories and, recently they were used as round-robin specimens in a project sponsored by the Committee on Data for Science and Technology (CODATA) described by Minges [2] and Berman, et al. [3]. Because of the new data, NBS decided to update the SRM's accordingly. Comparisons of the new data are given by Hust and Lankford [6]. Data from laboratories with established expertise are intercompared for the electrolytic iron SRM's in figure 2, and for the tungsten SRM's in figure 3. For the legend of the data sources the reader is referred to Hust and Lankford [6]. No research is currently underway at NBS to establish additional SRM's in this conductivity range. However, NBS has relatively large stocks of both well-characterized pure copper and pure aluminum that may be established as research materials (RM's) to establish

328

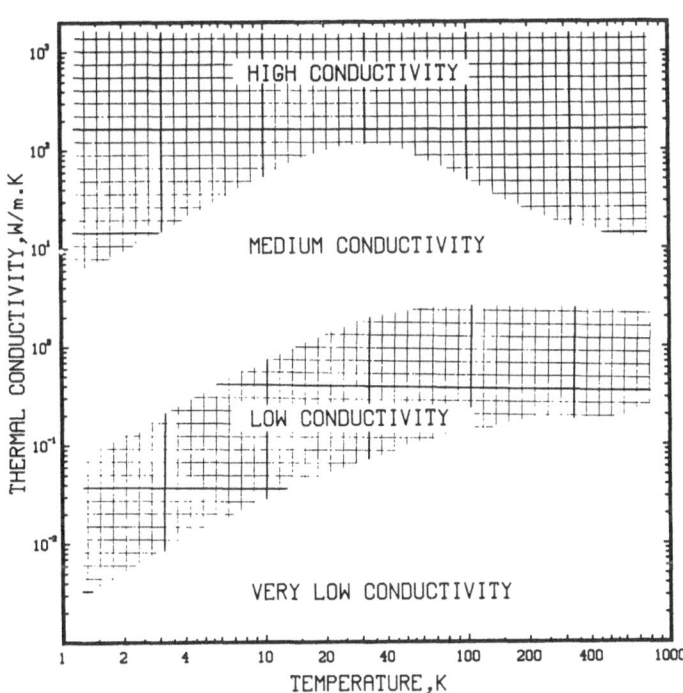

Figure 1. Approximate Ranges and Temperature Dependencies of Conductivity for the Four Ranges Described in the Text.

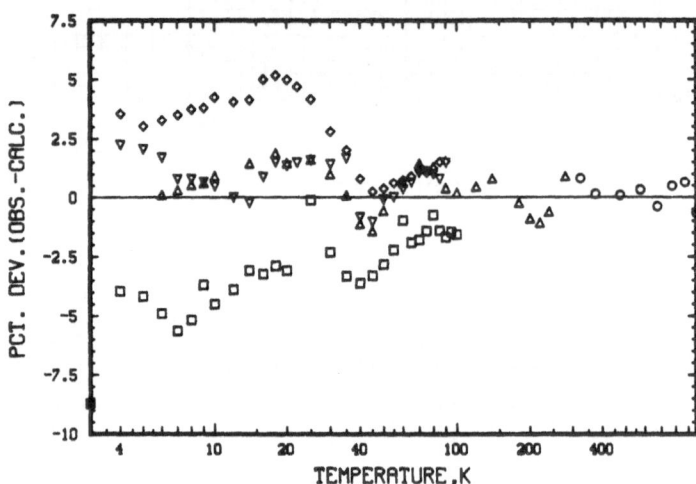

Figure 2.   Thermal Conductivity Deviations of Experimental Data
from the Recommended Values for the Electrolytic Iron
SRM.   For Details on the Sources of Data see Reference 6.

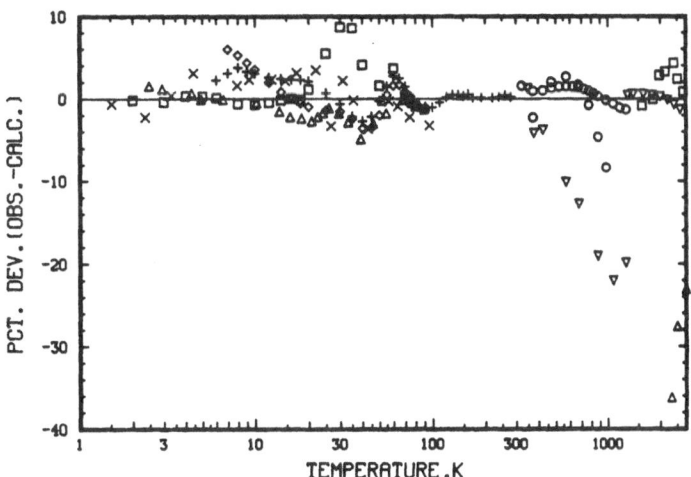

Figure 3. Thermal Conductivity Deviations of Experimentl Data from
the Recommended Values for the Tungsten SRM. For
Details on the Source of Data see Reference 6.

the basis for future SRM's. Because the NBS stock of tungsten is nearly depleted, consideration is being given to acquiring additional stock of similar material.

MEDIUM CONDUCTIVITY

The first SRM established in the medium conductivity range is a stainless steel (SRM's 1460, 1461, and 1462), Hust and Giarratano [7]. Note again that the SRM numbers have been changed since the original issue. This SRM was also updated by Hust and Lankford [6] as a consequence of the CODATA and other measurements. All of the data for this SRM are intercompared in figure 4. The legend for the data sources is given by Hust and Lankford [6].

A second NBS thermal conductivity SRM that falls in this category is AXM-5Q1 graphite. This material was investigated in an earlier project sponsored by the Advisory Group for Aerospace Research and Development (AGARD-NATO) as well as the CODATA project (see Hust [1] and Minges [2]). Several properties have now been measured and NBS is in the process of establishing this lot of material as an SRM of thermal conductivity and electrical resistivity. This lot of material has been found to be considerably more inhomogeneous than the metals; however, most of the effect of these inhomogeneities can be accounted for by simple room temperature electrical resistivity and density measurements. The intercomparison of the data, corrected for these room temperature variations is illustrated in figure 5. For a more detailed description of this extensive effort see reference [8].

LOW CONDUCTIVITY

NBS has not established any thermal SRM's in this range. However, preliminary research was initiated on Pyroceram 9606* by D. R. Flynn of NBS during the early 1960's (see earlier proceedings of these conferences for additional information). Currently, NBS has a low-level effort to establish this material or one of similar thermal characteristics as an SRM. It is recognized that a strong need for such an SRM exists. The CODATA task group has also initiated studies on similar materials. It is anticipated that at least two years will elapse before this SRM is established.

VERY LOW CONDUCTIVITY

During the past few years NBS has been relatively active in the area of insulation SRM's and CTS's. This interest and activity stems mainly from the existence of mandatory test methods requiring the use of reference standards. The principal test methods,

---

*The use of this tradename is necessary for clarity and no product endorsement is intended by NBS.

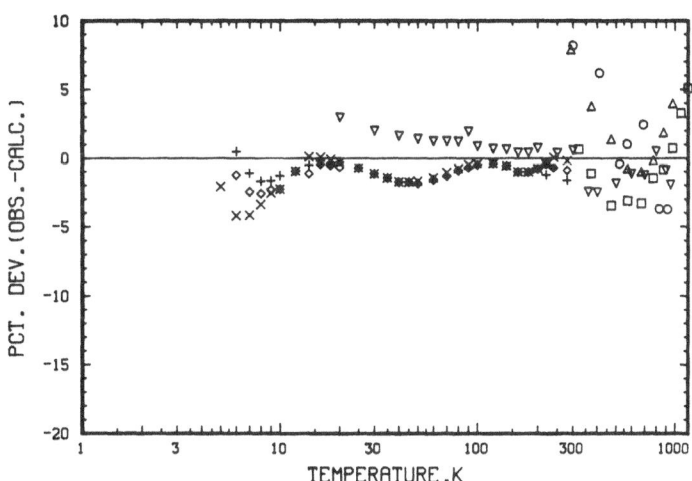

Figure 4. Thermal Conductivity Deviations of Experimental Data
from the Recommended Values for the Stainless Steel SRM.
For Details on the Sources of Data see Reverence.

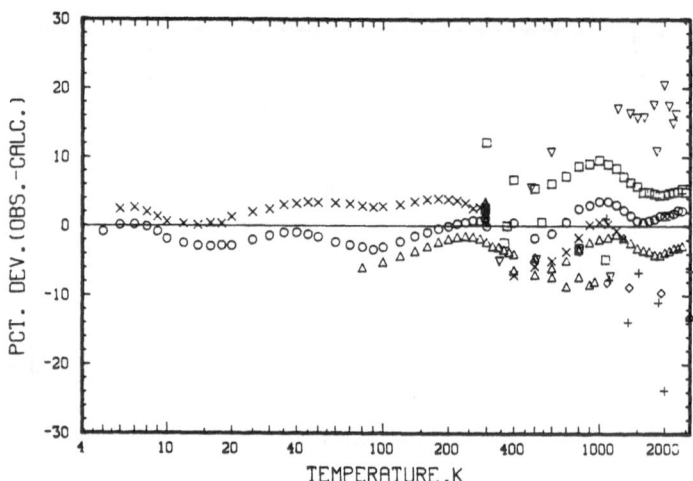

Figure 5. Thermal Conductivity Deviations of Experimental Data
from the Recommended Values for the Graphite SRM. For
Details on the Sources of Data see Reference 8.

produced by the American Society for Testing and Materials (ASTM), are C-177 and C-518. Accurate SRM's are especially important in the building insulation industry because of the large monetary impact, both to the consumer and to the producer, that may arise from a relatively small change in required thermal transmission properties.

As mentioned earlier, NBS supplied and continues to supply CTS's for the standardization of insulation products. The present stock of CTS's is glass fiber blanket of 2.54 and 7.62 cm (1 and 3 in) thickness. The specimens are also calibrated in other thicknesses up to 15.24 cm (6 in) by stacking combinations of these specimens. CTS calibrations are performed only in the vicinity of room temperature for specific apparatus as requested by the user. The density of this material is 9.6 ± 1.0 kg/m^3 (0.6 ± 0.07 pcf). The specimens are available up to 1.2 m squares (48 in squares) with guard masks. Further details are given by Rennex, Jones, and Ober [9].

Several years ago, NBS recognized the need for SRM's in this range of conductivity and established a glass fiber board as a thermal SRM (SRM 1450). This SRM was quickly exhausted and another lot was established (SRM 1450a). Again the lot was quickly sold out and the current lot was established (SRM 1450b). This type of material had been used as CTS for a long time and therefore its characteristics had been well studied. The nominal density of this material is 120 kg/m^3 (8 pcf) and the specimens are available as 70 cm squares (24 in squares). This 2.54 cm thick material is currently certified for the temperature range from 255 to 330 K. This certification will soon be extended down to 100 K. The apparent thermal conductivity of this SRM is shown in figure 6.

NBS is currently measuring the thermal transmission characteristics of a lower density glass fiber material (blanket) for use as an SRM. Its nominal density is 14 kg/m^3 (0.9 pcf) and its nominal thickness is 2.54 cm. It is anticipated that this SRM will be available within one year. The apparent thermal conductivity of this material is illustrated in figure 6. The reader is referred to references [10] to [13] for additional details on these studies.

Both the glass fiber board and glass fiber blanket described above contain phenolic binder and therefore are limited to use below about 350 K. Because very low conductivity SRM's are needed at higher temperatures, NBS has initiated research on candidate materials for high temperature applications. Current research is directed toward an upper temperature of 800 K. A guarded-hot-plate apparatus for this temperature range will soon be in operation at NBS (Boulder). This effort to establish higher temperature, very low conductivity SRM's will be conducted in cooperation with the

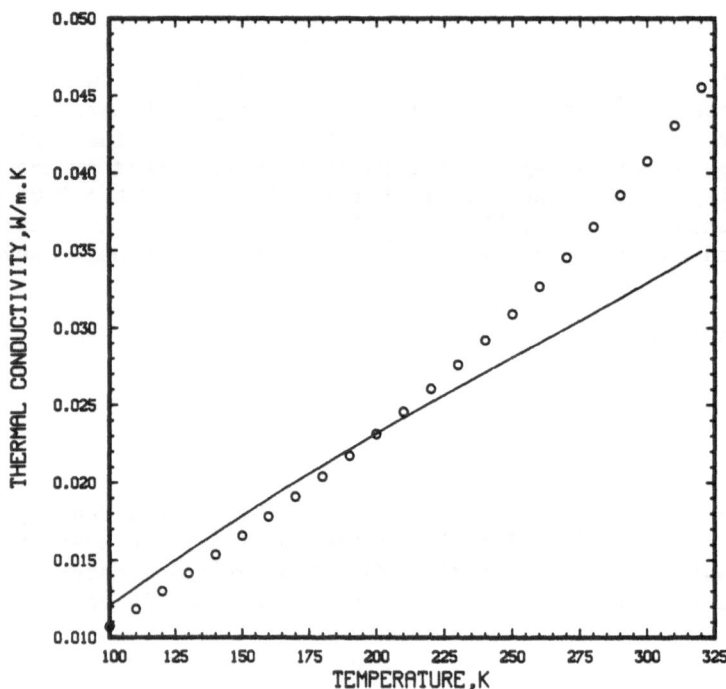

Figure 6. Apparent Thermal Conductivity of Glass Fiber Board and
Blanket SRM's. See Reference 10.

O - Blanket   ——— Board

effort by the insulation industry to establish in-house reference materials. It is anticipated that a higher temperature SRM with a conductivity similar to the glass fiber board will be available within two years.

REFERENCES

[1]   Hust, J. G., Standard Reference Materials for Thermal Conductivity and Electrical Resistivity, Proceedings of the 14th International Thermal Conductivity Conference, ed. P. G. Klemens and T. K. Chu, Plenum Press, New York (1976) pp. 221-231.

[2]   Minges, M. L., The Standard Reference Materials and Data Programs of the CODATA Task Group on Thermophysical Properties, Proceedings of the 17th International Thermal Conductivity Conference, ed. J. G. Hust, Plenum Press, New York (1983) pp. 73-92.

[3]   Berman, R., Hardy, N. D., Sahota, M., Hust, J. G., and Tainsh, R. J., Standard Reference Materials for Thermal Conductivity Below 100 K, Proceedings of the 17th International Thermal Conductivity Conference, ed. J. G. Hust, Plenum Press, New York (1983) pp. 105-116.

[4]   Hust, J. G. and Giarratano, P. J., Thermal Conductivity and Electrical Resistivity Standard Reference Materials: Electrolytic Iron, SRM's 734 and 797, from 4 to 1000 K, Nat. Bur. Stand. (U.S.) Spec. Publ. 260-50 (1975).

[5]   Hust, J. G. and Giarratano, P. J., Thermal Conductivity and Electrical Resistivity Standard Reference Materials: Arc Cast and Sintered Tungsten, SRM's 730 and 799, from 4 to 3000 K, Nat. Bur. Stand. (U.S.) Spec. Publ. 260-52 (1975).

[6]   Hust, J. G. and Lankford, A. B., Update of Thermal Conductivity and Electrical Resistivity SRM's of Electrolytic Iron, Tungsten, and Stainless Steel, Nat. Bur. Stand. (U.S.) Spec. Publ. 260-90 (1984).

[7]   Hust, J. G. and Giarratano, P. J., Thermal Conductivity and Electrical Resistivity Standard Reference Materials: Austenitic Stainless Steel, SRM's 735 and 798, from 4 to 1200 K, Nat. Bur. Stand. (U.S.) Spec. Publ. 260-46 (1974).

[8]   Hust, J. G., Characterization of a Fine-Grained, Isotropic Graphite for Use as NBS SRM's for Thermophysical Properties from 4 to 2500 K, Nat. Bur. Stand. (U.S.) Spec. Publ. 260-89 (1984).

[9]   Rennex, B. G., Jones, R. R., and Ober, D. G., Development of Calibrated Transfer Specimens of Thick, Low-Density Insulation, Proceedings of the 17th International Thermal Conductivity Conference, ed. J. G. Hust, Plenum Press, New York (1983) pp. 419-426.

[10] Smith, D. R. and Hust, J. G., Effective Thermal Conductivity
      of Glass-Fiber Board and Blanket Standard Reference
      Materials, Proceedings of the 17th International Thermal
      Conductivity Conference, ed. J. G. Hust, Plenum Press,
      New York (1983) pp. 483-496.

[11] Van Poolen, L. J., Hust, J. G., and Smith, D. R., A Model of
      Apparent Thermal Conductivity for Glass-Fiber Insulations,
      Proceedings of the 17th International Thermal Conductivity
      Conference, ed. J. G. Hust, Plenum Press, New York (1983)
      pp. 777-788.

[12] Siu, M. C. I., Comparison of Results of Measurements Made on
      a Line-Heat-Source and a Distributed-Heat-Source Guarded-Hot-
      Plate Apparatus, Proceedings of the 17th International
      Thermal Conductivity Conference, ed. J. G. Hust, Plenum
      Press, New York (1983) pp. 413-418.

[13] Siu, M. C. I., Fibrous Glass Board as a Standard Reference
      Material for Thermal Resistance Measurement Systems, STP 718,
      eds. D. L. McElroy and R. P. Tye, American Society for Test-
      ing and Materials, Philadelphia, PA (1980) pp. 343-360.

APPARENT THERMAL CONDUCTIVITY MEASUREMENTS

BY AN UNGUARDED TECHNIQUE*

R. S. Graves, D. W. Yarbrough, and D. L. McElroy

Martin Marietta Energy Systems, Inc.
Metals and Ceramics Division
Oak Ridge National Laboratory
Oak Ridge, Tennessee 37831

An unguarded longitudinal heat flow apparatus for measuring the apparent thermal conductivity (λ_a) of insulations was tested with mean specimen temperatures from 300 to 330°K on samples up to 0.91 m wide, 1.52 m long, and 0.15 m thick. Heat flow is provided by a horizontal electrically heated Nichrome screen that is sandwiched between test samples that are bounded by temperature controlled copper plates and 9 cm of mineral fiber insulation. A determinate error analysis shows λ_a measurement uncertainty to be less than ±1.7% for insulating materials as thin as 3 cm. Three-dimensional thermal modeling indicates negligible error in λ_a due to edge loss for insulations up to 7.62 cm thick when the temperature difference across the sample is measured at the screen center. System repeatability and reproducibility were determined to be ±0.2%.

Differences of λ_a results from the screen tester and results from the National Bureau of Standards were 0.1% for a 10-kg/m^3 Calibration Transfer Standard and 0.9% for 127-kg/m^3 fibrous glass board (SRM 1450b). Measurements on fiberglass and rock wool batt insulations showed the dependence of λ_a on density, temperature, temperature difference, plate emittance, and heat flow direction. Results obtained for λ_a as a function of density at 24°C differed by less than 2% from values obtained with a guarded hot plate.

These results demonstrate that this simple technique has the accuracy and sensitivity needed for useful λ_a measurements on thermal insulating materials.

*Research sponsored by the Office of Building Energy Research and Development, U.S. Department of Energy, under contract DE-AC05-84OR21400 with the Martin Marietta Energy Systems, Inc.

INTRODUCTION

This paper describes a large unguarded apparatus for measuring the apparent thermal conductivity (λ_a) of thermal insulations from 300 to 330 K that was tested at the Oak Ridge National Laboratory (ORNL). The ORNL apparatus uses an electrically heated 0.91 m by 1.52 m nichrome wire screen as the heat source and hot boundary for vertical, one-dimensional steady-state heat flow. The large dimensions of the test specimens and the low thermal conductivity of the screen heater make one-dimensional heat flow attainable in the central region of the tester without edge guarding and permit use of Fourier's law in finite difference form to calculate λ_a:

$$\lambda_a \approx Q \cdot \ell / A \cdot \Delta T \tag{1}$$

Equation (1) is exact for samples with λ_a that are constant or that depend linearly on temperature, in which case the measured value is taken to be at the mean specimen temperature. The idea of an unguarded system for λ_a measurements has previously been used by Jury et al.[1] for pipe insulations. The concept of an unguarded tester with longitudinal heat flow for thermal insulations has been discussed by Moore et al.,[2,3] Niven and Geddes,[4] Gilbo,[5] and Hager.[6,7,8]

The ORNL tester has been thermally modeled with a computer program, HEATING5[9] to determine the conditions for which one-dimensional heat flow will be realized. Measurements on a U.S. National Bureau of Standards (NBS) Calibration Transfer Standard[10] and a standard reference material[11] provide direct validation of the equipment design and procedure.

The ORNL tester was developed to demonstrate that a simple unguarded system can be used to obtain accurate λ_a values for a variety of insulations including batts, boards, loose-fills, and rectangular insulating systems. Furthermore, the small heat capacity of the heater reduces time to achieve steady-state and may provide an opportunity for transient measurements that is precluded in massive hot plate systems.

Three-dimensional thermal modeling with HEATING5 was used to find conditions for which one-dimensional heat flow is established and to show that heat exchange along the edges does not significantly affect the experimentally measured λ_a. Figure 1 is a schematic of the apparatus constructed and modeled. Modeling simulated the performance of the tester with sample in place and power applied to the nichrome screen. The computer-simulated temperature distributions were analyzed to determine the λ_a that would be measured at specific locations in the system. Simulations were

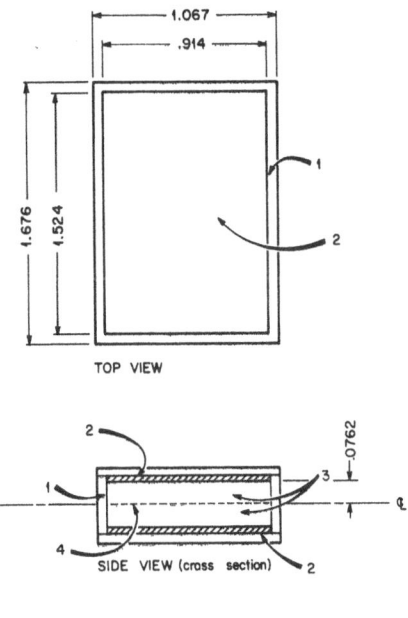

1. INSULATION   2. COPPER PLATES

3. SAMPLE   4. SCREEN HEATER

DIMENSIONS ARE IN METERS — NOT TO SCALE

Figure 1:   Schematic diagram of ORNL tester.

used to study the effect of sample λ_a, sample thickness, thermal resistance of edge insulation, and screen power on the ratio λ_a(simulation)/λ(input). For measurements made with mean sample temperatures from 295 to 330 K, the ratio is greater than unity and increases as the ΔT across the test sample is taken at points displaced from the center of the screen. Calculated values of percent error for sample thicknesses of 0.0762 m, 0.1524 m, and 0.3048 m are shown in Fig. 2 as a function of the position of the simulated ΔT measurement. The fiberglass batt edge insulation was assigned a thermal conductivity of 0.043 W/m·K.

The modeling calculations show that samples up to 0.076 m thick can be measured with edge-induced error below 0.2% if at least 0.076 m of edge insulation is used and ΔT is determined within 0.15 m of the center of the screen. Similarly, the modeling results show edge-loss induced errors above 1% will be encountered with 0.1524 m samples, and over 10% edge-loss error must be associated with 0.3048 m samples. The results in Fig. 2 also show that doubling the edge insulation thickness has a small effect on edge-loss error.

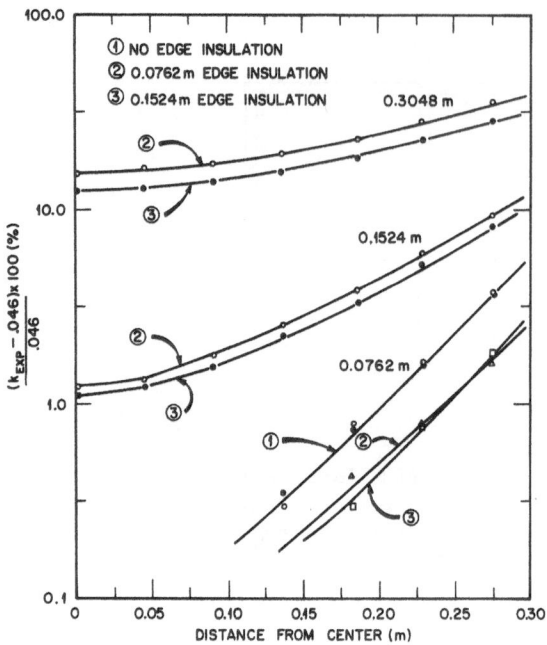

Figure 2:   Simulation results for ORNL tester. Percent error
[λ(simulation) - λ(input)] · 100/λ(input), versus
distance from the center of the screen

ESSENTIAL FEATURES OF THE EXPERIMENTAL APPARATUS

The ORNL tester consists of a pair of 1.14 m by 1.75 m temperature-controlled copper plates fabricated from 0.012 m thick
copper stock. An instrumented rectangle of nichrome wire screen[12]
0.00064 m thick, 0.91 m wide, and 1.52 m long, with electrical
power cables attached, is positioned between the horizontal copper
plates. Samples to be tested are placed between the copper plates
and the screen with spacing maintained by eight micarta spacer
tubes of known length ($\pm 1 \times 10^{-5}$ m). A measurement of electrical
power input to the central region of the screen is obtained from
the voltage drop across electrical taps on the edge of the screen
spaced 0.305 m on each side of the centerline. The dc current is
determined from the voltage drop across a 0.01Ω standard resistor
in series with the screen.[13] Electrical power is supplied from a
stable dc power supply.[14]

Screen and plate temperatures are measured with forty 36-gauge
Type E thermocouples referenced to an ice-water bath and read with
a Leeds and Northrup K-5 potentiometer.[15] Fifteen thermocouples
are located on each copper plate and 10 thermocouples are thermally

bonded[16] to the screen. The temperature of the copper plates is controlled by circulating water from a pair of Braun constant temperature baths[17] through coils attached to the plates on the sides away from the screen and monitored by rotameters.[18] Details of the screen heater and positioning frame are shown in Fig. 3 and a photograph of the apparatus is shown in Fig. 4. Additional information about the design and construction of the ORNL tester are contained in a paper by McElroy et al.[19]

Operation of the ORNL tester with two-sided heat flow involves installation of a pair of samples to form a cold plate-sample-screen-sample-cold plate vertical stack. The thermal resistances of the samples above and below the screen are matched as closely as possible so that total power input provides equal heat flow through each sample. Calculation of λ_a is accomplished with Eq. (2) after steady-state has been attained with constant power applied to the screen heater and the copper plates maintained at constant temperature.

$$\lambda_a = I \cdot \Delta V \cdot \ell / 2 \cdot \Delta T \cdot A \qquad (2)$$

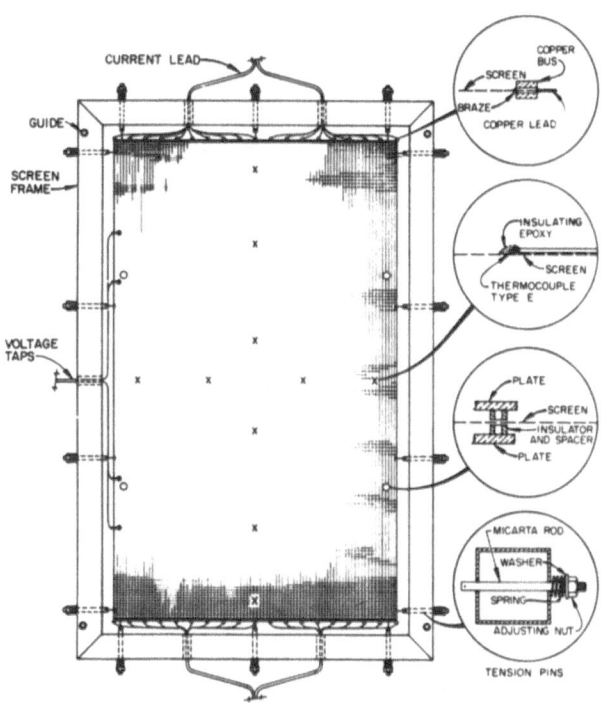

Figure 3: Details of screen heater and frame.

Figure 4:   Photograph of ORNL tester.

Specimen thicknesses on both sides of the screen are determined from
the lengths of the micarta spacers between the screen and plates.
A complete set of measurements for a pair of insulation samples
normally includes measurements at five to ten mean specimen temperatures resulting from plate temperatures near 297 K and screen
temperatures from 300 to 360 K and at three or more sample
thicknesses obtained by changing spacers. Thus, λ_a values are
determined as a function of temperature and density with density
determined from the thickness of the test sample. If the thermal
conductivity versus temperature is nonlinear, then λ_a versus mean
temperature will depend on the bounding temperatures.

The ORNL tester can also be used with heat flow up or down through a single sample of insulation in which case λ_a is calculated using

$$\lambda_a = (I \cdot \Delta V - \lambda_B \cdot \Delta T(B) \cdot A / \ell(B)) \cdot \ell / \Delta T \cdot A \qquad (3)$$

For one-sided heat flow measurements, the test samples are installed as before. The steady-state temperatures of one plate and the screen metering area are controlled to produce a minimum difference $\Delta T(B)$. The term $\lambda_B \cdot \Delta T(B) \cdot A / \ell(B)$ in Eq. (3) is a correction for heat flow across region B.

VALIDATION OF THE TEST METHOD

A determinate error analysis for the ORNL tester shows a maximum measurement uncertainty of 1.7% and a most probable error of 1.2% when sample ΔT is 5°C. Table 1 lists the contributions to the measurement uncertainty and it can be seen that the uncertainty in ΔT dominates the total. The error in temperature of ±0.058 K is based on previous calibration errors for these thermocouples using a platinum resistance thermometer to span the range 78 to 400 K.[20] In most cases λ_a measurements are made with ΔT greater than 5°C and the determinate error is correspondingly reduced. The reproducibility and repeatability of the λ_a measurements have been determined to be ±0.2%.

Validation of the ORNL tester and experimental procedure is provided by measurements on two standards from NBS. Table 2 shows ORNL results for λ_a of a low density fiberglass batt, NBS Certified Transfer Standard,[10] to be within 0.1% of NBS values at 303 and 313 K. A comparison of λ_a for heat flow up from screen to plate shows a 0.6% difference between ORNL and NBS values.

A comparison of ORNL and NBS results for NBS Standard Reference Material 1450b based on ORNL λ_a-measurements at mean sample temperatures from 295 to 333 K shows a difference of 0.9% between measurements by the two laboratories at 297.13 K. The difference is reduced to 0.5% if the lowest ΔT data point (No. 5) is deleted. The ORNL λ_a-data for SRM 1450b with point No. 5 deleted are described with an average absolute difference of 0.32% by Eq. (4) with coefficients evaluated using the procedure described in the following section. The appendix contains a listing of experimental data for the standard materials.

$$\lambda_a = 0.93559 \times 10^{-2} + 0.69073 \times 10^{-4} T_m + 0.18229 \times 10^{-9} T_m^3 \qquad (4)$$

Table 1. Fractional Uncertainties for the ORNL Tester

	Value	Error	Error (%)
Thickness, cm	5.1	0.007	0.14
Length, cm	61.0	0.05	0.08
Width, cm	91.0	0.05	0.05
Total uncertainty in length measurements			0.27
Uncertainty in area due to thermal expansion			0.14
Voltage, V	0.4	6×10^{-6}	0.0015
Current, A	0.4	6×10^{-6}	0.0015
Voltage divider			0.02
Standard resistor			0.01
Power supply instability			0.02
Total power uncertainty			0.053
Temperature, K			
Calibration		0.039	
Potentiometer		0.003	
Wire to wire		0.016	
ΔT, low	5.0	0.058	1.16
ΔT, high	30.0	0.058	0.19
Total temperature measurement uncertainty			1.16 or 0.19
Property/Degree	0.006	0.12	0.072

    Total uncertainty with low ΔT[a] = 1.70
    Total uncertainty with high ΔT = 0.73
    Most probable uncertainty with low ΔT[b] = 1.20
    Most probable uncertainty with high ΔT = 0.37

[a]Total uncertainty is:

$$\left| \Delta\lambda_a/\lambda_a \right| = \pm\left| \Delta Q/Q \right| \pm \left| \Delta A/A \right| \pm \left| \Delta(\ell)/\ell \right| \pm \left| \Delta(\Delta T)/\Delta T \right| \pm \left| d\lambda_a/dT \cdot \Delta T/\lambda_a \right|$$

[b]Square root of the sum of the squared errors (%).

Table 2. A Comparison of ORNL Tester and NBS Results

Specimen	Mean Sample Temperature (K)	Sample Density (kg/m^3)	λ_a-ORNL (W/m·K)	λ_a-NBS (W/m·K)	λ_a[a] (%)
Certified Transfer Standard					
Two-sided	303.14	9.255	0.04831	0.04827	0.08
	313.14	9.270	0.05165	0.05166	-0.02
One-sided	303.23	9.350	0.04811	0.04809	0.04
	303.23	9.340	0.04808	0.04835	-0.56
SRM 1450b					
Two-sided (14 points)	297.13	127.0	0.03454	0.03485	-0.89
(13 points)			0.03466	0.03485	-0.55

[a] $100 \cdot [(\lambda_a\text{-ORNL})-(\lambda_a\text{-NBS})]/(\lambda_a\text{-NBS})$

MEASUREMENTS ON COMMERCIAL PRODUCTS

The λ_a of fiberglass batt samples from a lot previously tested with a guarded hot plate (GHP) have been measured as a function of temperature from 305 to 330 K and thicknesses from 0.0508 m (2.0 in.) to 0.1524 m (6.0 in.). A comparison of ORNL results obtained with blackened copper plates, ε=0.74,[19] and GHP results[21] is shown in Fig. 5. A comparison of the ORNL results with GHP results requires an extrapolation of the ORNL data to 297.04 K (75°F). This was done by subtracting the thermal conductivity of air[22] from the measured λ_a and describing the difference with $A+BT_m^3$ using the Method of Least Squares. The resulting equations for $\lambda_a(T_m)$ at three densities are shown below.

$$\lambda_a = 0.13093\times10^{-2}+0.69073\times10^{-4}T_m+0.83407\times10^{-9}T_m^3 \quad @ \quad \rho = 11.28 \quad (5)$$

$$\lambda_a = 0.23126\times10^{-2}+0.69073\times10^{-4}T_m+0.68968\times10^{-9}T_m^3 \quad @ \quad \rho = 13.54 \quad (6)$$

$$\lambda_a = 0.31345\times10^{-2}+0.69073\times10^{-4}T_m+0.55197\times10^{-9}T_m^3 \quad @ \quad \rho = 16.92 \quad (7)$$

$$\lambda_{air} = 0.54818\times10^{-2}+0.69073\times10^{-4}T \quad (8)$$

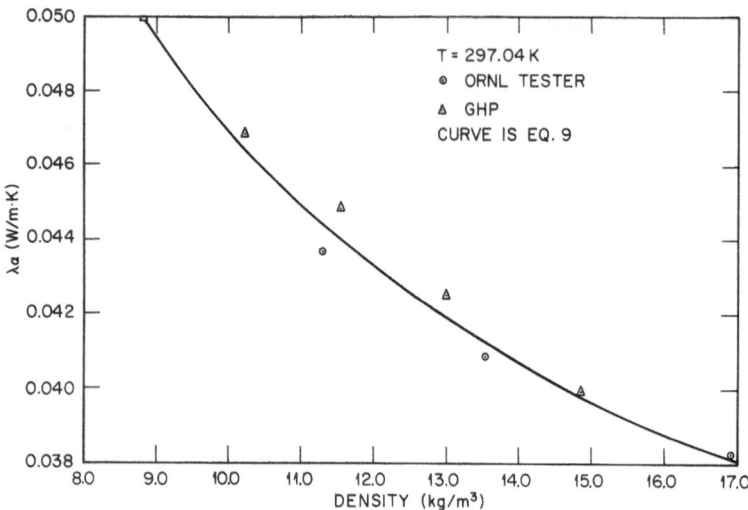

Figure 5: Apparent thermal conductivity of fiberglass sample 1101-1 at 297.04 K.

Equation (9) for λ_a as a function of density at 297.04 K (75°F) was obtained by the Method of Least Squares using three points calculated from Eqs. (5–8), and five points from the GHP.[21]

$$\lambda_a = 0.23185 \times 10^{-1} + 0.75743 \times 10^{-4}\rho + 0.23039/\rho \tag{9}$$

Figure 6 shows the deviation of the eight data points from Eq. (9). The maximum deviation is observed to be 1.8%.

Apparent thermal conductivity measurements for two samples of fiberglass batt insulation were made with the copper plates as machined, (low ϵ), and painted black to give $\epsilon = 0.74$. The results in Table 3 show increases in measured λ_a from 0.7 to 2.2% due to the increase in ϵ. As expected, the smaller λ_a increases occur with the highest density samples. Eqs. (5-7) were used to calculate λ_a for 1101-1[21] fiberglass batts measured with black plates. The data used to construct the table are in the appendix. A data reduction procedure like that used to obtain Eqs. (5-9) was used to determine λ_a at specific temperatures and densities.

A series of λ_a measurements for rockwool batt samples were completed using material 1302-1.[21] The data were analyzed as above to obtain $\lambda_a(T_m)$ at densities of 28.3, 34.0, and 42.5 kg/m^3 and λ_a at 297.04 K. The resulting expressions for λ_a are

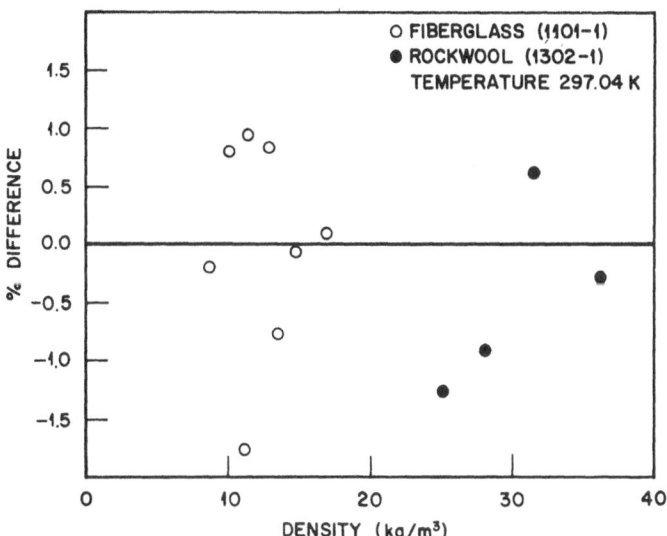

Figure 6: Differences between experimental λ_a and equations for samples 1101-1 and 1302-1 at 297.04 K.

$$\lambda_a = 0.12130 \times 10^{-2} + 0.69073 \times 10^{-4} T_m + 0.10509 \times 10^{-8} T_m{}^3 \quad @\rho = 28.3 \quad (10)$$

$$\lambda_a = 0.23313 \times 10^{-2} + 0.69073 \times 10^{-4} T_m + 0.85647 \times 10^{-9} T_m{}^3 \quad @\rho = 34.0 \quad (11)$$

$$\lambda_a = 0.45981 \times 10^{-2} + 0.69073 \times 10^{-4} T_m + 0.64887 \times 10^{-9} T_m{}^3 \quad @\rho = 42.5 \quad (12)$$

$$\lambda_a = 0.81506 \times 10^{-2} + 0.27777 \times 10^{-3} \rho + 0.94198 / \rho \quad (13)$$

The average λ_a difference between the ORNL tester results and those reported by Tye et al.[21] is 0.8%. Figure 6 shows the deviation of the λ_a for rockwool (1302-1) at 297.04 K from Eq. (13).

Measurements of λ_a were made for double thicknesses of fiberglass batts (1101-1) at three densities in order to determine the magnitude of the "thickness effect". A comparison of λ_a measured at thicknesses of 0.1524 m, 0.1270 m, and 0.1016 m with λ_a calculated from Eqs. (5–7) is in Table 4. As expected, the thickness effect is greatest for the low density specimen where increasing the measurement thickness from 0.0762 m to 0.1524 m at constant density and temperature decreases the R-value per unit thickness by about 5%.

Table 3. A Comparison of λ_a Measurements Made with Bright Copper and with Black Plates

Sample	Density (kg/m^3)	Temperature (K)	λ_a(Bright)[a] (W/m•K)	λ_a(Black)[b] (W/m•K)	% Increase
FG BATT c	10.14	297.04	0.04398	0.04479	1.84
		298.15	0.04430	0.04513	1.87
		323.15	0.05215	0.05330	2.21
	12.17	297.04	0.04116	0.04184	1.65
		298.15	0.04145	0.04213	1.64
		323.15	0.04842	0.04915	1.51
	15.21	297.04	0.03838	0.03890	1.35
		298.15	0.03863	0.03915	1.35
		323.15	0.04462	0.04511	1.10
FG BATT d	11.28	297.04	0.04325	0.04369	1.02
		298.15	0.04357	0.04401	1.01
		323.15	0.05109	0.05178	1.35
	13.54	297.04	0.04038	0.04091	1.31
		298.15	0.04066	0.04119	1.30
		323.15	0.04741	0.04791	1.05
	16.92	297.04	0.03776	0.03812	0.95
		298.15	0.03801	0.03836	0.92
		323.15	0.04377	0.04408	0.71

[a]As machined copper surface.

[b]$\varepsilon \approx 0.74$.

[c]Commercial fiberglass batt insulation produced by CertainTeed Corporation.

[d]Commercial fiberglass batt insulation code number 1101-1 (see Refererence 21.

Table 4. A Comparison of λ_a for Fiberglass Batts (1101-1) at
Two Thicknesses and Constant Temperature and Density

Sample Thickness (m)	Density (kg/m^3)	T_m (K)	λ_a (W/m·K)	R (m^2·K/W)
0.0508	16.92	313.12	0.04171	1.218
0.1016	16.92	313.12	0.04254	2.388
0.0635	13.54	313.09	0.04511	1.408
0.1270	13.54	313.09	0.04660	2.725
0.0762	11.28	313.02	0.04851	1.571
0.1524	11.28	313.02	0.05120	2.977

CONCLUSIONS

The ORNL flat screen tester has been demonstrated to have a λ_a measurement accuracy of better than ±2% by a determinate error analysis and measurements on two NBS reference materials. The ORNL tester has been used to obtain λ_a for materials up to 0.1524 m thick and at mean sample temperatures from 300 to 330 K. The effect on λ_a of increasing cold plate emissivity to 0.74 has been shown to be 1-2% in the case of low density fiberglass batts. A comparison of λ_a for fiberglass batts obtained at 0.0762 and 0.1524 m showed a thickness effect of 5%. The ORNL tester has been demonstrated to have the accuracy and sensitivity needed for useful λ_a measurements on thermal insulating materials.

REFERENCES

1. S. H. Jury, D. L. McElroy, and J. P. Moore, "Pipe Insulation Testers," Thermal Transmission Measurements of Insulation ASTM STP 660, R. P. Tye, Ed., American Society for Testing and Materials, pp. 310–326 (1978).
2. J. P. Moore, D. L. McElroy and S. H. Jury, A Technique for Measuring the Apparent Thermal Conductivity of Flat Insulations, ORNL/TM-6494, Oak Ridge National Laboratory, Oak Ridge, TN (October 1974).
3. J. P. Moore, D. L. McElroy and S. H. Jury, "A Technique for Measuring the Apparent Thermal Conductivity of Flat Insulations", Thermal Conductivity 17, J. G. Hust, Ed., Plenum Publishing Corporation, pp. 727–735 (1983).

4. C. Niven and A. E. M. Geddes, "On a Method for Finding the Conductivity for Heat," Proc. Roy. Soc. (London) A87, 535–539 (1912).

5. C. F. Gilbo, "Experiments with a Guarded Hot Plate Thermal Conductivity Set," ASTM Special Technical Publication No. 119, 45, (1951).

6. N. E. Hager, Jr., "Thin Heater Thermal Conductivity Apparatus", Rev. Sci. Instrum. 31(2), 177–185, (Feb. 1960).

7. N. E. Hager, Jr., U.S. Patent No. 3,045,473 (July 24, 1962).

8. N. E. Hager, Jr., "Miniature Thin-Heater Thermal Conductivity Apparatus," ISA Transactions 8(2), 104–109 (1969).

9. W. D. Turner, D. C. Elrod and I. I. Siman-Tov, HEATING5 - An IBM 360 Heat Conduction Program, ORNL/CSD/TM-15, Oak Ridge National Laboratory, Oak Ridge, TN (March 1977).

10. Report of Test on Thermal Resistance of Glass-Fiber Insulation, U.S. Department of Commerce, National Bureau of Standards, F. J. Powell to D. L. McElroy, Purchase Order 21X-48695, May, 10, 1983.

11. Excerpt from Ref. 10, "The uncertainty in the thermal resistance values is estimated to be not more than ±0.5 percent and includes apparatus systematic error and apparatus repeatability."

12. Phoenix Wire Cloth, Inc., 40x40 Per Inch Nichrome V Wire, 0.010 Inch, P. O. Box 610, Troy, Michigan 48084.

13. Leeds and Northrup, Philadelphia, PA, 0.01Ω Standard Resistor Model No. 4361, 100 amps dc current rating.

14. Hewlett Packard DC Power Supply Model 6260B (10V, 100A).

15. Leeds and Northrup K-5 Potentiometer with 1.6 Volt Accuracy of ±(0.001%+2μV).

16. Astrodyne, Inc., Thermal Bond 312, Burlington, MA.

17. B. Braun Instruments, Thermomix 1480 and Frigomix 1495, San Mateo, CA.

18. Fisher and Porter Company, Warminister, PA 18974, Flowrator Meter Model 10A1755XZ.

19. D. L. McElroy, R. S. Graves, D. W. Yarbrough, and J. P. Moore, "A Flat Insulation Tester that Uses an Unguarded Nichrome Screen Wire Heater," Forum on the Guarded Hot Box Plate and Heat Flow Meter State-of-the-Art, Quebec City, Quebec, Canada, (October 7–8, 1982).

20. J. P. Moore, R. K. Williams, and R. S. Graves, "Precision Measurements of the Thermal Conductivity, Electrical Resistivity, and Seebeck Coefficient from 80 to 400 K and Their Application to Pure Molybdenum," Rev. Sci. Instrum. 45 (1), 87–95 (1974).

21. R. P. Tye, A. O. Desjarlais, D. W. Yarbough and D. L. McElroy, An Experimental Study of Thermal Resistance Values (R-Values) of Low-Density Mineral-Fiber Building Insulation Batts Commercially Available in 1977, ORNL/TM-7266 (April 1980).

22. Y. S. Touloukian, P. E. Liley, and S. C. Saxena, Thermophysical Properties of Matter 3, Plenum Publishing Corp., New York, p. 512 (1970).

APPENDIX

Summary of Experimental λ_a Data

Material	Density (kg/m³)	Thickness (m)	T_m (K)	λ_a (W/m·K)	Comment
1101-1					
1.	16.92	0.0508	304.58	0.03980	$\varepsilon = 0.74$
2.	"	"	306.94	0.04028	
3.	"	"	312.94	0.04164	
4.	"	"	324.28	0.04437	
5.	13.54	0.0635	304.75	0.04297	
6.	"	"	307.72	0.04361	
7.	"	"	315.00	0.04556	
8.	"	"	328.12	0.04938	
9.	11.28	0.0762	305.05	0.04612	
10.	"	"	308.47	0.04707	
11.	"	"	316.36	0.04949	
12.	"	"	329.64	0.05399	
13.	16.92	0.1016	313.12	0.04254	
14.	13.54	0.1270	313.09	0.04660	
15.	11.28	0.1524	313.02	0.05120	
16.	"	0.0762	304.96	0.04571	as machined plates
17.	"	"	308.35	0.04639	
18.	"	"	316.08	0.04875	
19.	"	"	329.78	0.05331	
20.	13.54	0.0635	304.84	0.04242	
21.	"	"	307.77	0.04316	
22.	"	"	315.12	0.04511	

APPENDIX (Continued)

Material	Density (kg/m³)	Thickness (m)	T_m (K)	λ_a (W/m·K)	Comment
23.	"	"	327.56	0.04871	
24.	16.92	0.0508	304.61	0.03944	
25.	"	"	307.32	0.04007	
26.	"	"	313.34	0.04139	
27.	"	"	325.45	0.04435	
CTS					
1.	9.30	0.0762	304.91	0.04890	
2.	"	"	308.18	0.04983	
3.	"	"	315.46	0.05228	
4.	"	"	328.09	0.05698	
5.	11.16	0.0635	304.71	0.04528	
6.	"	"	307.73	0.04606	
7.	"	"	314.43	0.04799	
8.	"	"	326.45	0.05177	
9.	9.48	0.0762	308.22	0.04934	one-sided, heat flow up
10.	"	"	316.73	0.05214	one-sided, heat flow up
11.	"	"	302.79	0.04764	one-sided, heat flow up
12.	"	"	303.30	0.04766	one-sided, heat flow up
13.	9.13	0.0762	303.26	0.04883	one-sided, heat flow down
14.	"	"	308.19	0.05056	one-sided, heat flow down
15.	"	"	316.71	0.05338	one-sided, heat flow down

Material	Density (kg/m³)	Thickness (m)	T_m (K)	λ_a (W/m·K)	Comment
16.	13.96	0.0508	304.24	0.04147	
17.	"	"	307.23	0.04234	
18.	"	"	312.96	0.04366	
19.	"	"	323.76	0.04648	
SRM 1450 b					
1.	129.8[a]	0.0508	304.77	0.03591	
2.	"	"	307.84	0.03601	
3.	"	"	330.46	0.03870	
4.	"	"	315.43	0.03669	
5.	"	"	295.15	0.03359	
6.	"	"	297.32	0.03425	
7.	"	"	304.61	0.03551	ΔT = 3.64 K
8.	130.9	"	297.56	0.03460	
9.	128.8	"	297.29	0.03472	
10.	129.8	"	297.33	0.03464	
11.	"	"	301.15	0.03528	
12.	"	"	299.15	0.03491	
13.	"	"	303.32	0.03545	
14.	"	"	322.76	0.03780	
15.	"	"	333.43	0.03914	one-sided, heat flow down
16.	"	"	308.32	0.03610	one-sided, heat flow up

[a] Nominal density 127.0 kg/m³.

CALIBRATION OF HEAT FLOW METER APPARATUS USED FOR

QUALITY CONTROL OF LOW-DENSITY MINERAL FIBER INSULATIONS

Wendy S. Newton - Manville Canada Inc., Innisfail, Alta.
C.M. Pelanne - Manville Corp. of U.S., Denver, Colorado
M. Bomberg - National Research Council of Canada
Ottawa, Ontario

ABSTRACT

This paper provides industrial laboratories with guidelines for verifying precision in thermal resistance determinations. It deals with both equipment and test procedures, and discusses the development of a calibration program for the heat flow meter (HFM) apparatus used in quality control of low-density mineral fiber insulations manufactured at the Innisfail Plant of Manville Canada Inc. To ensure agreement between the HFM apparatus used in the plant and the guarded hot plate (GHP) apparatus used by National Research Council of Canada, the following elements were incorporated in the calibration program: selection, fabrication and verification of low-density glass fiber reference systems, calibration of apparatus, error analysis of apparatus, maintenance of calibration.

INTRODUCTION

A properly calibrated heat flow meter (HFM) apparatus provides fast, accurate measurement of thermal resistance and is therefore useful for industrial quality control. The Manville Corporation has installed an HFM apparatus* in each of its insulation manufacturing plants to monitor the thermal properties of its products, thereby ensuring that the product conforms to the required standards [1]. This paper discusses a cooperative program undertaken by Manville Canada Inc. (MCI), Innisfail, Alberta, the Research and Development Center (R/D) of Manville Corporation, USA,

---

* R-matic, manufactured by Dynatech R/D Company, Cambridge, Mass.

and the National Research Council of Canada (NRCC) to establish a calibration procedure and develop reference specimens for use by MCI. The calibration program contained the following elements:

1. Development of reference specimens
   - preliminary selection of candidate specimens by means of light transmission measurements,
   - determination of thermal characteristics on HFM apparatus,
   - selection of reference specimens,
   - tests of reference specimens in GHP apparatus at NRCC,
   - final tests on HFM apparatus.

2. Equipment calibration
   - calibration of the two HFM apparatuses at MCI,
   - evaluation of the characteristics of the equipment,
   - development of secondary reference specimens,
   - development of procedures to ensure consistent performance of equipment.

Prior to undertaking the calibration program the mechanical performance of each HFM apparatus was verified. Parallelism of the two surface plates was checked at various specimen thicknesses as well as the thickness measurement technique. The paper deals with the specific issue of thermal resistance measurements in an industrial quality control laboratory. The more general problems associated with the HFM apparatus are discussed elsewhere [2].

PRIMARY REFERENCE SPECIMENS

Selection of uniform specimens and closely matched pairs of specimens is imperative for success in establishing reference specimens [3]. The following criteria were defined:

1. Density and thermal resistance of pairs of specimens must match.
2. Structure and density of the specimen must match that of the material being manufactured.
3. Specimen density distribution must be uniform and free of any visible defects.

Prior to choosing the final reference specimens, a preliminary selection was made. A number of 610- × 610-mm pieces were cut from a batt of glass fiber insulation; the upper and lower surfaces were removed, using a horizontal bandsaw; the specimens were trimmed to a thickness slightly greater (by as much as 6 mm) than the thickness intended for testing; the pieces were pre-screened over a light box containing a series of fluorescent lights covered by white opaque plexiglas. (The box provides a very intense light over which the specimens can be observed for uniform density, with

particular attention to the central (metering) area.) Only then
are density and thermal resistance determined for specimens having
acceptable uniformity.

The reference specimens were selected on the basis of best
match density and thermal conductivity. Values of apparent thermal
conductivity and density (Fig. 1) were used to select the reference
specimens. Two 25-mm thick specimens and four 76-mm thick
specimens were selected. The 76-mm specimens were paired to form
two 152-mm thick specimens. Six additional specimens 25 mm thick
were selected for evaluating experimental errors introduced by
thick, low-density specimens.

In any comparative study of equipment having different
metering areas (as is the case with both the HFM and GHP), the
details of specimen preparation should be decided prior to testing
in order to avoid the necessity for adjustments during the test.
At NRCC the specimens are placed in 25-mm wide extruded polystyrene
frames; prior to testing at Innifail, therefore, the specimens
should have been cut to 560 × 560 mm and placed in frames supplied
by NRCC. As this was not done originally, testing at Innisfail
before the specimens were placed in frames had to be disregarded.

Table la shows average thermal conductivity for two test
series: an initial test on 610- × 610-mm specimens; one in which
specimens were placed in a smaller frame by cutting 25 mm from one
side, thereby displacing the metering area. There was a small
difference in the thermal conductivities determined by the two

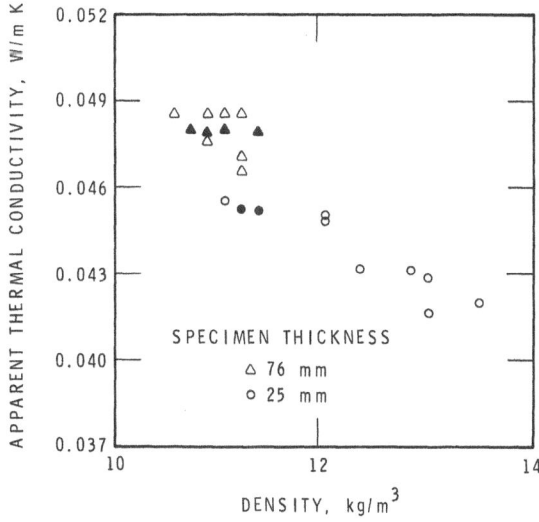

Fig. 1. Apparent thermal conductivity determined on HFM apparatus
and density of the pre-selected specimens

series. To check whether this was due to specimen preparation or to malfunction of the apparatus, four other specimens originally tested at the same time were retested. The results shown in Table 1b indicate good performance of the HFM apparatus. The difference shown in Table 1a is probably associated with the shift of metering area, and the initial tests were therefore disregarded. Each of the specimens was retested three more times and the results on specimens placed in 25-mm deep frames are given in Table 2.

Table 1a. Apparent thermal conductivity* determined on HFM 1 at Manville Canada Inc. Averages for two series: (1) initial, (2) after return from NRCC where the metering area was shifted by 25 mm in one direction

Specimen Code	Nominal Thickness mm	Initial 4 Tests λ, W/m K	After Return 3 Tests λ, W/m K	% Difference
106A	25	0.04540	0.04559	0.4
106B	25	0.04555	0.04506	1.1
305A	76	0.04800	0.04822	0.5
305B	76	0.04800	0.04793	0.1
306A	76	0.04800	0.04829	0.6
306B	76	0.04806	0.04828	0.4
305A&B	152	0.04987	0.04998	0.2
306A&B	152	0.05005	0.04991	0.3
				AVG = 0.5%

Table 1b. Repeatability of apparent thermal conductivity testing at Manville Canada Inc. during the same time without shift in metering area

Specimen Code	Nominal Thickness mm	Series (1) 4 Tests λ, W/m K	Series (2) 3 Tests λ, W/m K	% Difference
107A	25	0.04511	0.04519	0.2
107B	25	0.04514	0.04511	0.1
3½B2	89	0.04918	0.04925	0.2
6B2	152	0.05377	0.05384	0.1
				AVG = 0.1%

* Reported with four significant digits for the sake of error reduction in percentage calculations.

Table 2. Apparent thermal conductivity (λ) before and after calibration of HFM 1 at Manville Canada Inc.

Specimen Code	Average of 6 HFM Tests λ, W/m K	St. Deviation of λ on HFM, %	GHP at NRCC W/m K	λ on Recalibrated HFM
106A	0.04564	0.14		0.04638
			0.04611	
106B	0.04513	0.14		0.04585
305A	0.04823	0.20		0.04799
			0.04787	
305B	0.04800	0.27		0.04776
306A	0.04832	0.16		0.04793
			0.04790	
306B	0.04823	0.15		0.04786
305A,B	0.04996	0.32		0.04939
			0.04934	
306A,B	0.04986	0.47		0.04929

CALIBRATION OF HEAT FLOW METER APPARATUS

All the tests were performed according to ASTM C518 "Standard Test Method for Steady-State Thermal Transmission Properties by Means of the Heat Flow Meter." The apparatus was operated at a mean temperature of 24°C (75°F), with the hot surface plate at 37.8 ±0.6°C (100 ±1°F) and the cold surface plate at 10.1 ±6°C (50 ±1°F). Room temperature was maintained between 22°C and 26°C, and relative humidity of the ambient air was below 40%. No condensation on the cold plate was expected under such conditions.

After the thermal resistance values had been determined on GHP for each calibrated specimen, the calibration coefficent was calculated from the following equation:

$$C = \frac{T_h - T_c}{e} \frac{1}{R}$$

where C = heat flow meter calibration coefficient for a given temperature and specimen thickness, in units of heat flux per millivolt

T_h, T_c = hot and cold surface temperatures, in degrees

e = output of heat flow meter, in millivolts

R = thermal resistance measured on the GHP apparatus at the same thickness and mean temperature, with a known confidence interval.

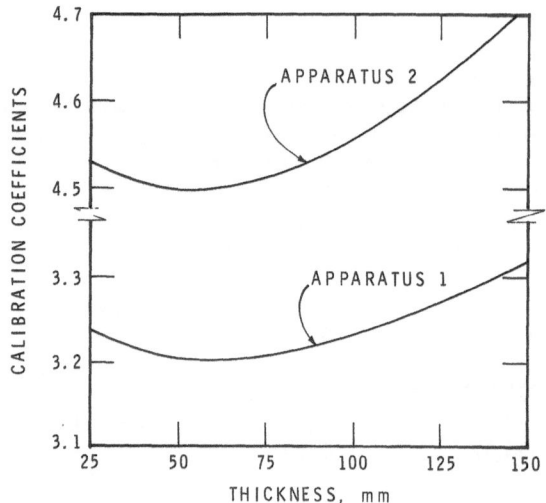

Fig. 2. Calibration coefficient (British units) versus specimen
thickness. Note that calibration curves for the two heat
flow meters are different, indicating need for calibration
covering the full range of thicknesses for which equipment
is to be used

Calculation of the calibration coefficient for HFM apparatus No. 1
was based on the test values obtained during six tests (Table 2).
For HFM No. 2, each specimen was tested at least twice. The
average calibration coefficients calculated for each apparatus and
thickness are shown on Fig. 2. Calibration coefficients are
relatively constant between 25 and 75 mm and increase for
thicknesses greater than 75 mm. This thickness effect appears to
be significant, particularly for HFM No. 2. Table 2 compares the
average values of the six HFM tests with the results obtained in
the GHP apparatus at NRCC on matched pairs of specimens.

For the comparison, values of the apparent thermal conductivity, λ_a, are used. They are calculated from thermal resistance R,
using the following equation:

$$\lambda_a = \frac{L}{R}$$

where L = thickness of specimen.

Either a single or a variable calibration coefficient may be
used for a specimen thickness range of 25 to 125 mm. Table 3 shows
percentage differences for apparent thermal conductivity values

determined with either a variable or single coefficient. The results suggest that a calibration coefficient varying with specimen thickness is preferred.

EVALUATION OF MEASUREMENT ERRORS

Radiation-absorbing membranes or septa were used to assess HFM apparatus measurement errors [4]. Radiation-absorbing barriers or septa (Kraft paper 0.06-0.08 kg/m^2) with an emittance similar to that of the plates of the HFM apparatus ($\varepsilon = 0.95$) were placed between 25-mm thick specimens of similar density and thermal conductivity to make up a specimen 152 mm thick. In effect, the septa absorb and re-emit radiant energy, thereby simulating a series of 25-mm thick specimens placed in the HFM apparatus. By comparing the results for single 25-mm thick specimens with those for the combined thicknesses of septa (Table 4), the apparatus error can be approximated.

One assumes that all test results with septa 25 mm apart should be equal to those obtained for a 25-mm thick specimen. Although this may not be strictly correct, it is applicable to low-density glass fiber products [5]. The difference between thermal conductivity measured for the insulation stack with septa and thermal conductivity measured for 25-mm thick specimen may be called "apparatus error." "Effect of thickness" of the material is obtained from test results without septa, corrected for instrument error.

It is the purpose of this section to show the error associated with each apparatus, and it is reasonable to assume that each has a systematic bias that increases with increase in specimen thickness. Figure 3 shows the variation in error with test thickness. The

Table 3. Estimate of error in thermal conductivity with four calibration coefficients

Specimen			Error		
Code	Thickness mm	Apparatus Number	Variable Coeff.	Coeff. Average	Coeff. Median
106A	25	1	0	0.3	0.4
		2	0	0.6	0.8
305A	76	1	-0.3	0.4	0.6
		2	0.1	0.8	1.1
305A,B	152	1	0	-0.9	-0.8
		2	0.1	-1.3	-1.1

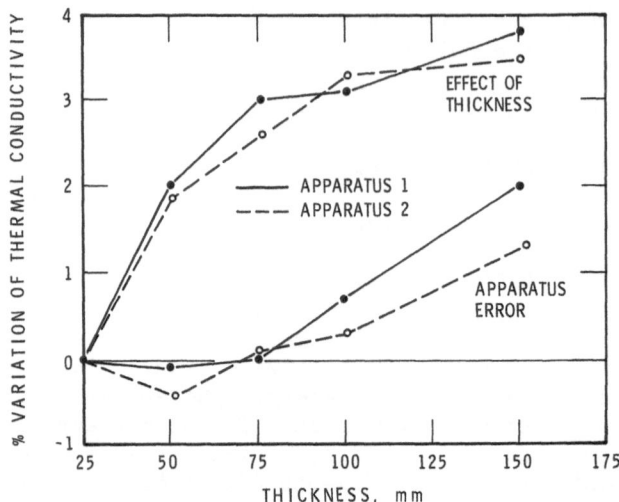

Fig. 3. Apparatus error and effect of thickness versus specimen
thickness for both apparatuses. The differences are
within the expected uncertainty limits of the HFM
apparatus

Table 4. Apparent thermal conductivity, λ, of low-density glass
fiber specimens tested with and without paper septa,
HFM 1 and HFM 2

Specimen Code mm	From 25 mm Testing λ, W/m K	Apparatus 1				Apparatus 2			
		λ Measured, W/m K		Error %	EOT** %	λ Measured, W/m K		Error %	EOT** %
		Septa	No Septa			Septa	No Septa		
50 AB*	0.0430	0.0430	0.0439	0.20	2.12	0.0428	0.0436	0.57	1.82
CD	0.0430	0.0430	0.0438	0.13	1.81	0.0429	0.0437	0.24	1.95
EF	0.0428	0.0427	0.0436	0.27	2.02	0.0432	0.0434	0.47	1.92
				0.20	1.98			0.43	1.90
76 ABC	0.0431	0.0432	0.0444	0.23	2.88	0.0433	0.0443	0.43	2.43
DEF	0.0428	0.0427	0.0441	0.17	3.16	0.0427	0.0438	0.17	2.43
				0.20	0.30			0.30	2.43
102 ABCD	0.0430	0.0433	0.0448	0.77	3.35	0.0433	0.0446	0.63	3.00
CDEF	0.0429	0.0432	0.0448	0.67	2.86	0.0429	0.0441	0.00	2.83
				0.72	3.10				2.92
152 ABCDEF	0.0430	0.0438	0.0454	2.01	3.63	0.0434	0.0449	1.07	3.37
CDEFAB	0.0430	0.0438	0.0452	1.94	3.39	0.0435	0.0452	1.23	3.87
EFABCD	0.0430	0.0439	0.0458	2.25	4.43	0.0437	0.0451	1.80	3.07
				2.07	3.83			1.37	3.44

* Last letter denotes specimen adjacent to the hot side
** EOT – Effect of thickness for the particular low-density glass fiber product

measurements show excellent agreement for the two apparatuses. Apparatus error, which is small up to 76 mm (3 in.), increases significantly beyond this thickness. The experiments were performed with and without paper septa, and the results were calculated using the variable coefficients shown in Fig. 2.

For practical purposes a reasonable range of uncertainty should be assigned to the results obtained at different thicknesses. The tolerances should take into account any measurement uncertainty and variation in specimen structure as well as factors related to the testing procedure, for example, specimen preparation and conditioning. Uncertainty values will increase as a function of thickness. The following uncertainty ranges were deemed appropriate for the HFM's reported: for 25 mm, ±0.5%; for 76 mm, ±0.8%; and for 152 mm, ±1.57%.

MAINTENANCE OF CALIBRATION

Calibration of an apparatus must be monitored on a continuous basis. Secondary reference specimens were therefore selected using the same criteria except that they were not paired. Each secondary reference specimen was tested a minimum of eight times and control limits were developed. These were later established once more, based on values obtained over a period of time and taking into account the variability due to different operators.

Any one of the secondary reference specimens may be used for control purposes. If a test result is outside the control limits, another secondary reference specimen should be checked before any calibration adjustment is considered. Only if such tests confirm that the results fall beyond the control limits and the cause of the discrepancy is not immediately obvious (for example, an error in thickness determination), new calibration coefficients must be developed based on the primary reference specimens.

There is no set rule for frequency of control checks. The procedure at the Innisfail plant requires a control check each time an apparatus is adjusted for a new specimen thickness, and each time the apparatus has not been used for some time (i.e., when a secondary reference specimen is tested before a new test series).

Primary and secondary reference specimens are identified using an alpha-numeric code. Marks are placed on the upper front surface to facilitate orientation of the specimen during test. Internal company certificates are issued for each calibrated specimen to signify the importance of the calibration program. These specimens are stored in rigid containers made of corrugated cardboard to protect them from damage.

CONCLUSION

This paper describes the procedure established by an insulation manufacturer to calibrate the heat flow meter apparatus used for its quality control operation. It identifies the characteristics of apparatus and testing procedure most important in calibration and calibration maintenance. The development of these procedures increased the confidence that quality control could meet the thermal resistance requirements specified by building codes and materials standards.

REFERENCES

1. Pelanne, C.M., Development of a company wide heat flow meter calibration program based on the N.B.S. certified transfer specimens; Presented, Forum on the Guarded Hot Plate and the Heat Flow Meter, Quebec City, Canada, October 1982. (To be published by ASTM)

2. Bomberg, M., Pelanne, C.M., Newton, W., Analysis of uncertainties involved in heat flow meter apparatus calibration; Prepared for 18th International Thermal Conductivity Conference, Rapid City, South Dakota, October 1983.

3. Pelanne, C.M., The development of low density glass fiber insulation as thermal transmission reference standards; Thermal Conductivity 17, Proc., International Thermal Conductivity Conference, Plenum, 763-775, 1983.

4. Pelanne, C.M., Discussion on experiments to separate the 'effect of thickness' from systematic equipment errors in thermal transmission measurements; Thermal Insulation Performance, ASTM, STP 718, 322-334, 1980.

5. Albers, M.A. and Pelanne, C.M., An experimental and mathematical study of effect of thickness in low density glass fiber insulation; Thermal Conductivity 17, Proc., International Thermal Conductivity Conference, Plenum, 471-482, 1983.

INSULATIONS WITH LOW THERMAL CONDUCTIVITY*

G. L. Copeland, D. L. McElroy, R. S. Graves,
and F. J. Weaver
Martin Marietta Energy Systems, Inc.
Metals and Ceramics Division
Oak Ridge National Laboratory
Oak Ridge, Tennessee 37830

H. A. Fine and T. W. Tong
Department of Metallurgical Engineering and Mechanical
  Engineering
University of Kentucky
Lexington, Kentucky 40506

Appliances would be more energy efficient if thermal
insulations could be developed with apparent thermal conductivity
(λ) below 0.01 W/m·K. In an effort to identify candidate
materials, λ was measured from 300 to 335 K for unevacuated and
evacuated systems of angstrom- and micron-size particles.

A radial heat flow apparatus provided λ-values for unevacuated
systems as a function of density, particle size, and composition.
For amorphous fumed silica particles nominally 100 Å diameter in
air the λ at 300 K shows a minimum near 10% solid of about
0.020 W/m·K, or 80% of λ of air. A simplified analysis was used to
describe the dependence of heat transfer on particle system characteristics and to suggest further tests.

A large, unguarded longitudinal heat flow apparatus was used
to measure λ of three sealed evacuated flat panels that contained
compacted amorphous fumed silica particles less than 100 Å
diameter. Each panel was originally evacuated to about 1 mm of Hg.
The panel λ-values increase between 300 and 325 K, are in the range
0.008 to 0.016 W/m·K, and can be described by $\lambda(T) = a + bT^3$.

---

*Research sponsored by the Office of Building and Community
Systems, U.S. Department of Energy, under contract
DE-AC05-840R21400 with the Martin Marietta Energy Systems, Inc.

INTRODUCTION

A major portion of the energy used in some residential and commercial appliances is lost through the thermal insulation system. Table 1 provides a comparison of thermal resistance (R-values) of two currently used insulations (polyurethane and fiberglass) and air (without convective or radiative transfer). Substantial energy savings would result for appliances if insulation systems could be developed having a metric R-value of 3.5 m^2K/W for 0.0254 m thickness (an R-value of 20 $h \cdot ft^2 \cdot °F/Btu$ for one inch).

The current study is based on recommendations made in an examination of several candidate concepts for technical feasibility, potential for saving energy, and economics in systems where conductive losses are large.[1] Tests were recommended to measure the R-value of evacuated systems and of air-filled systems containing small particles.

One way to produce improved insulation is based on reducing the gas conduction by making the mean free path of the gas large compared to the size of the voids between the solid particles or fiber. This may be accomplished either by making the effective pore size in the medium physically smaller or by increasing the mean free path of the gas. In practice the former could be accomplished by using very small particles so that the effective pore size is smaller than the mean free path of air (700 Å at atmospheric pressure and room temperature). The latter could be accomplished by reducing the pressure so that the mean free path of the gas is greatly increased. The relative importance and trade-offs between gas conduction, solid conduction, and radiation between gas conduction, solid conduction, and radiation were discussed in detail in a previous article.[2]

Table 1. Comparison of Metric and Standard Thermal Resistance Values for Thermal Insulations 0.0254 m Thick at 300 K

| Material | Thermal Resistance for 0.0254 m | | Apparent Thermal Conductivity | |
	$\dfrac{m^2K}{W}$	$\dfrac{h \cdot ft^2 \cdot °F}{Btu}$	$\dfrac{W}{m \cdot K}$	$\dfrac{Btu \cdot in}{h \cdot ft^2 \cdot °F}$
Fiberglass	0.48	2.7	0.053	0.370
Polyurethane	1.21	7	0.021	0.143
Air (Conduction only)	0.98	5.6	0.026	0.180
Target	3.52	20	0.0072	0.05

368

Thermal resistance values greater than the target value of 3.5 m^2K/W have been reported for evacuated and unevacuated samples of fumed silica particles when tested with isothermal surfaces at 77 and 300 K.[3] Table 2 provides results of some previous studies at temperatures more representative of operating appliances that obtained R-values for 0.0254 m in the range 0.7 to 1.2 m^2K/W for air-filled systems.[4-8] These studies were for fumed silica particles about 100 Å in diameter, while the best L'Air Liquide results[3] were obtained with fumed silica particles approximately 30 Å in diameter. Furthermore, the importance of temperature is demonstrated for Cab-O-Sil M-5 fumed silica (average particle diameter of 140 Å) whose thermal conductivity increases from 0.029 W/m·K at 20°C to 1.053 at 600°C.[5] The effect of increasing the mean free path of the gas molecules by decreasing pressure was demonstrated by Glazier et al.,[8] who obtained thermal resistivity-values on the order of 700 m·K/W for Cab-O-Sil M-5 at pressures less than 10^{-6} mm Hg.

High temperature insulations formed from fibrous media and small heat-resistant particles are available commercially.[9,10] These insulations exhibit a thermal conductivity in air lower than the molecular conductivity of still air. Evacuated powder insulations for high temperature battery applications have obtained a λ below 0.01 W/m·K at 300°C.[11] This requires optimization of the infrared extinction coefficient of the powder and minimization of the solid conduction component.

In the current work, the thermal properties of evacuated and unevacuated insulation systems made from angstrom- and micron-sized particulates were studied in the range 295 to 340 K. These data were used with a model for heat transfer to establish the requirements for developing an insulation system with a thermal resistance of 3.5 m^2K/W for 0.0254 m (20 h·ft^2·°F/Btu for one inch).

EXPERIMENTAL APPARATUS

Determinations of apparent thermal conductivity were made using two simplified radial heat flow apparatuses and one large longitudinal heat flow apparatus. Figure 1 is a schematic of the radial heat flow apparatus with a guarded core heater, ORNL-3. ORNL-3 allows loose powders to be tested in air, gas or vacuum. Figure 2 shows the radial heat flow apparatus with an unguarded core heater, ORNL-4. ORNL-4 was constructed so that powders could be compacted and tested in air as a function of density. Reference 2 gives details of ORNL-3 and ORNL-4 which operate in the range 295 to 340 K. Heat flow tests using a large unguarded longitudinal heat flow apparatus yielded λ values on evacuated panels accurate to ±2% in the range 295 to 320 K.[12,13]

Table 2. Previously Reported Apparent Thermal Conductivity Values for Fumed Silica Systems Including the Effects of Temperature, Density and Atmosphere [4]

Cab-O-Sil M-5 or Aer-O-Sil 200[a] at Atmospheric Pressure
Apparent Thermal Conductivity, W/m·K

Density $\frac{g}{cm^3}$	Temperature (°C)								Source
	-85	0	20	60	100	120	160	600	
0.051	-	-	0.029	-	0.053	-	0.091	1.053	(5)
	0.019	-	-	-	-	-	-	-	(6)
0.067[a]	-	0.026	0.027	0.035	0.049	0.059	-	-	(7)
0.096	-	-	-	-	-	-	-	0.981	(5)
0.176	-	0.027	-	-	-	-	-	1.096	(5)

Cab-O-Sil M-5 Evacuated to 10^{-6} mm of Hg
Apparent Thermal Conductivity at Mean Temperature (°C)

Materials	Density $\frac{g}{cm^3}$	W/m·K x 10^3		Source
Fumed Silica	0.056	2.21	(-85°C)	(6)
Fumed Silica	0.080	1.44	(-93°C)	(8)
75 F.S./25 C.B.[b]	0.080	0.66	(-93°C)	(8)
60 F.S./40 C.B.[b]	0.080	0.82	(-93°C)	(8)
Commercial Perlite	0.160	1.51	(-93°C)	(8)
Santocel Silica	0.096	1.79	(-93°C)	(8)

[a] Aer-O-Sil 200

[b] C.B. = Carbon Black

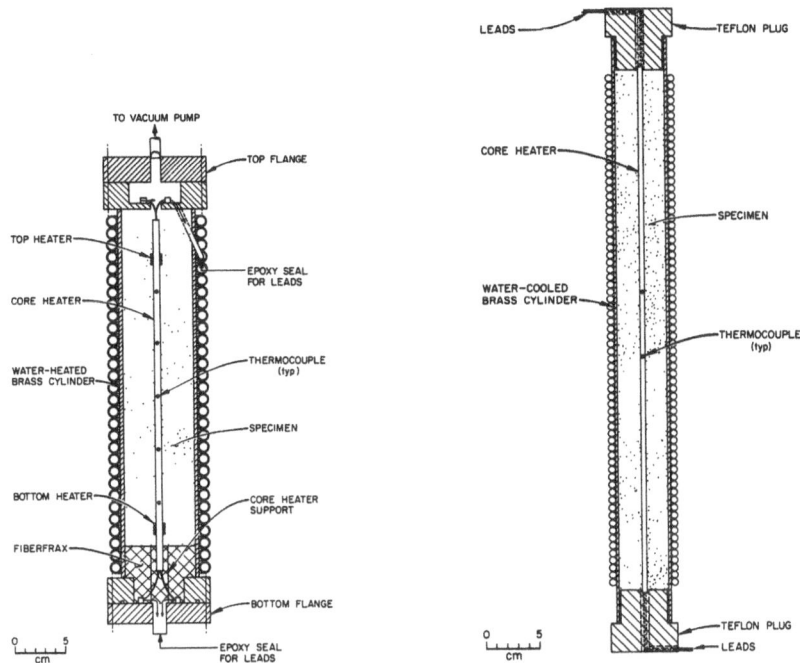

Figure 1: Schematic drawing of ORNL-3, a guarded radial heat flow apparatus for testing powders.

Figure 2: Schematic drawing of ORNL-4, an unguarded radial heat flow apparatus for testing powders.

MATERIALS

Thermal conductivity measurements were conducted on three materials: Al_2O_3 microspheres, commercially available fumed silica, and evacuated panels containing fumed silica particles.

The Al_2O_3 microspheres were used as a thermal conductivity standard to compare ORNL-3 and ORNL-4 to previous measurements. A sol-gel process was used to produce uniformly sized microspheres of theoretical density, 3.78 Mg/m^3.[14] The microspheres had a diameter range of 450 to 550 μm, a mean diameter of 508.3 μm, and contained less than 1 weight percent impurities. Beds of these microspheres pack reproducibly to 60.5 ± 1 volume percent solid.

Commercially available fumed silica was selected as the primary candidate for studies on insulation systems formed from small particles. Tests were conducted on Cab-O-Sil S17[15] and Aer-O-Sil

R974[1][6] which have nominal particle diameters (BET surface area method) of 0.007 μm (70 Å) and 0.012 μm (120 Å), respectively. Reference 2 describes additional tests conducted on fumed silica—carbon black mixtures, hollow microspheres, and perlite.

Fumed silica panels manufactured by L'Air Liquide were also tested. These panels were fabricated from Degussa 70 Å diameter fumed silica. The evacuated panels were fabricated in the fall of 1981 prior to the production process being shut down. The encapsulating material is an aluminized plastic which is heat sealed. The panels were originally evacuated to about 1 mm Hg. This evacuation compacted the powder by about a factor of three. Density varies from panel to panel because of the lack of a standardized manufacturing procedure. The plastic envelope for the panels is slightly permeable to air so that the internal pressure increases with time. L'Air Liquide calculations indicate that the pressure will increase to about 100 mm Hg after 5 years. Their best estimate was that the pressure was between 10 and 20 mm Hg at the time the λ measurements were made. Characteristics and thermal conductivity results for three panels are given in Table 3.

RESULTS

Figure 3 shows that λ results on the Al_2O_3 microspheres from ORNL-3, ORNL-4 and ORNL-2 agree to better than ±5% from 295 to 340 K. ORNL-2 is described in reference 17 and yielded λ values in nitrogen at a pressure of 0.1 MPa of 0.3361, 0.4365, 0.4687 and 0.5222 W/m·K at temperatures of 414.5, 631.5, 721.9 and 887.4 K, respectively. A quadratic fit to these data: $\lambda = 0.806 \cdot 10^{-1} - 0.721 \cdot 10^{-3}T - 0.252 \cdot 10^{-6}T^2$ is shown in Figure 3. The agreement is consistent with a computer modeling analysis that predicted 1—3% errors in λ results for ORNL-4 due to axial heat losses.[2]

Table 4 and Figure 4 contain the λ results on beds of Cab-O-Sil S17 and Aer-O-Sil R974 as a function of temperature and density at 0.1 MPa (atmospheric pressure) of air. For this temperature range the results may be described by a linear function of temperature with the temperature coefficient depending on density. Figure 5 shows the thermal conductivity at 300 K as a function of density. The thermal conductivity decreases rapidly as the volume fraction solid increases, passes through a minimum near 10% solid and then rises as the fraction solid continues to increase. Results for both types of powder seem to be on the same curve at 300 K.

Figure 6 shows that the λ results on three panels produced by L'Air Liquide increase with temperature but are very low from 295 to 320 K. These λ values are about four times larger than values

Table 3. Characteristics and Thermal Conductivity Results on
          Three Evacuated Panels Produced by L'Air Liquide

Mean Temperature (K)	Thermal Conductivity W/m•K
a. Panel 12 Density: 174 kg/m³; 96 x 48 x 5.3 cm	
297.24	0.00993
297.78	0.01007
306.57	0.01090
310.56	0.01150
311.02	0.01143
321.03	0.01277
b. Panel 22 Density: 183 kg/m³; 97.5 x 48.5 x 5.7 cm	
297.31	0.01208
306.54	0.01321
306.62	0.01358
320.60	0.01535
c. Panel 20 Density: 190 kg/m³; 97.5 x 48 x 5.5 cm	
297.45	0.00865
306.79	0.00946
311.30	0.00953
321.07	0.01084
321.22	0.01033

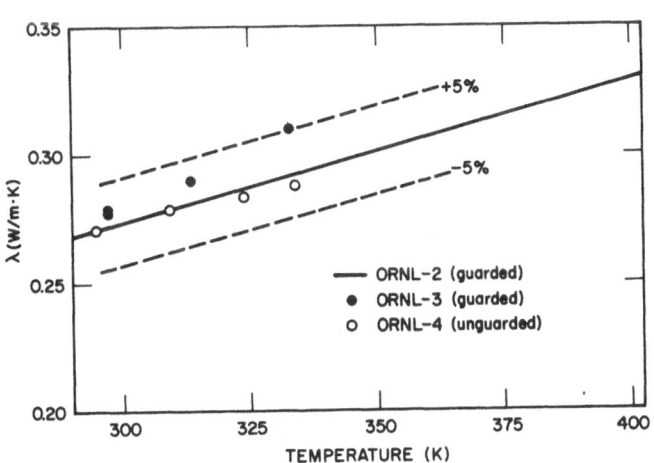

Figure 3: Comparison of thermal conductivity results from ORNL-2,
           -3, and -4 on a bed of nominally 500 μm diameter
           microspheres of alumina in air and N_2.

Table 4. Thermal Conductivity Results in Air at 0.1 MPa for Cab-O-Sil S17 and Aer-O-Sil R974

| Density | λ at 300 K (W/m·K) | From Least Squares Fit, $\lambda = A + BT$ | |
		$A_{coefficient}$ x 10^2	$B_{coefficient}$ x 10^4
1. Cab-O-Sil S17			
0.099	0.02334	-0.3784	0.9041
0.162	0.02130	0.0332	0.6990
0.280[a]	0.02291	0.8051	0.4951
0.280[b]	0.02179	0.0392	0.7131
0.496	0.04158	1.997	0.7203
2. Aer-O-Sil R974			
0.33	0.02498	0.5013	0.6654

[a]Run #1 not evacuated.

[b]Run #2 evacuated and backfilled with N_2.

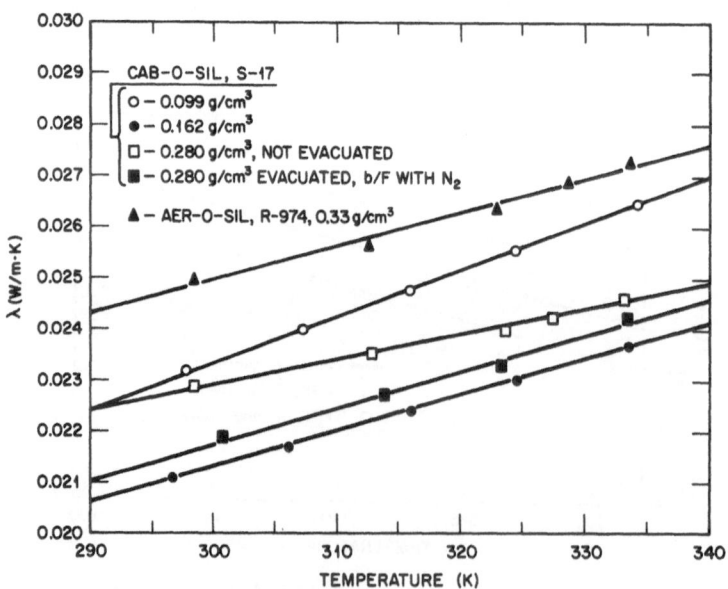

Figure 4: Thermal conductivity results in air at 0.1 MPa for Cab-O-Sil S17 and Aer-O-Sil R974 as a function of temperature and density.

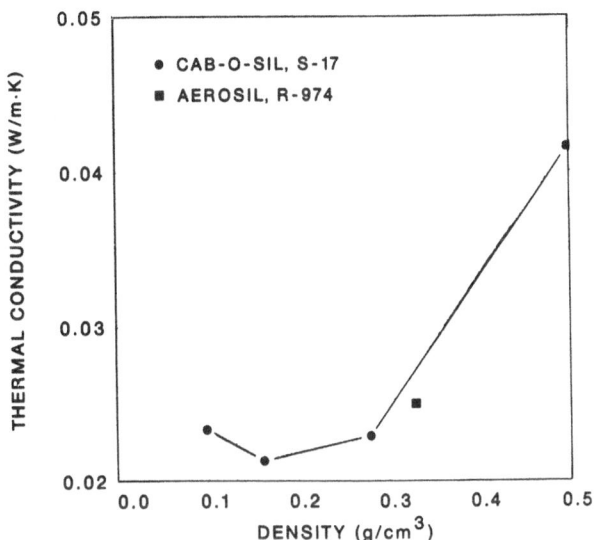

Figure 5: Thermal conductivity results at 300 K in air at 0.1 MPa for Cab-O-Sil S17 and Aer-O-Sil R974 as a function of density.

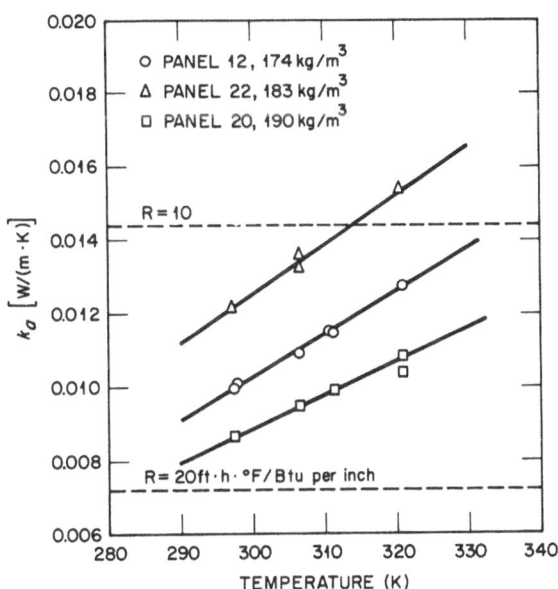

Figure 6: Thermal conductivity results on three evacuated panels produced by L'Air Liquide as a function of temperature.

reported in reference 3. The internal pressure is unknown but is thought to vary from panel to panel since the results do not correlate with panel density.

DISCUSSION

Heat transfer may occur in the insulation systems of interest by convection, conduction and radiation in the continuous gas phase and by conduction and radiation in the discontinuous solid phase. Simplified analyses were employed in establishing the approximate dependence of the thermal properties of the insulation system on solid volume fraction and particle size.[2] These models show that convection can be ignored. Furthermore, the analysis shows that for systems made only of angstrom-sized silica particles, radiation shine-through is significant at 300 K and dominates at temperatures above 300 K, thereby making the system less effective as a thermal insulator. Finally, the models predict the dependence of thermal conductivity on the solid fraction, as shown in Figure 5. Thus the target R-value can only be obtained by evacuating the material and reducing the radiative heat transfer.

ACKNOWLEDGEMENTS

The authors wish to thank J. P. Moore and D. W. Yarbrough for reviewing the manuscript and Brenda Hickey and Carolyn Whitus for typing the text.

REFERENCES

1. W. T. Lawrence and F. E. Ruccia, "Development of Advanced Insulation for Appliances Task 1," ORNL/Sub-81-13800/1 (June 1981).
2. G. L. Copeland, D. L. McElroy, H. A. Fine and T. W. Tong, "Development of Advanced Insulation for Appliances," ORNL/TM-9121 (June 1984).
3. P. Pellaux-Gervais and D. Goumy, "Insulating Material with Low Thermal Conductivity, Formed of a Compacted Granular Structure," United States Patent 4,159,359 assigned to L'Air Liquide, Paris, France, June 26, 1979.
4. Private communication from H. Cochrane, Cabot Research Center, Tuscola, Illnois to D. L. McElroy, ORNL, January 1983.
5. Private communication from S. Spinney, Dynatech R/D Company, to E. Wagner, Cabot Corporation, September 5, 1975 (Cab-O-Sil M-5).
6. Private communication from M. Fulk, National Bureau of Standards to H. Cochrane, Cabot Corporation, February 20, 1958 (Cab-O-Sil M-5).

7. Degussa Brochure dated July 2, 1950 on Aer-O-Sil 200.
8. P. E. Glazier, U.S. Patent No. 3,151,365 issued October 6, 1964 (Cab-O-Sil M-5).
9. MinK 1300 and 2000, Manville Corporation, P. O. Box 5108, Denver, Colorado 80217.
10. Microtherm2OCR, Micropore International, LTD., Droitwich, Wors., WR97DJ, England.
11. H. Reiss, "An Evacuated Powder Insulation for a High Temperature Na/S-Battery," AIAA-81-1107, AIAA 16th Thermophysics Conference, Palo Alto, CA (June 1981).
12. D. L. McElroy, R. S. Graves, D. W. Yarbrough and J. P. Moore, "A Flat Insulation Tester that Uses an Unguarded Nichrome Screen Wire Heater," in Forum on the Guarded Hot Plate and Heat Flow Meter State-of-the-Art, Quebec, Canada (October 7-8, 1982).
13. R. S. Graves, D. W. Yarbrough and D. L. McElroy, "Apparent Thermal Conductivity Measurements by an Unguarded Technique," in Thermal Conductivity 18, T. Ashworth, ed., Plenum Press, New York (in press).
14. Private communication from O. D. Erlandson, General Atomic Company, San Diego, CA to R. Roberts, Union Carbide Corporation Nuclear Division, Oak Ridge, TN, March 17, 1980.
15. Cabot Corporation Bulletin, "Cab-O-Sil Properties and Functions," Headquarters: 125 High Street, Boston, MA 02110.
16. Degussa, Inc., Pigments Division, Route 46 at Hollister Road, Teterboro, NJ 07608.
17. J. P. Moore, R. J. Dippenaar, R. O. A. Hall and D. L. McElroy, "Thermal Conductivity of Powders with UO_2 or ThO_2 Microspheres in Various Gases from 300 to 1300 K," ORNL/TM-8196 (June 1982).

THERMAL DIFFUSIVITY OF HETEROGENEOUS MATERIALS AND

NON-FIBROUS INSULATORS

R. R. Bittle[†] and R. E. Taylor

Thermophysical Properties Research Laboratory
Purdue University School of Mechanical Engineering
West Lafayette, Indiana

ABSTRACT

The flash technique has proven to be a fast and accurate method for determining the thermal diffusivity (and hence conductivity of a wide range of homogeneous materials from cryogenic temperatures into the molten region. The applicability of this technique has been extended to many heterogeneous materials including certain layered, dispersed and fiber-reinforced composites. Extensions of its use for some types of insulators has been limited due to the larger temperature rise which occurs on the front face of highly insulating materials, partial in-depth absorption of the laser energy by porous or translucent samples and problems associated with rear face temperature transient measurements on such materials. There are also difficulties involved in measuring the thermal diffusivity of large-grain heterogeneous materials where the grain size is of the order of the usual sample thicknesses used. Substituting step heating for the laser pulse tended to overcome problems associated with both large-grain heterogeneous materials and many insulating materials.

A step-wise heating apparatus was developed and tested using several materials including solid insulators and large-grain carbon magnesite brick. Results were compared to the standard flash technique where applicable, and procedures were developed which permitted these materials to be measured quickly and accurately. For example, diffusivity measurements were made on refractory brick materials from room temperature to 500C (seven data points) in less than four hours. Typical results are presented.

---

† Presently at General Electric Corporation, Evendale, OH

INTRODUCTION

Use of variable-state methods for measuring thermal transport properties has become increasingly popular in the last decade. The advantages of variable-state methods over steady-state methods such as the Guarded Hot Plate include a relatively short measurement duration. Steady-state techniques can take up to several hours to achieve a necessary steady temperature distribution for a single data point. Variable-state measurements, on the other hand, usually vary in duration from less than a second to 3 - 5 minutes [1,2]. Relative specimen size requirements also often favor variable-state techniques. Specimen sizes on the order of a small coin are used in many variable-state methods, whereas steady-state methods require much larger samples.

Variable-state measurement techniques include continuous heating, semi-infinite plate methods. These methods derive thermal diffusivity from the temperature response of a specimen subjected to a continuous heat flux. The term "semi-infinite" implies that the heat flow is unidirectional and normal to the surface on which the heat flux is imposed. Much of the early development of variable-state measurement techniques during the 1950's and 1960's was carried out using continuous heating, semi-infinite plate methods. The obvious advantages of variable-state techniques over steady-state methods, such as a shorter measurement duration, were realized but further development and refinement have awaited technological advancements.

The objective of this work was to develop a continuous heating, semi-infinite plate method for measuring thermal diffusivity values of specimens that are relatively thin compared to those used in the guarded hot plate or hot wire methods, but that are relatively thick compared to samples used for the laser flash technique. The method is referred to as step heating and involves subjecting one face of a generally disk-shaped specimen to a constant and uniform heat flux at time zero, and recording the surface temperature response on the opposite face. Incorporated into this step heating method are ideas proposed in earlier works [3,4], but in addition it utilizes computerized data acquisition and analysis.

The developed step heating technique is characterized in Table 1 along with the laser flash method for comparison. Laser flash is a pulsed heating, variable-state measurement technique first proposed by Parker, et. al. [5]. Pulsed heating differs from step heating in that the imposed heat flux is an instantaneous deposition of heat on the specimen surface. One primary advantage of step heating over laser flash is its application to larger specimens. The larger specimen allows for more accurate measurement of heterogeneous materials since the specimen is more representative of the bulk material. Another important advantage of the step heating method is the relatively low intensity level of the imposed constant heat flux compared with the instantaneous heat pulse necessary with the laser flash. Specimens are thereby less likely to begin a phase transformation or decompose as a result of a sudden rise in temperature of the front surface caused by a heat pulse.

MATHEMATICAL DEVELOPMENT

1. Derivation of the Working Equation

The model used in the mathematical development of this step heating technique is a semi-infinite slab of finite thickness. Initially, the slab is at a constant temperature T_0. At time zero, one surface ($x = 1$) is exposed to a constant and uniform heat flow. It is assumed that there are no heat losses from the slab. The resulting temperature distribution in the slab is given by Carslaw and Jaeger [6] and the resulting mathematical developments are given by Bittle and Taylor [7]. The observed parameter V is a ratio of rear face surface temperature changes and is defined as

$$V = \frac{T_1(t_1) - T_0}{T_2(t_2) - T_0} \tag{1}$$

where $t_2 > t_1$. This results in the working equation given by

$$V = \frac{\dfrac{\alpha t_1}{l^2} - \dfrac{1}{6} - \dfrac{2}{\pi^2} \sum_{n=1}^{\infty} \dfrac{(-1)^n}{n^2} \exp\left[-\alpha\left(\dfrac{n\pi}{l}\right)^2 t_1\right]}{\dfrac{\alpha t_2}{l^2} - \dfrac{1}{6} - \dfrac{2}{\pi^2} \sum_{n=1}^{\infty} \dfrac{(-1)^n}{n^2} \exp\left[-\alpha\left(\dfrac{n\pi}{l}\right)^2 t_2\right]} \tag{2}$$

Table 1. Characterization of Laser Flash and Step Heating Techniques (Typical Values).

	LASER FLASH	STEP HEATING
Specimen Size	1.25×10^{-2} m dia. 0.2-0-5×10^{-2} m thick	4.0×10^{-2} m dia. 0.3-1.25×10^{-2} m thick
Measurement Duration	30 μsec. 5 sec	3–60 sec
Elapsed Time for Measurements 20–500C (6–7 values)	3 hours	4 hours
Heat Input Source	Laser	Halogen Projector Lamp, 600 W
(Cost)	($20,000)	($20)
Rear Surface Temperature Rise	1–2°C	1–2°C
Front Surface Temperature Rise	20–100°C	5–30°C

Once the experimental temperature response is obtained, the only unknown left in Equation 2 is thermal diffusivity (α).

The secant iteration method is used to determine diffusivity values. For a given measured value of V, the diffusivity (α) is varied until the absolute value of the difference of the left and right hand sides of Equation 2 becomes less than 0.0001. When this condition is met, the corresponding diffusivity value is taken to satisfy Equation 2. Thermal diffusivity is calculated at several different times, t_1 for each set of temperature response data, and if the values are consistent, the analysis is considered valid.

2. Heat Loss Correction

The derivation of the working equation, Equation 2, assumed no heat losses from the slab. Cowan derived a solution for this same problem assuming radiative heat losses from both faces of the slab due to the temperature difference between the slab surroundings [8]. Cowan's analytical procedure for determining thermal diffusivity not only requires specimen dimensions and rear surface temperature response data but also knowledge of a surface heat loss parameter. The solution derived by Cowan was used to generate normalized rear surface response curves which included heat losses from the specimen. In each case, the specimen thickness and thermal diffusivity value (α) were held constant. Only a heat loss parameter was varied. For each generated response curve, thermal diffusivity values were calculated at several times using Equation 2. The calculated diffusivity values increased with time over the measurement time interval. A higher heat loss parameter was accompanied by a more rapid increase in the calculated diffusivity values over time. It was concluded that heat losses could be recognized by the shape of the normalized temperature response curve, and from the increase in diffusivity values calculated over the measurement duration time.

It was apparent that a straight line extrapolation of any set of calculated diffusivity values to zero time intersects at $\alpha(t)/\alpha$ (actual) = 1.0 Thus, the heat loss correction procedure used was to extrapolate the calculated diffusivity values for a given response curve to zero time.

DESCRIPTION OF MEASUREMENT APPARATUS

A schematic of the equipment arrangement for this study is diagramed in Figure 1. The heat flux source is directed at the sample and the step input controlled using a rotating shutter mechanism. The sample's rear surface temperature response is detected using either a thermocouple or an infrared (IR) detector. The output from either detector is amplified, filtered if necessary, sent to the computer through the analog-to-digital (A-D) converter then stored in data files.

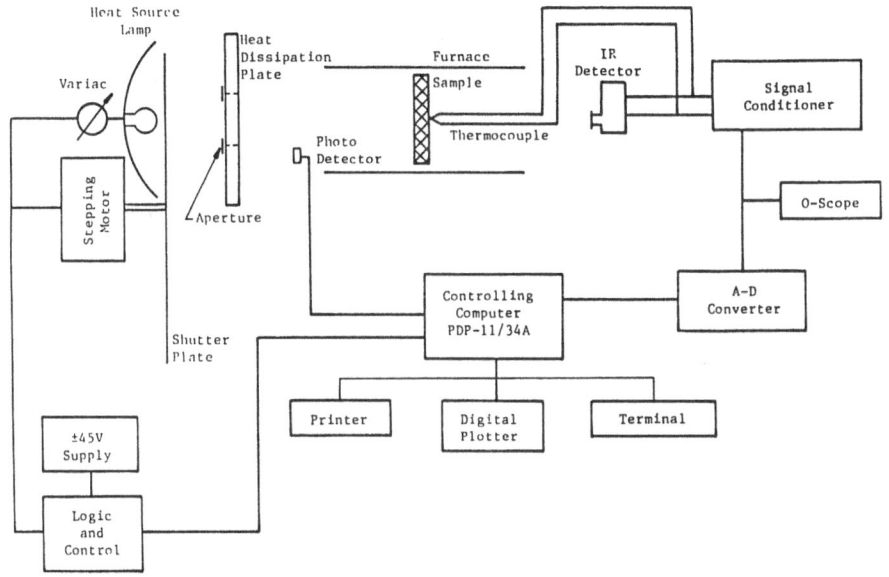

Figure 1. Schematic of Step Heating Apparatus

A 600 Watt, quartz-iodide tungsten element bulb mounted within an aluminum parabolic reflector is the heat flux source. The reflector is cooled using both a water cooling system and forced convection to an air stream. The source has been experimentally verified to be reasonably constant over a thirty minute period and requires approximately two seconds to reach maximum output. The heat flux can be controlled between zero and maximum output using a variac. Thirty centimeters from the heat source the flux intensity was measured and found to vary less than ±2% across a 0.08 m diameter.

The shutter mechanism controlling the heat flux step input consists of a plate concentrically attached to the shaft of a stepping motor. The driving circuit and logic that control the motor are powered by a 45 volt, bipolar supply. The input signal to begin the shuttering sequence is sent to the stepping motor controller from the controlling computer. This same input signal also energizes the heat source. The shutter remains in the normal closed position for a ten second delay, shielding the sample from the heat source. During the delay, the heat source becomes stabilized and the initial sample temperature data are collected. The heat flux is imposed on the front face of the specimen when the stepping motor rotates the shutter plate 90 degrees. When the data collection is completed, the shutter is returned to the closed position.

The initial specimen, temperature T, is controlled above room temperature with a tubular air furnace. The heating element is a chrome-aluminum iron alloy wire embedded in Al_2O_3 cement and can reach a maximum temperature of 1200C. Both ends of the furnace are closed using 1.5 mm thick fused-quartz optical flats which transmit much of the infrared radiation. This decreases convective losses from the sample, but still permits viewing the sample by the IR detector and heat lamp.

A thermocouple or an IR detector is used to detect the temperature response at the rear surface. The voltage output signal from both detectors changes linearly with temperature over a small temperature range. In determining thermal diffusivity, the analysis requires the temperature response to be in the form of a ratio of temperature change (Equation 2). Because of the linear relationship between output voltage and temperature, the ratio of temperature change has the same value as the corresponding ratio of voltage change. Hence, the recorded output voltage from the thermocouple or IR detector can be directly used in the analysis procedure.

An advantage of the IR detector is in the application to heterogeneous materials such as dispersed composites. In this case, the surface of a specimen machined for a diffusivity measurement would be composed of two or more materials: the dispersed particles and the main matrix. During diffusivity measurement, the temperature distribution may not be uniform across the rear surface due to the presence of two different materials. A major concern is viewing by the temperature sensing device of an area of the rear surface which is not representative of the bulk materials. Since the area viewed by the IR detector is controllable, an average response can be obtained by trial and error.

The output from the thermocouple or IR detector is routed through a signal conditioner having high gain, very low internal noise, fast response and an amplification range of 250,000. It has ten levels of low-pass filtering between 10 and 100,000 Hz as well as a notch filter for 60 Hz noise. The controlling computer used for data acquisition and analysis is the DEC PDP 11/34A.

SAMPLE DESCRIPTIONS AND PREPARATION

The present study employed six different materials on which thermal diffusivity measurements were made over a limited temperature range. Five of the materials are characterized in Table 2. They cover a wide range of thermal diffusivities from moderately good thermal conductors to very good thermal insulators. Material makeup ranges from a very homogeneous structure as with the stainless steel, to a coarse granular structure of one of the insulating refractories. The sixth material, a large-grain heterogeneous carbon magnesite brick, is treated separately [7]. Some of the materials are available commercially while others are still in the developmental stage.

Table 2. Materials Characterization

Material No.	Material	Color	Make-Up	Bulk Density (kg/m^3)	Thermal Diffusivity @ 100°C (m^2/s)	Remarks
1	Stainless Steel 303	Metallic	Machined Steel	7830	3.70×10^{-6}	All samples machined from the same round bar stock.
2	Zircar Alumina, Type SALI	White	Porous, Homogeneous appearance	472	0.230×10^{-6}	Relatively rigid, easily machined.
3	Insulating Refractory-2300	Beige	Porous, non-homogeneous appearance	476	0.340×10^{-6}	Voids randomly dispersed within the material - machinable.
4	Insulating Refractory-2800	White	Porous, non-homogeneous appearance	698	0.440×10^{-6}	Very coarse, granular, and abrasive. Significant air space within. Not machinable.
5	Space Shuttle Insulation, LI-900	White	Porous, homogeneous appearance	145	0.375×10^{-6}	Very soft, breaks apart easily. Machinable.

Thermal diffusivity specimen size was dependint on the material being measured. Samples of Materials 1, 2 and 5 were machined to a diameter of 0.038m (1.5 in) and the thickness ranged between 0.003 and 0.0075 m. Specimens of the insulating refractories, materials 3 and 4, were machined to a 0.0445m (1.75 in) diameter with a thickness ranging between 0.004 and 0.0095. The larger sample size for Materials 3 and 4 was necessary due to their relatively heterogeneous makeup compared with Materials 1, 2 and 5, since larger specimens are more representative of the bulk material.

A common characteristic of the insulators, Materials 2, 3, 4 and 5, is their high porosity. The large voids within the insulating refractories, Materials 3 and 4, resulted in numerous holes in the surface of specimens. The chemical composition and porosity of the Zircar insulation and LI-900 (Materials 2 and 5) gave these specimens a translucency to the heat flux. Both the surface holes and translucency would have allowed the heat flux to penetrate to the interior of a specimen during diffusivity measurements. This would cause significant volume heating, which is a violation of the boundary condition requiring the constant heat flux to be deposited on the sample's front surface.

The volume heating caused by translucence to the heat flux in Materials 2 and 5 was eliminated by painting the sample's front surface. This also gave the surface a uniform emissivity. The paint used was a black, heat resistant spray paint made by Derusto. It is commonly applied to charcoal grills and wood-burning stoves. In all cases a thin coating approximately 5×10^{-5} m thick was satisfactory. A visual inspection indicated absorption of the paint by the material was insignificant.

The insulating refractory specimens with surface holes required a front layer cover which was smooth and uniform. A graphite-based sealer made by Devcon Industries was used to achieve

this condition. The sealer is rated to 800C, has a thermal diffusivity of 0.085 cm /sec at room temperature, and forms a good bond
with the insulators. Surface preparation consisted of spreading a
thin coating evenly over the sample surface and then allowing it to
dry. Next, excess sealer was machined off until a layer on the
order of 1×10^{-4} m remained leaving a smooth surface with a relatively uniform emissivity. There was still some non-uniform heating
over the surface due to the surface holes being partially filled,
but not a volume heating problem which would occur if no surface
cover were used.

A mathematical analysis was made to determine the limiting
front layer thickness for the sealer which would ensure no adverse
effects on the measured diffusivity values. Using a finite-element
heat transfer analysis program, it was determined that a limiting
sealer-to-specimen thickness ratio l_1/l_2 of 0.025 would result in
less than 5% error in measured diffusivity values due to the presence of the sealer layer. Experimental verification of the limiting thickness ratio was made by comparing measured diffusivity values over a limited temperature range for the two cases showed no
significant difference.

The porosity of the insulating materials would not permit accurate use of the IR detector during diffusivity measurements. The
mathematical model for this step heating technique utilizes the
rear surface temperature change in the diffusivity analysis. Internal temperature changes were detected when the IR detector was
used to monitor the surface temperature of specimens with surface
holes or which were translucent to IR. A solution to the problem
of detecting the internal temperature was to use a thermocouple
arrangement on the sample's rear face which would respond to the
surface temperature rise.

Direct placement of a thermocouple at a point on the surface
of a low conductivity material could lead to a non-representative
transient temperature response. The problem is the thermocouple
would act as a heat sink due to the relatively high thermal conductivity of the thermocouple wire compared to that of an insulator. To overcome the heat sink problem, the thermocouple should
be distributed uniformly over the sample's rear surface. This was
accomplished by spot-welding the thermocouple to a thin (0.005 cm)
sheet of stainless steel approximately .02 x .02 m square which was
then attached to the rear surface of the sample. A thin layer of
Devcon sealer was used to fasten the steel sheet to the specimen.

MEASUREMENT RESULTS

1. Verification of the Step Heating Technique

The first diffusivity measurements were made to verify the

step heating technique for thermal conductors and thermal insulators. The measured thermal diffusivity values for Materials 1, 2 and 5 over a limited temperature range were compared with values obtained using the laser flash method. These materials can be accurately measured using the laser flash technique [2] for maximum variation in the results for the two methods was 5% for Material 2 (Zircar alumina), and 4% for Material 5 (LI-900, Figure 2).

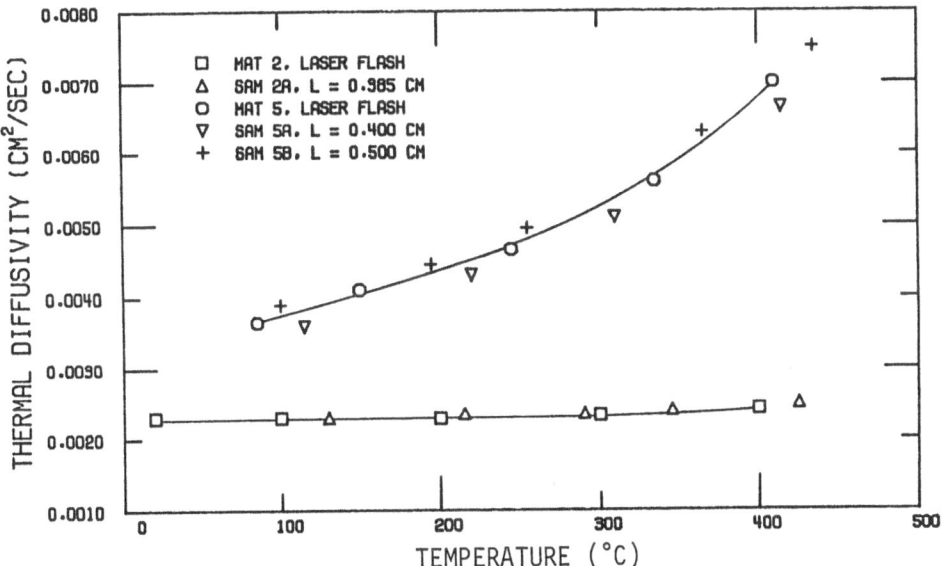

Figure 2. Technique Verification

2. Correction for Heat-Loss

The heat-loss correction procedure was verified using specimens of Materials 2 and 3. For both materials, the correction procedure lowered measured diffusivity values which were too high as a result of heat losses. Measurements of a Material 2 specimen, 1 = 0.005 m, were made through 500C and compared with the laser flash results. The corrected diffusivity values for the Material 2 specimen were in close agreement with the values obtained using the laser flash method (Figure 3). Measurements of a Material 3 specimen 1 = 0.0095 m, were made through 457C and compared with measured values for a specimen, 1 = 0.00385 m (Figure 3). The corrected diffusivity values of the Material 3 specimen, 1 = 0.0095 m, were in good agreement with the measured diffusivity values for the specimen, 1 = 0.00385 m.

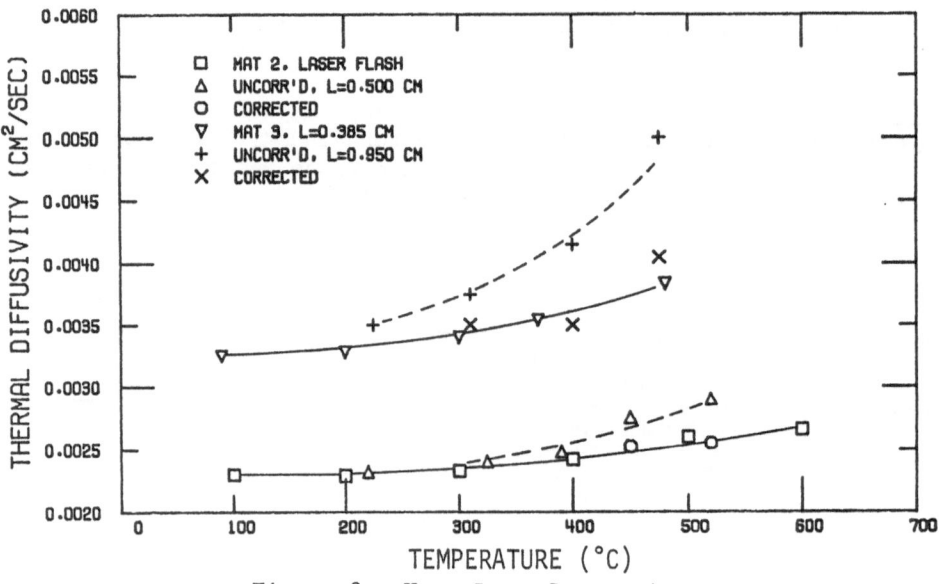

Figure 3. Heat Loss Corrections

SUMMARY AND CONCLUSIONS

A step heating technique was developed for measuring thermal diffusivity values. The method involves subjecting one face of a generally disk-shaped specimen to a constant and uniform heat flux and recording the surface temperature response on the opposite face. The advantages of this method include the application to specimens of relatively large size compared to the limited size of specimens used in pulse heating methods. A larger sample allows for more accurate measurement of heterogeneous materials since the specimen is more representative of the bulk material. Although a larger specimen can be accommodated, its size is still much smaller than that required for a steady-state method such as the guarded hot plate. Another advantage of the step heating method is the relatively low intensity of the imposed heat flux compared with that necessary for pulse heating techniques. Specimens are thereby less likely to begin a phase transformation or decompose as a result of a sudden large temperature increase at the front surface caused by an intense heat pulse.

Thermal diffusivity measurements were made through approximately 500C on thermal insulators and conductors. The step heating technique was shown to be applicable over at least the range of thermal diffusivity values given in Table 2. Measurement duration at a given temperature depended on the material and specimen size, but varied between 3 seconds and 1 minute. The total time for measurement of even diffusivity values from room temperature to

500C was on the average four hours. Verification of the step heating method was made by comparing measured diffusivity results for several materials with those obtained using the laser flash method. A maximum variation of 5% was found between the results of the two techniques.

Two major problems were encountered with diffusivity measurements of the insulating materials resulting from their relatively high porosity. First, volume heating of specimens occurred during diffusivity measurements since the imposed heat flux could penetrate inside the specimen. This is a violation of the boundary condition requiring the heat flux to be deposited on the sample's front surface. Second, the mathematical analysis to determine diffusivity values for this step heating technique utilizes the temperature response at the rear surface. Internal temperature changes were detected in the porous insulator specimens when the IR detector was used to monitor rear surface temperature during diffusivity measurements.

The problem of volume heating was solved by either painting the front surface of the specimen, or covering the surface with a thin, conductive graphite-based sealer. Both methods insured deposition of the heat flux on the front surface. It was mathematically and experimentally verified that a front layer could be applied which did not significantly affect measured diffusivity values.

A solution to the problem of detecting internal temperature changes was to place a thermocouple arrangement on the sample's rear surface to measure changes in the surface temperature. The thermocouple wire was spot-welded to a thin sheet of stainless steel which was then attached to the sample rear surface. For some of the insulation materials, internal temperature detection could be eliminated by painting the rear surface. This produced a secondary effect of increasing heat losses from the specimen, which influenced measured diffusivity values. Consequently, all diffusivity measurements of the insulating materials were made using a thermocouple to monitor the rear surface temperature.

The rise in temperature for the rear surface of a specimen during diffusivity measurements is approximately 1 or 2°C above the initial temperature T_0. A heat loss analysis demonstrated that convective radiative heat losses from the rear surface as the specimen temperature increases can affect significantly the measured diffusivity values.

Heat losses can be minimized by limiting the duration of the measurement. This is accomplished by measuring a specimen which is as thin as possible, yet still representative of the material. Losses from the specimen can also be decreased with a low rear surface emissivity and by keeping the rear-surface temperature rise small. A heat loss correction procedure was developed to correct measured diffusivity values affected by heat losses. The procedure was verified with the specimen for which measured diffusivity values were most affected by heat losses. In this case, the measured diffusivity values were as much as 1.25 times the corrected value. The limits to the applicability of the procedure were not determined.

ACKNOWLEDGEMENTS

The authors wish to thank W. Dennis and R. Jefferess, the American Iron and Steel Institute, for their support and R. Padfield, Bethlehem Steel Corporation, for discussions anf for acting as technical monitor.

REFERENCES

1. A.W. Pratt and J.M.E. Ball, "Thermal Conductivity of Building Materials, Methods of Determination and Results", J. Inst. Heating Ventilating Engrs., 24, P. 201, 1956.

2. R.E. Taylor, "Heat-Pulse Thermal Diffusivity Measurements", High Temperatures-High Pressures, 11, P. 43, 1979.

3. C.P. Butler and E.C.Y. Inn, "Thermal Diffusivity of Metals at Elevated Temperatures", Thermodynamic and Transport Properties of Gases, Liquids and Solids, Trans. ASME, N.Y., 377-90, 1959.

4. G. Sonnenschein and R.A. Winn, "A Relaxation Time Technique for Measurement of Thermal Diffusivity", WADC Tech. Rept. 59-273, 1-23, Feb. 1960.

5. W.J. Parker, R.J. Jenkins, C.P. Butler and G.L. Abbott, "Flash Method of Determining Thermal Diffusivity Heat Capacity and Thermal Conductivity", J. of Appl. Phys., 32(9) 1679-84, Sept. 1961.

6. H.S. Carslaw and J.C. Jaeger, Conduction of Heat in Solids, 2nd ed., Oxford University Press, P. 112, 1959.

7. R.R. Bittle and R.E. Taylor, "Step-Heating Technique for Thermal Diffusivity Measurements of Large-Grained Heterogeneous Materials", J. Amer. Ceramic Soc., 67(3), P. 186-90, Mar. 1984.

8. R.D. Cowan, "Pulse Method of Measuring Thermal Diffusivity at High Temperatures", J. of Appl. Phys., 34(4), P. 926, Apr. 1963.

SESSION N

Insulations II

SESSION CHAIRMAN

C. M. Pelanne
Littleton, CO.

OPTICAL PROPERTIES AND RADIATIVE HEAT TRANSPORT

IN POLYESTER FIBER INSULATION

[1]N.L. McKay, [1]T. Timusk and [2]B. Farnaworth

[1]McMaster University, Hamilton, Ontario, Canada
[2]Defense Research Establishment
 Ottawa, Ontario, Canada

ABSTRACT

We present calculations of combined radiative and conductive
heat flow in two types of polyester fiber insulation. The radiative transport was represented by both a diffusion model and a two-
flux model, and the results compared. Optical parameters in both
cases were derived from realistic scattering calculations using
the frequency-dependent complex refractive index of polyester and
the measured fiber diameter. We found that anisotropy in the
scattering must be included in the model to obtain accurate results. Model calculations agreed very well with direct measurements of thermal resistance for samples of several thicknesses,
with no adjustable parameters in the theory.

INTRODUCTION

Thermal radiation has long been recognised as an important
mechanism of heat transfer in glass-fiber and other low-density
materials used as insulation in buildings [1,2], in high-temperature applications, and even in animal fur [3]. Recently it has
been shown that radiative transfer is equally important in several
natural and synthetic materials used as insulation in sleeping bags
and cold-weather clothing [4,5]. Convection, previously thought
to be a significant component of the thermal performance of these
materials, was shown to be negligible. Differences in apparent
thermal conductivity among goose down, large-diameter polyester
fiber insulation, and polypropylene microfiber insulation result

instead from differences in the radiative component of the heat flow. In this paper we show how the radiative transport in two types of polyester fiber insulation may be calculated from their optical properties.

Exact solutions of the radiative transfer problem can only be obtained numerically and are not necessary to describe heat flow adequately in insulation materials. Generally a simple model is used and fits the data well if the optical parameters are properly chosen. Most common is the two-flux model, either with a single (absorption) parameter, or with separate absorption and scattering parameters. A diffusion model is also frequently used, usually with absorption only, or with isotropic scattering. Below we use the results of realistic scattering calculations to include the anisotropy in a diffusion model and a two-flux model of radiative transport. Calculated thermal resistances are compared with heat-flow measurements.

All of these models use absorption and scattering parameters which are averaged over incident and scattered angles according to different prescriptions. For thermal transport problems the model equations can be averaged over wavelength if thermal weight functions are used to integrate the wavelength-dependent cross-sections.

One approach for obtaining the required optical parameters is to measure diffuse transmittance and reflectance of optically thick samples, either as a function of wavelength [6] or averaged over a blackbody spectrum [7]. These multiple-scattering phenomena can be described by a two-flux model, so a fit to the data yields the two-flux parameters. Assuming the same optical constants enter into the model for heat transport, these parameters can be used to predict the radiative contribution to the thermal conductivity.

A second approach is to calculate the scattering and absorption cross-sections directly from measured or assumed fiber properties. Both the fiber geometry - size, shape and orientation of the fibers - and the optical properties of the bulk material are needed. Hager and Steere [8] used such a procedure for very thick fibers, simplifying the optical problem by assuming the fibers were black. Davis and Birkebak presented similar calculations for various fiber orientations, using an adjustable emissivity parameter to represent the unknown optical properties [3]. Both of these calculations were for fibers large compared to the wavelength of thermal radiation, so diffraction effects could be ignored. In general such microscopic calculations are possible only for very regular and well-defined fiber geometries.

The materials "Hollofil", made by Dupont, and "Polarguard", made by Celanese, do have a simple geometrical structure. Both consist of cylindrical polyester (polyethylene terephthalate) fibers of uniform diameter, with orientations which are approximately random. "Hollofil" fibers are in fact hollow tubes with about 10% of the material removed, a complication which we have not included in the scattering calculations. Recently "Polarguard" has been modified to include a binder. All the measurements in this paper were done on the older version, with no binder.

Scattering and absorption are of comparable magnitude in these battings, and the scattering is highly anisotropic. Diffraction effects are important, as the fiber diameter (about 25 μm) is comparable to a thermal wavelength (5-50 μm). The optical constants of the bulk material vary strongly with wavelength in this range. We include all of these effects in our theory and obtain good agreement with the thermal measurements, with no adjustable parameters in the theory.

HEAT TRANSPORT

In the absence of convection the total heat flux may be written

$$\dot{q} = F - \lambda \, dT/dx \tag{1}$$

for a uniform slab perpendicular to the x-direction, which approximates the conditions of our measurements. The radiative flux is denoted by F, and T is the temperature, assumed to vary only in the x-direction. The heterogeneous mixture of fibers and air is represented by a single thermal conductivity λ whose value can be calculated from the conductivities of the two separate media. By solving Laplace's equation for an isolated fiber and averaging over fiber orientations we have

$$\lambda = \lambda_{air} + f(\lambda_f - \lambda_{air})[(\lambda_f + 5\lambda_{air})/(3\lambda_f + 3\lambda_{air})] . \tag{2}$$

Bhattacharyya [9] obtained an expression which reduces to Eq. 3 for small volume fractions f by considering a fiber as the limiting case of a prolate spheroid.

A complete description of the radiation involves specifying the angular intensity function $I(x,\nu,\theta)$ as a function of frequency ν and angle θ (measured from the x-axis), even when the radiation is assumed to be unpolarised. However, we are interested only in the radiative flux $F(x)$, which is the integral over both angle and frequency of $I\cos(\theta)$. The diffusion model approximates $I(\theta)$ by a linear function of $\cos\theta$ to integrate the equation of transfer over angle [10]. Assuming the radiation has nearly a blackbody spectrum, the resulting coupled differential equations are further integrated over frequency to give

$$dF/dx = \rho\bar{\kappa}_A (4\sigma T^4 - U) \tag{3}$$

$$dU/dx = -3\rho\bar{\kappa}_T F \tag{4}$$

where U is a quantity proportional to the radiative energy density, ρ is the density of the insulation, and σ is the Stefan-Boltzmann constant. The optical properties of the material are contained in the frequency-averaged cross-sections per unit mass,

$$\bar{\kappa}_A = 2/(\pi r \rho_f) <Q_A(\nu)> \tag{5}$$

$$1/\bar{\kappa}_T = \pi r \rho_f/2 <1/(Q_A(\nu) + Q_S'(\nu))> \tag{6}$$

The brackets indicate an average over frequency with the thermal weight function $dB(\nu,T)/dT$ where B is the blackbody spectral energy density. The factor $(2/\pi r \rho_f)$ is the geometrical cross-section per unit mass of a fiber of radius r and density ρ_f, and the dimensionless absorption efficiency Q_A is the absorption cross-section divided by the geometrical cross-section. In the diffusion model the scattering is represented by the average efficiency

$$Q_S' = \int_{4\pi} dQ_S/d\Omega \ (1-\cos\theta_S)d\Omega \tag{7}$$

where θ_S is the scattering angle and $dQ_S/d\Omega$ the differential scattering efficiency as a function of angle.

A commonly-used simpler version of this model treats the scattering as isotropic. In this case the weighted scattering efficiency Q_S' in the definition (6) of $\bar{\kappa}_T$ should be replaced by the total scattering efficiency Q_S. Such a change considerably simplifies the problem as both the absorption and the total scattering may be measured directly [12]. Unfortunately the isotropic-scattering model performed poorly in predicting the heat transfer in our test materials.

Grey boundary surfaces of temperature T_H and T_C and emissivity ε are represented by the usual condition,

$$F(x=0) = \frac{\varepsilon}{2-\varepsilon} (2\sigma T_H^4 - U(x=0)/2) \tag{8}$$

and a similar condition, with a minus sign, at x=L.

Equations (1), (3) and (4) with these boundary conditions form a system of nonlinear differential equations which may be solved by standard numerical techniques. Since in our thermal measurements the temperature differences across the samples were small (10-20°K) we linearise the equations in T and obtain the approximate analytic solution

$$T(x) = C_0 + C_1 x + C_2 \exp\{-px\} + C_3 \exp\{-p(L-x)\} \tag{9}$$

where

$$p = [3\rho\bar{\kappa}_T)(\rho\bar{\kappa}_A)(1 + \lambda_R/\lambda)]^{1/2} \tag{10}$$

and the constants are determined from the boundary conditions. We have defined the "radiative conductivity"

$$\lambda_R = 16\sigma T_M^3/(3\rho\bar{\kappa}_T) \tag{11}$$

where T_M is the mean temperature. In the case of identical boundaries this leads to a thermal resistance given by

$$R = \frac{L}{\lambda+\lambda_R} [1 + \frac{\lambda_R}{\varepsilon/(2-\varepsilon)(\lambda+\lambda_R)(3\rho\bar{\kappa}_T L/4)+\lambda pL/2 \; \coth(pL/2)}] \tag{12}$$

In general the radiative and conductive parts of the heat flux cannot be separated in eq. (12). Energy is transferred between the two modes over a coupling distance $1/p$. Only when the absorption cross-section is very small (so that p goes to zero) do the radiative and conductive terms become independent.

The widely-used two-flux model divides the radiation into two hemispheres and treats the left and right fluxes F_L and F_R separately. A pair of differential equations are derived by considering the change in each partial flux due to scattering, absorption, and emission in a thin slab of the medium. Written in terms of the variables $F = F_R - F_L$ and $U = 2(F_R + F_L)$, the frequency-averaged equations are

$$dF/dx = \rho\bar{\kappa}_A(4\sigma T^4 - U) \tag{13}$$

$$dU/dx = -4\rho\bar{\kappa}_2 F \tag{14}$$

where $\bar{\kappa}_2$ shares the definition (6) with $\bar{\kappa}_T$ except that the scattering parameter Q_S' is now obtained by calculating the flux backscattered from a thin slab uniformly illuminated from one side. Normalisation factors have been kept the same, so the extra factor 4/3 in (14) as compared to (6) is a significant difference. It arises from the different assumptions for the angular variation of the intensity in the two models, and leads to different predictions. Otherwise the equations (13) and (14) are identical to (5) and (6) and may be solved in the same way.

SCATTERING AND ABSORPTION FUNCTIONS

The models of heat transport described above require the differential cross-section as a function of angle as well as the absorption cross-section of the batting. These were obtained from a combination of optical measurements on Mylar films (made from the same polymer, polyethylene terephthalate, as the insulation fibres) with theoretical calculations based on the structure of the materials. We shall briefly outline the procedure below. A detailed description will be published elsewhere [12].

Our calculations were restricted to the materials "Polarguard" and "Hollofil", made from cylindrical fibers of diameter 23.4 μm and 26.4 μm respectively. The "Hollofil" fibers are hollow tubes with an inside diameter of 5 μm. These dimensions were obtained from electron microscopy and have an accuracy of about 1 μm.

On a scale of 1 mm or so the fibers are straight and randomly oriented. We used the theory of electromagnetic scattering from infinite dielectric cylinders at oblique incidence [13] to calculate the absorption and scattering functions. For this purpose the hole in the centre of the Hollofil fibers was ignored, although it must contribute some additional scattering. The complex refractive index as a function of frequency was obtained from transmission measurements on several different thicknesses of Mylar film by an iterative scheme. Finally the scattering functions were averaged over angle and frequency with appropriate weighting to give the parameters of the diffusion and two-flux models.

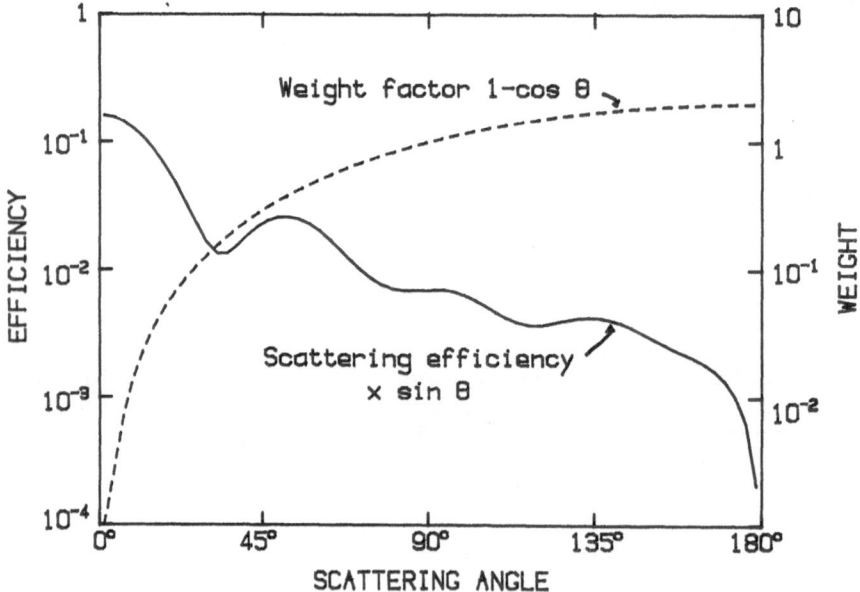

Fig. 1. Scattering anisotropy. The solid curve is the differential scattering efficiency calculated for polyester fibers of 23.4 μm diameter. Also shown is the weight function used in averaging over scattering angles in the diffusion model.

RESULTS AND DISCUSSION

In Fig. 1 we show a typical calculated differential scattering efficiency for Polarguard at a frequency of 500 cm^{-1}. The solid curve is $\sin\theta \, dQ_S/d\Omega$, the efficiency for scattering into the annular region between θ and $\theta + d\theta$, plotted against the scattering angle. Note the predominance of forward scattering. This is the reason that an isotropic-scattering model using the total scattering as a parameter fails to describe heat transport in these materials. Scattering at larger angles is more effective in impeding the radiative flow of energy. The dashed curve represents the weight function $1 - \cos\theta$ which appears in the diffusion model. It is seen at least qualitatively that this weight function will emphasize scattering at large angles and reduce the contribution of forward scattering.

Averaging a scattering function such as that in Fig. 1 over angle gives the weighted scattering efficiency for a particular frequency. Figure 2 shows the sum of weighted scattering and ab-

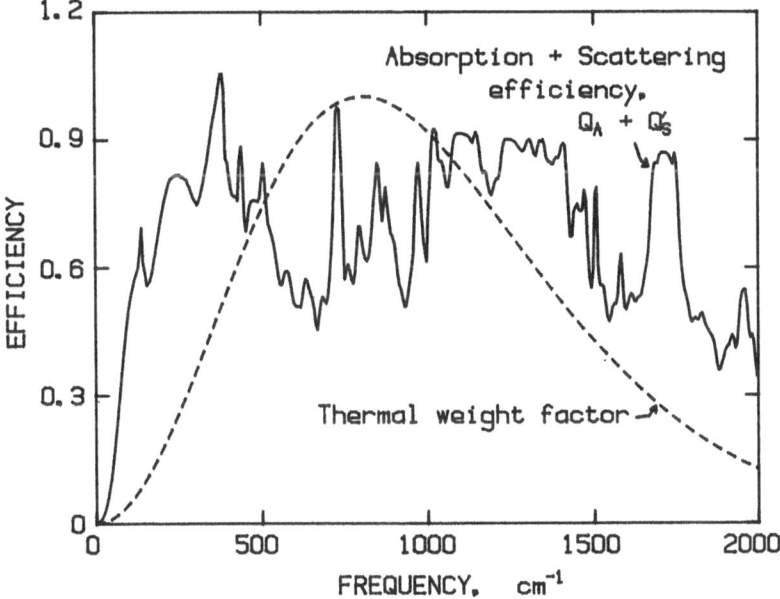

Fig. 2 Frequency dependence. Shown is the sum of absorption and weighted scattering efficiencies calculated for 23.4 μm fibers, along with the thermal weight function.

sorption efficiencies which appears in the definition (6) of $\bar{\kappa}_T$ plotted against frequency. Also shown is the thermal weight function (derivative of the blackbody function) for a temperature of 30°C.

Parameters of the two insulation materials are listed in Table 1. We use the fiber density essentially to determine the geometrical cross-section per unit mass of the fibers. Therefore the value listed for Hollofil has been adjusted to account for the fibers being lighter than solid fibers of the same size.

A comparison of calculated and measured thermal resistances appears in Table 2. For Polarguard, the diffusion model gives excellent agreement in all cases, while the two-flux model gives resistances which are too large. We emphasize that no adjustable parameters or optical measurements on the fibers directly were used in either theory. Heat flow was calculated from the measured fiber diameter and the complex refractive index of bulk polyester.

The better agreement of the diffusion model is to be expected as it is based on an angular intensity distribution which is realistic for thick specimens and moderate temperature gradients. In these materials unit optical depth corresponds to a mass/unit area of about 0.03 kg/m^2, so the batts are about 5–10 path lengths thick. For a thin sample the two-flux model should represent the radiation more accurately. Possibly this effect is seen in the results for the thinnest batt, Polarguard #3.

Table 1. Parameters of sample materials

Material	Fiber radius μm	Fiber density kg/m^3	Fiber conductivity W/m-°K	$\bar{\kappa}_A$ (at 30°C) m^2/kg	$\bar{\kappa}_T$
Polarguard	11.7	1.38×10^3	0.145	18.7	26.6
Hollofil	13.2	1.33×10^3	0.145	18.4	26.2

Density of polyethylene terephthalate from [14];
thermal conductivity from [15].
Density of Hollofil is reduced 4% to account for hole.
The two-flux mean $\bar{\kappa}_2$ has the same value as $\bar{\kappa}_T$.

Table 2. Measured and calculated thermal resistances

| Sample | Mass/area | Thickness | Thermal resistance | | |
| | | | measured | diffusion model | two-flux model |
	kg/m^2	mm	$m^2-°K/W$	$m^2-°K/W$	$m^2-°K/W$
Polarguard					
1	0.220	20.2	0.40 ±.02	0.39	0.44
2	0.295	25.0	0.485±.01	0.49	0.55
3	0.155	12.4	0.27 ±.01	0.26	0.29
4	0.255	24.2	0.46 ±.01	0.455	0.51
5	0.270	26.3	0.48 ±.01	0.49	0.55
Hollofil					
1	0.185	12.2	0.28 ±.01	0.27	0.29
2	0.260	22.0	0.49 ±.02	0.43	0.485
3	0.350	27.4	0.60 ±.02	0.55	0.62

Thermal measurements are from [4].
Mean temperature in all cases was 30°C.
Wall emissivities $\varepsilon_H = \varepsilon_C = 0.92$

For the three Hollofil samples the diffusion model gives resistances which are too low, and the two-flux model fits the data better. We believe this is in part the result of ignoring the hole in the fiber in the scattering calculations. The hole is a scattering surface with about 20% of the area of the fiber itself. A 20% increase in scattering (an upper limit; the actual effect is surely smaller) would reduce the radiative conductivity by 20% and increase the calculated total resistance by about 10%. This would bring the diffusion-model results into better agreement with the thermal measurements, and make the two-flux results too high.

CONCLUSION

We have shown that the radiative heat transport in fibrous insulation materials can be calculated directly from the complex refractive index spectrum of the bulk material from which the fibers are made. Expressions are given relating the parameters of two widely-used models of radiative transfer to the fundamental scattering and absorption cross-sections of the fibres. It is necessary to include the effects of the anisotropy of the scattering. Excellent agreement is obtained for the solid-fiber material Polarguard between the measured thermal resistance and the value calculated from the diffusion model. The poorer agreement of the two-flux model reflects the fact that this model begins with an approximation to the angular variation of the intensity which is not valid for optically thick samples.

REFERENCES

[1] Verschoor, J.D. and Greebler, P., Trans. ASME 74 961 (1952)
[2] Bankvall, C., J. Test. and Eval. 1 235 (1973)
[3] Davis, L.B. and Birkebak, R.C., Biophys. J. 14 249 (1974)
[4] Farnworth, B., Crow, R.M., and Dewar, M.M., DREO report #818, DREO, Ottawa, 1979
[5] Farnworth, B., Textile Research J. 53 715-725 (1983).
[6] Cabannes, F., Maurau, J-C, Hyrien, M., and Klarsfeld, S.M., High Temperatures - High Pressures 11 429 (1979)
[7] Timmermans, G., Van Paemel, O., and Myncke, H., XV Int. Conf. on Refrigeration, Venice, Sept. 1979
[8] Hager, N.E., and Steere, R.C., J. Appl. Phys. 38 4663 (1967)
[9] Bhattacharyya, R.K., in Thermal Insulation Performance, (ASTM STP 718) edited by D.L. McElroy and R.P. Tye, ASTM, 1980, pp. 272-286
[10] Pomraning, G.C., The Equations of Radiation Hydrodynamics, Pergamon Press, Oxford, 1973
[11] Modest, M.F., and Azad, F.H., J. Heat Transfer 102 92 (1980)
[12] McKay, N., Timusk, T., and Farnworth, B., J. Appl. Phys. (to be published in June 1984)
[13] Wait, J.R., Can. J. Phys. 33 189 (1955)
[14] Hefflinger, C.J., and Knox, K.L., in The Science and Technology of Polymer Films, edited by O.J. Sweeting, John Wiley & Sons, New York, 1971, pp. 587-639
[15] Steere, R.C., J. Appl. Phys. 37 3338 (1966)

SPECTRAL TRANSMISSION AND REFLECTION PROPERTIES OF HIGH
TEMPERATURE INSULATION MATERIALS AND THEIR RELATION TO
RADIATIVE HEAT FLOW

R. Caps, A. Trunzer, D. Büttner and J. Fricke

Physikalisches Institut der Universität Würzburg
D-8700 Würzburg, FRG

H. Reiss

Brown, Boveri & Cie. AG., Zentrales Forschungslabor
D-6900 Heidelberg, FRG

ABSTRACT

This paper describes two approaches for a determination of
radiative properties of high temperature insulations: a) Experimental: calorimetric measurements of total thermal conductivity and
optical transmission and reflection studies are applied for determination of extinction coefficients and for verification of the increase of radiative transfer in cases of forward scattering media;
b) Theoretical: application of rigorous Mie scattering theory,
Monte Carlo simulations of both calorimetric and optical measurements, and five-flux formulas obtained from the method of discrete
ordinates. The findings from both approaches establish the wavelengths resolved extinction coefficient and the first order term of
the scattering phase function series expansion to be indispensible
parameters for an optimization of the insulation properties of
anisotropically scattering media. The paper demonstrates the validity of the diffusion approximation for radiative transfer also in
these cases.

INTRODUCTION

The development of a high temperature insulation of lowest
thermal conductivity requires a large optical thickness τ_0 through-

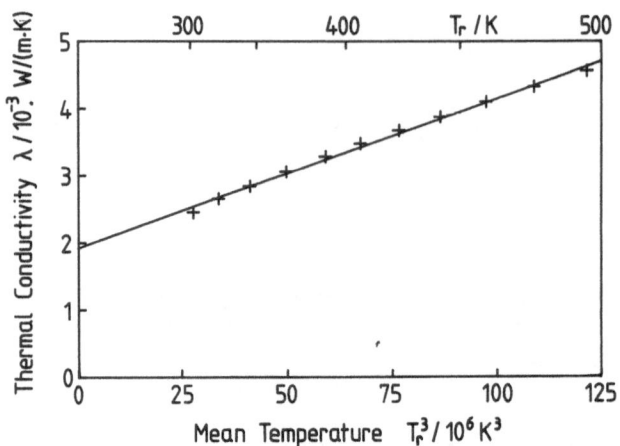

Fig.1. Calorimetric determination of total thermal conductivity, λ, as a function of effective temperature, T_r.

out the ir-spectrum. For an evacuated insulation, this usually allows the total heat flux $\dot{q}$ to be calculated as the algebraic sum of two heat flow components, i.e. a solid conduction part, $\dot{q}_{sc}$, and a radiation heat flux, $\dot{q}_{rad}$, which are, if evaluated with the boundary temperatures, considered as independent from each other:

$$\dot{q} = \dot{q}_{sc} + \dot{q}_{rad} .$$

The aim of this paper is to report on our investigations how to determine and minimize $\dot{q}_{rad}$. These investigations include: a) measurements of transmission and reflection spectra for fiber insulations in the range of wavelengths $2.5 < \Lambda < 14.5$ µm; b) determination of extinction coefficients, albedo, and the amount of forward scattering from experimental data; c) calculation of extinction properties and scattering phase functions from rigorous Mie theory.

DETERMINATION OF EXTINCTION COEFFICIENTS FROM CALORIMETRIC MEASURE-MENTS USING THE DIFFUSION MODEL

Fig.1 shows the results of total thermal conductivity measurements of a load-bearing evacuated glass fiber insulation as function of the cube of radiation temperature, T_r. The guarded hot plate device (with the very large metering section of 0.25 m²) is described elsewhere [1]. From the total conductivity, λ, a mean calorimetric radiation extinction, E_{cal}, of the insulation can be obtained, according to the equation (diffusion model [2])

$$\lambda = \lambda_{sc} + \lambda_{rad}$$

$$= \lambda_{sc} + 16n^2 \sigma T_r^3/(3E_{cal}),$$

where n is the effective index of refraction (real part), σ the Stefan-Boltzmann constant and E_{cal} the extinction coefficient.

If λ_{sc}, n and E_{cal} are independent of temperature, the total thermal conductivity, λ, should be a linear function of T_r^3. This is demonstrated by the example given in Fig.1. The slope of the graph yields $E_{cal} = 14900$ /m (assuming with respect to the high porosity of the sample an effective value of n^2 of 1.1).

The above diffusion model solution for λ_{rad} is in this simple form valid only for isotropic scattering, where the scattering phase function, $p(\Theta)$, is a constant ($\leqslant$ 1), or in the case of a purely absorbing medium. In the case of anisotropic scattering, the extinction coefficient E has to be corrected with respect to the scattering phase function: Forward or backward scattering would lead to an effective value E^* which is smaller or larger than E, respectively, and to a corresponding increase or decrease in λ_{rad}.

Like in neutron diffusion theory, the effective extinction coefficient E^* is related to E by

$$E^* = E(1-\bar{\mu})$$

where
$$\bar{\mu} = \cos \Theta = 1/2 \int_{-1}^{1} p(\mu)\mu d\mu$$

is the weighted mean of the scattering angle $\mu = \cos \Theta$ with respect to the phase function. Since these relations hold also for any value of the wavelength, Λ, it is thus sufficient for a calculation of the total radiative heat flux to know the asymmetry factor $\bar{\mu}_\Lambda$ and the extinction coefficient, E_Λ.

In order to verify the validity of this first order approximation also for strong anisotropic scattering, we have performed Monte Carlo simulations[2] using, first, a model phase function ($p(\Theta) = 6$ for $0 \leqslant \Theta \leqslant \pi/6$ and $p(\Theta) = 0$ for $\Theta > \pi/6$ for all Λ) yielding $\bar{\mu} = 0.955$, and, second, the phase function from Mie theory for perpendicular incidence of radiation onto a glass fiber with diameter D = 2 μm and Λ = 2 μm, which results in $\bar{\mu}_\Lambda = 0.84$. If the transmitted flux, t, is plotted in Fig.2 and compared to the equation

$$t = 1/(1 + (3/4) \cdot \tau_0 \cdot (1 - \bar{\mu}))$$

(linear anisotropic model[4]), which for large optical thicknesses, τ_0, is asymptotically equal to the diffusion approximation, we find a very good agreement between both procedures, especially at high τ_0, i.e. in the domain of optical thicknesses where the diffusion model is valid. Thus the diffusion solution using the above defined

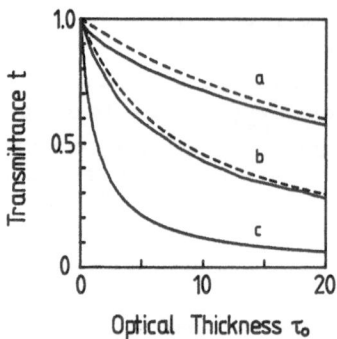

Fig.2. Total transmitted flux t from Monte-Carlo calculations (solid lines) and from the linear anisotropic model (diffusion approximation, dashed lines) for a diffuse radiation field impinging onto the probe; curve a) asymmetry factor μ = 0.955; curve b) μ = 0.84; curve c) isotropic phase function.

effective value E^* of the extinction coefficient is demonstrated as a good approximation for radiative heat transfer in a plane parallel, optically thick insulation also for strong anisotropic scattering.

Not only the phase function (and thus the asymmetry factor), also the spectral extinction coefficient E_Λ can be calculated numerically from Mie theory for cylindrical and spherical geometry (see e.g. [5]), if the complex index of refraction, $m(\Lambda)$, and the particle or fiber diameter, D, are known. From E_Λ and μ_Λ, a mean $E_R(T)$, the Rosseland mean [2], of E_Λ for a given thermal spectrum at temperature, T, can be calculated. For this calculation we used the complex index of refraction of silicate glass from 1 µm to 30 µm[6]. As a first step, Fig.3 shows calculated specific extinction coefficients, E/ρ, (ρ: insulation density) and effective extinction coefficients, E^*/ρ, of glass fibers (D = 2 µm) for perpendicular incidence following rigorous Mie theory. It demonstrates that due to strong forward scattering which occurs at wavelengths Λ < 5 µm, the (conventional) extinction coefficients E/ρ are strongly reduced by the anisotropy factors to rather small effective values E^*/ρ.

In the second step of the calculation of $E_R(T)$, the E^*/ρ of Fig. 3 are weighted by the Rosseland weight function for temperatures T between 300 and 700 K. Fig. 4 shows the temperature dependent specific extinction coefficient $E^*/\rho(T)$ versus temperature, T, and fiber diameter, D. The decrease of the extinction off the maximum at D ≈ 1 µm is below 20%, for 0.6 µm < D < 2 µm, so the fiber size distribution is not of great importance within these limits. Temperature - averaged specific extinctions for D = 0.5 µm and D = 0.7 µm vary between 0.060 and 0.076 m²/g. This is in good agreement with

406

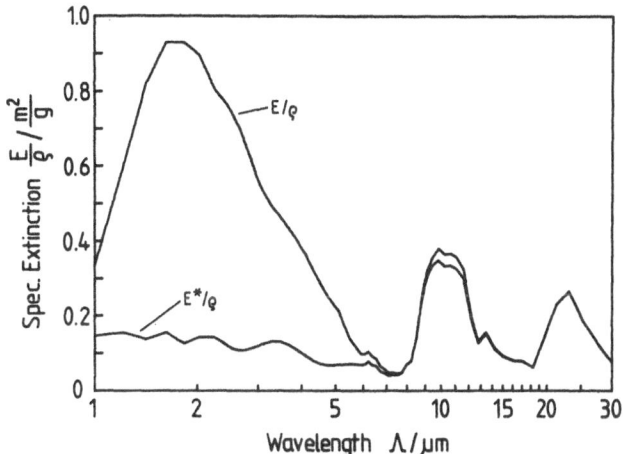

Fig.3. Specific extinction coefficient, E/ρ, of glass fibers (D = 2 μm) predicted by Mie theory for perpendicular incidence, and effective extinction, E*/ρ.

the experimental value given earlier (E_{cal}/ρ = 0.053 m²/g).

This agreement between measured and calculated extinction coefficients indicates that the radiative flux in an insulation can in principle be deduced from theory, i.e. from the optical constants and the internal geometry of the material. Nevertheless it is essential to get experimental data on the extinction and amount of forward scattering in additional experiments, especially if the index of refraction is unknown.

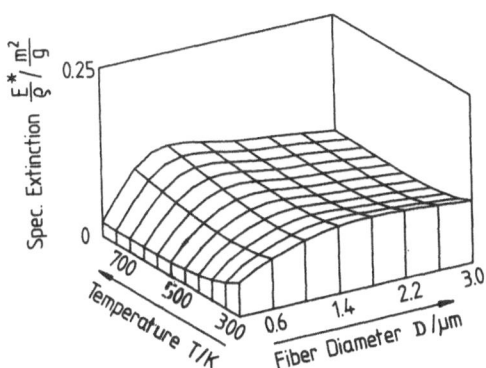

Fig.4 Rosseland mean E*/ρ of glass fibers as a function of temperature, T, and diameter, D.

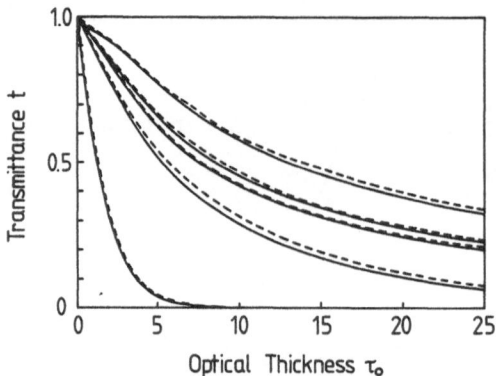

Fig.5. Comparison of transmittance t derived from Monte Carlo calculations (solid lines) and five flux model (dashed lines)
as a function of optical thickness, τ_0. Fiber diameter D =
2 μm and wavelengths Λ = 2.5, 3.0, 4.0, 5.0, 6.0 μm (from
top to bottom) are assumed for Mie theory calculations
(yielding $\bar{\mu}$ = 0.86, 0.79, 0.78, 0.72 and 0.26, respectively).

The experiments should reveal the total transmittance and reflectance, t and r, of an approximately parallel beam falling onto
a sample at room temperature, which depend on the optical thickness, τ_0, the albedo, ω_0, and the phasefunction $p(\Theta)$. Then the
experiment can be compared with the predictions of Mie theory and
with solutions of the equation of transfer (for details see[7]). A
two-flux approximation is frequently used to solve this equation,
where the radiation field is divided into a forward and backward
hemisphere. It was shown[8] that a three-flux approximation including the μ = 0 direction can be reduced to the formalism of two-
flux calculations. We found that extending this method to anisotropic scattering still yields sufficiently accurate results if five-
flux approximations and, besides ω_0, the first three moments of the
phasefunction $\omega_1 \ldots \omega_3$ are used. As a demonstration of this finding,
five-flux calculations of transmittance of 2 μm glass fibers at
several wavelengths, Λ, were performed numerically and compared to
Monte Carlo calculations (Fig.5). The good agreement between both
theoretical approaches indicate that as Monte Carlo calculations
are very time-consuming, the five-flux method can be used to compare
quickly optical transmissivity measurements with the predictions
of combined Mie theory and radiation transfer theory.

EXTINCTION COEFFICIENTS FROM TRANSMISSION MEASUREMENTS

The thermal radiation of a globar (Fig.6) is focused by a

408

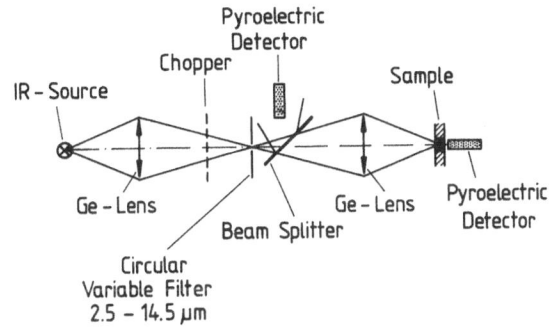

Fig.6 Experimental setup for ir-transmission measurement.

germanium lens on a circular variable filter of wavelengths range
2.5 to 14.5 µm. A second germanium lens focuses the chopped beam
onto the sample. A pyroelectric detector is in close contact with
the sample and collects about 75% of the hemispherical ir transmission. The signal is passed to a lock-in-amplifier and correlated
with the chopper-frequency of 12.5 s^{-1}. A semi-reflecting ZnSe-
mirror and a second pyroelectric detector enable us to monitor the
original intensity. Both signals are digitized and stored in a TRS-
80 microcomputer, which is also used to automatically turn the
variable filter and a movable detector holder as well as to process
and plot the data. In the arrangement described here, the total ir
transmission through a sample can be measured.

If the extinction of a sample is to be determined, it is
positioned in front of the second lens. Since the scattered radiation is distributed over the hemisphere, only a small fraction
arrives at the detector. From the primary intensity, I_0, and the
unscattered intensity, I_1, the optical thickness of the sample is
derived in the usual manner: $\tau_0 = \ln(I_0/I_1)$. Since the scattered
intensity hitting the detector is below $0.03\% \cdot I_0$, the τ_0-values are
very accurate, and no correction has to be made with respect to a
possible source of scattered radiation. Furthermore from the mass
and surface area of the sample we get the specific extinction, E/ρ.

Fig.7 shows the specific extinction spectrum E/ρ of Microglass
1000 (Manning Corp., Troy, N.Y.) and a comparison with Mie-calcula-
tions for D = 0.5 µm, D = 1.0 µm and D = 1.5 µm. A mean diameter
of 0.5 µm is claimed by the manufacturer. SEM-pictures, however,
show an appreciable amount of additional larger fiber sizes in the
range of 1 µm. For this diameter the figure shows an excellent
agreement between measured and calculated spectral extinction coefficients. Note that the figure contains no adjustable parameters,
and that it uses only experimental refractive indices for the calculations. Two different regions have to be distinguished: one of

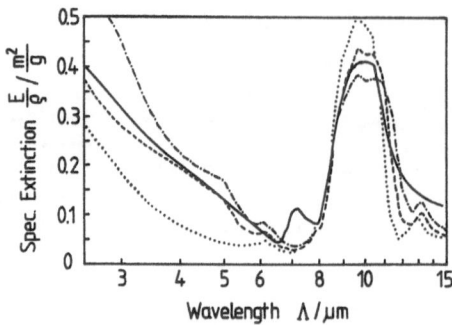

Fig.7 Comparison between experimental specific extinction, E/ρ, of Microglass 1000 (solid line), and predictions of Mie theory assuming $D = 0.5$ μm (dotted line), $D = 1.0$ μm (dashed line) and $D = 1.5$ μm (dashed-dotted line).

high albedo ω_0 (mainly scattering, $\Lambda < 6.5$ μm), and second, a strong absorption band around $\Lambda = 10$ μm, which is also reproduced by the calculations.

Following the discussion in the first paragraph of this paper on the effective extinction coefficient, E^*, and in view of the strong reduction of E/ρ to E^*/ρ at smaller wavelengths shown in Fig.3 because of strong forward scattering, it is to be expected that transmittance through even several layers of Microglass should be decreasing only very slowly with the number of layers. This is experimentally verified by Fig.8: Measured transmittance values t

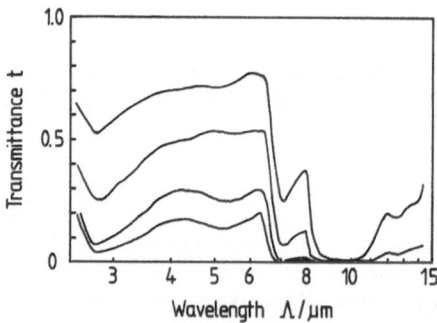

Fig.8 Measured transmittance, t, versus wavelength for Microglass 1000 with surface densities $\sigma_s = 14.2$, 28.4, 56.8 and 71.0 g/m², proceeding from top to bottom.

410

are large at smaller wavelengths although Fig.7 predicts a large extinction coefficient E/ρ. Once again, Fig. 8 demonstrates the importance of the anisotropy factor for a reliable prediction of transmittance (and radiative heat flow).

In the example given here, this importance is made particularly clear by the high albedo which follows from the agreement found between measured and calculated scattering cross sections at smaller wavelengths in Fig.7. In regions of wavelengths, Λ, where absorption dominates (small albedo, e.g. at Λ = 10 μm, Fig. 7), anisotropic scattering is of no importance since absorption/reemission processes are isotropic. This is reflected by the small transmittance, t, at Λ = 10 μm in Fig.8.

REFERENCES

1. D. Büttner, J. Fricke, R. Krapf, and H. Reiss, Proc. 8th ETPC Baden-Baden (1982), High Temp.-High Press. 15:115 (1983)
2. R. Siegel, J. R. Howell, "Thermal Radiation Heat Transfer", McGraw-Hill, Tokyo (1972)
3. S. Glasstone, M. C. Edlund, "Kernreaktortheorie", Springer Verlag, Wien (1961)
4. T. W. Tong, C. L. Tien, Trans. ASME 105:70 (1983)
5. M. Kerker, "The Scattering of Light", Academic Press, New York (1969)
6. C. K. Hsieh, K. C. Su, Solar Energy, 22:37 (1979)
7. S. Chandrasekhar, "Radiative Transfer", Dover Publications, New York (1960)
8. M. G. Kaganer, Optika i Spektroskopika, 26:443 (1969)

TEMPERATURE-DEPENDENT EXTINCTION COEFFICIENTS AND SOLID THERMAL CONDUCTIVITIES OF GLASS FIBER INSULATIONS

H. Reiss and B. Ziegenbein

Brown, Boveri & Cie AG.
Central Research Laboratory
Eppelheimer Str. 82
6900 Heidelberg F.R.G.

ABSTRACT

An experimental method has been derived for determining temperature-dependent total extinction coefficients and solid conductivities from measured total thermal conductivities of glass fiber insulations. The method uses a stationary calorimetric measurement of thermal conductivities at temperatures $100 \leq T \leq 430°C$ and a simultaneous measurement of the temperature profile within the insulation. Experimental data were fitted by calculated temperature profiles using the diffusion model for combined radiative/solid conduction and assuming linear temperature-dependent values of extinction and solid conductivity.

INTRODUCTION

Only advanced high temperature batteries provide sufficiently high energy and power densities for propulsion of electric vehicles to be an economic alternative to gasoline powered cars. Such a battery is maintained within a specified temperature range. The battery requires an improved insulation to prevent excessive heat loss during long idle periods, a heating system to raise the battery cells to their operating temperature and a cooling system to prevent overheating during high-rate-discharge (Fischer [1]).Thermal model calculations for a 50 kWh-passenger car battery lead to a maximum heat loss of less than 60 W/m^2 at 320°C, or a thermal conductivity lower than 2.5 x 10^{-3} W/mK, that can be tolerated. The need for a lightweight vacuum insulation of high performance is obvious. A light-weight vacuum insulation calls for a thin leak-tight metal vacuum casing and an insulation material with very low degassing rates. The insulation system must be capable of withstanding the load due to atmospheric pressure.

Recent experimental results of lead-bearing vacuum insulation development indicate that for powder insu-lations the solid conduction seems to be too high to meet the above design value if the powder density cannot be reduced to about 200 kg/m³ (Reiss [2]). Nowobilski [3] measured thermal conductivities for the best continuous load-bearing fiber board system of 3.2 x 10^{-3} W/mK between 300 and 325°C. A quantitative knowledge of the relative amounts of combined radiative and solid-conduction heat transfer rates in fiber board insulations significantly facilitates further improve-ments of lower-conductivity materials.

The present paper describes an experimental method to separate the two heat transfer mechanisms in evacuated fiber board insulations leading to temperature dependent extinction coefficients and solid conduc-tivities. Application of this method to plain glass fiber insulations clearly shows that increasing the extinction by using higher density boards cannot sufficiently reduce radiative heat transfer for the above high-temperature application.

EXPERIMENTAL PROCEDURE

Thermal Conductivity Measurement

Measurement of thermal conductivities is performed in a guarded hot plate calorimeter with two cold re-ference plates (Figure 1). Testing was performed under varying residual gas pressure from less than 10^{-2} to 10^5 Pa, compressive load from less than 0,01 MPa to 0.3 MPa and hot plate temperatures from 120°C to 475°C (Ziegenbein [4]). The fact that the lateral heat flow in compressed multilayer insulation may be a factor 10^2 to 10^3 greater than that perpendicular to the layer surface requires a uniform lateral temperature, i.e. a very precise matching of test section plate (11 cm dia-meter) and guard ring temperatures. An auxiliary heater attached to the outer guard ring minimizes additional

414

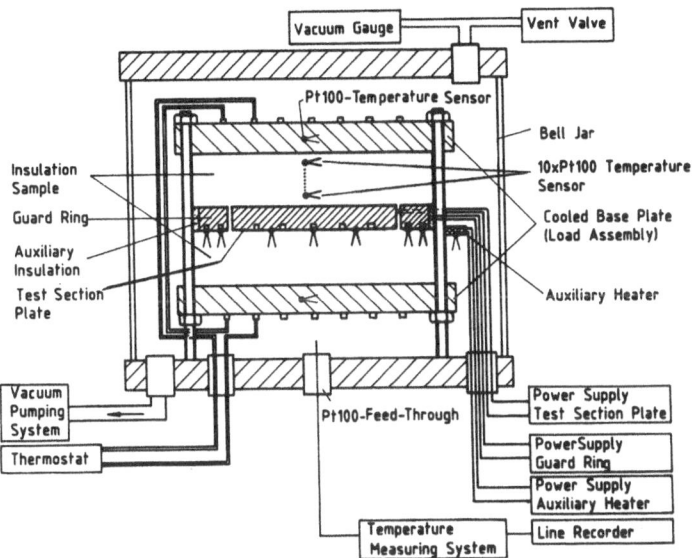

Fig.1   Schematic of Guarded-Hot-Plate Calorimeter

heat losses through temperature sensor and resistance
heater wires. The typical dimensions of insulation
samples to be tested are 25 x 25 cm^2 x 1 cm. Depending
on the type of insulation it generally took about 12
hours to reach steady-state conditions.

Plain and opacified fibrous system insulations
were made from binderless borosilicate glass fiber
materials with fibers of 0,69 to 1,1 µm diameter
(Johns Manville; 475-Tempstran; Code 108 A). For an
Al_2O_3 insulation sample a 99%-Al_2O_3 ceramic fiber with
5 µm diameter has been used (Didier GmbH, Wiesbaden,
FRG).

Figure 2 shows effective thermal conductivities, λ_{eff},
for three different insulation samples. For mean temperatures between 90°C and 210°C, the λ-values could be
fitted within the experimental error by a linear curve
if plotted versus $T*^3$, a quantity defined below.

Temperature Profile Measurements

Determination of solid conductivities and extinction coefficients from temperature profiles in the
insulation requires highly accurate temperature
measurements. Up to 10 calibrated Platinum resistance

415

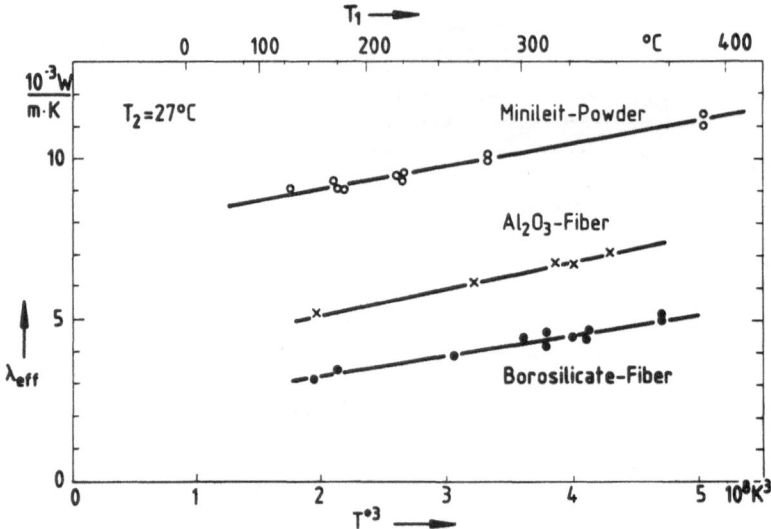

Fig.2   Measured Effective Thermal Conductivities vs Radiation
        Temperature

sensors (Heraeus GmbH, Hanau, FRG, Type FKG 1030,6)were
inserted at different positions in the upper insulation
sample (Figure 1). The exact distance of each sensor
from the cold plate is given by the layer number.
Uniform horizontal temperature distribution was controlled by inserting up to three sensors in the same
layer. A disturbance of the temperature profile by the
sensors could be minimized by increasing the sample
thickness to 35 mm and using 0,1 mm diameter wires.

The shapes of the measured temperature profiles in
Figure 3 clearly indicate the different solid conduc-
tion/radiation heat transfer ratios. For the low-density insulation a high rate of radiative transfer leads
to a strongly curved temperature distribution, where
∿ 4/5 of the insulation thickness is required to
decrease the local temperature to 50 % of the hot side
value. A nearly linear slope for the lowest curve
results from prevailing gas conduction (residual
pressure 10^3 Pa) and high radiation extinction.

METHOD OF ANALYSIS

Based on the standard diffusion model, the
following solution for the total thermal conductivity
λ of a fibrous insulation operating in vacuum and

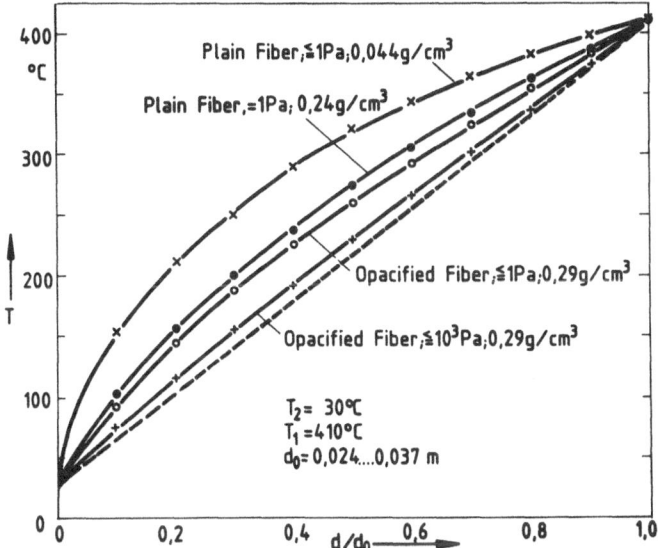

Fig. 3 Measured Temperature Profiles in Fiber Insulations vs
Dimensionless Position

bounded by opaque surfaces can be derived:

$$\lambda = \lambda_{SC} + 4/3 \; \sigma n^2 \; T_{1,2}^{*3}/E \qquad (1)$$

with $T_{1,2}^{*3} = (T_1^2 + T_2^2) \cdot (T_1 + T_2)$, where T_1 and T_2 are
the hot and cold boundary temperatures, σ the Stefan-
Boltzmann constant, n the real part of the effective
index of refraction and E the extinction coefficient
(for the diffusion model to be applicable, E must be
sufficiently large). According to this equation, λ
should be linear in $T_{1,2}^{*3}$, if λ_{SC} and E are independent
of temperature. The high optical thickness of the insulations studied here indicates that the above additive
approximation is valid. Thus from $(\lambda, \; T_{1,2}^{*3})$-plots in
Figure 2, λ_{SC} and E can be extracted from intercept and
slope for different insulation materials in the
(comparatively small) temperature range $10^8 \; K^3 \leq T_{1,2}^{*3}$
$\leq 5 \cdot 10^8 \; K^3$. The obtained λ_{SC} and E must be interpreted
as <u>mean</u> values averaged over the total insulation
thickness.

In contrast to the above frequently-used procedure
the method described here allows the calculation of
<u>local</u> solid conductivities and extinction coefficients
by combination of experimental calorimetric λ-data with
measured temperature profiles in the insulation. For one

set of boundary temperatures T_1, T_2, the heat flux in the insulation, $\dot{Q}/A$, is constant (A: area of test section plate). Local thermal conductivities $\lambda_{(x_i)}$ can therefore be derived following the usual relation $\lambda = (-\dot{Q}/A) \cdot \Delta X / \Delta T$, i.e.

$$\lambda_{(x_i)} = -\dot{Q}/A \cdot (x_i - x_{i+1})/(T_{x_i} - T_{x_{i+1}})$$

where T_{x_i} are temperatures at the insulation layer positions x_i.

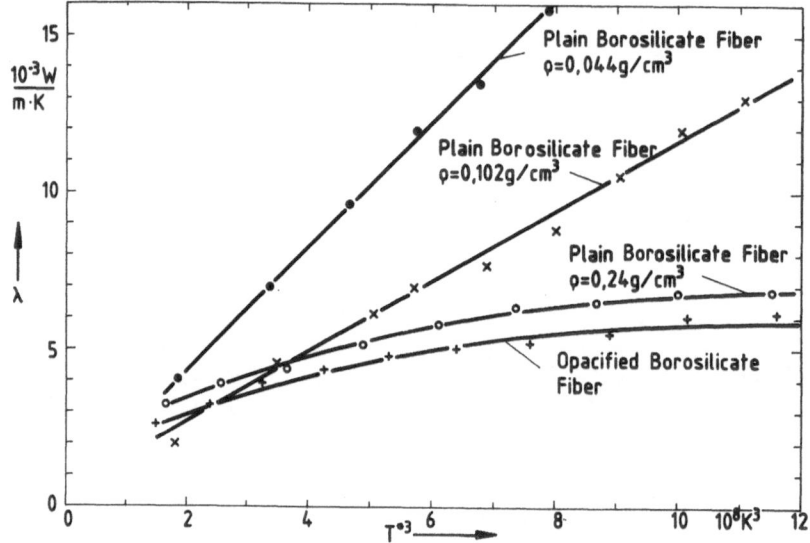

Fig. 4 Measured Local Thermal Conductivities vs Radiation Temperature

A plot of local conductivities, $\lambda(x_i)$, versus $T^{*3}_{x_i,i+1} = (T^2_{x_i} + T^2_{x_{i+1}}) \cdot (T_{x_i} + T_{x_{i+1}})$ in Figure 4 shows that for high-density evacuated fiber insulations there is no longer a linear relationship within the whole range of T^{*3}_x. This observation can be interpreted as follows: The terms λ_{sc} and E are in reality <u>not</u> constant with respect to the temperature in the insulation. Furthermore, it should be noted that the T^{*3}_x-values in Figure 4 extend to considerably higher values than the usual $T^{*3}_{1,2} = (T_1^2 + T_2^2) \cdot (T_1 + T_2)$ in Figure 2.

418

Temperature Variation of Extinction and Solid Conductivity

Assuming temperature-dependent $\lambda_{sc}(T)$ and $E(T)$ differentiation of eq. (1) yields

$$\frac{d\lambda}{dT^{*3}} = \frac{d\lambda_{sc}(T)}{dT^{*3}} + \frac{4\sigma n^2}{3} \left[\frac{1}{E(T)} - \frac{T^{*3}}{E(T)^2} \cdot \frac{dE(T)}{dT^{*3}} \right] \qquad (2)$$

Obviously, λ is linear in T^{*3} (index 1,2 omitted) if both λ_{sc} and E are constant or linear in T^{*3}. A linearity of λ has been observed only in low density fiber insulations (Figure 4) with relatively small solid conductivities and extinction coefficients. For high-density or opacified samples the nonlinearity of λ must be explained by a temperature dependence of E and/or λ_{sc}.

A possible temperature variation of E was investigated by calculating for a given temperature distribution, the Rosseland mean $E_{R(T)}$ (Siegel [5]) using measured spectral attenuation data of specimens

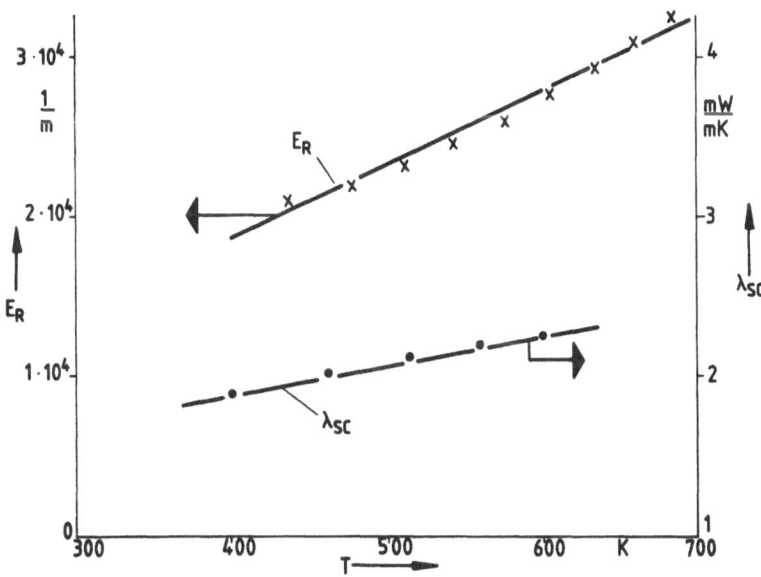

Fig. 5   Calculated Rosseland Extinction Coefficient and Solid Conductivity of Borosilicate Fiber Board.

of low optical thickness in the wavelength region $1 \le \lambda \le 50$ μm. Except for the low temperature region, $E_R(T)$ in Figure 5 can be quite well approximated by $E_R(T) = a_1 \cdot T$.

A temperature dependence for λ_{sc} is derived in a formula given by Kaganer [6].

$$\lambda_{sc}(T) = \frac{16(1-m)^2}{\pi^2} \cdot \frac{\lambda_T}{\dfrac{1}{1.86 \ A \ p^{1/3}} + \dfrac{1}{4(1-m)}}$$

Here m denotes the porosity of the fibers, λ_T the thermal conductivity of the solid material the fibers are made from, and p the mechanical pressure on the sample. A is the Hertz contact radius according to

$$A = \left(\frac{1-\mu^2}{Y(1-m)^2}\right)^{1/3},$$

Y is the Young's modulus of elasticity and μ the Poisson coefficient. It is shown by Espe [7] that Y is proportional to the dynamic viscosity which in turn decreases exponentially with temperature. Therefore, it is not at all obvious that assumptions such as $\lambda_{sc}(T) = $ const are justifiable. Using Y-, μ- and λ_T-data for borosilicate glasses, the Kaganer formula yields an almost linear relationship of λ_{sc} and T in the interval $400 \le T \le 600$ K as shown in Figure 5:

$$\lambda_{sc}(T) = b_o + b_1 \cdot T$$

Inserting the approximations for $E_R(T)$ and $\lambda_{sc}(T)$ tentatively in eq. 1, the thermal conductivity is given by:

$$\lambda = b_o + b_1 T + \frac{16}{3} n^2 \sigma T^2 / a_1$$

The radiation temperature T^* has been substituted by the local temperature $T = (T^{*3}/4)^{1/3}$ in the interval $T_{x_i} \le T \le T_{x_{i+1}}$. The temperature coefficient a_1 of the extinction was calculated from the slope of a $(d\lambda/dT, T)$-plot. Subtraction of the radiative component from the measured total conductivity according to eq. 1 gives the solid thermal conductivity $\lambda_{sc}(T)$. Figure 6 contains experimental extinction coefficients and solid conductivities for a plain borosilicate fiber insulation. It should be noted that the nonlinear (λ, T^{*3})-curves in Figure 2 are explained quite well by temperature-dependent extinction coefficients and solid conductivities but that a linear relationship as shown in Figure 6 is only one possible assumption which might be justified by the theoretical consideration according to Figure 5.

420

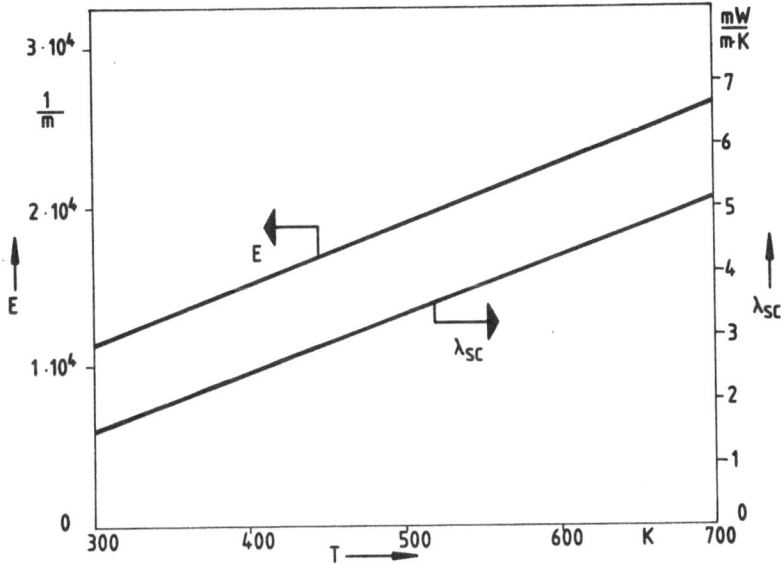

Fig. 6　Experimental Extinction Coefficients and Solid Conductivities for Plain Borosilicate Fiber Insulation.

RESULTS

Measured temperature profiles of various plain and opacified borosilicate fiber insulations have been analysed by fitting their thermal conductivities by the following four assumptions for a possible temperature dependence:

$$\lambda = \lambda_{sc}(T) + \frac{4\sigma}{3} T^{*3}/E(T)$$

with

(a)　$\lambda_{sc}(T) = \text{const}$　and $E(T) = \text{const}$
(b)　$\lambda_{sc}(T) = b_0 + b_1 T$ and $E(T) = \text{const}$
(c)　$\lambda_{sc}(T) = b_0 + b_1 T$ and $E(T) = a_1 T$
(d)　$\lambda_{sc}(T) = b_0 + b_1 T$ and $E(T) = a_1 T^2$

Systematically, a temperature dependence of λ_{sc} and E according to ⓑ and ⓒ yielded far better fits to experimental data than ⓐ . It must be concluded, that there is a definite increase in the solid conductivity of fibrous insulations with temperature. It also seems probable that the extinction coefficient varies with temperature because of the theoretical $E_R(T)$ in Fig. 5. Inserting ⓓ into eq. (2), we have $d\lambda/dT^{*3} = c_1/x^{2/3}$ or $\lambda = c_2 x^{1/3} + c_3$, where $x = T^{*3}$ and c_1, c_2 and c_3 are constants. This result explains the (λ, T^{*3})-curves

of the high density insulations in Fig. 4. A decision
between ⓒ and ⓓ from calorimetric λ-data, however,
can be made only after further experiments provided the
total experimental error is reduced clearly below 5%.

DISCUSSION

The measured calorimetric $E(T)$ in Figure 6 can be
attributed to absorption and/or scattering mechanisms.
Pure or dominating scattering would lead to a more or
less linear temperature profile if $\lambda_{sc}(T)$ = const
because the radiation field is completely decoupled
from the temperature field for the case of pure scattering. Since Figure 6 shows that $\lambda_{sc}(T)$ depends on temperature calorimetric measurements alone cannot reveal
whether scattering or absorption dominates. However,
evaluation of measured specific extinction coefficients
$E/\rho = \tau_0/\kappa$ show scattering as the prevailing extinction
mechanism at short wavelengths, i.e. in the high temperature regions of the insulation sample ($\tau_0 = E \cdot d$:
optical thickness; d: sample thickness; κ: specific
surface density in g/cm^2).

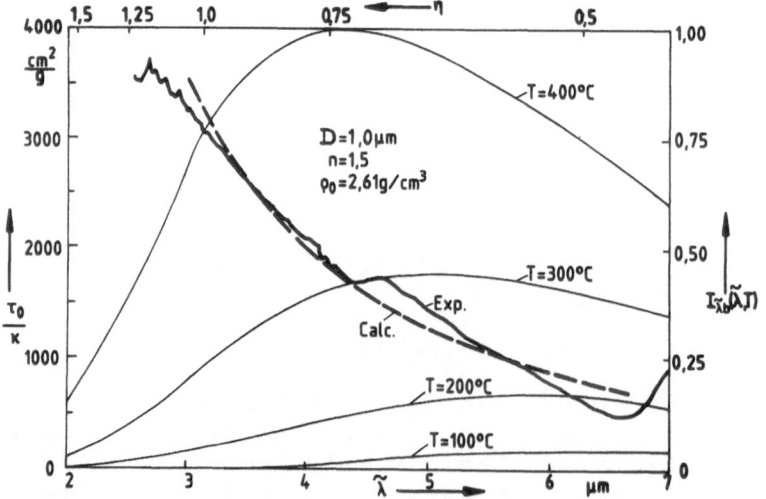

Fig. 7   Measured (thick solid curve) and Calculated (dashed curve)
Specific Optical Thickness $E/\rho = \tau_0/\kappa$ of a Plain Glass Fiber Insulation versus Wavelength (the thin solid curves denote the spectral
emissive power of the black body, $I_{\lambda b}^{\sim}$, normalized to unity).

422

Figure 7 shows a comparison of measured and calculated E/ρ-values for glass fiber samples with mean fiber diameter D = 1 μm and solid glass density ρ_O = 261 kg/m³. Micrographs of the fiber board in Figure 8 confirm that individual fibers in the compressed board are randomized in planes perpendicular to heat flow.

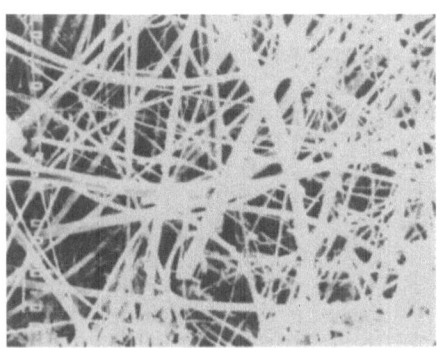

View Parallel to Temperature Gradient

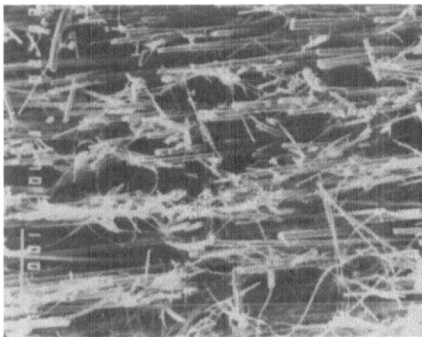

View Perpendicular to Temperature Gradient

Fig.8 Micrographs of Plain Borosilicate Fiber Board

Measured τ_O/κ are taken from Caps [8]. Theoretical values have been calculated using a constant index of refraction n = 1.5 and an appropriate expression for the extinction cross section for perpendicular incidence of unpolarized radiation on very long cylinders. Van de Hulst [9] has shown that the efficiency factor for extinction, Q_{ext}, for a real refractive index near 1, approaches for small η the value $Q_{ext}(\eta)$ = $2/3\eta^2$ for both orientations of the electric vector with respect to the fiber axis, where η is given by $\eta = 2(n-1)\alpha$ and $\alpha = \pi \cdot D/\lambda$

This expression for Q_{ext} and the following equation for the specific extinction coefficient $E/\rho = 4/\rho_O \cdot Q_{ext}/(\alpha \cdot \lambda)$ yield the theoretical (dashed) curve in Fig. 7. Multiple scattering is considered to be negligible in the transmission measurement since the optical thickness of this sample was very low. No account has been made for the distribution of the real fibers with respect to their diameter and orientation to the incident beam. The good agreement between measured and calculated extinction coefficients leads to the conclusion that at least for T > 200°C, (anisotropic) scattering is the predominant extinction mechanism. This is in accordance with measurements of the spectral albedo for fiber insulations from Cabannes [10] and extended calculations of Caps et al [8].

REFERENCES

1. W. Fischer, F. Gross, D. Hasenauer, H. Kahlen, K. Liemert, "A Passenger Car...", in Proceedings of the 30th Intern. Power Sources Symposium, Atlantic City,1983,The Electrochem. Soc., Inc. (Publ.), Princeton NJ, 59 (1983).
2. H. Reiss, "Evacuated, Load Bearing Powder Insulation for High Temperature Applications", Journal of Energy 7: 152 (1983).
3. J.J. Nowobilski, "Insulation Development for High-Temperature Batteries for Electric Vehicle Application", Final Report, Union Carbide, Linde Div., Tonawanda NY, Contract No.EM-78-C-01-5160 (1979).
4. B. Ziegenbein, "Evacuated High Temperature Insulations for Electrochemical Batteries", in Proceedings of the 8th Europ.Conf. Thermophys.Prop., Baden-Baden FRG, 1982 High Temp.-High Press., 15: 211 (1983)
5. R. Siegel, J.R. Howell, "Thermal Radiation Heat Transfer",McGraw-Hill Kogakusha Ltd., Tokyo 487 (1972).
6. M.G. Kaganer, "Thermal Insulation in Cryogenic Engineering", Transl. from Russian by A.Moscona, Israel Progr. Sci. Transl., Jerusalem, (1969).
7. W. Espe, "Werkstoffkunde der Hochvakuumtechnik", Vol. II, VEB Deutscher Verlag der Wissenschaften, Berlin (1960).
8. R. Caps, A. Trunzer, D. Büttner, J. Fricke, H. Reiss, "Spectral Transmission and Reflection Properties of High Temperature Insulation Materials and their Relation to Radiative Heat Flow", Proc. 18th Intern. Thermal Cond. Conf., Rapid City (1983), pp. 403-411.
9. H.C. Van de Hulst, "Light Scattering by Small Particles", Dover Publ. Inc., New York, 297 (1981).
10. F. Cabannes, J.-C. Maurau, M. Hyrien, S.M. Klarsfeld, "Radiative Heat Transfer in Fiberglass Insulating Materials as related to their Optical Properties", High Temp.-High Press. 11: 429 (1979).

THERMAL BEHAVIOUR OF INSULATING BUILDING MORTARS: EXPERIMENTAL

APPARATUS, CARBONATATION AND AGEING

[1]A. Mourtada, [2]A. Bloch, [1]G. Menguy and [3]M. Laurent

[1]Lyon 1, Laboratoire d'Etudes Thermiques et Solaires
69622 Villeurbanne, France
[2]Weber et Broutin, 02140 Saint Paul de Varax, France
[3]Insa Lyon, Laboratoire de Physique Industrielle
69621 Villeurbanne, France

ABSTRACT

Thermal properties of insulating building mortars are dependent on those of their components.

Some properties of mortars change with age as well as on carbonatation phenomena. Carbonation develops only when moisture and carbon dioxide gas are present. The carbonation is due to the diffusion of carbon dioxide gas into the air in the pores; Fick's laws are applicable in a first approximation.

In this study the thermal conductivity of the insulating mortars is determined as a function of ageing, moisture and components

The thermal conductivity is measured in the conditions similar to these of utilisation.

In order to determine the variation of the thermal conductivity of insulating mortars as a function of humidity ratio, a series-parallel arrangement consisting of two layers is used. The thermal conductivity of the mortars is calculated. Values calculated in this way provided good correlations with test values for insulating mortars and a linear relation between the thermal conductivity and the initial dry density is obtained.

It is shown finally, that acting on the composition (nature of components, cement/lime ratio) it is possible to improve the quality of the mortars.

This study deals with the influence of ageing (carbonatation) moisture and composition on the thermal conductivity of an insulating mortar.

1. THE INSULATING MORTAR

The Components : a mixture of :

- lime and cement,
- expanded polystyrene in granular form stabilized by resin treatment,
- additives : retainer of water, air carrier, water repellent.

Fabrication :

The components are mixed with water. The density depends on mixing time and water quantity. The mortar is characterized by its initial dry density. The characteristics of the samples are shown in Fig. 1.

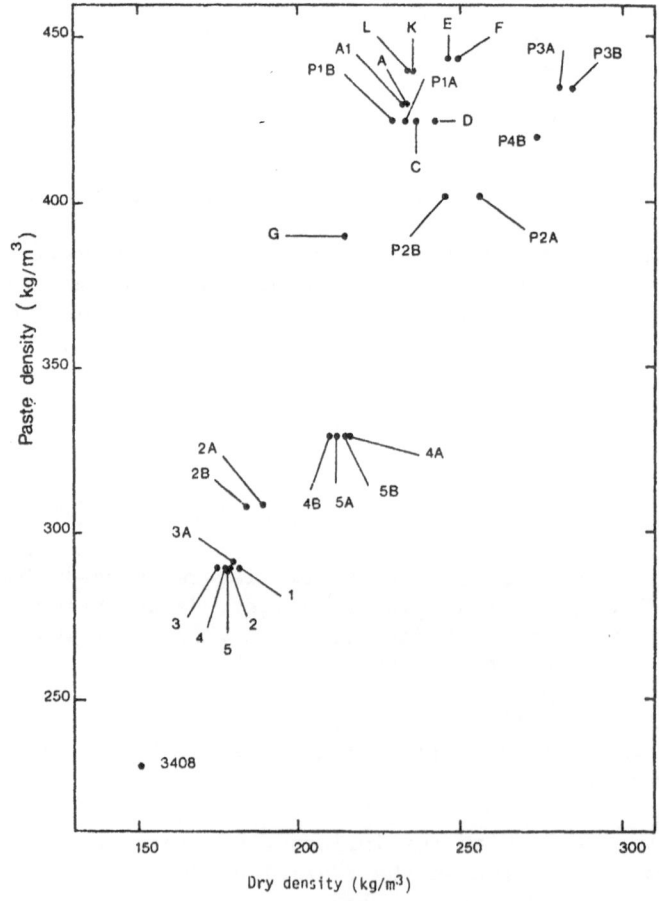

Figure 1.   Sample Properties

2. EXPERIMENTS

2.1. Thermal conductivity measurements

The thermal conductivity measurements are based upon steady heat conduction. The sample is set between two atmospheres (fig. 2). Their temperature corresponds to real cases. As the surroundings are at the same temperature as the upper boxes, the power $\dot{q}$ supplied by the electrical resistance goes through the sample and the thermal conductivity is given by :

$$\lambda = \frac{\dot{q}_e}{S(T_2 - T_1)}$$

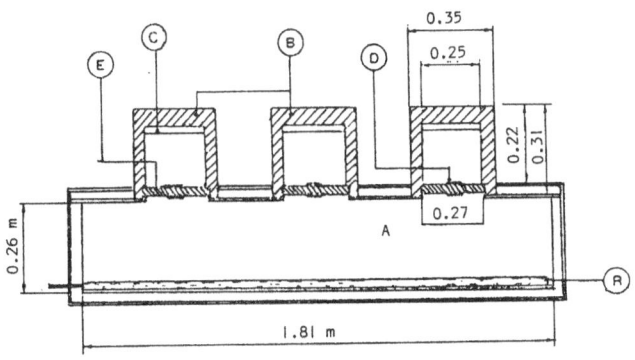

A - Isothermal enclosure   D - Platinum resistances

B - Insulating box   E - Sample

C - Heater   F - Heat exchanger

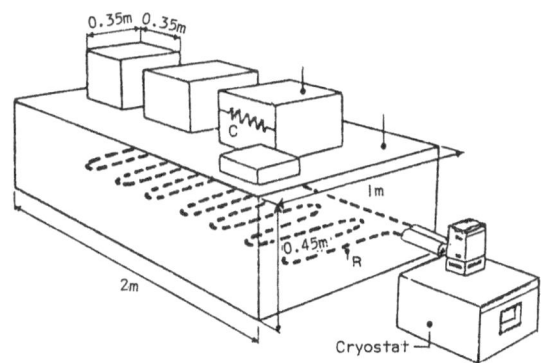

Figure 2. Thermal Conductivity Measurement Apparatus

On the two sides of the samples, temperatures T_2 and T_1 are measured with platinium resistance thermometers.
Three boxes on the same apparatus enable us to achieve many thermal conductivity measurements. The temperature difference between the opposite surfaces of the sample is about 15°C (16 to 1°C) ; the accuracy is about 5%.[4]

2.2. Moisture content

The difference of weight of the sample before and after drying to a constant weight at 70°C gives us the moisture content*

2.3. Carbonatation

The samples are broken and we vaporize a phenolphthalein solution (0.1%). The outside carbonated parts don't change color and the inside parts are colored in red.

2.4. Quantitative analysis of carbonate of calcium

The sample is broken and some parts are taken off at different distances from the surface. By hydrochloric acid action, the carbonate of calcium is decomposed, and from weight loss we obtain the quantity of calcium carbonate :

$$CaCO_3 + 2\ HCl \rightarrow CaCl_2 + H_2O + CO_2$$

3. RESULTS

3.1. Carbonatation

The lime $Ca(OH)_2$ which is in the mortar dissolves in pore-water and reacts with the carbon dioxide of the air to give calcium carbonate and water. This reaction takes place quickly at the surface and is retarded inside the mortar because the carbon dioxide gas must diffuse into the interior.

This is shown by the non-colored region near the surface with phenolphthalein; and we found that the non-colored depth X was proportional to the square root of the time :

$$X = a' \sqrt{t}$$

Then we can interpret the carbonatation as a diffusion phenomenon.

From the analysis of carbonate of calcium we get the curves quantity of carbonate versus depth (fig. 3). And by using Fick's

* The dry weight is obtained at 66 % of relative humidity during fifteen days

law, we determine the diffusion coefficients and the times for 90% carbonatation (table 1).

Figure 4 shows the density increases of the samples; they can be up to 20 per cent (sample E).

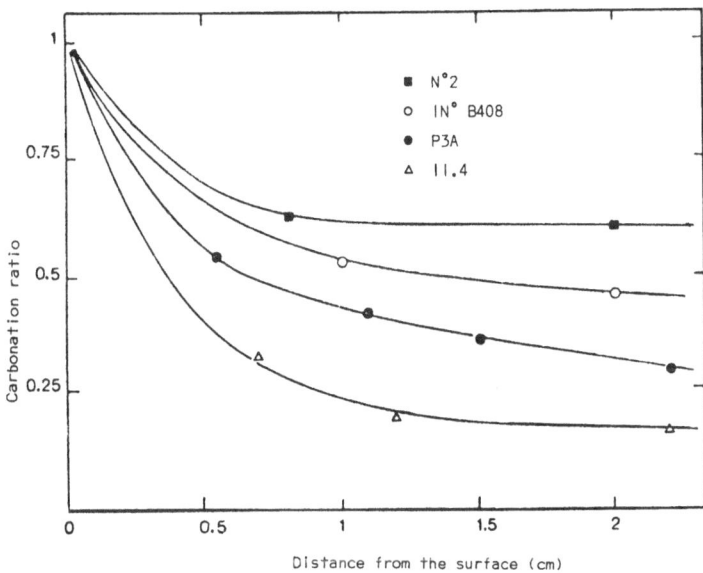

Figure 3. Carbonation Ratio Versus Distance from Surface

Table 1. The diffusion coefficients and the times
for 90% carbonatation

Sample ref.	Age (year)	D (mm²/year)	Time for 90% carbo-natation	dry initial density
P3A	2.25	76	6.6	280
B408	0.33	654	0.7	151
II 4	1.33	65	7.8	1124
2	0.75	339	1.35	179

$$\Delta \varrho_S = \frac{\varrho_S - \varrho_{S_0}}{\varrho_{S_0}} \%$$

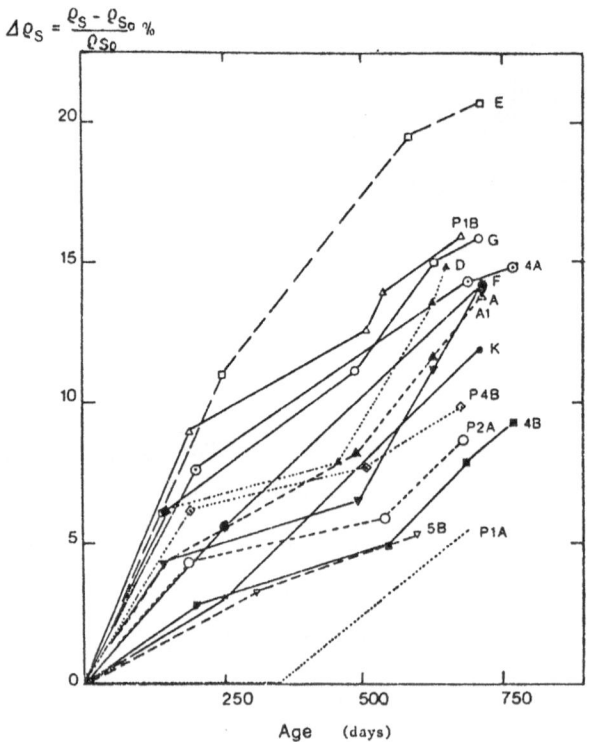

Figure 4.   Density Increase Versus Age of Samples

3.2. Influence of carbonatation on the thermal conductivity

For the same moisture, figure 5 shows the thermal conductivity curves vesus dry density at different ages ; they are different. But if we draw thermal conductivity versus the initial dry density (fig. 6) we get one curve only. We conclude that carbonatation does not influence the thermal conductivity, and that is corroborated by results on dry mortars (table 2). This is an important result because simultaneously the carbonate of calcium makes the mortar stronger[1,2,3]

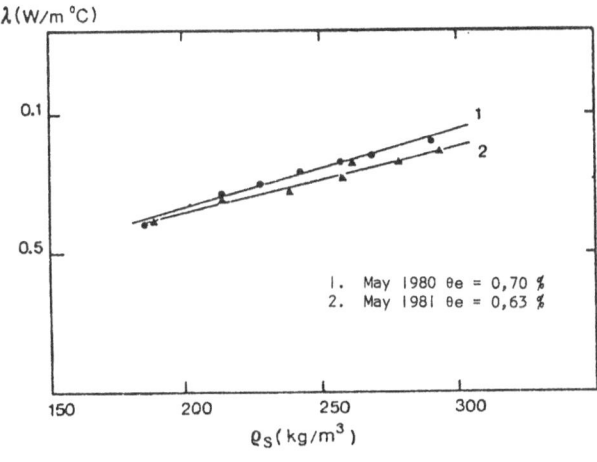

Samples: (D, 3A, P2B, P4B, C, IB, G)

Figure 5.   Thermal Conductivity Versus Dry Density at Different Ages

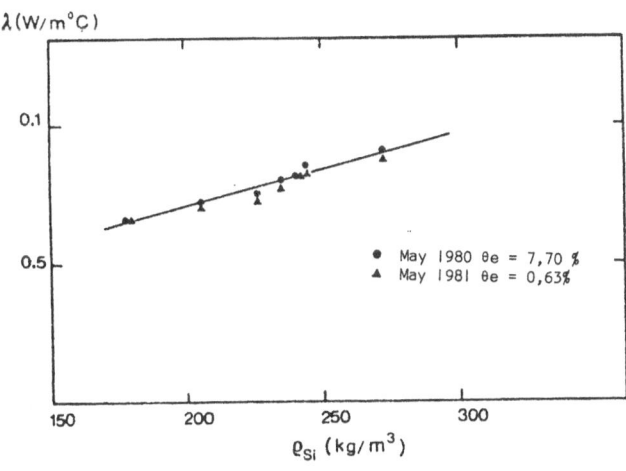

Samples: (D, 3A, P2B, P4B, C. IB, G)

Figure 6.   Thermal Conductivity Versus Initial Dry Density at Different Ages

431

Table 2. Thermal Conductivity and Density Versus Age for Dry
Samples

Reference samples	ρs kg/m³	Date	λ W/m°C	Date	ρs kg/m³	Date	λsec W/m°C	Date	ρs kg/m³	Date	λsec W/m°C	Date
D	242	02.80	0,079	03.80	257	06.80	0,077	05.80	261	4.81	0,077	05.81
3A	180	01.80	0,063	"	185	"	0,063	"	189	"	0,063	"
P2B	245	01.80	0,079	"	267	"	0,080	"	277	"	0,076	"
P4B	273	01.80	0,085	"	290	"	0,084	"	294	"	0,081	"
C	236	02.80	0,077	"	242	"	0,075	"	248	"	0,073	"
1B	207	01.80	0,070	"	214	"	0,068	"	214	"	0,066	"
G	214	02.80	0,072	"	227	"	0,070	"	238	"	0,072	"
A	233	02.80	0,076	"	243	"	0,077	"	252	"	0,073	"
P2A	256	01.80	0,081	"	271	04.81	0,080	05.81				
I	213	02.80	0,071	"	223	04.81	0,071	05.81				
5A	212	01.80	0,071	03.80	218	08.80	0,070	09.80				
P3A	280	01.80	0,088	"	296	"	0,087	"				
4					179,8	06.81	0,063	07.81	185,6	01.82	0,062	01.83
5					179,8	06.81	0,063	07.81	191,3	01.82	0,062	01.83

3.3. Influence of carbonatation on mortar

Another interesting result is the quantity of water in the sample with age (fig.7). The older the mortar is, the less water we find in it. That means the carbonate of calcium is less hygroscopic than lime.

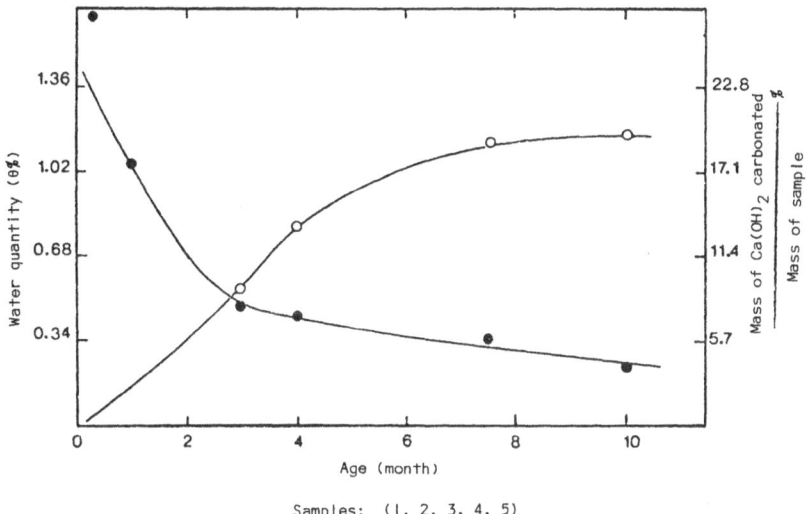

Figure 7.   Quantity of Water and Carbonate Versus Age

Then it seems worth increasing the quantity of carbonate of calcium because :
  - first, the rise in density does not alter the thermal conductivity with the same moisture ratio,
  - then, with the same outside conditions there is less moisture. Consequently we obtain a better thermal insulator.

4. THERMAL CONDUCTIVITY EVALUATION

We assume the mortar to be described by the scheme of figure 8, after Krischer[7].
The porous solid is assumed to be made of plates. Plates e represent the water and plates s represent the other components. The thermal conductivities are λ_e and λ_s ; θ_e and θ_s are volume ratios ; and we have the part d in series with the part 1-d, of two parallel components. [5,6]

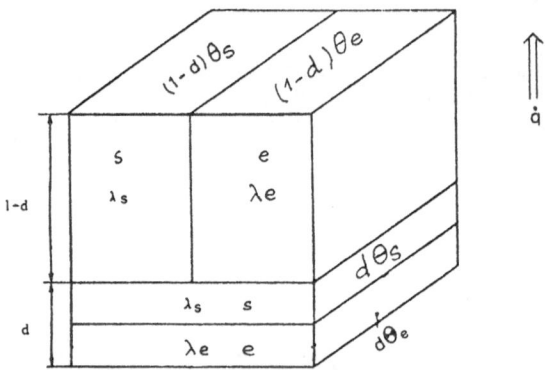

Figure 8.   Series-Parallel Model

Then the thermal conductivity is given by :

$$\lambda = \frac{1}{\frac{1-d}{\lambda'} + \frac{d}{\lambda''}}$$

with $\quad \lambda' = \lambda_s\lambda_s + \theta_e\theta_e = (1 - \theta_e)\lambda_s + \theta_e\lambda_e$

$$\lambda'' = \frac{1}{\frac{\theta s}{\lambda_s} + \frac{\theta_e}{\lambda_e}} = \frac{1}{\frac{1 - \theta_e}{\lambda_s} + \frac{\theta_e}{\lambda_e}}$$

For many samples with two measured conductivity and with two moisture ratios, (fig. 9) we calculated d. We always found the same value d = 0.24. Therefore we can use this value to obtain the thermal conductivity of dry samples from the conductivity measurements of moist samples with the above equation. And for smaller quantities of water, we get :

$$\lambda_s = \lambda(1 - 9 \theta_e)$$

By comparing the calculated dry conduct with measurements of moist and dry specimens (fig. 10) we can conclude they correspond to one another.

The thermal conductivity of dry insulating mortars is given by :

$$\lambda_s = 0.00022 \rho_{si} + 0.0236$$

ρ_{si} is the initial density.

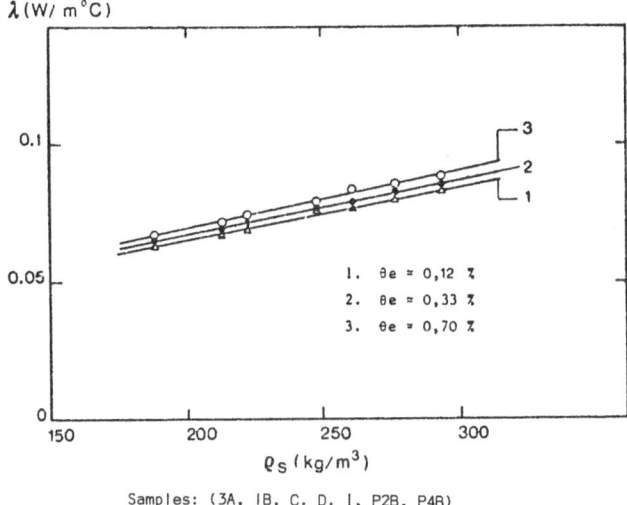

Samples: (3A, 1B, C. D. 1, P2B, P4B)

Figure 9. Thermal Conductivity of Mortars Versus Dry Density
with Various Quantities of Water

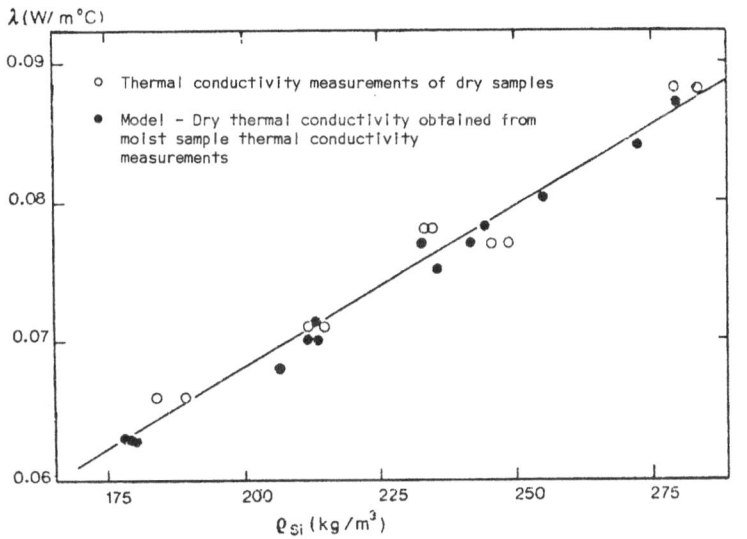

o Samples: (E, F, K. L. 2A, 2B, 5A, 5B, P3A, P3B)
● Samples: (5A, P3A, D, P2B, P4B, 3A, C, 1B, G, A, P2A, 1, 4, 5)

Figure 10. Thermal Conductivity Versus Initial Dry Density
(Experimental Results and Model Results)

435

By using the initial density ρ_{si} we do not have to take the carbonatation into account.

From this study and from the many conductivity measurements we did, we conclude :

- the cement quality does not influence the conductivity ;
- the quality ratio, % cement / % lime, does not influence the conductivity, but acts on carbonatation and this reaction must be improved ;
- the dry density must be lowered as much as possible by increasing the quantity of expanded polystyrene, and by acting on preparation : water quantities and mixing speed.

Following these conclusions, a new insulating mortar was developed : its dry conductivity is 20 % better ($\lambda = 0.054$ w/mK). Further improvements with expansed polystyrene as the insulating component will be more difficult to achieve because of its effect on mechanical properties.

REFERENCES

1. Tye, R.P. "Thermal conductivity" Academic Press, London and New York (1969)
2. Luikov A.V, "Heat and mass transfer in capillary porous bodies" Advance in heat transfer. Vol 1 (1964)
3. Philip J.P., De Vries D.A. "Moisture movement in porous bodies" Pergamon Press (1984)
4. Mourtada A. "Comportement thermique des mortiers d'isolation extérieure du bâtiment". Thèse Docteur Ingénieur INSA Lyon (1982)
5. Eyraud M.N.C, Lareal P., et Gielly J. "Propriétés physicochimiques de l'eau dans les milieux poreux. Caractéristiques texturales des milieux poreux et facteurs susceptibles de les modifier". Bull. Rilem. n°27 (1965)
6. Fauconnier R., Florence B., Laugier D., "Le transfert de l'humidité dans les matériaux isolants". Promoclime Tome E n°3 (1979)
7. Krischer O., Kroll K. "Bases scientifiques de la technique du sèchage". Traduction du Centre technique des industries aérauliques et thermiques.France.

THERMAL CONDUCTIVITY OF A POLYIMIDE FOAM

D. D. Burleigh

Materials and Process Laboratory
General Dynamics Convair Division
San Diego, California

Abstract

The thermal conductivity of a flexible polyimide closed-cell foam
has been measured from 100°K to 300° K by a guarded hot plate
technique (ASTM C 177).   Polyimide foam is useful as an insulator
from the cryogenic range through 590°K.   It shows great promise for
future use as a replacement for polyurethane foam.

Introduction

The thermal conductivity of a flexible "closed-cell" foam made from
a modified polyimide resin, Chem Lon 500FC, made by the Chemtronics
Co. of San Diego, has been measured from 100K to 300K by a guarded
hot plate technique   (ASTM C 177).

In this work, two thin circular concentric electrical heaters were
sandwiched between two specimen panels whose dimensions are approximately 15 by 15 by 1 centimeter thick.   The temperatures of the
specimen surfaces and of the center and guard heaters are measured
by twelve 36-gauge chromel-constantan (Type E) thermocouples.

At equilibrium, the electrical power to the center heater and the
average hot and cold face temperatures are recorded.   Thermal conductivity, λ, is calculated from the relationship:

$$\lambda \;\; = \;\; C \; (EI)t/A\Delta T$$

Where:

C   = geometry and unit factor

(EI)   = center heater power

t   = average center-section thickness

A   = effective center-section area

ΔT = temperature difference between hot and cold faces

The heat sinks used in this testing were a pair of aluminum plates with interior channels through which coolant can be pumped. For the lowest temperatures, liquid nitrogen was the coolant. Chilling the outer surfaces of the plates with blocks of dry ice produced a specimen temperature of ~210 °K. Circulating refrigerated water through the plates resulted in a specimen temperature of near ambient.

A plastic bag surrounded the specimen stack and a dry gas purge flowed around the specimen at a rate of about 0.05 liters/sec. Both nitrogen and helium were tried as purge gases. We found that with a helium purge the thermal conductivity measured was seven times higher than that measured with a nitrogen purge. This indicates that the "closed" cells are permeable to helium. We used a nitrogen purge for all data reported here. Tests were performed either with nitrogen flowing or with a static nitrogen environment. Results were identical for the two cases as long as the gas flow rate was low.

Several thermal conductivity tests showed that the orientation of the specimens did not affect the measured values. Loading was kept constant by the use of spacers and clamps.

The test apparatus is shown in Figure 1. The test stack with the purge bag and cooling plates is shown in Figure 2. The guarded hot plate itself is shown in Figure 3.

The thermal conductivity of both bulk and "skinned" foam samples were tested. The skin was produced by pressing the bulk foam against a hot metal plate. All specimens were the same thickness. Figure 4 shows the structure of the bulk foam. It is clear that the cells are not well formed and that few cells are closed. The average cell diameter is approximately 0.25 mm.

Figure 5 shows the difference in structure between the bulk foam and the skinned surface. The skin layer has a depth of approximately 0.5 mm. Skinned specimens have skin on both surfaces.

The data is plotted in Figure 6 for both specimens. The presence of the skin increased the conductivity by approximately 15%.

The measured density of the bulk foam was 13 Kg/m^3. Similar foams of densities up to 80 Kg/m^3 can be produced, as well as rigid foams, by modifying the processing parameters. Polyimide foam is useful as an insulator from the cryogenic range through 590°K.

This material shows great promise for future use. It is being considered by the FAA as a replacement for polyurethane foam in airplane seat cushions, and by the Navy for submarine hull insulation. Polyurethane foam has been shown to be a hazard during fires in enclosed spaces. When exposed to flame it burns, exuding a thick, black toxic smoke containing HCN and Cl_2. Polyimide foam, on the other hand, chars instead of burning, as long as the atmospheric oxygen content is below 38 percent, and is likely to produce CO and CO_2. The smoke density is in the range of 2 percent.

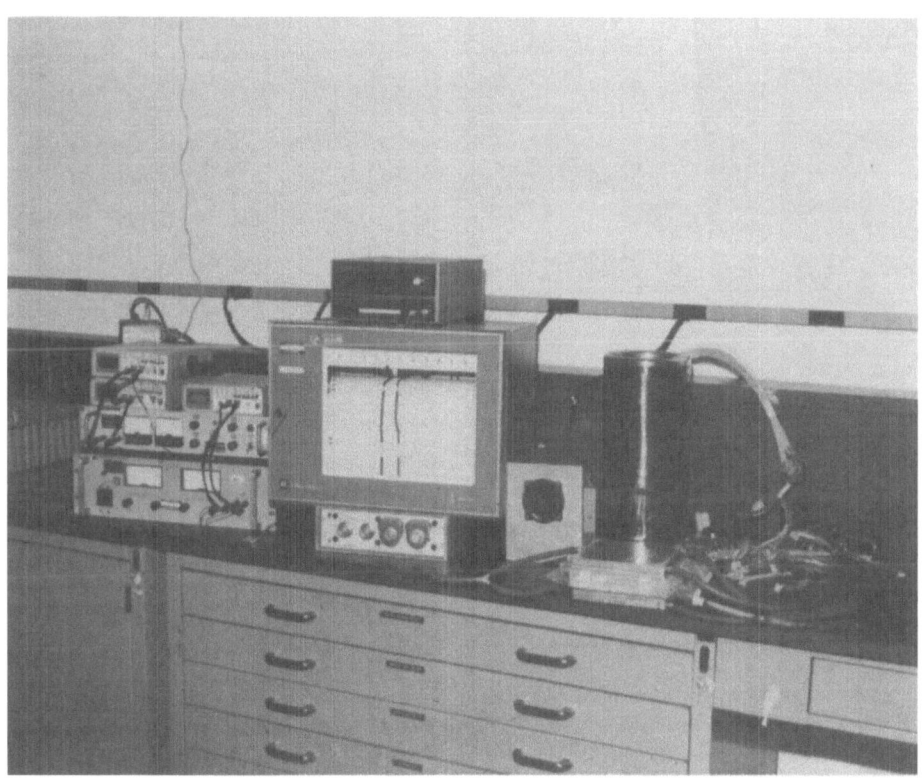

Figure 1. Thermal conductivity test apparatus.

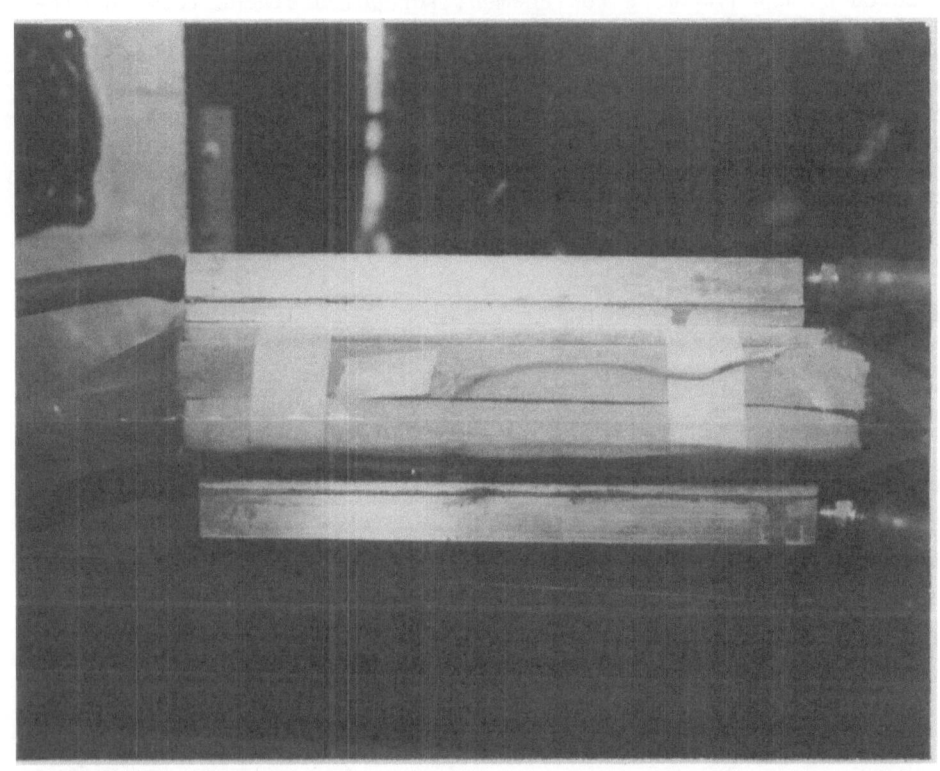

Figure 2. Detail of apparatus showing cooling plates, purge
bags, and specimens.

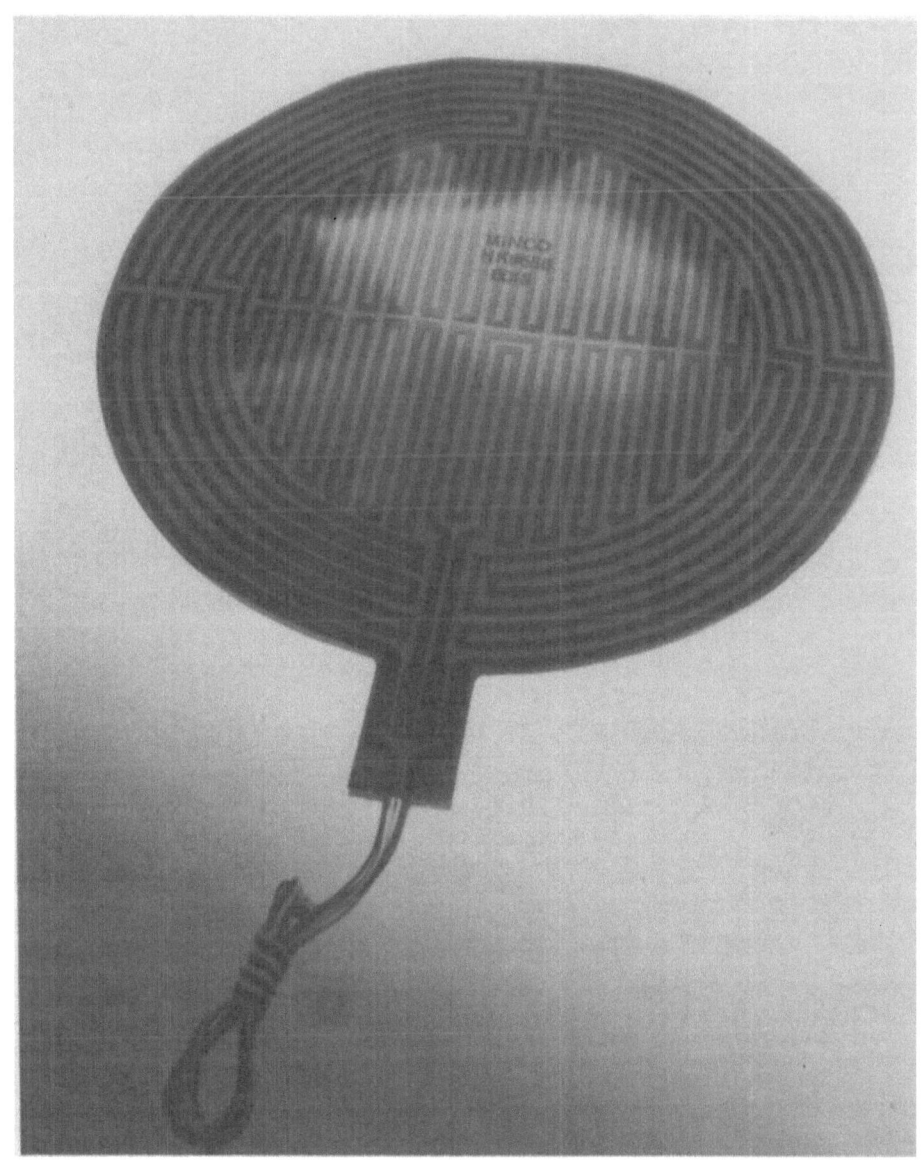

Figure 3. Guarded hot plate.

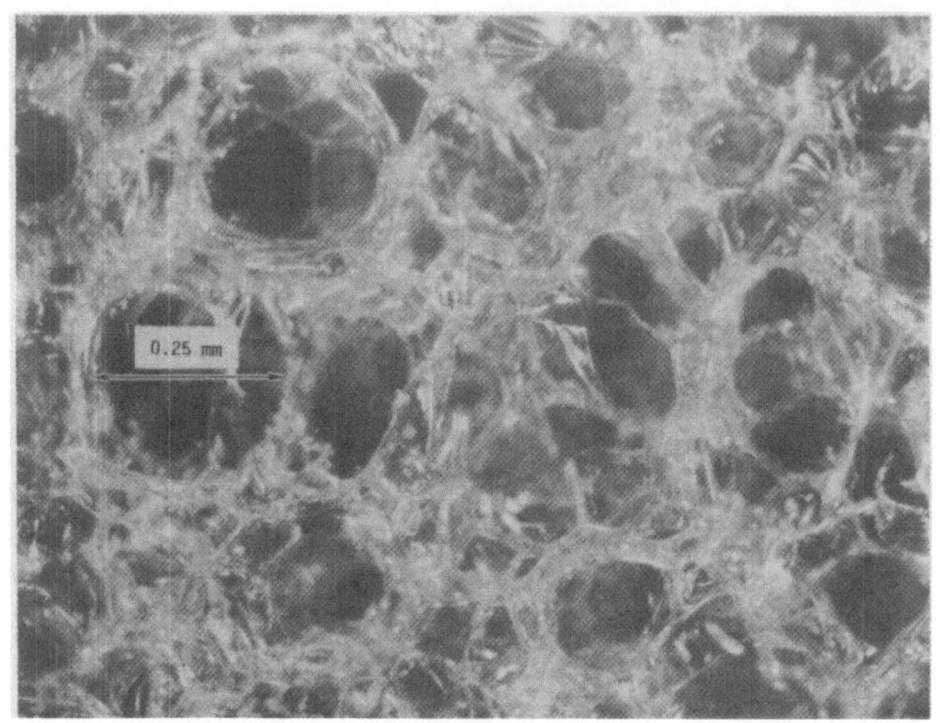

Figure 4. Bulk foam microstructure.

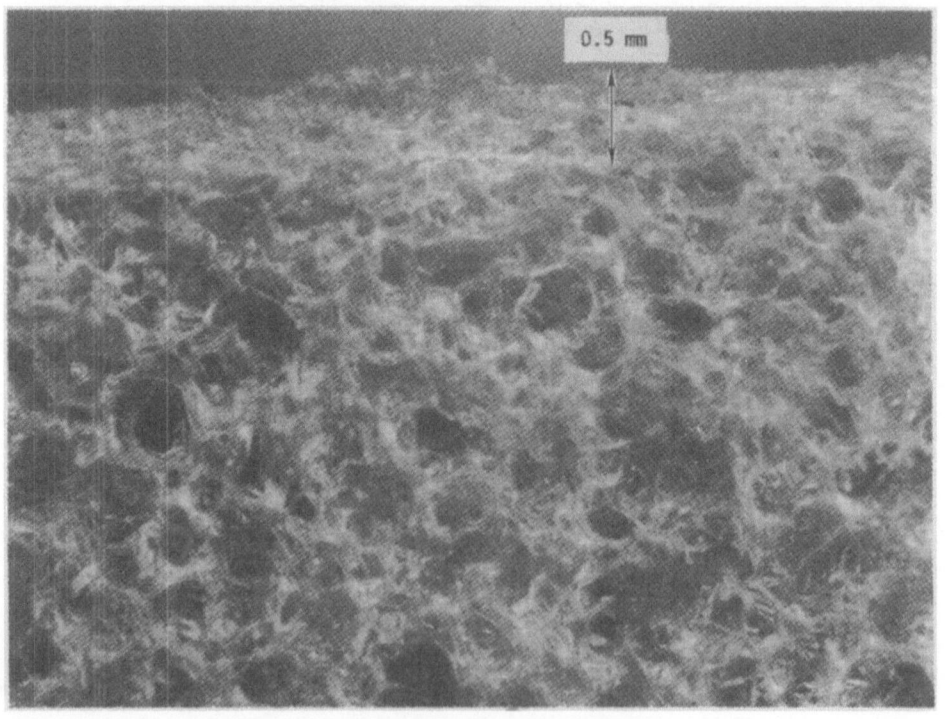

Figure 5. Microstructure showing skin.

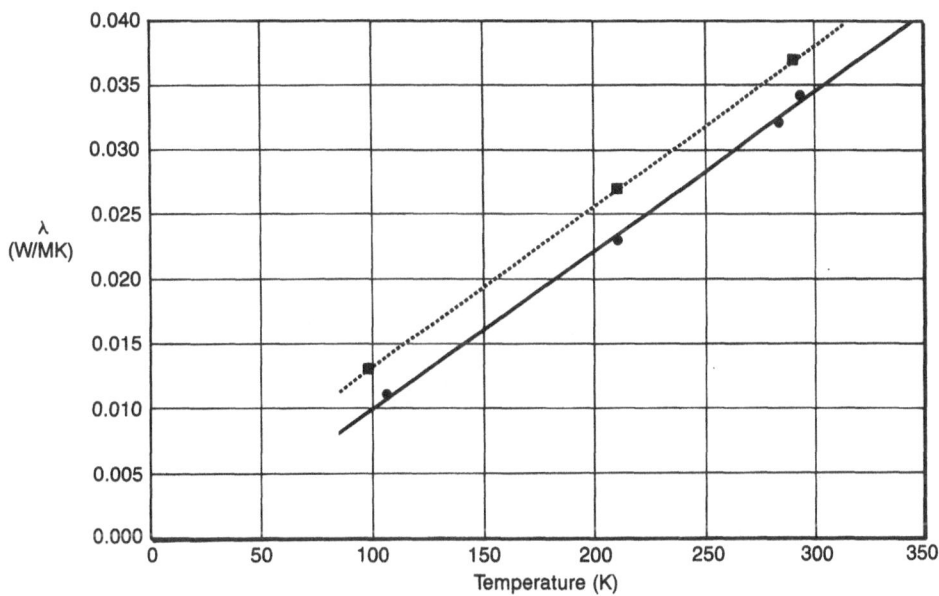

Figure 6. Thermal conductivity of Chem Lon 500 polyimide foam.

SESSION P

Invited Paper

SPEAKER

R. Taylor
UMIST
Manchester, England

SESSION CHAIRMAN

P. Wagner
Los Alamos National Laboratory
Los Alamos, NM.

THERMAL DIFFUSIVITY MEASUREMENTS OF THERMAL BARRIER

COATINGS AND CARBON FIBRE/CARBON COMPOSITES

Roy Taylor

Department of Metallurgy and Materials Science
University of Manchester/UMIST, U.K.

ABSTRACT

The thermal diffusivities of plasma-sprayed thermal barrier coatings and of carbon/carbon fibre composites manufactured by two different production routes are reported. The results are discussed and related to the microstructural features and processing variables.

INTRODUCTION

Since its inception some twenty years ago, the success of the flash method for measuring the thermal diffusivity of a wide range of materials such as metals and alloys, ceramics, semiconductors has been well established.

However most technologically significant materials are, in one way or another composites of two or more components. Within the regime where inhomogeneities on a microscopic scale are small compared to the macroscopic dimensions the specimen can be considered essentially homogeneous and reproducible results are obtained irrespective of sample size. This is particularly important for the flash method where samples of only a few millimetres thick are required. In an elegant experiment Lee and R.E. Taylor [1] showed that even when the particle size of a dispersion was 25% of the sample thickness the flash method gave reproducible results.

In recent years two technologically significant advances in

materials developments have been made with regard to the development of specific composites.

a) the production and application in a number of diverse areas of layered composites.

b) the use of directionally reinforced fibrous composites such as carbon-carbon, carbon-phenolic resin and more recently composites using ceramic fibres such as alumina, zirconia, silicon carbide etc.

In the case of layered composites it is essential to measure the thermal transport properties of the coated layer *in-situ* primarily because such layers are usually too friable and fragile to permit easy handling. Furthermore attempts to deposit thicker coatings to permit easier handling are likely to produce unrepresentative material. Hence it becomes necessary to measure the thermal diffusivity or conductivity of a coating that may be as thin as 0.1mm. Considerable effort has been devoted in recent years to solving the heat diffusion equations appropriate to the flash method for heat flow through two and three layers composites. Such analyses have led to the adaptation of the flash technique to measure the diffusivity of one layer of a two or three layer composite or even of liquids or powders in containers. To this extent the flash method is perhaps the ideal method to measure the thermal transport properties of one component forming part of a layered composite.

By way of a contrast more caution is needed when applying the flash techniques to a directionally reinforced composite, because of inhomogeneities arising from differences in the conductivities of the two components. For carbon fibre epoxy resin composites this conductivity ratio can, for example, be as high as 500:1. This can lead to diffusivities whose value depends on sample thickness.

The thermal diffusivity facility at UMIST has been actively concerned for a number of years in measuring the thermal diffusivity of a wide range of materials and two examples, one of a layered composite and one of a reinforced composite, will be considered in this paper.

1. The thermal diffusivity of certain plasma-sprayed coatings based on stabilised zirconia has been measured from 500-1500K.

2. The thermal diffusivity of orthogonally reinforced (2 and 3 dimensional) carbon/carbon fibre composites has been measured from 300-3000K.

In each instance we have attempted to explain the results

obtained and in this paper emphasis will be placed on relating the results obtained to the microstructure of the material under investigation and equally importantly to any changes that may be occurring. For the case of the carbon/carbon fibre composite a model developed to describe the derived thermal conductivity in terms of the conductivity of the components will be briefly referred to.

Theory and Apparatus

In the flash technique the front face of the sample is subjected to a short pulse of radiant energy and the resultant dimensionless temperature history of the rear face is recorded [2]. For a 1 layer sample the thermal diffusivity may be determined from the temperature change with time. The accuracy and reproducibility of the technique have been discussed in detail in a critical review [3]. Mathematical analyses for heat flow through two layers have been given by Larson and Koyama [4] Bulmer and R. Taylor [5] and for two or three layers by Lee, Donaldson and R.E. Taylor [6].

The UMIST apparatus consists of three basic units, pulse source, heating assembly and transient recording and analysis system [7]. The heat pulse is supplied by a solid state Neodymium glass laser having a dissipation time of ~6 x 10^{-4}s. The specimen in its holder is supported inside a susceptor which is heated to the desired measurement temperature using an induction coil. This is located inside a vacuum chamber/pressure vessel with a measurement capability from 300-3000K. Radiation from the rear face of the specimen is collected and focussed onto an infra-red detector. A range of detectors is available: 293-750K — $Hg_xCD_{1-x}Te$ or a thermoelectrically cooled PbS detector, 500-3000K using a room temperature PbS detector and >1500 using a Si photodiode. The amplified output from the detector is sampled and analysed using a PDP-11 minicomputer possessing full data reduction capabilities.

Thermal Barrier Coatings

The specific fuel consumption of a gas turbine engine can be decreased by more efficient burning of the fuel and this can be achieved by running the engine hotter. In the case of a jet engine this poses severe problems. For example maximum gas temperatures in the turbine section of a modern jet engine are around 1600°C and such high temperatures are only possible because of engineering developments such as blade cooling. The application of plasma-sprayed ceramic coatings to critical areas of a jet engine is being intensively investigated. For a material to be an effective thermal barrier several criteria have to be met of which the most important are; a low thermal conductivity, a thermal expansion coefficient that matches the superalloy substrate and a high thermal shock resistance. Many refractory ceramics have properties which fit many of the criteria but the requirement for a match of thermal expansion to that of the substrate has focussed attention on ZrO_2, which possesses a suitable thermal expansion coefficient. Zirconia however possesses three allotropic forms

Monoclinic	< 1170°C	a = 0.51477 nm
		c = 0.52030
Tetragonal	1170-230°C	a = 0.3640 nm
		c = 0.527 nm
		β = 99.26°
Cubic	> 2370-2690°C	a = 0.509 nm

It is necessary therefore to stabilise the cubic form of ZrO_2 and this may be achieved by adding either MgO, CaO or Y_2O_3. We have investigated the temperature dependence of the thermal diffusivity of a number of plasma-sprayed thermal barrier coatings of ZrO_2 stabilised with MgO and with Y_2O_3. Some of these results will be presented here.

The coatings were produced from commercially available powders which were plasma-sprayed by Rolls Royce Limited onto nickel-based superalloy substrate discs 1mm thick x 10mm diameter. The maximum temperature to which measurements can be made is thus limited by the substrate to 1500K. A flash layer of a NiCrAl bond coat 10-20μm thick was deposited on the substrate to improve adhesion of the ceramic. Successive passes were made with the plasma gun to build up a ceramic coating 0.2 to 0.5mm thick. Details of the powders used and the coatings obtained are given in Table 1.

Table 1. Characterisation of Thermal Barrier Coatings

Powder No.	Composition	Phase distribution in sprayed coating	Coating Density kg.m^{-3}x10^3
1	ZrO_2 + 18-25wt%MgO	–	5.15 to 5.40
2	ZrO_2 + 18-22wt%Y_2O_3	–	5.50 to 5.82
3	ZrO_2 + 8wt%Y_2O_3	97.5% Tetragonal ZrO_2 ~2.5% Monoclinic ZrO_2	5.78±0.1
4	ZrO_2 + 8wt%Y_2O_3	97.5% Tetragonal ZrO_2 ~2.5% Monoclinic ZrO_2	5.18±0.1
5	ZrO_2 + 6wt%Y_2O_3	98.5% Tetragonal ZrO_2 ~1.5% Monoclinic ZrO_2	5.29±0.1

The densities of the sprayed coatings were determined by the displacement method whereby the composite specimen is weighed both in air and in a liquid, in this case iodoethane. X-ray analysis of the constituent phases was carried out using a Philips X-ray diffractometer. The thermal diffusivity of the substrate alloy + bond coat was measured beforehand and all substrate discs were fabricated from the same batch of alloy. Since there are slight variations in the thickness but more particularly in the density of the coating 3 or 4 samples of each coating were prepared. At least two samples were measured to check reproducibility. In spite of variations in density thermal diffusivities for each particular coating material agreed to better than 10%. Changes in length of substrate and coating with temperature were corrected using known thermal expansion data. Specimen thicknesses were determined metallographically on a section of the specimen after all measurements were completed. All measurements were carried out under a pressure of 1.25 atmospheres of nitrogen.

Typical results for coating made from powder 1 are presented in figure 1. Initially the thermal diffusivity had a low value of ~2.5 x 10^{-7}m^{2}s^{-1} which increased significantly on heating above 1100K. Three measurement runs up to 1400K were carried out and hysteresis effects were observed, the diffusivity systematically increasing. The microstructure of our coatings has been studied

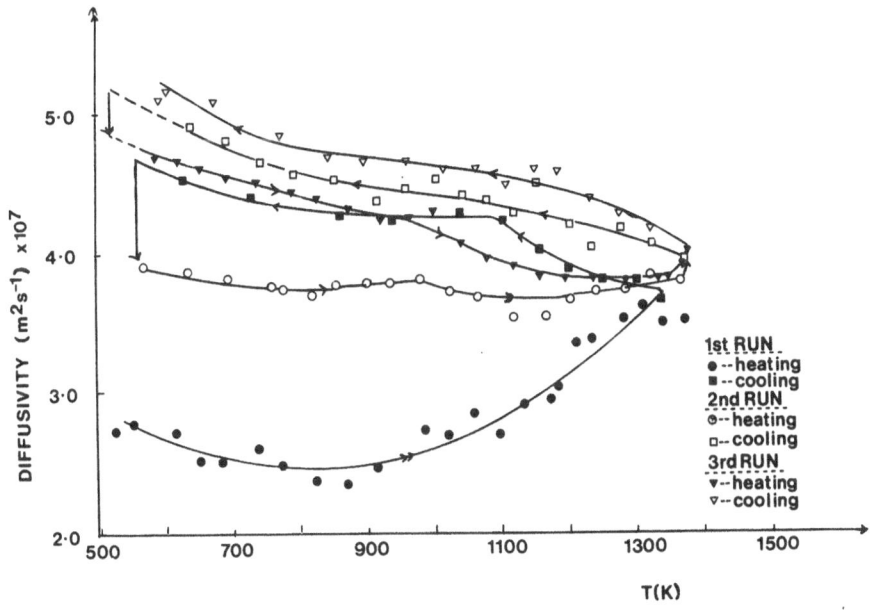

Fig. 1.   Thermal diffusivity of coating plasma sprayed using powder 1.

in some detail in order to understand this behaviour. We find that precipitation of free MgO occurred from the stabilised ZrO_2 solid solution during thermal cycling and we could distinguish monoclinic zirconia (MgO deficient), cubic zirconia and MgO particles [8]. The volume change associated with this phase change leads to an enhancement of the crack network which increases the toughness of the ceramic. Similar conclusions were deduced by Grot and Martyn [9] when they carried out burner life tests for Pratt and Whitney. They concluded that the MgO precipitates acted as point stress relief systems via the extended crack network and dissipate strain energy by stopping propagating microcracks. This behaviour may be readily understood by reference to the equilibrium diagram [10] which shows that the cubic form of ZrO_2 in the ZrO_2–MgO system is only stable at temperatures above the eutectoid temperature of $1400^{\circ}C$ and will therefore decompose with time at temperatures above 1100K to monoclinic ZrO_2 + MgO. This enhancement of the thermal diffusivity reduces the effectiveness of ZrO_2 coatings stabilised with MgO as thermal barriers.

In contrast coatings from powder 2, ZrO_2 highly alloyed with Y_2O_3, show a thermal diffusivity which is roughly constant with temperature and reproducible during repeated runs up to 1400K. Differences of up to 25% were detected between different specimens but for any one specimen the variation between heating and cooling during repeat runs is within ±5% (figure 2).

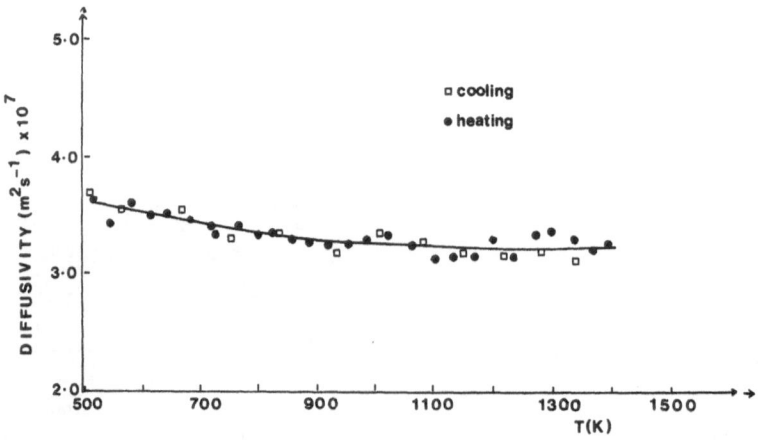

Fig. 2. Thermal diffusivity of coating plasma sprayed using powder 2.

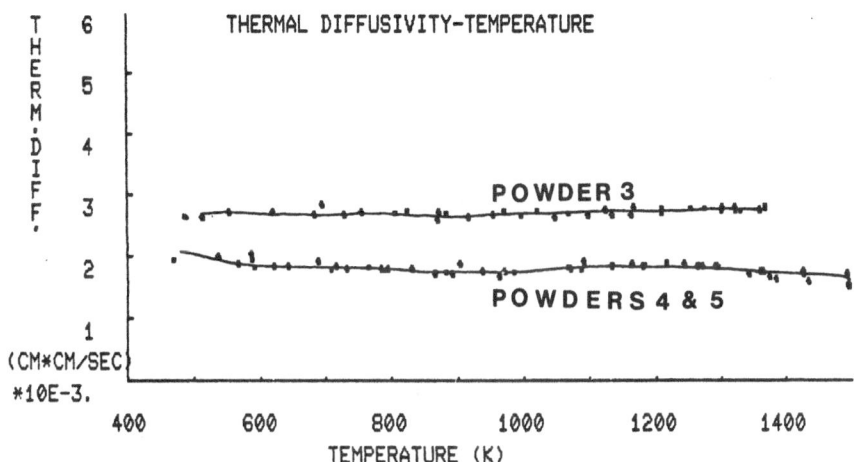

Fig. 3. Thermal diffusivities of coating plasma sprayed using
powders 3, 4, 5.

However the proportion of Y_2O_3 added to ZrO_2 is far greater
than that necessary to stabilise the cubic phase [11] so current
interest has centred on partially stabilised zirconia having Y_2O_3
contents in the range 6-12wt%. Coatings from three different
powders have been measured and the results are shown in figure 3.
The first of these from powder 3 showed a dense coating which had
a fairly high thermal diffusivity of ~3.0 x $10^{-7}m^2s^{-1}$ which
remained constant even on repeated heating up to 1490K. The
change between runs was marginal and the density showed no
detectable change. The data for coatings sprayed from powders 4
and 5 also show no change on thermal cycling up to 1495K and are
indistinguishable, having an average value for thermal diffusivity
of ~1.8 x $10^{-7}m^2s^{-1}$. The density and microcrack network of the
sprayed ceramic obviously have a significant effect on the thermal
diffusivity. For example the thermal diffusivity of coatings
sprayed using powder 4 is lower than that of coatings sprayed
using powder 3 which has the same composition. A typical example
of the microstructure is given in figure 4 which is a coating
obtained by spraying powder 3. This work is at an interim stage
and we are relating the thermal transport properties of coatings
to the microstructure and other variables. [12]. However further
detailed discussion lies outside the scope of this paper.

Fig. 4.   Optical micrograph of coating, plasma sprayed using
             powder 3.

Carbon-Carbon Fibre Composition

Carbon-carbon fibre composites have only limited commercial
use because of their high cost of manufacture. However because of
their   high   temperature   mechanical   and   thermal   property
capabilities they are essential in several specific areas such as
brakes for supersonic aircraft, solid-fuel rocket nozzles or
external heat shields for atmospheric re-entry.

Several types of carbon fibres are commercially available,
made from rayon, polyacrylicnitrile (PAN) or pitch precursors.
The fibre shape, diameter and other properties are influenced by
selection of precursor. The technology now exists to fabricate 1,
2,   3   or   even   more   complex   carbon/carbon   fibre   composites.
Initially the fibres are laid up in the desired orientation to
construct a fibre preform.   Then the preform is coated with
pyrolytic carbon by low pressure isothermal deposition of vapour
phase carbon (carbon vapour deposition or C.V.D.).   This stiffened
preform is then sufficiently rigid to be handled.   In the final
stage this CVD'd preform is densified either by impregnation with
pitch, a process which may be repeated up to 5 times, or by
multiple impregnations lasting several hundred hours using the
C.V.D. process. Finally the composite is graphitised by heating to
2750-2800°C.

There is a strong interest in obtaining data for the designer to the high temperatures such components may be required to sustain in service, as much as 2500°C. Additionally there is a requirement to develop a mathematical model to explain the effective thermal conductivity in terms of the properties of the constituents. In this respect the component properties particularly those of the fibre, are markedly affected by processing variables.

A number of fine-weave carbon-carbon fibre composites have been measured at UMIST and we have attempted to rationalise the observed behaviour by a detailed study of the microstructure. To illustrate the effect of processing and highlight the observed differences, data for composites prepared by two different production routes will be presented. For each composite we have also measured the thermal diffusivity of 1-D composites having similar densities produced from the same fibre.

Two materials reported here typify the two alternative methods of production. Composite A supplied by Wright-Patterson A.F.B. was a 3-D orthogonal carbon fibre/carbon composite (C.F.C.C.). This is referred to as a 2-2-3 composite made by laying up two yarns of 1440 filaments/yarn to form reinforcement in the X and Y directions and piercing this with 3 yarns of 1440 filaments/yarn to form reinforcement in the Z direction. Fibre volume fraction in the X and Y axis was 13%, and 22% in the Z axis. The fibres used were Thornel 50 having a mean diameter of 6.5µm. In this instance the preform, stiffened by CVD, was impregnated five times with pitch and finally graphitised at 2750°C. Compsite 1A was a 1-D composite made from the same fibres and containing a 52% fibre volume fraction. In this instance the fibres were impregnated with pitch directly and the CVD stage was omitted. A sample of matrix graphite was also provided.

Composite B which was supplied by Societe Europeene de Propulsion has 2-D reinforcement comprising alternative layers of 8 harness satin-weave carbon fibre mats. These were made from woven mats of fibres laid up with the warp of alternate layers at 90° to each other. The matrix of the composite is formed by successive cycles of CVD followed by high temperature graphitisations which serve to open up any porosity and permit more efficient densification by subsequent infiltrations. The 2-D composite presented here has been subject to three CVD infiltration stages.

400 hours CVD at 1000°C followed by graphitisation at 2800°C
100 hours CVD at 1000°C followed by graphitisation at 2800°C
100 hours CVD at 1000°C followed by graphitisation at 2800°C

A 1-D composite B-1 was prepared having a similar density. However this density was achieved after the first CVD treatment so further impregnations were omitted.

Since these materials are directionally inhomogeneous and composed of components, matrix and fibre reinforcement, whose thermal properties are expected to be different, a cautious approach is needed to ensure that a sample of sufficient thickness to ensure a homogeneous thermal response is used. A theoretical study to determine the thermal properties of reinforced composites using the flash method has been carried out by [13] and an experimental investigation reported [14]. The broad conclusion from these investigations is that the thermal diffusivity tends to the diffusivity deduced from steady state properties ($\alpha_e = \lambda_e/\rho_e c_{p_e}$) if the thickness of the specimen is large compared to the spatial period of the reinforcement. Typically the specimen will respond as a homogeneous medium if the thickness is some 4-5 times the unit cell dimensions. We have done a similar study on our materials and conclude that for these fine weave composites the thickness needs to be >3mm.

Disc-shaped specimens 10mm diameter were cut from each billet. In the case of the 1-D composites specimens were cut both parallel and perpendicular to the fibre axes. For the 3-D composite samples were cored from the Z and X directions. In the case of the 2-D woven cloth, composite samples were machined parallel and perpendicular to the cloth (X and Z directions). Orthogonal symmetry in X and Y directions was verified. For the matrix graphite appropriate to composite A samples were cored from the orthogonal directions to verify isotropy. Sample details are given in Table 2.

Table 2. Specimen Details for C.F.C.C. Composites

Material	Orientation	Sample length mm	Density kg m^{-3}x10^3
Composite A (3-D)	‖ X-axis	3.4	1.92-1.93
	‖ Z-axis	3.0-4.0	1.89-1.93
Composite A-1 (1-D)	‖ fibre-axis	4.5-4.6	1.90-1.91
	⊥ fibre-axis	2.0	1.90-1.91
Matrix graphite	-	3.0	1.37-1.39
Composite B (2-D)	‖ X-axis	4.0	1.78
	‖ Z-axis	4.0	1.80
Compsite B-1 (1-D)	‖ fibre axis	3.0	1.81
	⊥ fibre-axis	4.4	1.86

The densities of the specimens were again determined by the displacement method. Detailed metallographic examination was carried out on all materials using either optical microscopy or scanning electron microscopy (S.E.M.).

For all measurements the sample length was corrected for thermal expansion using a mean expansion coefficient of $9 \times 10^{-6}K^{-1}$. Data obtained were converted to thermal conductivity using the relationship $\lambda_e = \alpha_e \, Cp_e \, \rho_e$, using the above expansion coefficient to correct for density changes and a specific heat curve sythesised from data from 350-1000K on Poco graphite [15] and directly measured specific heat data for composite A [16].

$$Cp \; (J \; gm^{-1}K^{-1}) = -0.6853 + 5.9199 \times 10^{-3} \, T - 5.5271 \times 10^{-6}T^2$$

$$+ \; 2.6677 \times 10^{-9}T^3 - 6.4429 \times 10^{-13}T^4 + 6.1622 \times 10^{-17}T^5.$$

The thermal diffusivity for the two directions of composite A are presented in figure 5. The thermal diffusivity in the Z axis direction is some 40% higher than the thermal diffusivity in the X axis direction which is broadly in accord with the greater fibre reinforcement in this direction (22% compared with 13% for the X-Y

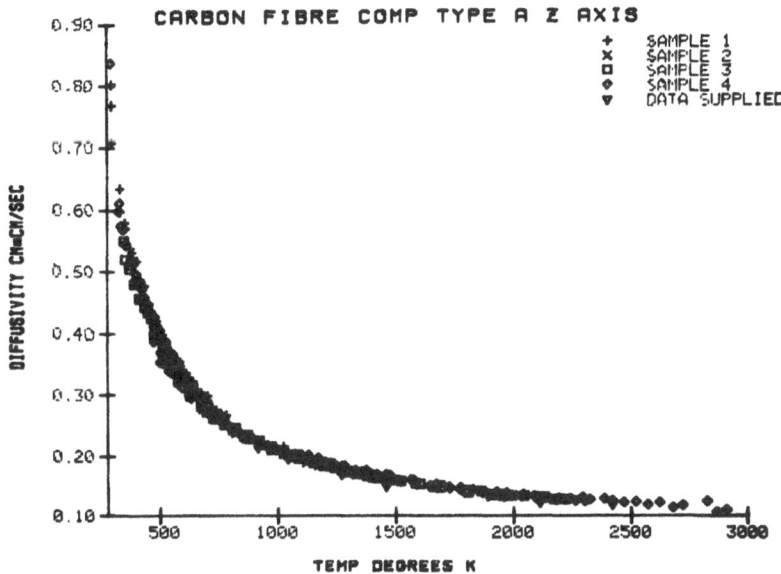

Fig. 5 a) Thermal diffusivity of Z-axis direction of composite A.

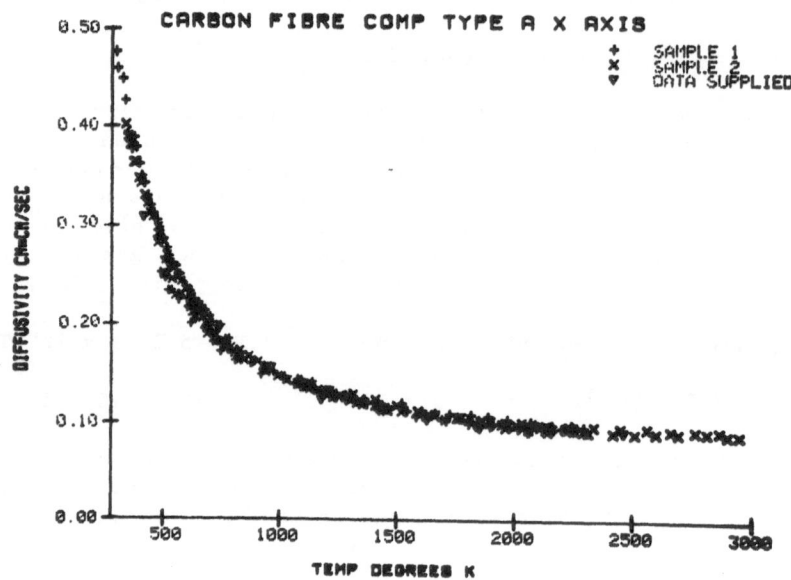

Fig. 5 b) Thermal diffusivity of X-axis direction of composite A.

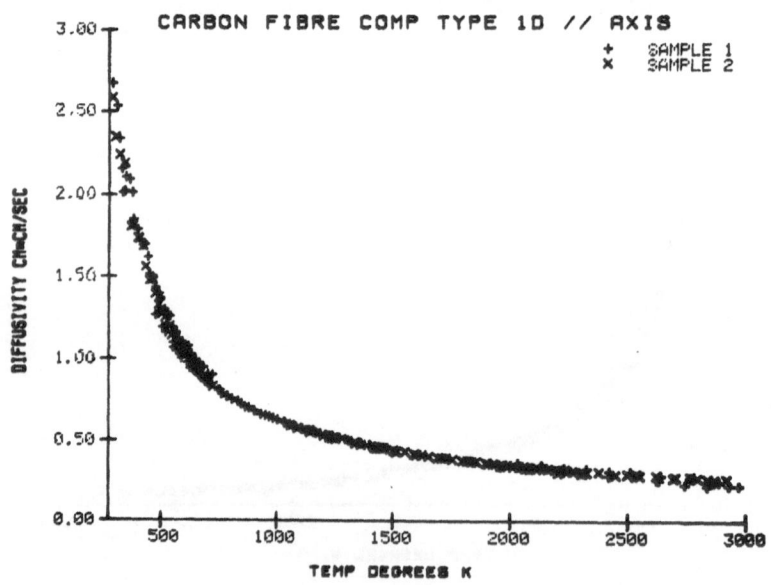

Fig. 6 a) Thermal diffusivity || to fibres of composite Al

458

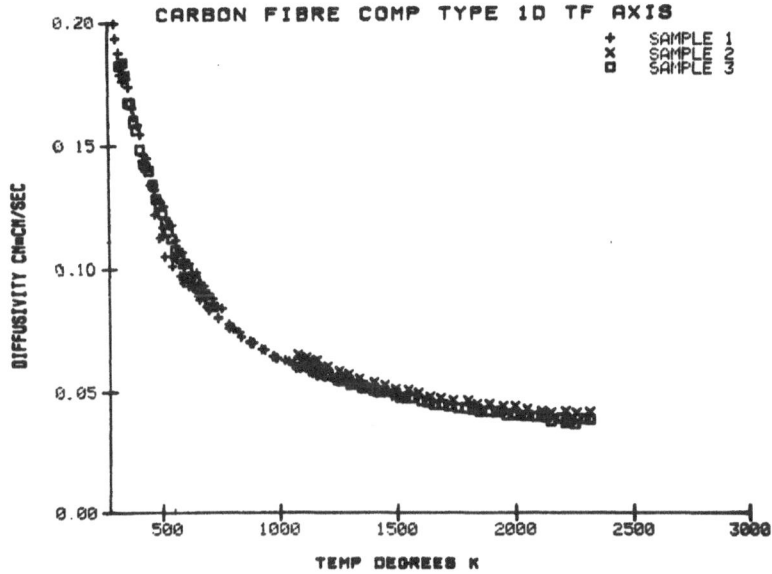

Fig. 6 b) Thermal diffusivity ⊥ to fibres of composite A1.

direction). In figure 6 data are presented for the thermal diffusivity of the two directions of the 1-D composite. The anisotropy ratio is approximately 10 over the whole temperature range. The thermal diffusivity of the matrix graphite decreased

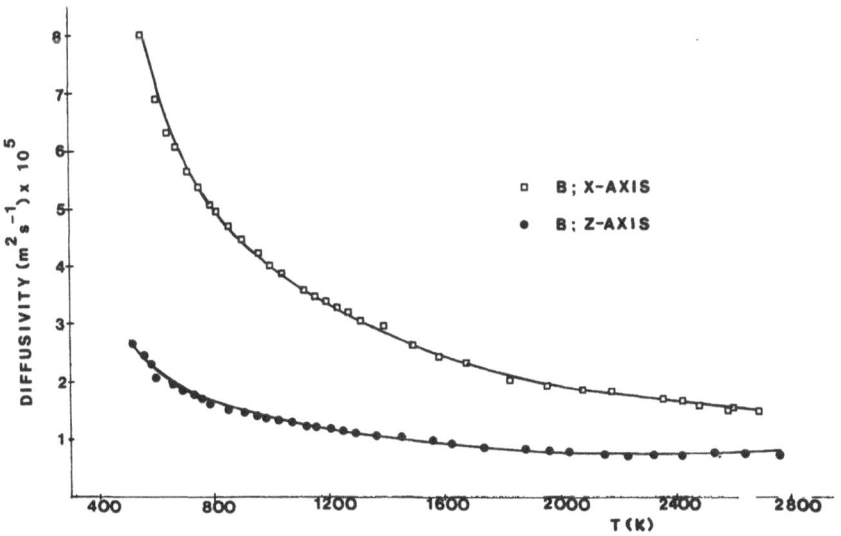

Fig. 7. Thermal diffusivity for 2 directions of composite B.

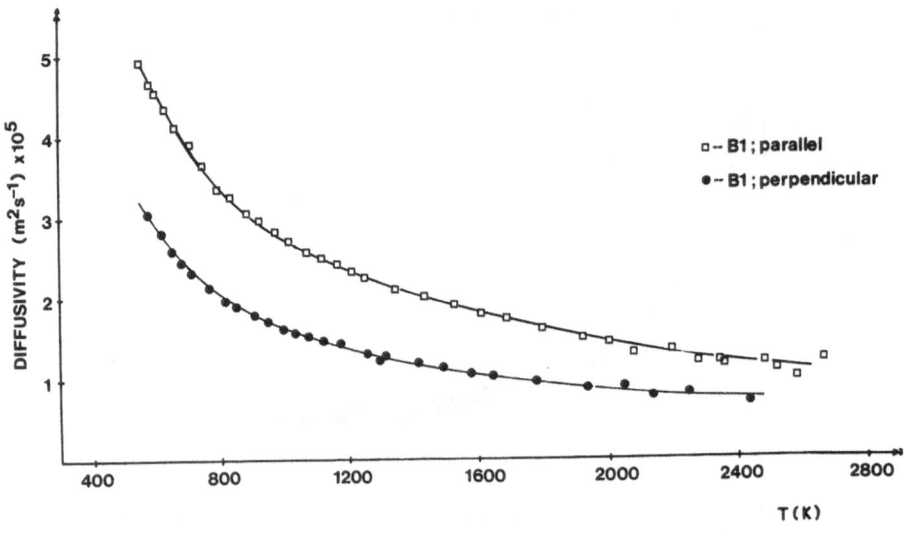

Fig. 8. Thermal diffusivity for 2 directions of composite B1.

from $9 \times 10^{-5} m^2 s^{-1}$ at 300K to $1 \times 10^{-5} m^2 s^{-1}$ at 2750K. These are high values for a material for which the density at $1.36 \times 10^{-3} kg$ m^{-3} is only 60% of the theoretical density.

In Figure 7 we present data on the thermal diffusivity of the two directions of composite B. The anisotropy ratio between parallel and perpendicular directions is 3 at 500K but decreases as the temperature is increased. Interesting data are presented in figure 8 for the thermal diffusivities of the two directions of the composite B-1. Although the density is similar to that of composite B the thermal diffusivity of this 1-D composite cut parallel to the fibres is *lower* than that of the 2-D composite and even more significantly the thermal diffusivity of the transversely oriented sample is as high as 2/3 that of the specimen oriented parallel to the fibre axis.

Smoothed data for all these composites have been converted to thermal conductivity using density data corrected for thermal expansion and the specfic heat equation quoted earlier. These are plotted in figures 9 (2 and 3-D composites) and 10 (1-D composites).

There have been a number of determinations of the thermal conductivity of some of the various types of carbon fibre. Kalnin [17] has reported room-temperature values for a variety of commercial fibres. These range from 142 W $m^{-1}K^{-1}$ (for high modulus fibre to 3 W $m^{-1}K^{-1}$ for a low modulus fibre. Thornel 50 has a thermal conductivity reported to be 60 W $m^{-1}K^{-1}$. Lee and R.E. Taylor [18] have reported the thermal diffusivity of

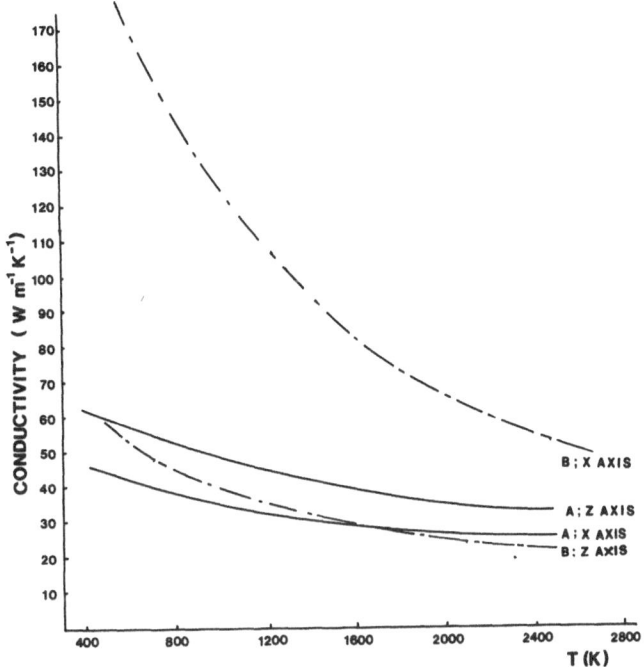

Fig. 9. Derived thermal conductivities of orthogonally reinforced composites A and B.

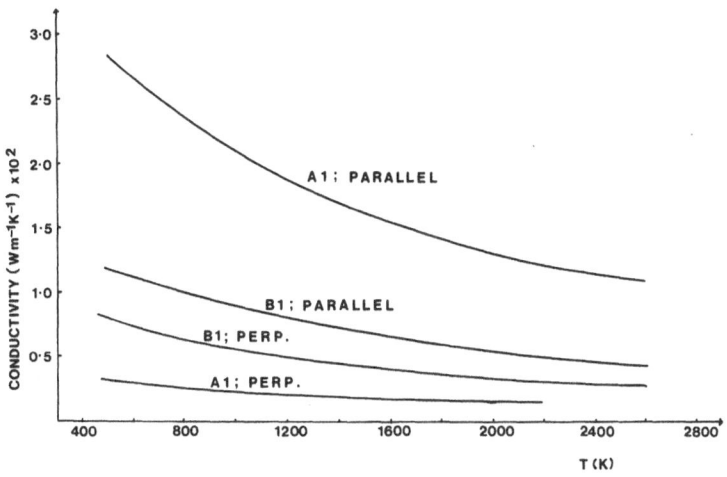

Fig. 10. Derived thermal conductivities of 1-D composites.

461

Morganite II,Thornel 50 and PX505 carbon fibres to be 118, 58.3 and 1.49 W $m^{-1}K^{-1}$ respectively. The values obtained by Kalnin and by Lee and Taylor are thus in excellent agreement for Thornel 50. Pilling et al [19] from measurements on carbon fibre epoxy composites have determined the thermal conductivities at 270K of a high modulus fibre after graphitisation at 2600°C to be 80 W $m^{-1}K^{-1}$ for the parallel orientation and 6 W $m^{-1}K^{-1}$ for the transverse orientation.

However such values refer to the conductivity of a carbon fibre in the virgin condition. When fibres are embedded in a carbon fibre composite certain things will happen:
a) during the graphitisation stage fibres will be subject to high temperatures which will increase the structural perfection of the fibre and increase its thermal conductivity;
b) the fibres will form nucleation sites and influence the growth of matrix graphite during the densification process.
Hence the use of reported thermal conductivity data on fibres is unrealistic since the thermal conductivity of the fibre itself will be increased. Furthermore the processing influences the properties of the constituent phases and the properties of

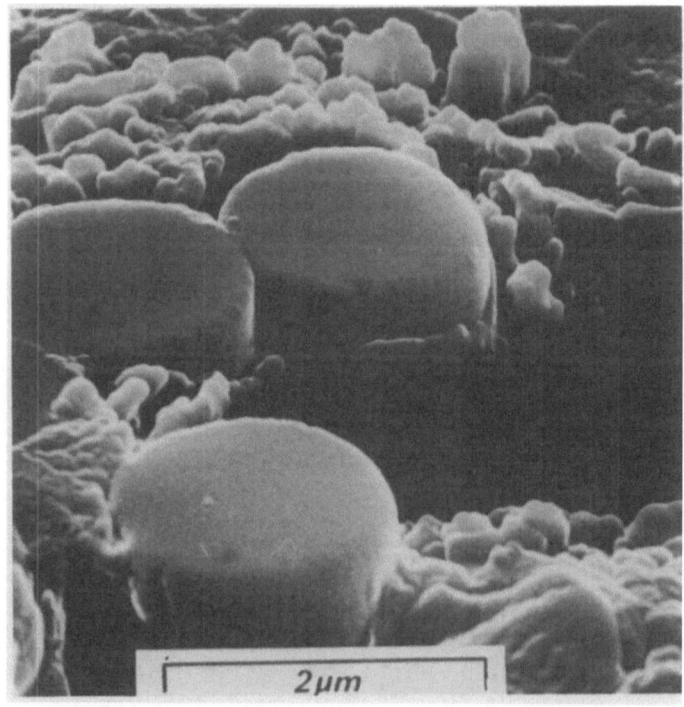

Fig. 11. S.E.M. micrograph of composite A.

C.F.C.C. materials derive from a complex interaction of the individual phases. To illustrate this effect we have carried out a detailed microscopical investigation of our materials. The following conclusions may be deduced:

1) The deposition of C.V.D. graphite produces a sheath-like coating around individual fibres. If the C.V.D. process is only used to rigidise the C.F.C.C. the interfibre space will be completely filled with C.V.D. graphite for separations less than $1^{1}/_{2}$ fibre diameters. This sheath like coating is illustrated in figure 11 for a 1-D carbon fibre composite.

2) C.V.D. graphite will tend to infill pores and heal cracks in individual fibres, thereby increasing fibre thermal conductivity. Increases in fibre density of up to 20% have been reported [20].

3) Prolonged densification by C.V.D. results in the formation of matrix graphite which gives an "onion peel" appearance to individual fibre filaments and eventually results in a matrix of graphite in which individual filaments are embedded. This is illustrated for composite B in figures 12. Prolonged densification by C.V.D. alone will result in the filling in of interfibre spacing by C.V.D. graphite which appears to have a complex orientation dependence related to the fibre orientation when viewed under polarised light [21].

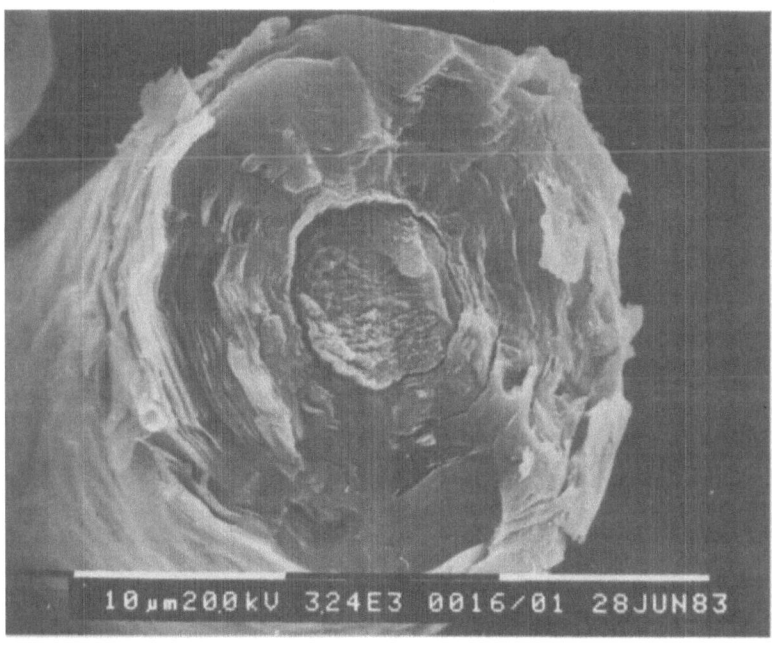

Fig. 12. S.E.M. micrograph of fibre & sheath coating composite B.

4) Where densification of rigidised fibre bundles is by pitch impregnation the matrix graphite crystals appear to be transversely oriented with respect to the fibre axis. This has also been observed by Gebhardt [22].

5) Matrix crystallites in the vicinity of fibres that have not been subject to the C.V.D. process have been reported to be parallel oriented [23]. Our own observations on composite A-1 have been unable to confirm this.

6) For multi dimensional composites cracks are frequently visible at interfaces between the various axial reinforcement directions (figure 13). These will also influence heat transfer within such composites and complicate any extrapolation from data for 1-D composites.

The examples given in this paper show that it is difficult to interpret the thermal transport properties of directionally reinforced carbon/carbon fibre composites. We have attempted to analyse the thermal conductivity behaviour of composites A and A-1 [24]. The thermal conductivity of a unidirectionally reinforced composite in a direction parallel to the fibres λ_c'' is usually expressed in terms of the volume-weighted conductivities of the constituents using an Ohm's law approach:

Fig. 13. Interfacial cracks in composite A.

$$\lambda_c'' = V_f \lambda_f'' + (1-V_f) \lambda_m \qquad\qquad (1)$$

where λ_f'' is the conductivity of the fibres

λ_m is the matrix conductivity

V_f is the fibre volume fraction (= 0.52 for composite A-1).

However the value obtained for the matrix graphite (85 W $m^{-1}K^{-1}$ at 300K) is inappropriate to use in the above equation in view of the density difference, 1.36 x 10^3 kg m^{-3} compared with 1.91 x 10^3 kg m^{-3} for the composite A-1. However applying the porosity/tortuosity factor postulated by Taylor et al [25] for polycrystalline graphite appropriate to the increase in density of matrix graphite in the 3-D composite would predict a room temperature conductivity for this matrix graphite of 180 W $m^{-1}K^{-1}$. This value is comparable with many values for well-graphitised material obtained by a number of authors. Inserting this value in equation 1 would yield a value of 410 W $m^{-1}K^{-1}$ for the conductivity of fibres parallel to the fibre axis at 300K. A similar value of 400 W $m^{-1}K^{-1}$ for Thornel 50 fibres has been deduced by R.E. Taylor [26] who measured the conductivity of numbers of 1-D composites with different fibre volume fractions and extrapolated the data fo 0% matrix. Luc-Bouhali et al [27] have estimated the conductivity of graphitised fibres to be 360-430 W $m^{-1}K^{-1}$. These values are consistently higher than for the virgin fibre and clearly reflect graphitisation during processing of the composite. Volga et al [28] have shown that the room temperature, thermal conductivity of fibres increases during graphitisation by as much as a factor of 15 (20 to 300 W $m^{-1}K^{-1}$) following heat treatment for 1 hour at 2800°C.

The data obtained for composite A-1 has been used to predict the conductivity of the 3-D composite A. A simple model was constructed (figure 14) whereby the composite was divided into four parallel conduction channels. Assuming a constant temperature difference along each channel the total heat flow Q is divided into 4 parts

$$Q = Q_1 + Q_2 + Q_3 + Q_4$$

and the thermal conductivity for each channel calculated using a weighted Ohm's law approach. The total conductivity of the unit cell is obtained from the conductivities of the individual channels weighted according to area. This will of course give an upper bound of conductivity. In a real composite, structural defects such as broken interfaces of the type shown in figure 12 must also be taken into account. When the 1-D data was applied to composite A the following modification was needed to be made in order to achieve an acceptable fit:

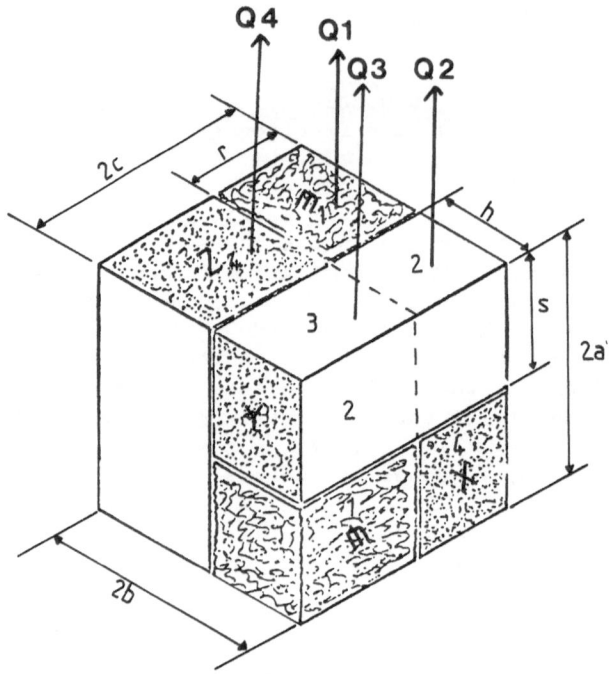

Fig. 14.   Model for heat flow in composite A (after Proctor
and Taylor)

a) A reduction factor of 0.66 was applied to the axial (|| fibre)
    data for composite A-1.
b) Transverse conductivity for composite A-1 was increased by a
    factor of 2.0.
c) Measured matrix conductivity was used unmodified since its
    porosity was estimated to be similar to that of the cross-over
    pockets in the 3-D composite.
d) A lower-bound solution corresponding to one broken yarn
    interface/unit cell gave the best fit with r.m.s. errors of
    11% and 9% for the X and Z direction over the temperature
    range 300-2500K.

    These assumptions seem justifiable in view of the known
differences in processing between composites A and A-1 especially
when a parallel is drawn between these and composites B and B-1.
It is clear that prolonged exposure to the C.V.D. process has a
dramatic effect on the conductivity of the composite.   It is
particularly noteworthy that the anisotropy ratio for the two

directions of composite B-1 is less than 2 (compared to ~10 for composite A-1) and for the 2-D composite the ratio of conductivities for the two directions with and without fibre reinforcement is less than 3. This strongly suggests that there are strong orientation affects in the graphite deposited by C.V.D. In view of the marked anisotropy in the thermal conductivity of graphite [29, 30], which can be as high as 200 this will have a pronounced effect on the conductivity components. However our investigations of this new material are still continuing.

CONCLUSIONS

Basically two investigations are covered in this paper. The common denominator is that data can only be explained from an understanding of the materials under investigation. This frequently necessitates comprehensive microstructural investigation. In the case of thermal barrier coatings, the observed changes can be explained in terms of the equilibrium diagrams following S.E.M. and X-ray diffraction characterisation. Carbon fibre/carbon composites present a more difficult problem to analyse. It would be presumptuous to imply that the work undertaken at UMIST has resolved the problem of analysing the thermal properties of directionally reinforced composites since we hve highlighted more problems than we have solved. However measurements have been reported up to 3000K and we have shown that a detailed knowledge of the microstructure and its relation to processing variables is an essential prerequisite to any understanding of the thermal transport properties.

REFERENCES

1. Lee H.J. and Taylor R.E. "Thermal diffusivity of dispersed composites" J. Appl. Phys. 4 148-54 (1976).
2. Parker W.J. Jenkins R.J. Butler C.P. and Abbott G.L. "Flash method for determining thermal diffusivity, heat capacity and thermal conductivity" J. Appl. Phys. 32 1679-34 (1961).
3. Taylor R.E. "Critical evaluation of flash method for measuring thermal diffusivity" Rev. Int. Htes. Temp. et. Refract. 12 141-5 (1975).
4. Larson K.B. Koyama K. "Measurement by the flash method of thermal diffusivity, heat capacity and thermal conductivity in two layer composite samples" J. Appl. Phys. 39, 4408-16 (1963).
5. Bulmer R.F., Taylor R. "Measurement by the flash method of thermal diffusivity of two layer composite samples" High Temp-High Press 8 469-78 (1974).

6. Lee T.Y.R., Donaldson A.B. and Taylor R.E. "Thermal diffusivity of layered composites" Thermal conductivity 15 ed. V.V. Mirkovich (New York, London-Plenum) $\underline{135}$-48 (1978)

7. Taylor R., "Construction of apparatus for heat pulse thermal diffusivity measurements from 300-300K". J. Phys. E. Sci Inst. $\underline{13}$, 1193-99 (1980).

8. Fitzgerald L. "Thermal properties of aerospace ceramic coating". M.Sc. Thesis UMIST (1979).

9. Grot A.S. and Martyn J.K. "Behaviour of plasma sprayed ceramic thermal barrier coatings for gas turbine applications" J. Am. Ceram. Soc. Bull., $\underline{60}$ 807-11 (1981).

10. Grain C.F. "Phase relations in the ZrO_2-MgO system" J. Am. Ceram. Soc. $\underline{50}$ 288-   (1967).

11. Stubican V.S. Hink R.C. and Ray S.P. "Phase equilibria and ordering in the system ZrO_2-Y_2O_3". J. Am. Ceram. Soc. $\underline{61}$ 17-   (1978).

12. Morrell P. "Thermal properties of ceramic thermal barrier coatings for Aero-engine applications" M.Sc. Thesis UMIST (1982).

13. Luc A.M. and Balageas D.L. "Compartement thermique des composites a renforcement oriente soumis a des flux impulsionnels". ONERA T.P. 1982-105 (1982) also presented at 8th E.T.P.C., Baden-Baden.

14. Taylor R.E. Groot H. and Shoemaker R.L. "Thermophysical properties of fine weave carbon-carbon composites"AIAA-81-1103 (1981).

15. Taylor R.E. and Groot H. "Thermophysical properties of POCO graphite" AFOSR-77-3280 (1978).

16. Cezairliyan A. and Miller, A.P. "Specific heat capacity and electrical resistivity of a carbon-carbon composite in the range 1500-3000K by a pulse heating method" Int. J. Thermophysics $\underline{1}$ 317-30 (1979).

17. Kalnin I.L. "Thermal conductivity of high modulus fibres" Composite Reliability ASTM STP 580 p460-73 (1975)

18. Lee H.J. and Taylor R.E. "Thermophysical properties of carbon/graphite fibres and MOD-3 fibre reinforced graphite" Carbon $\underline{13}$ 521-527 (1975).

19. Pilling M.W., Yates B., Black M.A., Tattersall P. "Thermal conductivity of carbon fibre-reinforced composites" J. Mat. Sci. $\underline{14}$ 1326-1338 (1979).

20. Jortner J. "Cracking in 3-D carbon-carbon composites during processing and affects on performance" McDonnell-Douglas Astronautics Co-West MDAC Paper WD 2694 (1978).

21. Whittaker A.J. "A study of the thermal transport properties of a series of carbon fibre carbon composites" M.Sc. thesis UMIST (1982).

22. Gebhardt J.J. "Surface effects in pyrolytic infiltration of carbon fibre preforms" Proc. 14th Int. Conf. on Carbon (State College Penn.) (1979).

23. Theibert L.S. Private Communication (1978).
24. Procter R.N. and Taylor R. "Measurement of thermal diffusivity of carbon fibre carbon composites from 300-3000" A.F. Report AFOSR-77-3449 (1980).
25. Taylor R. Gilchrist K.E. and Poston L.J. "Thermal conductivity of polycrystalline graphite" Carbon 6 537-44 (1968).
26. Taylor R.E. "Thermal diffusivity of composites" presented at 8th E.T.P.C. Baden-Baden (1982) also available as T.P.R.L. Report 253.
27. Luc-bouhali A.M. Pujola R.M. and Balageas D.L. "Measure in-situ de la diffusivite thermique des renforcements de composites carbone-carbone" presented at 18th AIAA Thermophysics conference Montreal May (1983).
28. Volga V.I., Foder V.I. and Usov V.K. Inorganic Materials 9 543- (1973).
29. Taylor R. "Thermal Conductivity of pyrolytic graphite" Phil. Mag. 13 157-66 (1966).
30. Kelly B.T. "Thermal conductivity of graphite" in Chemistry and Physics of Carbon (P.L. Walker Jr. ed.) M. Dekker N. York 5 119- (1969)

SESSION Q

Ceramics

SESSION CHAIRMAN

R. Taylor
UMIST
Manchester, England

THERMAL DIFFUSIVITY OF CEMENTED CARBIDES

W. Neumann

Austrian Research Centre Seibersdorf
Department of Materials Technology
Lenaugasse 10, A-1082 Vienna

ABSTRACT

The thermal diffusivity of cemented carbides was measured by the laser flash method up to 1470 K. All materials tested were based on WC-Co hardmetals. The influence of cobalt content, temperature and grain size was studied. In addition the effect of the substitution of WC by TiC, TaC and NbC and the substitution of cobalt by nickel was investigated.

INTRODUCTION

The lifetime of cutting tools strongly depends on the thermophysical properties of their constituent materials. The thermal diffusivity, which determines the temperature distribution within the tool is especially important. Because thermal diffusivity data obtained at ambient temperature does not describe the behaviour at working temperatures and an extrapolation to higher temperatures is insufficient it is necessary to measure the thermal transport properties up to high temperatures.

Therefore, in the present study different compositions of carbides used for cutting tools were investigated. To obtain a good overview of the different types, especially the influence of cobalt content, samples of various compositions, commercially available as well as special prepared ones, were tested. The thermal diffusivity was measured using the laser flash method.

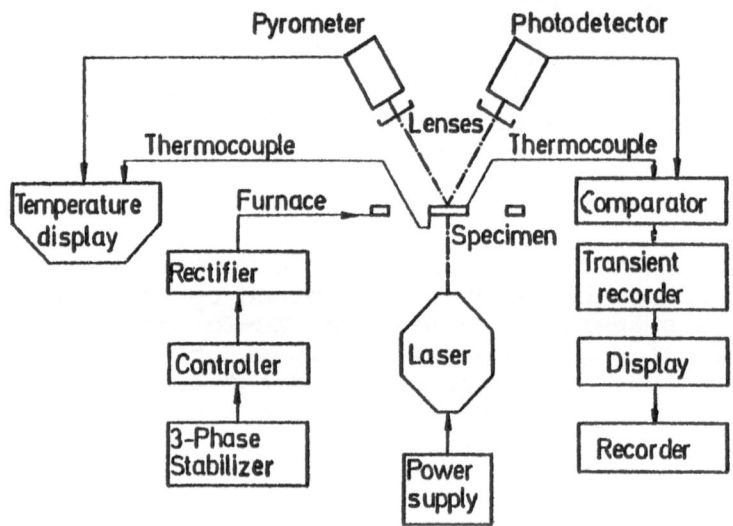

Fig. 1   Thermal diffusivity measurement system

EXPERIMENTS

The apparatus, based on the well-known laser flash method /‾1‾7 was designed and built in the workshop and laboratories of the Austrian Research Centre Seibersdorf /‾2‾7.

The flash energy source was a high energy (40 J max.) neodymium-glass laser, which has a pulse duration of 0,5 ms. The diameter of the laser beam was 10 mm, therefore the whole front surface of the cylindrical specimen (diameter 10 mm, height about 1mm) was illuminated by the flash.

A schematic diagram of the equipment is shown in the following figure.

To avoid heat loss of the specimen during heating the specimen was supported in a horizontal position on three thin insulating ceramic tubes, the specimen chamber was evacuated and radiation shields were installed. A short tantalum tube acted as a resistance heater. The temperature versus time history of the rear surface was detected by a Ge photodiode. The temperature signal was amplified and registered by a transient recorder. The temperature-time diagram could be shown graphically on a screen. Simultaneously the thermal diffusivity was calculated by a personal computer. The test temperature was measured by a thermocouple, which was in contact with the rear surface of the specimen. All experiments were carried out under vacuum ($\sim 1.10^{-5}$ bar).

The laser, specimen chamber and detector were mounted on a large, vertical steel tube.

All specimens were prepared by grinding them to the size required. To improve the energy adsorption on the front surface, the specimens were sand-blasted.

All specimens contained high concentrations of tungsten carbide and could be subdivided into 3 groups according to composition.
WC-Co composites
a) of different Co-content

b) with TiC, TaC, NbC (specimens 1-18) substituted for WC (specimens 19-23)

c) with Ni as substitute for Co.

The exact compositions were listed in Table 1. At least two specimens of each type were tested in the apparatus to check the reproducibility.

Table 1. Thermal Diffusivity and Composition of Cemented Carbides

Specimen No.	WC	TiC	TaC	NbC	Co	Ni	others	WC grain size	Therm. diff. [cm^{2}s^{-1}] 600 °C	1100 °C
1	100								0,177	0,144
2	97				3			M	0,160	0,132
3	95				5				0,168	0,137
4	94				6			F	0,142	0,126
5	94				6			C	0,174	0,137
6	92				8			M	0,175	0,138
7	92				8			M	0,170	0,136
8	90,5				9,5			C	0,170	0,135
9	90,5				9,5			C	0,171	0,134
10	90				10				0,151	0,128
11	89				11			M	0,158	0,129
12	89				11			C	0,170	0,131
13	85				15			M	0,140	0,117
14	85				15			C	0,152	0,125
15	85				15				0,142	0,115
16	80				20			C	0,147	0,118
17	80				20				0,140	0,111
18	74				26			C	0,134	0,107
19	18	60			2	8	12		0,042	0,053
20	55,5	19	12,2	3,8	9,5				0,061	0,071
21	68,6	11,9	6,0	4,0	9,5				0,081	0,082
22	77	4	6,1	1,9	11				0,111	0,101
23	74,8	12	3,0	2,0	8,2				0,090	0,086
24	91					8	1		0,135	0,120
25	95				4,6	0,4			0,135	0,122

F ... fine

M ... medium

C ... coarse

RESULTS AND DISCUSSION

The thermal diffusivity of all specimens was measured at 50°C intervals from 500 to 1200°C. The results obtained at 600°C and 1100°C are listed in Table 1. The diffusivity-temperature diagrams of selected specimens are shown in Fig. 2, 4 and 5.

- Temperature dependence

The thermal diffusivity of all WC-Co materials tested decreases with increasing temperature (Fig. 2). A significant peak is shown at about the Curie temperatures for all samples containing cobalt. Only at very low cobalt content and for very fine-grained specimens was this peak not readily detectable. The precise influence of the cobalt content on the shape of these curves could not be determined since measurements were taken only at 50°C intervals.

If Ni is used instead of Co the diffusivity still decreases with increasing temperature but to a smaller extent (Fig. 4). This can be explained by the lower temperature dependence of the Ni component. A peak at the Curie point could not be detected, because the Curie temperature is out of the investigated temperature range.

An increase of the diffusivity was only detected for materials with high TiC content.

- Influence of Co content

As known from published investigations $\underline{/}^-3\underline{\,/}$ the thermal conductivity decreases with increasing cobalt content in the composition range studied in this work (0 - 26 wt.%) (Fig. 3). This unexpected result -- the conductivity of pure Co being about as high as the conductivity of WC -- was adequately explained by the formation of a Co-W-C solid solution of low conductivity.

In addition the contiguity, i.e. the ratio of grain-boundary area to total interface of carbide grains, decreases rapidly with increasing content of the Co-binder phase, as shown in the literature $\underline{/}^-4\underline{\,/}$, and the lower the contiguity the lower the diffusivity because the degree of skeleton formation is reduced.

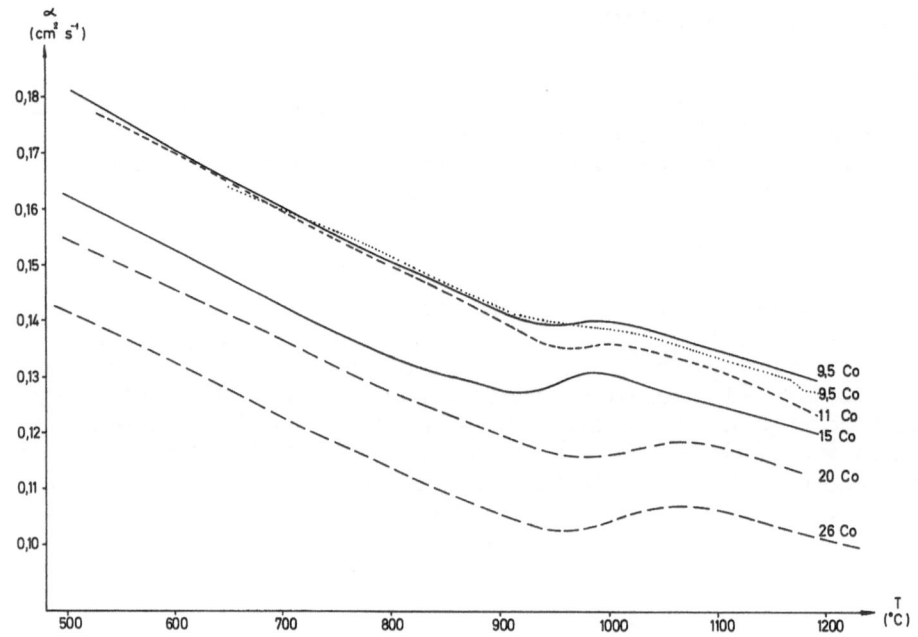

Fig. 2   Thermal diffusivity of WC-based cemented carbides.

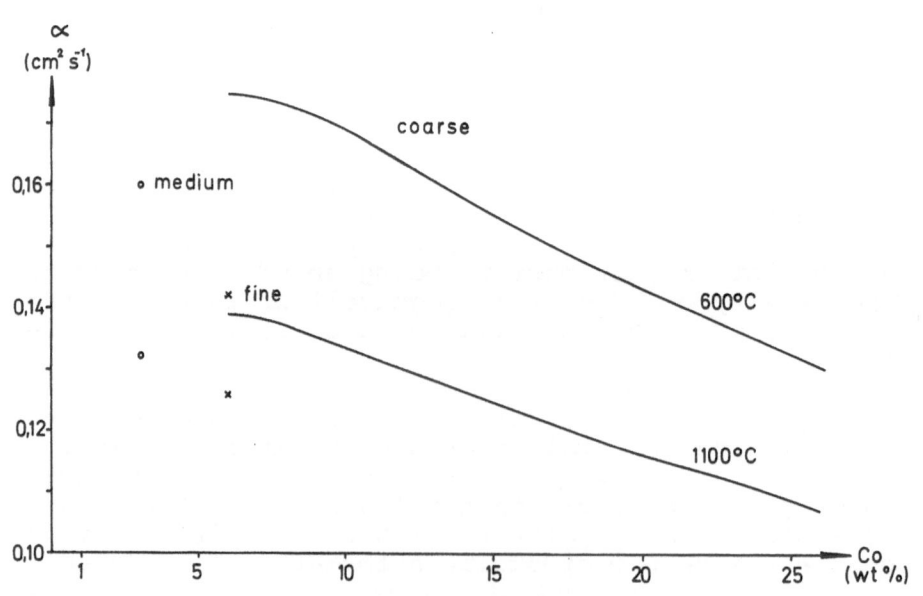

Fig. 3   Thermal diffusivity of WC-based cemented carbides
with different Co content.

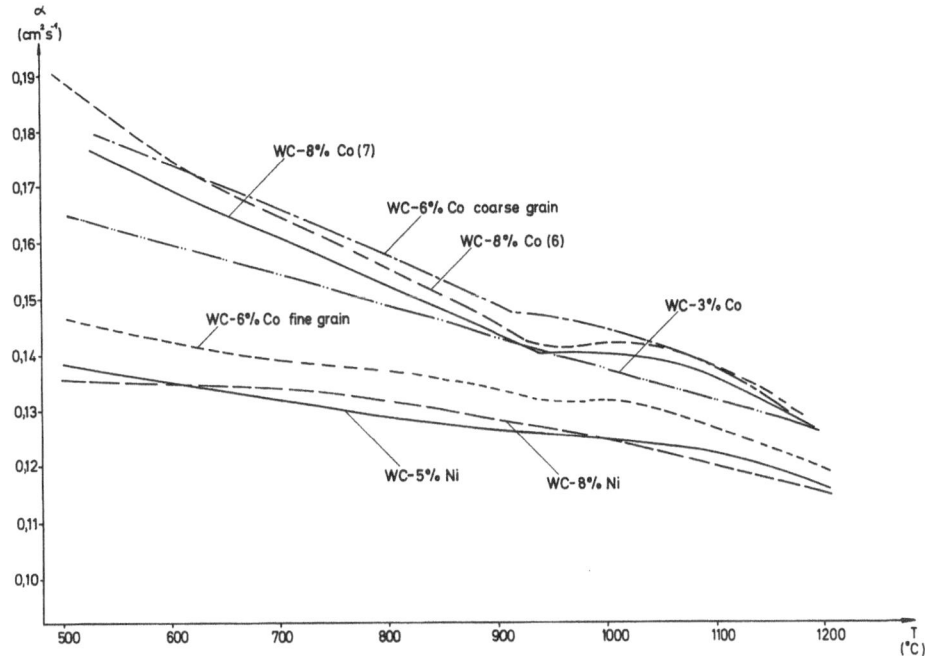

Fig. 4   Thermal diffusivity of WC-Co and WC-Ni.

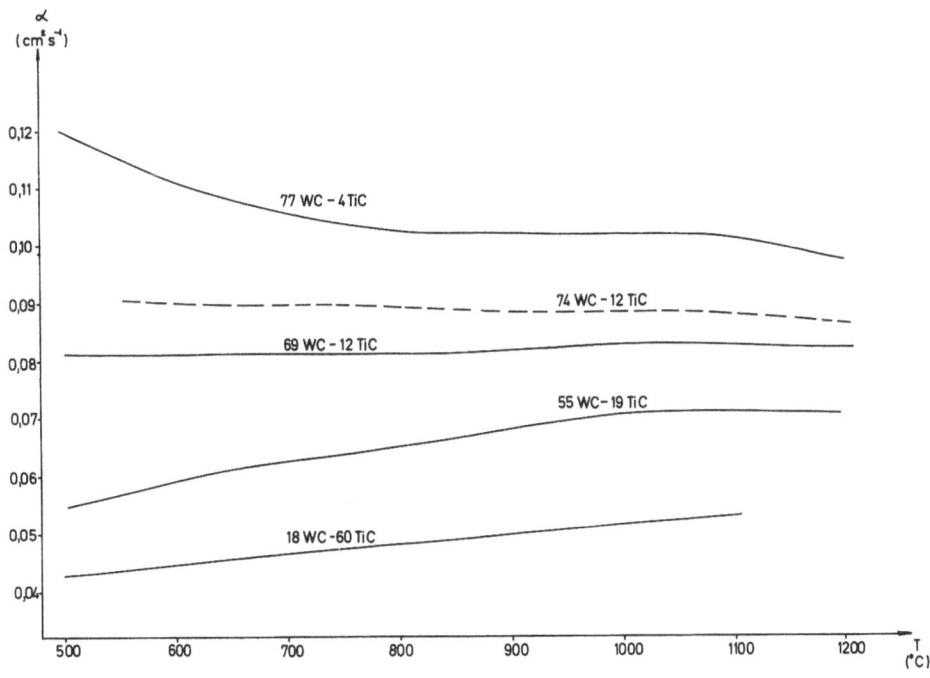

Fig. 5   Thermal diffusivity of cemented carbides
(WC-TiC-TaC-NbC)-Co.

The slope of the thermal diffusivity versus cobalt content does not depend on the temperatures, as is seen in the diagram.

- Influence of grain size

A remarkable decrease of diffusivity with decreasing grain size was observed. In specimens with fine grains -- at about 1/3 of the coarse grain size -- the diffusivity is reduced to 80% of that for the coarse-grained specimen of the same composition. This may be explained by the multiple interface between the phase containing cobalt and the WC grains with their additional thermal resistance.

- Influence of TiC and other carbides (TaC, NbC)

The low value of the diffusivity of TiC determines the data obtained on the specimen with high TiC content (Fig. 5). According to this result the lowest diffusivity value at 600°C (0,042 cm^2s^{-1}) was measured for the specimen with the highest TiC content. It is less than 1/4 of the highest values obtained in this test series at this temperature. Above a certain content of WC the diffusivity decreases again with increasing temperature.

A general interpretation of the effects of these multicomponent content carbides is not possible without a high number of tests and a very in-depth examination of their structure.

CONCLUSION

The laser flash method is a quick and reliable method to measure thermal diffusivity of different kinds of cemented carbides.

Different composition of WC-Co based materials influenced the diffusivity as follows

- the higher the cobalt content the lower the diffusivity

- the finer the grain size the lower the diffusivity

- nickel instead of cobalt reduces the diffusivity

- addition of TaC strongly reduces the diffusivity.

The results obtained support combined with other properties the selection of a suitable cemented carbide for special applications as well as a further development of these materials.

ACKNOWLEDGEMENT

Many thanks are given to Prof. B. Lux of the Technical University of Vienna for his cooperation, to Dr. Schubert, also of Technical University of Vienna, and to Sandvik Hard Metal R&D organisation for providing the materials.

REFERENCES

/⁻1_7 W.J. Parker et al, "Flash Method of Determining Thermal Diffusivity, Heat Capacity and Thermal Conductivity", J. Appl. Phys. 32, 1679 (1960)

/⁻2_7 W. Neumann, K. Wallisch, "Thermal Conductivity Determination of Graphite and High Temperature Alloys by the Laser-Flash-Method", Measurement (1984) in press

/⁻3_7 O. Rüdiger, A. Winkelmann, "Thermal Conductivity of WC-Co hardmetals", Techn. Mitt. Krupp 18, 19-24, (1960) (in German)

/⁻4_7 H.E. Exner, "Physical and Chemical Nature of Cemented Carbides", Int. Metals Review 4, 149 (1979)

THE HIGH-TEMPERATURE EX-REACTOR THERMAL

CONDUCTIVITY OF THORIA AND THORIA-URANIA

J. Belle and R.M. Berman

Bettis Atomic Power Laboratory

West Mifflin, Pennsylvania

ABSTRACT

Thermal conductivity of a material can be determined by the product of thermal diffusivity, heat capacity, and density. Each of these properties is reviewed for ThO_2 and for ThO_2-UO_2 solid solutions, and compared with similar data for UO_2. Between room temperature and 2950 K for ThO_2 and between room temperature and the temperature range 2850 - 2950 K for the solid solutions $(Th_{0.92}U_{0.08})O_2$ to $(Th_{0.70}U_{0.30})O_2$, thermal conductivity of fully dense $(Th_{1-M}U_M)O_2$, for $M \leq 0.3$, can be expressed by the following fitted equations:

$$1/\lambda_0 = A + BT;$$
$$A = 1/(46.947 - 112.072\ M);$$
$$B = 0.0001(1.597 + 6.736\ M - 21.56\ M^2)$$

where λ_0 is the thermal conductivity in W/m·K and M is the mole fraction of UO_2. Corrections for porosity (P) are made using the Maxwell-Eucken equation, $\lambda/\lambda_0 = (1-P)/(1+\beta P)$, with a constant value of 1.15 for the coefficient β. Since there is no significant electronic contribution to the heat capacity of ThO_2, unlike the case for UO_2, the thermal conductivity of ThO_2 continues to decrease with increasing temperature until a solid-solid phase transition at 2950 K is reached, at which point heat capacity and thermal conductivity increase about 50%. Thermal conductivity decreases again above the transition temperature. ThO_2-UO_2 solid solutions also show evidence for phase transitions from enthalpy data and thus show increases in calculated heat capacities and thermal conductivities. The pretransition thermal conductivities of the ThO_2-UO_2 solid solutions are intermediate between those of thoria and urania. Experimentally determined pretransition heat capacities for the solid solutions agree with values calculated from mole averages of ThO_2 and UO_2.

Further details of this work may be found in a report: WARD-TM-1530, Dec. 1982, by J. Belle and R.M. Berman.

ROLE OF POROSITY IN THE EFFECT OF MICROCRACKING ON THE THERMAL

CONDUCTIVITY OF BRITTLE CERAMIC COMPOSITES

L. D. Bentsen and D. P. H. Hasselman

Department of Materials Engineering
Virginia Polytechnic Institute and State University
Blacksburg, Virginia 24061

ABSTRACT

   Fracture-mechanical principles were used to predict the role
of porosity in the formation of microcracks in brittle ceramic
composites containing a dispersed phase with a coefficient of thermal expansion less than that of the matrix. It is shown that
microcracking will occur only if the amount of porosity lies within
a region $P_i < P < P_a$, and that all the microcracks will be of equal
size. Using this criterion, the effect of the microcracks and
pores on the thermal conductivity is examined. The expected behavior includes a sharp drop in the relative thermal conductivity
and diffusivity at P_i, followed by a porosity-independent value
between P_i and P_a. These conclusions were confirmed by measurements
of the thermal diffusivity of composites containing silicon carbide
dispersed in a matrix of magnesium oxide, beryllium oxide, and
aluminum oxide, each with a range of compositions and amounts of
porosity.

INTRODUCTION

   Many brittle materials such as geological formations and
structural ceramics for high temperature purposes can undergo
extensive microcracking. The principle causes for the formation
of microcracks include phase transformations, thermal expansion
anisotropy of the grains of polycrystalline aggregates, or mismatches in the thermal expansion of the individual components within a composite. These latter two effects result in the generation
of internal stresses upon changes in temperature of the polycrystal

485

or composite from the manufacturing temperature. The magnitude of these stresses is a function of the range of temperature change, the degree of thermal expansion anisotropy or mismatch, and the elastic properties of the grains or individual components. These internal stresses can be large enough to result in localized fracture and microcrack formation. Such microcracks can have a major effect on the continuum behavior of brittle materials, affecting the elastic moduli, strength, fracture energy, and toughness. Microcracking can also lead to significant improvements in machinability and thermal shock resistance.

Microcracks represent barriers to heat flow by phonon and electron transport and have a significant effect on the thermal conductivity and diffusivity at temperatures low enough that radiative heat transfer across the cracks does not make a major contribution. The continuum heat-transport properties of microcracked materials have received much theoretical and experimental attention. Microcracks in structural ceramics of technical interest were found to decrease the thermal conductivity and diffusivity by as much as a factor of three [1-4]. A number of theoretical expressions for the effect of cracks on thermal conductivity have appeared in the literature. For a matrix with randomly oriented penny-shaped cracks of radius b, the effective thermal conductivity, K_{eff} is [5]

$$K_{eff} = K_o (1+8Nb^3/9)^{-1} \qquad (1)$$

where K_o is thermal conductivity of the crack-free material and N is the number of cracks per unit volume.

Many ceramic materials also contain a pore phase. Porosity of 10 to 30% is desirable for improving the thermal insulating ability and thermal shock resistance of refractory ceramics used in furnace linings and related applications. A residual pore phase of only a few percent in structural ceramics for high performance applications, such as components of high temperature turbine engines, generally results from incomplete densification during sintering or hot-pressing of compacts made by powder-metallurgical techniques.

For temperatures at which radiation across the pores is negligible, a porous material represents a special case of a composite consisting of a continuous matrix with a dispersed phase of zero thermal conductivity. Theoretical solutions for the effective thermal conductivity of composites indicate that the pore phase should lead to a decrease in thermal conductivity, with the relative decrease determined by the volume fraction, shape, and orientation of the pores [6-7]. Such a decrease in thermal conductivity or diffusivity was confirmed by experimental data [7-8]. For the morphology and orientation of the residual pore phase found in sintered powder compacts, the relative decrease in thermal conductivity generally was found to be of the order of three times the pore-volume fraction. Because of the accompanying decrease in the specific heat per unit volume, the corresponding relative decrease in thermal diffusivity is expected to be about two times the fractional porosity. For spherical pores the relative decrease in thermal diffusivity is only one half the pore-volume fraction.

Microcracked brittle materials usually also contain a residual pore phase. For this reason, the thermal transport behavior of ceramics prone to microcracking should be affected by the combined presence of the microcracks and the pore phase. In general, at concentrations sufficiently low that the local temperature fields around second-phase inclusions or other thermal discontinuities are not affected by neighboring inclusions or discontinuities, the relative effects of microcracks and porosity on thermal conductivity are additive. At first sight, then, it appears that the combined effects of the cracks and the pore phase on the heat conduction behavior of microcracked materials could be ascertained from information on the separate effects for the cracks and pores, as determined either from theory or experiment.

For microcracked materials, however, this approach is complicated by the very role of the pore phase in the formation of microcracks. For any residual stress distribution, the magnitude of peak stress alone is not sufficient to describe the generation of microcracks. The appropriate criterion for microcrack formation is that the magnitude of the stress intensity factor, which is governed by the magnitude of the stress and size of the precursor crack, equals or exceeds the critical stress intensity factor (K_{IC}) appropriate to the preferred plane of crack propagation, which may be grain boundaries in polycrystalline materials or phase boundaries in composites. For materials prone to microcracking, pores represent the very source of microcrack precursors, especially the crack-like pores located at the triple points, grain boundaries, and interfaces.

The fracture mechanics of microcrack stability and propagation has received considerable theoretical attention [9-11]. Figure 1 illustrates schematically the stability of a circumferential microcrack (i.e., pore) around a spherical inclusion contained within a brittle matrix. The thermal expansion coefficient of the matrix is greater than that of the inclusion, resulting in a tensile tangential stress in the matrix near the inclusion when the composite is cooled from the fabrication temperature, at which the internal stresses are zero. Figure 1 indicates that the stress intensity factor (K_I) is a function of the ratio of the crack size to the inclusion size. Crack instability and propagation will occur only between crack sizes c_i and c_a. For $c < c_i$ no microcracking will occur since the crack size is so small that $K_I < K_{IC}$ for the

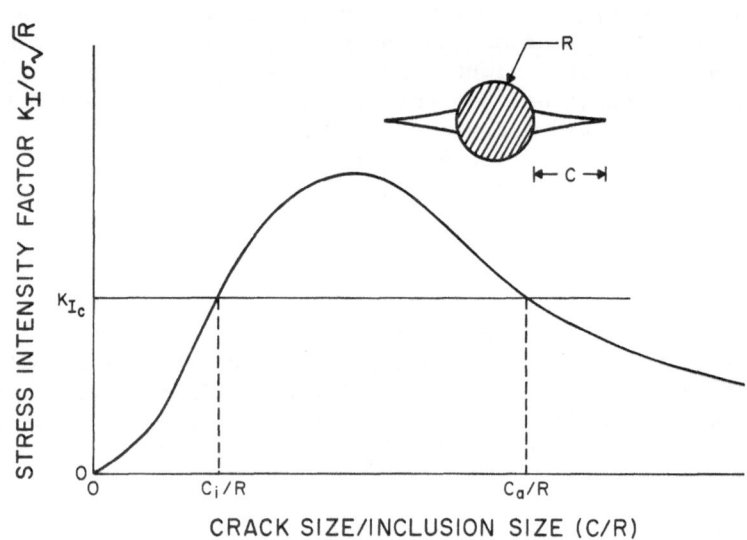

Fig. 1. Normalized stress intensity factor as a function of relative crack length for a radial crack produced by a spherical inclusion under compression.

maximum stress involved. The absence of crack propagation for $c>c_a$, the crack length for crack arrest, is due to the combined effects of the decrease in stress with increasing distance from the inclusion, and the stress relief due to the formation of the crack, which permits accommodation of the mismatch in thermal expansions of the matrix and inclusion. The conclusion obtained from Figure 1 is critical to the objectives of this study and the experimental results to be presented. The figure shows that the crack length for crack arrest, c_a, is independent of the initial crack length $c>c_i$. Furthermore, the value of the initial crack length is included in the value of crack length at crack arrest. It is for this reason that the combined effects of porosity and microcracks are expected to be coupled rather than simply additive.

A direct relationship generally exists between the pore content and the number of pores per unit volume, their geometry and their dimensions. For simplicity, it will be assumed that the dimensions and geometry of all pores are identical. The existence of a minimum crack size, c_i (i.e., pore size), implies that no microcracking will occur below a minimum value of porosity, P_i. Thus for values of porosity $O<P<P_i$, the thermal conductivity will be governed by the effect of the pore phase only. The same effect will prevail for $P>P_a$ for which no microcracks can form either.

For $P_i<P<P_a$, however, microcrack formation will occur, with all cracks being of equal size. Because the size of the microcrack precursor (i.e., the pore) is included within the dimensions of the microcrack at arrest, it is expected that for $P_i<P<P_a$, the thermal conductivity will be governed only by the microcracks and should be independent of pore content.

The implications of the above conclusions with regard to the effect of porosity on the thermal conductivity of brittle materials prone to microcracking are shown schematically in figure 2. For $P<P_i$ and $P>P_a$, the thermal conductivity decreases monotonically with increasing porosity. At P_i, the thermal conductivity decreases discontinuously to a value which remains constant for $P_i<P<P_a$.

Clearly, the dependence of thermal conductivity on pore content shown in figure 2 is valid for the idealized situation of all pores having equal size and geometry. In practice, however, the pores will exhibit a range of dimensions and geometries. Depending on the size and shape distributions, the dependence of thermal conductivity on pore content shown in figure 2 will show a more continuous variation preceded by an initial rapid decrease in thermal conductivity at $P\simeq P_i$. Furthermore, for $P_i<P<P_a$, a slight decrease in the thermal conductivity is expected for those small pores which did not contribute to the formation of microcracks.

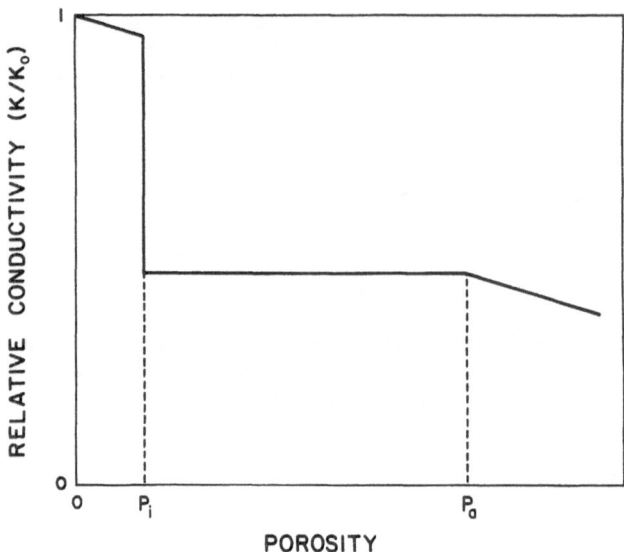

Fig. 2. Predicted effect of porosity on the relative thermal conductivity of brittle
composites prone to microcracking due
to thermal expansion mismatch.

It should be noted that microcrack formation itself also contributes to the creation of porosity. For this reason, the measured
pore content represents the sum of the original pore phase plus the
extra porosity created by microcracks. For two-component composite
materials, an upper bound on the pore content created by the microcracks, (P_{mc}), can be estimated assuming that the total thermal
expansion mismatch between the two components can be accounted for
by the volume of the cracks. This yields:

$$P_{mc} \simeq 3\Delta\alpha\Delta T$$

where $\Delta\alpha$ is the mismatch in the coefficients of thermal expansion
of the two components and ΔT represents the temperature range of
cooling from the processing temperature.

The above conclusions regarding the effect of porosity on the
thermal conductivity of microcracked materials are supported by
experimental data reported herein.

EXPERIMENTAL

The microcracked brittle composites consisted of aluminum oxide, beryllium oxide, or magnesium oxide with a dispersed phase of silicon carbide. Table I lists the coefficients of thermal expansion [12] for the three matrix materials and the silicon carbide, together with the values of thermal conductivity [13], density, specific heat [14], and thermal diffusivity at room temperature. The latter four values are required for the calculation of the composite thermal conductivity without microcracks. Because the coefficients of thermal expansion for the three oxide materials exceed the value for the silicon carbide, microcrack formation will occur in the matrix phase when the composites are cooled from the manufacturing temperature.

The specific process by which these composites were made consisted of sintering mixtures of the appropriate powders under pressure in graphite dies. The amount of residual pore phase could not be closely controlled during the pressurized sintering operation; therefore, a number of specimens were obtained to assure a sufficient range of pore content. Typical microstructures for composites consisting of magnesium oxide with a range of silicon carbide contents are shown in figure 3. The mean grain size of the magnesium oxide matrix was of the order of 1 μm, with similar values for the composites based on aluminum oxide and beryllium oxide. The

Table I. Thermal expansion and room temperature heat transfer properties of SiC and three oxide matrix materials.

	SiC	Al_2O_3	BeO	MgO
Coefficient of Thermal Expansion* (293–1600 K) (x 10^6/°C)	5.1	8.9	9.8	14.4
Thermal Conductivity (W/m·K)	168.0	36.4	272.0*	48.4*
Density (g/cm^3)	3.21	3.986	3.008	3.576
Specific Heat* (cal/g·C)	.161	.230	.246	.222
Thermal Diffusivity (cm^2/S)	.7770	.0950	.8785	.1457

---

*
Touloukian, et al.

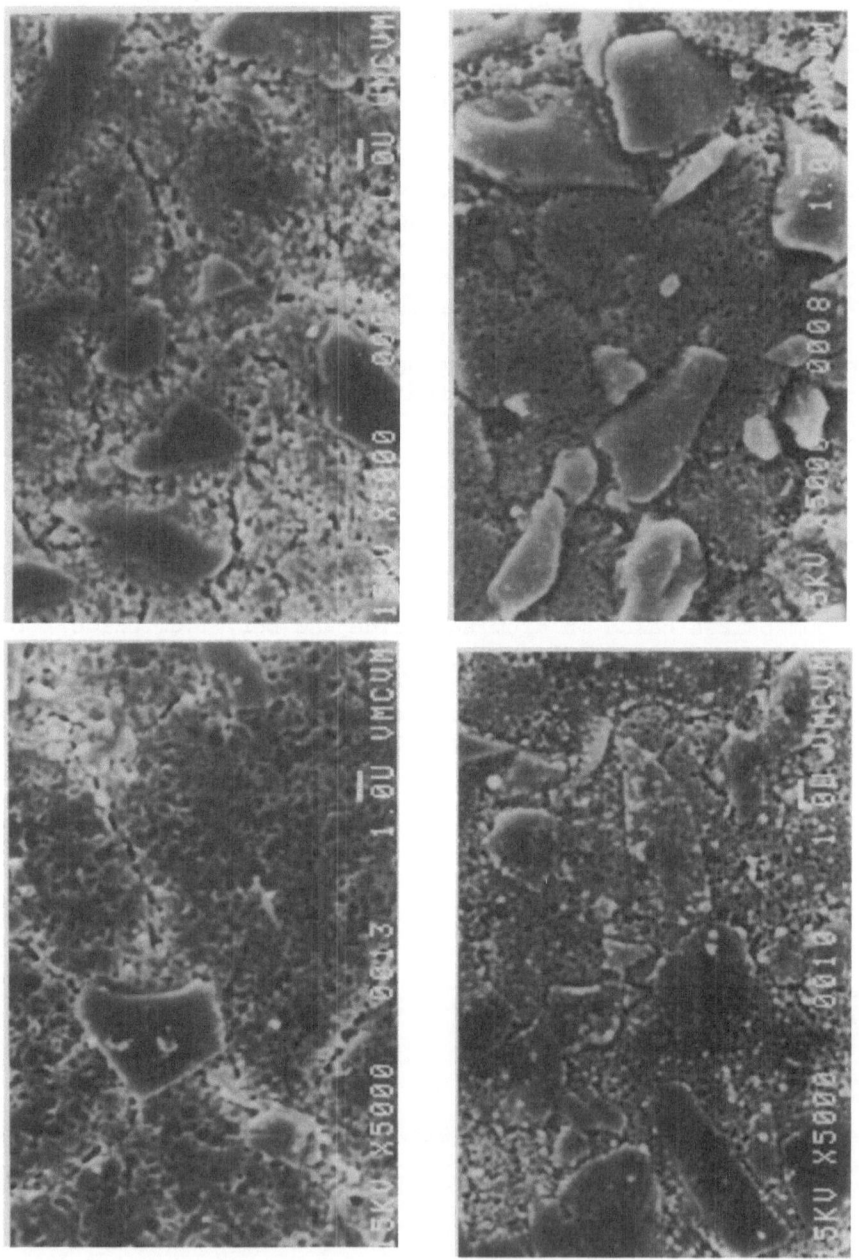

Fig. 3. SEM micrographs of MgO–SiC composites containing, clockwise from upper left, 10, 20, 30, and 40% SiC.

492

silicon carbide particles were somewhat disc-shaped with a diameter of about 6 μm and a thickness of about 1-2 μm.

The porosity of the samples was determined from the measured density of the specimens and the theoretical density of the pore-free composites as calculated from the values of density listed in table I. The effect of the pores and microcracks on the heat conduction behavior was determined by measurements of the thermal diffusivity by the flash method [15] using a glass-Nd laser as the flash source. All measurements were made at room temperature using a liquid nitrogen-cooled infrared detector to monitor the transient temperature of the specimen. The specimens were either circular disks about 12 mm in diameter or square plates about 1 cm on a side, each about 2 mm thick.

The specimens for all three series of composites varied in pore content as well as silicon carbide content. To facilitate reporting the data, the actual experimentally determined diffusivity data were multiplied by the density and specific heat to obtain the thermal conductivity. These values of thermal conductivity were then divided by the value of the conductivity without cracks. This latter value was calculated using the Bruggeman mixture equation and the appropriate values of the thermal conductivity of the individual components as listed in Table I.

RESULTS AND DISCUSSION

Figure 4 compares typical experimental data for the thermal diffusivity of the alumina-silicon carbide composites as a function of silicon carbide content to the values for the crack-free composities calculated from theory. The experimental data clearly fall far below the calculated values. The difference is attributable to the formation of the microcracks due to the thermal expansion mismatch. Dividing the experimental value by the calculated value for any given silicon carbide content yields the relative thermal conductivity, which reflects the effect of the microcracks alone, without the accompanying change in the thermal conductivity due to the silicon carbide phase. These relative values can then be plotted as a function of pore content in order to verify the role of porosity in the heat conduction behavior of microcracked composites.

Figures 5, 6, and 7 show the data for the relative thermal conductivity of the composites of silicon carbide dispersed in alumina, beryllia, and magnesia, respectively, as a function of the pore content. All three sets of data, especially the extensive set of data for the MgO-SiC system, are in general agreement with the behavior shown in figure 2, predicted from fracture-mechanical

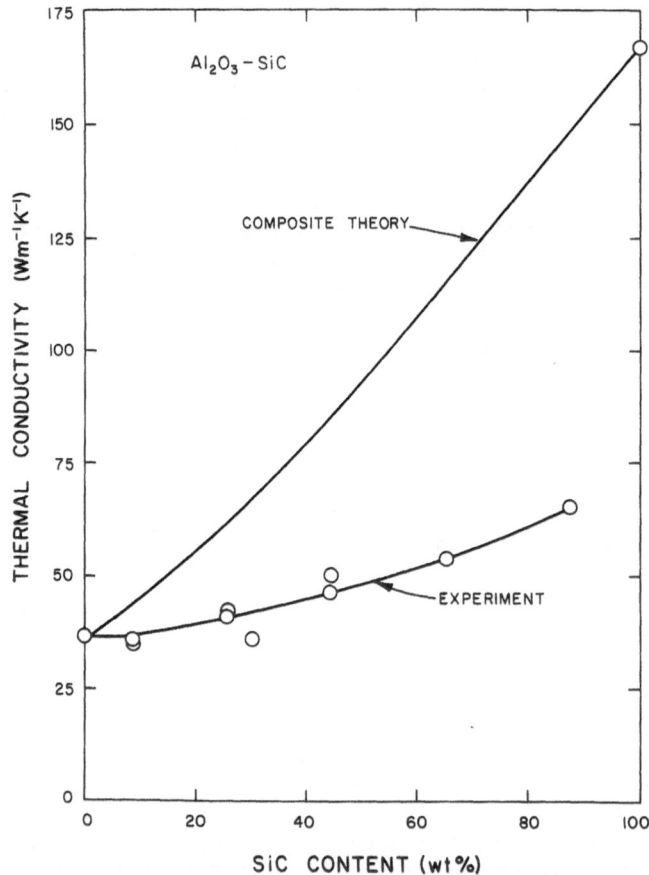

Fig. 4. Room temperature thermal conductivity of
Al₂O₃-SiC composites as compared to values
expected from the theory of mixtures.

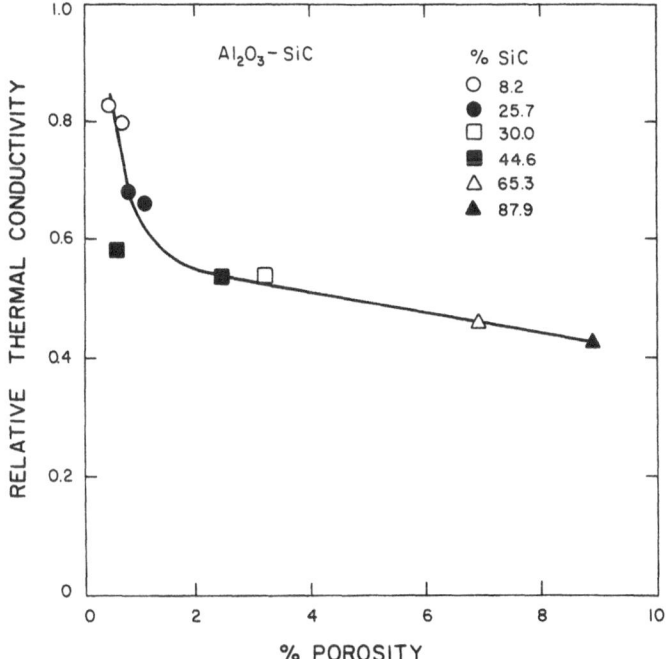

Fig. 5. Effect of pore content on the relative thermal conductivity of microcracked Al_2O_3-SiC composites relative to crack-free values.

principles. The very rapid decrease in thermal conductivity with increasing porosity at the lower values of porosity is clearly evident. The invariance of the thermal conductivity with porosity at the higher pore contents is compatible with the conclusion that the final crack size is independent of the initial pore size (i.e., initial pore content at a given crack density). The data scatter for the thermal conductivity at the lowest values of porosity is too large to make a reliable extrapolation; nevertheless, the general trend of the data suggests the existence of a minimum value of porosity below which no microcracking occurs.

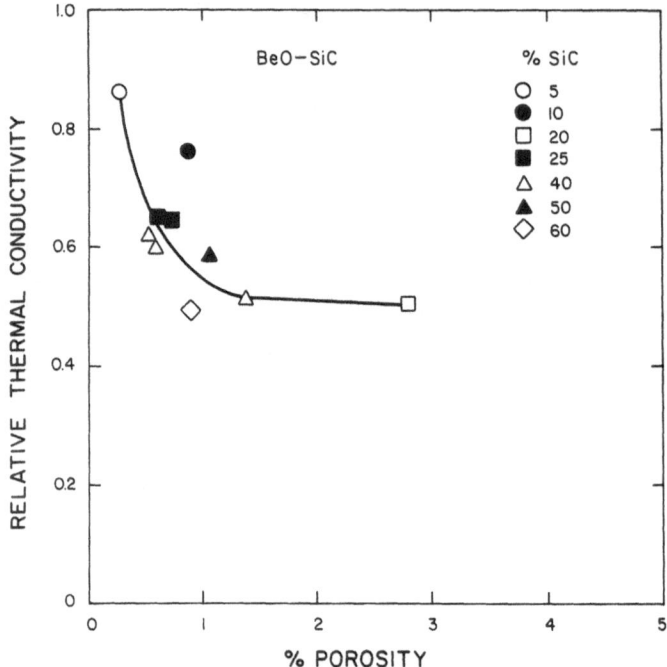

Fig. 6. Effect of pore content on the relative
thermal conductivity of microcracked
BeO-SiC composites relative to crack-
free values.

The amount of porosity created by the formation of the microcracks cannot be ascertained from the present data. However, the
differences in porosity at a given value of silicon carbide content must represent the amount of original pore phase prior to
the formation of the microcracks.

Comparison of the data shown in figures 5, 6, and 7 indicates
that the relative thermal diffusivity at the higher pore contents
is lowest for the magnesium oxide matrix, followed by the beryllium
oxide and then the aluminum oxide matrix. This is expected from
the differences in the mismatch in the coefficients of thermal expansion between the silicon carbide and the three matrices. The
extent of crack propagation and resulting decrease in thermal

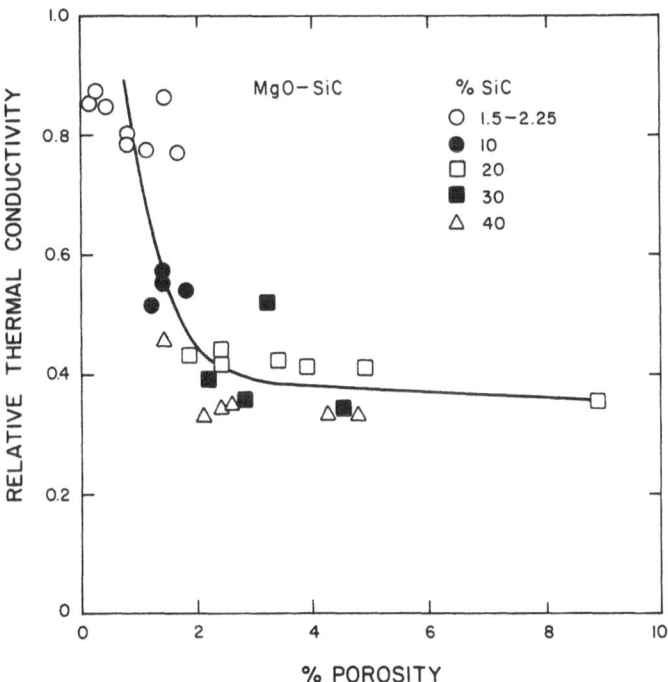

Fig. 7. Effect of pore content on the relative
thermal conductivity of microcracked
MgO-SiC composites relative to crack-
free values.

diffusivity is expected to increase with increasing thermal expansion mismatch.

In summary, on the basis of the stability of microcracks in
internal stress fields produced by thermal expansion heterogeneity,
a prediction was made concerning the relative effect of the pore
content on the thermal diffusivity and conductivity of brittle
materials subjected to microcracking and was confirmed by experiment.

ACKNOWLEDGEMENT

This study was conducted as part of a research program on the thermal and thermomechanical behavior of structural materials for high temperature applications funded by the Office of Naval Research under contract N00014-78-C-0431. Ceradyne Corporation, Santa Ana, CA, and Max-Planck Institute, Stuttgart, FRG are acknowledged for providing many of the specimens used in this study.

REFERENCES

1. H. J. Siebeneck, D. P. H. Hasselman, J. J. Cleveland and R. C. Bradt, J. Am. Ceram. Soc., 59:241 (1976).
2. H. J. Siebeneck, D. P. H. Hasselman, J. J. Cleveland and R. C. Bradt, J. Am. Ceram. Soc., 60:336 (1977).
3. J. P. Singh, D. P. H. Hasselman, W. M. Su, J. A. Rubin, R. Palicka, J. Mat. Sci., 16:141 (1981).
4. D. P. H. Hasselman, pp. 417-33 in Thermal Conductivity, Vol. 16, Ed. by D. C. Larson, Plenum Press, New York (1983).
5. D. P. H. Hasselman, J. Comp. Mat., 12:403 (1978).
6. A. E. Powers, Conductivity in Aggregates, Knolls Atomic Power Laboratory Report KAPL-2145, March 6, 1961.
7. S. C. Cheng and R. I. Vachon, Int. J. Heat and Mass Trans., 12:249 (1969).
8. J. Francl and W. D. Kingery, J. Am. Ceram. Soc., 37:99 (1954).
9. A. Evans, Acta. Met., 26:1845 (1978).
10. D. Clarke, Acta. Met., 28:913 (1980).
11. M. V. Swain, J. Mat. Sci., 16:151 (1981).
12. Y. S. Touloukian, R. K. Kirby, R. E. Taylor, and T. Y. R. Lee, "Thermal Expansion-Nonmetallic Solids", (Thermophysical Properties of Matter, Vol. 13) IFI/Plenum Press, New York-Washington, 1977.
13. Y. S. Touloukian, R. W. Powell, C. Y. Ho, and P. G. Klemens, "Thermal Conductivity-Nonmetallic Solids", (Thermophysical Properties of Matter, Vol. 2) IFI/Plenum Press, New York-Washington, 1970.
14. Y. S. Touloukian and E. H. Buyco, "Specific Heat-Nonmetallic Solids", (Thermophysical Properties of Matter, Vol. 5) IFI/ Plenum Press, New York-Washington, 1970.
15. W. J. Parker, R. J. Jenkins, C. P. Butler and G. L. Abbott, J. Appl. Phys., 32:1679 (1961).

RADIATIVE CONTRIBUTION TO THE THERMAL DIFFUSIVITY AND CONDUCTIVITY

OF A SILICON CARBIDE FIBER REINFORCED GLASS-CERAMIC

L. D. Bentsen and D. P. H. Hasselman
Department of Materials Engineering
Virginia Polytechnic Institute and State University
Blacksburg, Virginia 24061 USA

J. J. Brennan
United Technologies Research Center
East Hartford, Connecticut 06108 USA

ABSTRACT

    The thermal diffusivity and conductivity of a silicon carbide
fiber-reinforced lithium aluminosilicate glass-ceramic was measured
using the laser-flash method. As indicated by the effect of specimen thickness on thermal conductivity and its positive temperature
dependence, heat transfer by radiation makes a significant contribution to the total thermal conductivity. The observed dependence
of the effect of specimen thickness on thermal conductivity suggests that radiative heat transfer between the carbon coated
surfaces of dielectric materials for the laser-flash technique may
be partially governed by the view factor between the specimen surfaces.

INTRODUCTION

    Electrons, phonons, and photons represent the primary mechan-
isms of heat transport in solids. Each of these mechanisms dis-
plays a unique temperature dependence [1]. In dielectric materials
the thermal conductivity due to electron transport is generally
not significant, and above room temperature the conductivity is
approximately inversely proportional to the temperature as the
result of increased phonon collisions.

    The thermal conductivity due to photon transport is a function
of the optical absorption and emission. For highly transparent
materials, heat transfer can occur by direct radiation between the

surfaces of the material. In this case the effective radiative thermal conductivity for surfaces with equal emissivity is [2]:

$$K_R = \left(\frac{e}{2-e}\right) 4\sigma n^2 T^3 d \tag{1}$$

where e is the emissivity of the specimen surface, σ is the Stefan-Boltzmann constant, n is the index of refraction, T is the absolute temperature, and d is the thickness of the material in the direction of heat flow.

Heat transport by radiation can also occur by emission, absorption, and re-emission. When the distance between sites of absorption and re-emission of a photon is small compared to the specimen dimensions, the radiative conductivity becomes [2]:

$$K_R = \frac{16}{3} \sigma n^2 T^3 \ell \tag{2}$$

where ℓ is the photon mean-free-path which is equal to the reciprocal of the extinction coefficient.

Equations 1 and 2 indicate that the thermal conductivity due to radiation between the specimen's surfaces, and the thermal conductivity due to absorption and re-emission of radiation are proportional to T^3. This implies that the radiative thermal conductivity should exhibit a strong positive temperature dependence. Equations 1 and 2 also show that the thermal conductivity due to radiation directly across the material is directly proportional to the specimen thickness, whereas radiative heat transfer by photon absorption and re-emission is independent of the specimen size.

In general, the conduction of heat in dielectric materials occurs by the combined contributions of phonon and photon transport. Above room temperature, the negative temperature dependence of the thermal conductivity due to phonon transport will be offset by the positive dependence of the radiative contribution. In crystalline dielectric materials which have a large phonon mean-free-path, such as aluminum oxide, beryllium oxide, and diamond, the negative temperature dependence of phonon transport generally overrides the positive temperature dependence of radiative heat transfer up to temperatures of some 1000°C. In highly amorphous dielectrics, phonon transport tends to be suppressed. For these materials, the relative contribution of radiative heat transfer can be such that the combined thermal conductivity due to phonons and photons produces a significant positive temperature dependence.

Such a positive temperature dependence was observed in a feasibility study of the composite method for determining the thermal conductivity of amorphous silicon carbide fibers contained within a continuous matrix of a lithium aluminosilicate (LAS)

500

glass-ceramic [3]. The experimental data indicated that radiative heat transfer within the fibers as well as the glass-ceramic matrix made significant contributions to the total thermal conductivity of the composite. A few exploratory measurements indicated the existence of the effect of specimen thickness on the thermal conductivity, suggestive of radiative heat transfer between the specimen surfaces as described by equation 1. The purpose of this study was to investigate the effect of specimen size on the thermal conductivity of the SiC-LAS composites in order to determine in further detail the nature of the radiative contribution to the total thermal conductivity.

EXPERIMENTAL PROCEDURES

Materials

The silicon carbide fibers from a commercial source* were made from an organometallic polymer by the method of Yajima [4]. Chemical analysis indicated that the fibers consisted of about 65% SiC, 25% SiO_2, and 10% C. The fibers were in the form of tows of yarn containing about 500 fibers per tow. The average fiber diameter was approximately 10 μm and the density was about 2.55 $g \cdot cm^{-3}$. The degree of peak broadening in x-ray analysis showed that the fibers were nearly amorphous with a crystallite size of 2.5 to 3.0 nm.

The matrix material consisted of a lithium aluminosilicate glass-ceramic with a composition nearly identical to that of a commercial glass-ceramic**; however the 3% TiO_2 nucleating agent was replaced by about 2% ZrO_2 in order to eliminate the reactivity between the TiO_2 and the SiC. The density of the LAS was 2.52 $g \cdot cm^{-3}$.

Composites were fabricated by passing the SiC yarn through a slurry of the uncrystallized glass powder dissolved in isopropyl alcohol. After drying, the coated yarn was cut to fit a graphite die, and was hot-pressed in vacuum for 5 to 60 minutes at 1400 to 1500°C and 14 MPa. Following hot-pressing, the glass was crystallized by heat treating the composite for 1 to 2 hours at temperatures of 880 to 1100°C. Both uniaxial and 0/90 cross-ply composites were fabricated with SiC contents up to 50 vol.%. Figure 1 shows a photomicrograph of a composite with uniaxial fibers, and figure 2 shows a scanning electron fractograph of a 0/90 cross-ply composite.

---

*Nippon Carbon Company, Japan.
**C-9608, Corning Glass Works, Corning, New York.

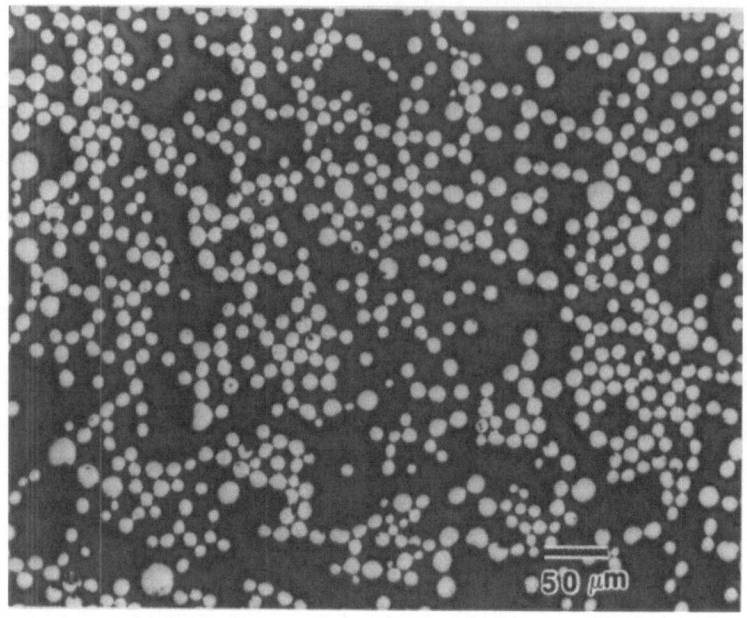

Fig. 1. Photomicrograph of a composite of LAS glass-
ceramic containing 48 vol.% uniaxially
aligned SiC fibers.

Measurement and Evaluation of Thermal Properties

The laser-flash method [5] was used to measure the thermal
diffusivity of the composites and the matrix. The specimens of
the composites cut from the hot-pressed blanks were in the form
of nearly square plates about 9.5 mm on a side with thicknesses
ranging from 0.5 to 4.5 mm. Samples were cut to permit measurement of the thermal diffusivity both parallel and perpendicular to
the fibers. Specimens of the LAS matrix were circular disks 12.8
mm in diameter with thicknesses of 0.3 to 4.0 mm. For all samples,
direct transmission of the laser beam was prevented by coating the
specimen surfaces with carbon. The transient temperature response
of the specimen was monitored with remote optical sensors. For
measurements above room temperature, the specimens were supported
within a carbon resistance furnace containing a nitrogen atmosphere.

In the analysis of the data, corrections for heat loss were
taken into account using the method of Heckman [6]. Changes in
the thickness and density of the specimens due to thermal expansion
were also accounted for. The specific heat of the fibers and the
glass-ceramic matrix was determined with differential scanning
calorimetry.

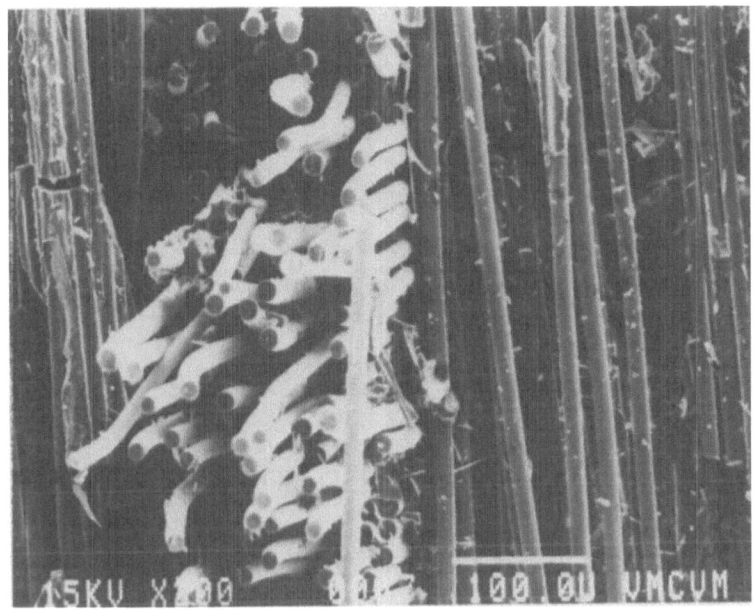

Fig. 2. Scanning electron fractograph of a composite
of LAS glass-ceramic containing 45 vol.% of
crossplied SiC fibers.

From the measured value of the thermal diffusivity of either
the fiber-free matrix or the composite, the corresponding value of
the thermal conductivity can be calculated from:

$$K = \kappa \rho c \qquad (3)$$

where κ is the thermal diffusivity, ρ is the density, and c is the
specific heat. The specific heat of the composite was calculated
using the rule of mixtures and the measured specific heats of the
fibers and the matrix shown in Figure 3.

The thermal conductivity and diffusivity of the fibers were
evaluated from the corresponding data for the matrix and the composites using the theory for the thermal conductivity of mixtures
with a dispersed phase in the shape of cylinders [7]. For heat
flow parallel to uniaxially aligned fibers, the thermal conductivity
of the composite is:

$$K_c = K_m V_m + K_f V_f \qquad (4)$$

where K is the thermal conductivity, V is the volume fraction, and
the subscripts c, m, and f refer to the composite, matrix and

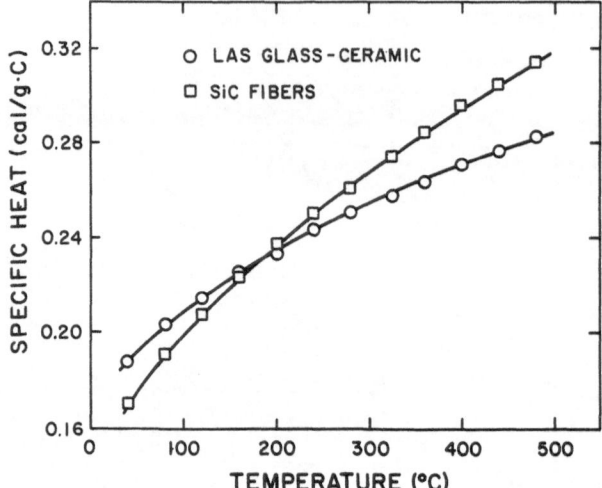

Fig. 3. Specific heat of LAS glass-ceramic and of amorphous SiC fibers.

fibers, respectively. For heat flow perpendicular to the fibers, the thermal conductivity is:

$$\left(\frac{K_m - K_c}{K_m + K_c}\right) V_m = \left(\frac{K_c - K_f}{K_c + K_f}\right) V_f \qquad (5)$$

After determining the thermal conductivity of the fibers using equations 4 and 5, the thermal diffusivity can be calculated by means of equation 3.

RESULTS AND DISCUSSION

Figure 4 shows the data for the effect of specimen thickness on the thermal conductivity and diffusivity of the LAS glass-ceramic matrix and two composites with heat flow parallel and per-pendicular to the fiber axes. The thermal conductivity was cal-culated using Eq. 5 and the experimental data for the thermal diffusivity, density, and specific heat. The data indicate a significant specimen size effect for these three materials, sug-gestive of radiative heat transfer between the carbon-coated surfaces of the specimens.

A similar size effect is observed for the values of thermal conductivity and diffusivity of the SiC fibers at room temperature

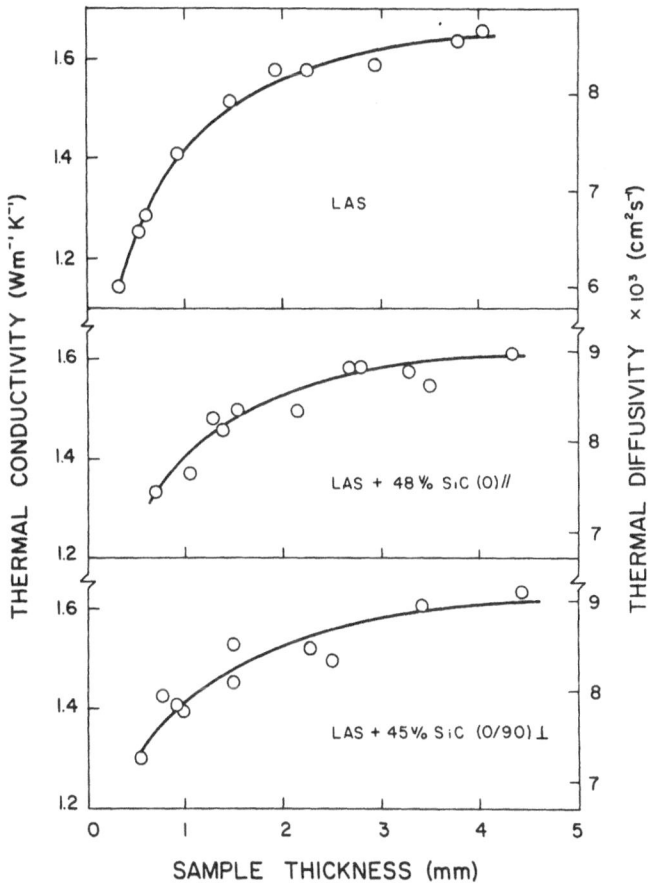

Fig. 4. Thickness dependence of the thermal diffusivity and conductivity at 25°C for LAS glass-ceramic and two composites of LAS with SiC fibers oriented parallel (middle) and perpendicular (bottom) to the direction of heat flow.

as shown in Figure 5. These values were calculated from the experimental data for the composite and matrix samples using equations 4 and 5. The values are essentially independent of the direction of heat flow relative to the fibers. This suggests that any exixting interfacial effects do not influence heat flow perpendicular to the fibers.

For a quantitative interpretation of the size effect, it is important to note that the thermal conductivity does not rise linearly with specimen thickness, which is expected from Eq. 1 if all the heat transfer through the specimen occurred by radiation between the surfaces. This indicates that additional mechanisms of heat transfer independent of the specimen thickness also contribute to the thermal conductivity.

Such mechanisms include heat transfer by phonons and by continuous emission and absorption of photons with short mean-free-path. Both processes are independent of specimen thickness. However, an additional factor to explain the observed size effect should be considered as well. First it should be noted that equation 1 for the effective thermal conductivity due to radiative heat transfer between sample surfaces is valid when the ratio of the length of the surface to the distance between the surfaces

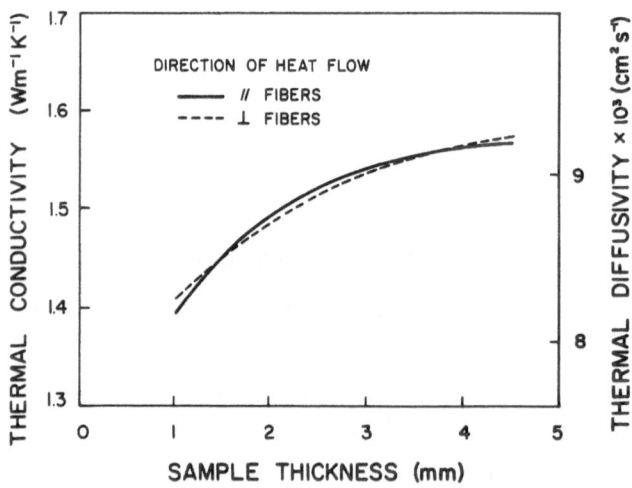

Fig. 5. Thermal diffusivity and conductivity of SiC fibers at room temperature determined from experimental values for the LAS matrix and the composites.

approaches infinity. Under this condition, all the radiation emitted by one surface is absorbed by the second surface. This condition is not met for the specimens used in this study, where in the case of the thicker specimens, the ratio is only two to three. Thus an appreciable fraction of radiation emitted hemispherically from the hotter surface will not be intercepted and absorbed by the colder surface. This condition is governed by the view factor from one surface to another, which is defined as the fraction of the total radiation emitted by the one surface that is directly incident on the other [8]. For parallel circular disks of equal diameter positioned directly opposite each other, the view factor (F) can be written [9]:

$$F = \frac{1}{2} \left[\frac{d^2}{r^2} + 2 - \left(\frac{d^4 + 4r^2 d^2}{r^4} \right)^{\frac{1}{2}} \right] \tag{6}$$

where r is the radius of the disks and d is the distance between them.

The effect of specimen geometry on the radiative thermal conductivity of typical carbon-coated laser-flash specimens can be taken into account by incorporating the view factor of equation 6 into equation 1, which yields:

$$K_R = \left(\frac{e}{2-e} \right) 4F\sigma n^2 T^3 d \tag{7}$$

Figure 6 shows the normalized radiative conductivity, $K_R (2-e)/4e\sigma n^2 T^3$, as a function of thickness for a range of specimen diameters, which for the LAS matrix in this study is 12.8 mm. Comparison of the data for the thermal conductivity as a function of specimen thickness shown in Figure 4 with the curve for specimen diameter of 12.8 mm shown in Figure 6 indicates that the variation in the observed thermal conductivity can be partially accounted for by the change in specimen shape with specimen thickness. Such an effect should be taken into account in the interpretation of the data for the thermal diffusivity of transparent materials measured by the laser-flash method.

Figure 7 shows the temperature dependence of the thermal conductivity for two values of thickness of specimens of the glass-ceramic matrix and two composites with heat flow parallel and perpendicular to the fiber axes. The increase in thermal conductivity with temperature for all three materials is indicative of the contribution of radiation to the total heat transfer. The relative increase in thermal conductivity with temperature, however, is well below the T^3 dependence expected if radiation were the only contributing mechanism of heat conduction. This indicates that in spite of the observed specimen size effect and the positive temperature dependence, phonon transport still makes the primary

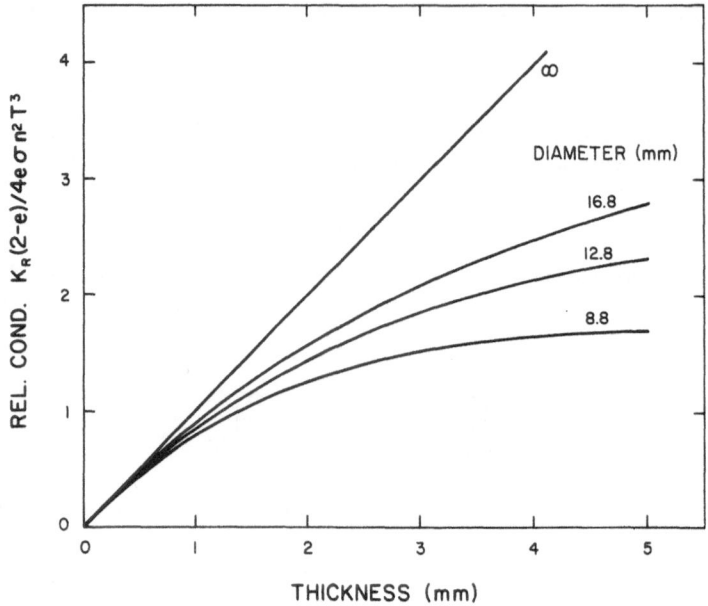

Fig. 6. Normalized thermal conductivity due to radiation between the surfaces of samples of
various diameters as a function of the
sample thickness.

contribution to the thermal conductivity. Of course, the temperature and spectral dependence of the absorption coefficient may also
play a role. However, at this time no data for these effects
appear to be available for these composites.

The data presented in Figure 7 indicate that the curves for
the temperature dependence of the thermal conductivity for the
two values of specimen thickness are nearly parallel for all three
materials. This suggests that the relative effect of specimen
thickness on thermal conductivity decreases with increasing temperature. This in turn suggests that heat transfer by photon emission
and absorption makes a relatively larger contribution to the total
thermal conductivity with increasing temperature.

It is interesting to note that compared to the present materials, the contribution of radiation to the thermal conductivity of
such highly transparent materials as quartz and sapphire appears to
be much higher, as observed by Howlett [10]. This suggests that
the suppression of radiative heat transfer by photon scattering at
optical discontinuities could play a significant role in the

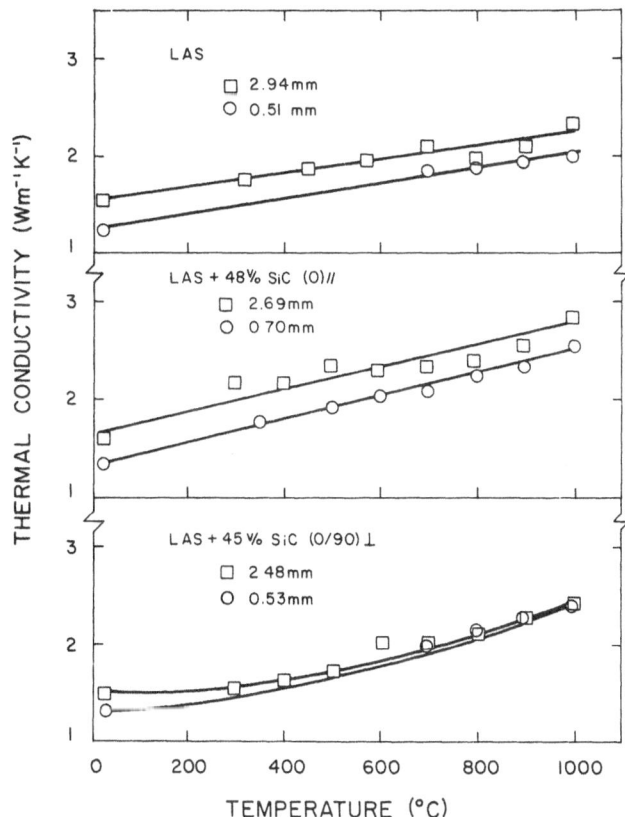

Fig. 7.   Temperature dependence of the thermal con-
          ductivity of LAS glass-ceramic matrix and
          two composites of LAS with SiC fibers
          oriented parallel and perpendicular to
          the direction of heat flow. Data for
          samples of two different thicknesses
          are shown for each material.

conduction of heat in the materials investigated in this study.

In summary, it was found that radiative heat transfer between the specimen surfaces contributes to the overall thermal conductivity of a silicon carbide fiber-reinforced lithium aluminosilicate glass-ceramic. The dependence of thermal conductivity on specimen thickness suggests that the radiative heat transfer is partially governed by the view factor between the carbon coated laser-flash specimens.

ACKNOWLEDGEMENTS

This study was supported by the Office of Naval Research under contract Numbers N00014-78-C-0431 and N00014-78-C-0503. Financial assistance for L.D. Bentsen was provided by a Cunningham Fellowship from Virginia Polytechnic Institute and State University.

REFERENCES

1. R. Berman, "Thermal Conduction in Solids", Clarendon Press, Oxford (1976).
2. D. W. Lee and W. D. Kingery, J. Amer. Ceram. Soc., 43 (1960) 594.
3. J. J. Brennan, L. D. Bentsen and D. P. H. Hasselman, J. Mater. Sci., 17 (1982) 2337.
4. S. Yajima, K. Okamura, J. Hayashi and M. Omori, J. Amer. Ceram. Soc., 59 (1976) 324.
5. W. J. Parker, R. J. Jenkins, C. P. Butler and G. L. Abbott, J. Appl. Phys., 32 (1961) 1679.
6. R. C. Heckman, J. Appl. Phys., 44 (1973) 1455.
7. A. E. Powers, Knolls Atomic Power Laboratory, Report KAPL-2145, 1961.
8. W. A. Gray, and R. Muller, "Engineering Calculations in Radiative Heat Transfer", (Pergamon Press, New York, 1974), p. 29.
9. R. Siegel and J. R. Howell, "Thermal Radiation Heat Transfer", (McGraw-Hill, New York, 1972) p. 787.
10. S. P. Howlett, R. Taylor, and R. Morrell, "Thermal Conductivity-17", (Plenum Press, New York, 1983), p. 447, J. G. Hust, Ed.

SESSION R

Pulse Measurement Techniques

SESSION CHAIRMAN

M. L. Minges
Air Force Materials Laboratory
Wright-Patterson AFB, OH.

MEASUREMENT OF THERMAL CONDUCTIVITY FROM HIGH TEMPERATURE PULSE DIFFUSIVITY AND CALORIMETRY MEASUREMENTS*

E. P. Roth

Sandia National Laboratories
P. O. Box 5800
Albuquerque, NM   87185

ABSTRACT

The measurement of the thermal conductivity of small samples over an extended temperature range is discussed. Experimental data on thermal diffusivity obtained by the laser pulse technique are combined with specific heat data, obtained either by differential scanning calorimetry or by vaporization calorimetry, to derive the thermal conductivity. An an example of the use of this technique, data are reported on the conductivity of an ASTM A517 steel over the temperature range 100-1200°C. The thermal diffusivity was derived from temperature vs time data obtained from a laser pulse diffusivty system designed and constructed at Sandia. A data reduction technique was used which simultaneously corrects for finite pulse time effects and sample heat losses by radiation. The specific heat was measured directly at temperatures below 680°C using a commercial differential scanning calorimeter. For temperatures up to 1200°C, the specific heat was determined by differentiating a curve fit to the measured sample enthalpy. The enthalpy was measured using a liquid argon vaporization calorimeter developed at Sandia. The thermal conductivity of the A517 steel was then calculated over the entire temperature range. A classical Curie transition was observed during measurements of both the diffusivity and enthalpy. The effect of this transition on the measured thermal conductivity is discussed.

*This work performed at Sandia National Laboratories supported by the U.S. Department of Energy, under Contract Number DE-AC04-76DP00789.

INTRODUCTION

The determination of the thermal conductivity of small samples over extended temperature ranges poses many difficult problems. A technique often used is the flash diffusivity method whereby the thermal conductivity is determined using the expression

$$\lambda = \alpha \rho c_p \qquad \qquad (1)$$

where λ = thermal conductivity, α = thermal diffusivity, ρ = density, and c_p = specific heat. As this formula shows, in addition to the measurement of sample thermal diffusivity, measurements must also be made of the sample density and specific heat over the same temperature ranges. The measurement of specific heat over extended temperature ranges itself presents some problems. Commercial instruments which perform direct heat capacity measurements such as the Differential Scanning Calorimeter (DSC) are limited in temperature range (725°C) and in the size of samples that can be measured. Specific heat can be determined to higher temperatures from enthalpy measurements obtained using drop calorimetry techniques. However, these techniques require that the specific heat be determined by differentiating a curve fit to the enthalpy data. Care must be taken in applying this method to the determination of thermal conductivity, especially for materials that undergo transformations. This technique may require many data points in the region of the transformation in order to properly determine its form.

As an example of the application of these techniques in determining thermal conductivity, results are reported on a study of an ASTM A517, grade D steel measured from 100-1200°C.

METHOD

The thermal diffusivity was measured using the laser flash method whereby the front face of a small disk-shaped sample was subjected to a short laser pulse and the temperature rise of the back face was recorded as a function of time. This method was proposed by Parker et al. and recently reviewed by Taylor. [1,2] The analytical solution of the temperature rise for an infinite slab with parallel faces subjected to an instantaneous energy pulse is well known. However, this analytical solution must be corrected for experimental conditions in which the energy pulse is not instantaneous. Corrections must also be made for radiative losses from the sample faces which can be significant at elevated temperatures. Several authors have treated the case of correcting the diffusivity for finite pulse times [3,4] or for radiative losses [5,6]. However, previous works have only corrected for these effects individually. In this study, the thermal diffusivity was simultaneously corrected for both effects with data reduction methods developed at Sandia by J. A. Koski [7].

In this program, simultaneous laser pulse time and radiative heat loss corrections were obtained for the experimentally measured laser pulse shape. A large digital computer was initially used to obtain correction values as a function of the measured temperature rise waveform and the laser pulse width. Surface fitting techniques were then applied to produce polynomial expressions for the correction factors which could be programmed on the laboratory mini-computer for rapid data analysis. This program was checked by performing a set of measurements on AXM-5Q POCO Graphite. The data obtained agreed to within 5% of the results of a round-robin study on this material [8].

The sample specific heat was measured from 25-680°C using a Perkin-Elmer DSC-2 Differential Scanning Calorimeter. The DSC is a differential power device whose operation is described in the manufacturer's literature [9]. The specific heat was determined at 0.5°C intervals using a continuous scan data analysis program developed at Sandia. This program calculates specific heat based on NBS sapphire standards and is calibrated at the time of use for every set of measurements. This program automatically corrects for baseline offsets in the DSC and any shift in the slope between the initial and final isotherms. The data are taken in 200-300°C scan intervals in order to minimize non-linear corrections to the DSC baseline. A scan rate of 20°C/min was chosen to insure sample thermal equilibrium during the measurement. At the conclusion of every sample measurement NBS standards were again checked to insure the accuracy of the initial calibration. These measured specific heat values were within ± 3% of the recommended NBS values. Temperature calibrations were made at the melt points of known materials and were accurate to ± 1°C.

The material specific heat was determined at temperatures greater than 680°C from enthalpy measurements obtained using a liquid argon vaporization calorimeter developed at Sandia [10,11]. The sample was first encapsulated in a platinum crucible to prevent oxidation at elevated temperatures. The encapsulated sample was heated in a tube furnace and then dropped into a receiving well which was in good thermal contact with a dewar of liquid argon. Heat released by the sample while thermally equilibrating with the liquid argon vaporized a portion of the liquid. The resulting gas flow was measured by calibrated flow units and the flow data recorded by digital computer. The heat content of the sample was determined from the thermodynamic properties of the liquid argon and the total amount of gas released. The heat content of the encapsulated sample was corrected for radiative and convective losses which occurred during the drop. This temperature-dependent correction factor was obtained by first performing a series of calibration measurements using encapsulated NBS sapphire standards. The measured heat content of the standards was compared to calculated values obtained from the specific heats of the platinum and sapphire. The heat-loss corrections were less than 1% of the measured sample heat content. Corrections were also made for systematic measurement errors (<1%) based on electrical calibrations. The overall system accuracy for enthalpy measurement was ± 1%. The sample heat content was obtained by subtracting the heat contribution of the platinum capsule. The sample enthalpy referenced to 85K, H_T-H_{85}, was computed using this result and the sample mass. The sample enthalpy was referenced to room temperature (298.15K) by performing a measurement at room temperature and using the relation

$$H_T - H_{298} = (H_T-H_{85}) - (H_{298}-H_{85}) \tag{2}$$

Specific heat was obtained using the relation

$$c_p = \frac{d(H_T - H_{298})_p}{dT} . \tag{3}$$

An analytical expression for the enthalpy was obtained by fitting the enthalpy data to a curve using a third order B-spline data-fitting program [12]. Enthalpy measurements were performed from room temperature to 1150°C in order to obtain a smooth transition in the specific heat measured with the DSC and the specific heat obtained from the fitted enthalpy curve.

The thermal conductivity was calculated using a computer program based on Eq. 1 at temperatures corresponding to the thermal diffusivity data points. The corresponding specific heats at those temperatures were obtained by interpolating between the closest data points obtained from the DSC or by using an analytical expression obtained from the enthalpy curve fit.

SAMPLE CHARACTERIZATION AND PREPARATION

A compositional analysis of the sample material was performed at Sandia using atomic emission spectroscopy. The sample was identified as an ASTM standard A517, grade D steel. Table I lists the measured composition along with the nominal composition [13].

The samples prepared for the laser pulse diffusivity apparatus were disks measuring 1.27 cm in diameter. Three thicknesses were used: 0.296 cm, 0.196 cm, and 0.101 cm. Data for the three sample thicknesses were compared as a check for possible systematic errors in the measurement. The surfaces were sandblasted to improve emissivity. The DSC samples were small disks weighing approximately 0.1 g. Sample geometry did not affect the DSC measurements. The samples for the vaporization calorimeter weighed 14.8 g and were cylindrical in shape, measuring 1.9 cm in length and 1.2 cm in diameter. The enthalpy sample was then vacuum sealed in a platinum capsule.

TABLE 1. Compositional Analysis

Element	A517 Grade D Composition (w/o)	Sample Composition (w/o)
C	0.13 - 0.20	0.17
Mn	0.40 - 0.70	0.56
P	0.035	0.004
S	0.040	0.025
Si	0.20 - 0.35	0.25
Cr	0.85 - 1.20	0.87
Ni	---	0.12
Mo	0.15 - 0.25	0.21
V	---	0.01
Ti	0.04 - 0.10	0.10
Cu	0.20 - 0.40	0.27
B	0.0015 - 0.005	0.0020
Fe	Balance	Balance

The sample density measured at room temperature was 7.81 g/cm^3. Corrections were made for thermal expansion in performing the thermal conductivity calculation.

EXPERIMENTAL EQUIPMENT

The thermal diffusivity apparatus used in this study consisted of a pulsed ruby laser, a high temperature furnace/vacuum system, and an InSb infrared detector. A schematic of the apparatus is shown in Fig. 1. The laser was a KORAD K1 ruby laser (λ = 6943Å) which delivers approximately 35J of energy in 10^{-3}s [14]. The sample was held in a vacuum/furnace system using a low-thermal-contact tantalum holder with tantalum shields on both the front and back sample faces to eliminate laser flashby. The temperature rise at the back sample face was monitored using an InSb infrared detector system [15]. The output of the IR detector was conditioned through an amplifier and biasing network and then recorded by a digital transient recorder. The output of the transient recorder was read by a Hewlett-Packard HP1000 mini-computer which then performed the data analysis.

Direct specific heat measurements were made with the Perkin-Elmer DSC System which was modified for computer control by the HP1000. The laboratory mini-computer records the DSC output and automatically starts and finishes the scan sequence after recording isothermal baselines before and after each scan. The computer then performs all data analysis. The principle and operation of the DSC are described in the manufacturer's literature.

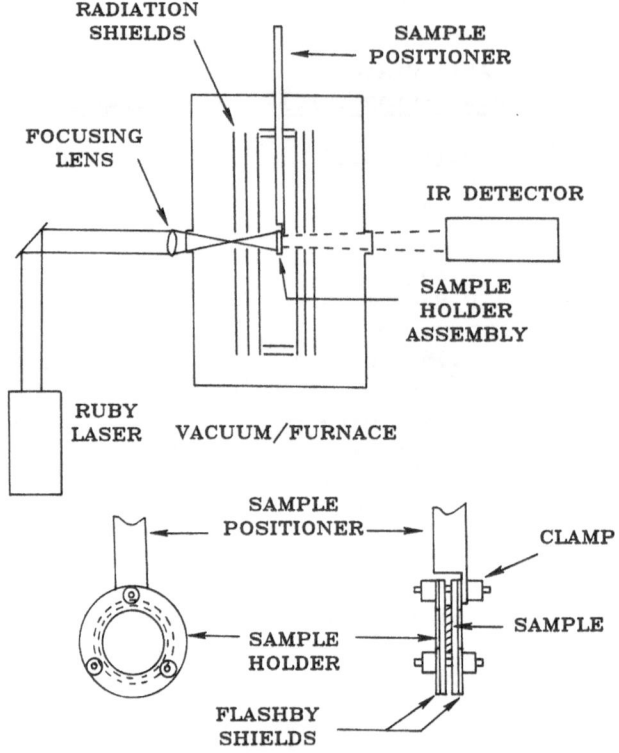

Fig. 1. Thermal diffusivity apparatus.

A portion of the liquid argon vaporization calorimeter system
is shown in Figure 2. The encapsulated sample was suspended by a
platinum wire in a tube furnace which was capable of reaching 1200°C.
Sample temperature was measured by thermocouples placed near the
capsule. The sample drop speed was controlled by an air piston
brake during the drop into the receiving well. The receiving well
was in contact with the liquid argon working fluid in the calorimeter
dewar. This dewar was placed in a liquid argon thermostating dewar
(not shown) to reduce heat leaks to the system. The vaporized argon
was brought to a known temperature by passing through thermostating
coils in a temperature-controlled water bath (not shown). The flow
rate was determined from the pressure drop across calibrated flow
units measured with a capacitance manometer. A control system maintained the pressure in the calorimeter dewar. All gas flow data
were recorded by the HP1000 laboratory mini-computer. The data acquisition program monitored the background gas flow from the system
both before and after each measurement to determine the residual
heat leak to the system. Corrections were then made to the measured
heat release to compensate for this residual heat input. This system
is described in more detail in Ref. 10 and 11.

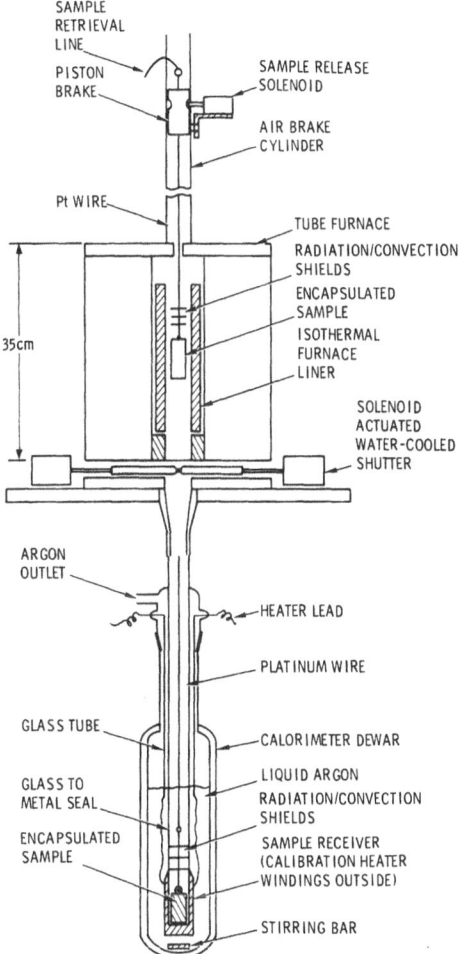

SAMPLE
RETRIEVAL
LINE

PISTON
BRAKE

SAMPLE RELEASE
SOLENOID

AIR BRAKE
CYLINDER

Pt WIRE

TUBE FURNACE

RADIATION/CONVECTION
SHIELDS

ENCAPSULATED
SAMPLE

ISOTHERMAL
FURNACE
LINER

35cm

SOLENOID
ACTUATED
WATER-COOLED
SHUTTER

ARGON
OUTLET

HEATER LEAD

PLATINUM WIRE

GLASS TUBE

CALORIMETER DEWAR

GLASS TO
METAL SEAL

LIQUID ARGON
RADIATION/CONVECTION
SHIELDS

ENCAPSULATED
SAMPLE

SAMPLE RECEIVER
(CALIBRATION HEATER
WINDINGS OUTSIDE)

STIRRING BAR

Fig. 2.   Liquid argon vaporization calorimeter.

RESULTS

The thermal diffusivity data for the three sample thicknesses
are shown in Figure 3.   The data for these three thicknesses agreed
to within ± 5% indicating that there were no systematic errors attributable to sample size.   The sample half-rise times were in the
range 0.01 - 0.1 s which gave ratios of pulse time to half-rise time
of 0.1 - 0.01.   These small ratios resulted in small pulse time
corrections to the diffusivity values.   Radiative loss corrections
were also small.

The thermal diffusivity was observed to decrease with increasing temperature until reaching a minimum at approximately 765°C.
At this temperature the diffusivity increased sharply and continued

to rise with increasing temperature. This sharp change in diffusivity was attributed to a magnetic Curie transition during which the material underwent a transition from the ordered ferromagnetic to the disordered paramagnetic state.

Figure 4 shows the enthalpy data referenced to room temperature. The Curie transition is seen as an increase in enthalpy at approximately 760°C. The solid line in this figure is the curve obtained from the B-spline fitting routine. The first derivative of this function with respect to temperature gives the specific heat. Figure 5 shows the specific heat data obtained from the enthalpy curve starting at approximately 680°C. The specific heat data shown at temperatures below this point were obtained directly using the DSC. The DSC data were obtained in four separate temperature scans: 25-310°C, 310-480°C, 480-630°C, and 630-680°C. No adjustments were made to match the data between sequential scans. The data from sequential scans agreed within the ± 5% accuracy limit of the DSC. The data obtained from the DSC were also in good agreement with the specific heat data obtained from the enthalpy curve. The sharp peak in the specific heat is characteristic of a Curie transition. However, the shape of the specific heat curve in this transition region is only an approximation to the true specific heat function. A Curie transition is a transition of the second kind which gives a discontinuity in the slope of the enthalpy function. A curve fitting routine with a limited number of data points in the transition region smooths over this discontinuity. The asterisks in Fig. 5 represent literature values for the specific heat of a similar alloy [16]. These data are in good agreement with the data shown for the A517 steel except in the narrow range of the transition. The points represented by the + symbols are discussed later.

The resulting thermal conductivity values based on these data are shown in Figure 6. The thermal conductivity was seen to decrease with increasing temperature until reaching a minimum at approximately 900°C. The thermal conductivity then increased until reaching a constant value at approximately 1050°C. The narrow peak in the conductivity at approximately 760°C is believed to be an artifact resulting from the curve fit used for the specific heat in Eq. 1. The thermal conductivities of similar alloys measured using steady-state methods show no discontinuities through the Curie transition region [17] and there is no known theoretical model that would predict such a large discontinuity. This discrepancy is discussed in the following section.

DISCUSSION

The calculation of thermal conductivity using Eq. 1 imposes stringent requirements on the accuracy of the measured thermal parameters α, ρ, and c_p, especially in regions where the parameters

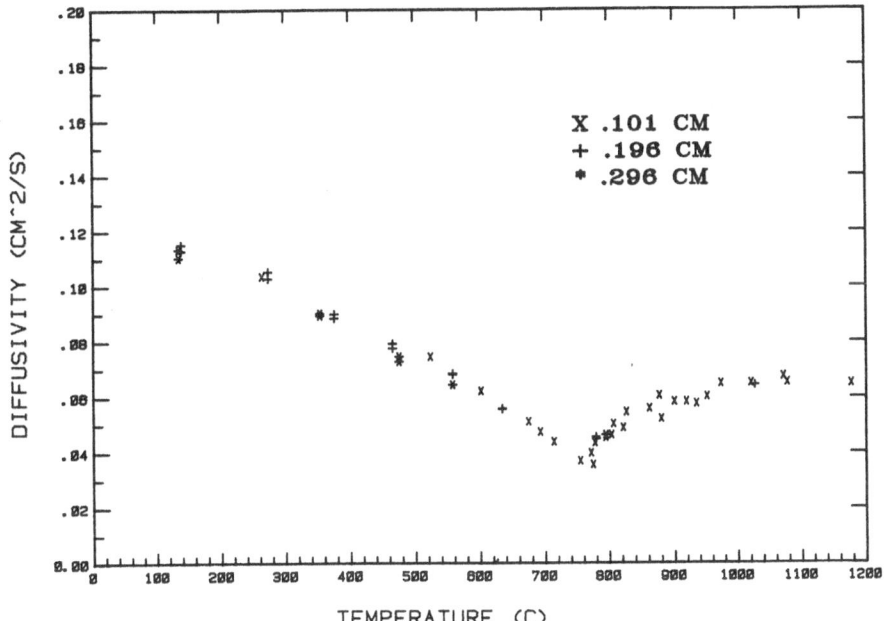

Fig. 3. Thermal diffusivity data.

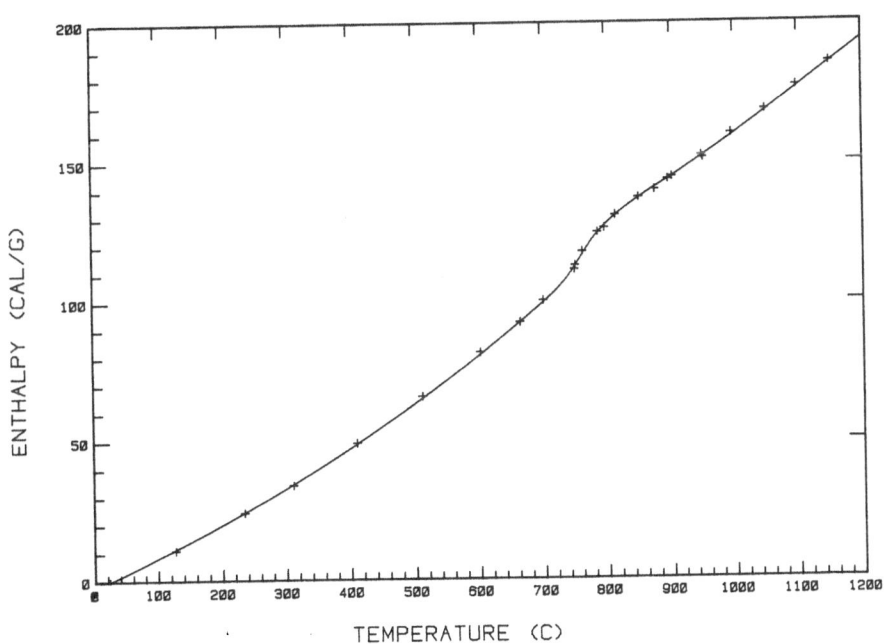

Fig. 4. Enthalpy data.

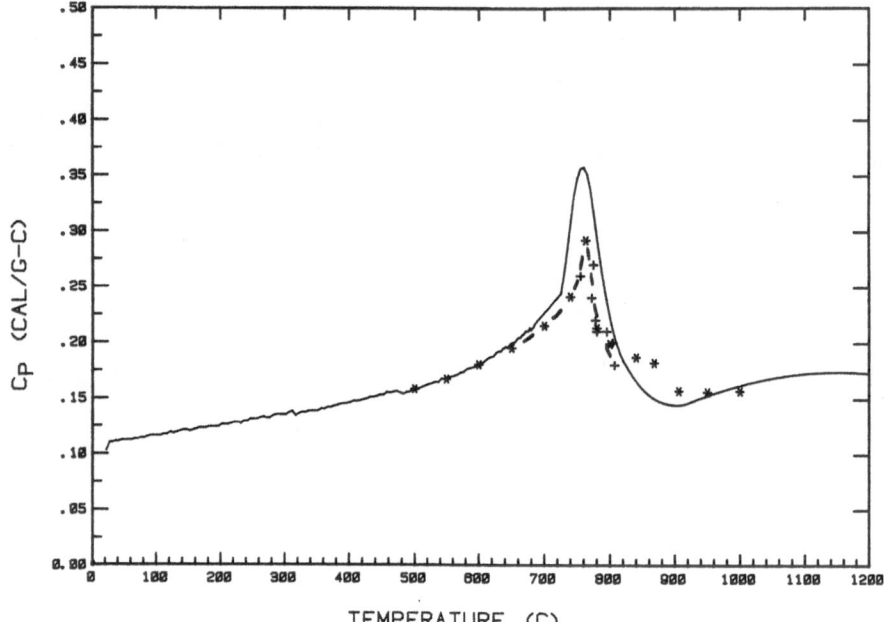

Fig. 5.   Specific heat data.

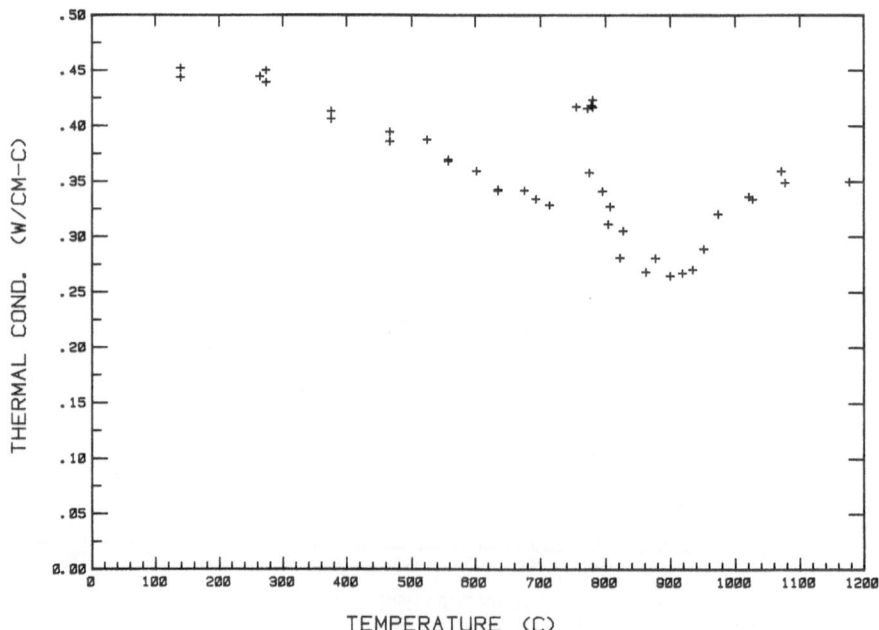

Fig. 6.   Thermal conductivity data.

are changing rapidly with temperature. The narrow peak in the thermal conductivity shown in Fig. 6 results from the approximate form of the specific heat used in the transition region. A more accurate representation of the specific heat in this region might eliminate the discontinuity. A calculation was performed to determine the specific heat values that would be necessary to eliminate the sharp peak in the calculated thermal conductivity. These values are indicated by the + symbols in Fig. 5 and agree closely with the literature values shown in that region. The dashed line shows the expected form of the specific heat in the transition region. However, examination of Fig. 4 shows that if the transition region were narrower, the peak in the specific heat would be higher. In order for the thermal conductivity to remain continuous in this narrower transition region, the thermal diffusivity would have to be correspondingly lower in that region. Although no sharp drop in diffusivity was measured, more measurements in this transition region might have resolved the effect of the transition. The overall confidence limits on the calculated thermal conductivity, excluding the transition region, is 8% based on the uncertainty in the measured values of α, ρ, and c_p.

SUMMARY

The determination of thermal conductivity requires the use of several experimental techniques. This paper has described how three separate experimental systems were used to determine thermal conductivity from 100-1200°C, while maintaining reasonable accuracy and precision limits. The flash diffusivity method has been shown to be a useful method for determining thermal conductivity for small samples at elevated temperatures. This method has the advantage of determining thermal conductivity based on small perturbations in the sample temperature which is important for materials whose properties vary sharply with temperature. One limitation of the flash diffusivity method is that measurements must also be made of density and specific heat over the same temperature range. Care must be taken that these measurements accurately represent the temperature-dependent structure of the specific heat function.

REFERENCES

1. W. J. Parker, R. J. Jenkins, et al, "A Flash Method of Determining Thermal Diffusivity, Heat Capacity, and Thermal Conductivity", U. S. Navy Technical Report USNRDL-TR-424, May, 1960.
2. R. E. Taylor, "Heat Pulse Thermal Diffusivity Measurements", High Temperature-High Pressures, 11, 213 (1979).
3. J. A. Cape and G. W. Lehman, "Temperature and Finite Pulse-Time Effects in the Flash Method for Measuring Thermal Diffusivity", J. Appl. Phys., 34, 1909 (1963).

4. R. C. Heckman, "Finite Pulse-Time and Heat-Loss Effects in Pulsed Thermal Diffusivity Measurements", J. Appl. Phys., 44, 1455 (1973).
5. R. D. Cowan, "Pulse Method of Measuring Thermal Diffusivity at High Temperatures", J. Appl. Phys., 34, 926 (1963).
6. L. M. Clark and R. E. Taylor, "Radiation Loss in the Flash Method for Thermal Diffusivity", J. Appl. Phys., 46, 714 (1975).
7. J. A. Koski, "Improved Data Reduction Methods for Laser Pulse Diffusivity Determination with the use of Minicomputers", Proceedings of the Eighth Symposium on Thermophysical Properties, Vol. II: Thermophysical Properties of Solids and of Selected Fluids for Energy Technology, ASME, New York, (1982), pp. 94-103.
8. E. Fitzer, "Results of the Cooperative Measurements on Heat Transport Properties up to 2800 K", AGARD-R-606, Technical Editing and Reproduction Ltd., Harford House, 7-9 Charlotte St., London WIP 1HD, 1973.
9. Perkin-Elmer Corp., Norwalk, Conn.
10. H. P. Stephens, "The Enthalpy and Specific Heat of a Nickel-Cobalt Steel to 2000 K", High Temp. Sci., 10, 95 (1978).
11. E. P. Roth, "Measurement of the Enthalpy and Specific Heat of a Be_2C-Graphite-UC_2 Reactor Fuel Material to 1980 K", Int. J. of Thermophysics, 3, 45 (1982).
12. J. F. Lathrop, D. L. Crawford, and R. J. Hanson, "DATFIT-User's Manual, An Interactive Least Squares Data Fitting Program", Sandia National Laboratories Report SAND80-8204.
13. Metals Handbook 9th Edition, Vol. 1, Properties and Selection: Irons and Steels, American Society for Metals, Metals Park, OH 44073, p. 187.
14. Korad Laser Systems, 2520 Colorado Avenue, Santa Monica, CA 90406.
15. Properties Research Laboratory, Box 2224, West Lafayette, Indiana 47906.
16. Thermophysical Properties of Matter, the TPRC Data Series, Vol. 4, Y. S. Touloukian and E. H. Buyco, IFI/Plenum, New York-Washington, (1970) p. 647.
17. Thermophysical Properties of Matter, the TPRC Data Series, Vol. 1, Y. S. Touloukian, R. W. Powell, C. Y. Ho, P. G. Klemens, IFI/Plenum, New York-Washington, (1970), p. 1182.

SENSITIVITY AND ACCURACY ANALYSIS OF PULSE

DIFFUSIVITY MEASUREMENTS ON LAYERED SAMPLES*

J. A. Koski

Sandia National Laboratories
Albuquerque, New Mexico   87185

ABSTRACT

Measurement by the laser pulse technique of the thermal diffusivity of an unknown layer in a planar layered sample consisting of two or three materials requires knowledge of the thermal
properties and thicknesses of the remaining sample layers. This
paper examines the sensitivity of the technique to uncertainties
in the known parameters of the adjacent layers, and applies the
results of the sensitivity analysis to the topic of sample selection and assessment of the overall accuracy of the calculated thermal diffusivity. Methods of estimating and bounding
the error in the calculated diffusivity are also discussed.
Finally, some typical cases for two- and three-layer samples are
presented, and some observations regarding the cases are discussed.

INTRODUCTION

With the laser pulse diffusivity technique[1], the thermal
diffusivity of a material is determined by subjecting one side of
a disk-shaped sample to a pulse of laser energy while recording
the temperature versus time history on the opposite side of the
sample. Analytical models are then used to relate the measured
temperature history to the thermal diffusivity of the sample.
The necessity for measurement of layered samples with this
technique occurs for several different circumstances. For

---

*This work was supported by the U.S. Department of Energy (DOE)
 under contract #DE-AC04-DP00789.

example, to measure the thermal diffusivity of a liquid with the laser pulse method, the walls of a container must be considered during data analysis. Similarly, measurements of coatings which are not easily removed from their base or substrate, and the measurement of transparent media require analysis of samples composed of two or three layers.

Unlike single-layer samples, which often require only knowledge of the sample thickness and the half-rise time to the peak temperature in order to determine thermal diffusivity, analysis of multilayer samples requires the additional knowledge of the density-specific heat products and thicknesses of all sample layers. The thermal diffusivity of all materials but the unknown one must also be known. Errors in measurement of these additional parameters are propagated through the data reduction analysis and result in errors in the diffusivity determined for the unknown material.

ANALYTICAL BASIS

The one-dimensional analytical solutions presented by Lee[2] have been chosen as a basis for sensitivity studies and data analysis algorithms. These solutions, with adiabatic boundaries assumed following the input pulse, represent the temperature rise at the back face of the two or three layer samples shown in Figure 1.

Three Layer Solution

For the three layer sample shown in Figure 1 (b), the normalized temperature rise at the back of the sample may be written with the use of Lee's notation as

$$T/T_\infty = 1 + 2 \sum_{k=1}^{\infty} \frac{(\omega_1 X_1 + \omega_2 X_2 + \omega_3 X_3 + \omega_4 X_4) \ Q(\gamma_k, \tau_p, \tau)}{\omega_1 X_1 \cos(\omega_1 \gamma_k) + \omega_2 X_2 \cos(\omega_2 \gamma_k) + \omega_3 X_3 \cos(\omega_3 \gamma_k) + \omega_4 X_4 \cos(\omega_4 \gamma_k)} \tag{1}$$

where,

$$\begin{aligned}
X_1 &= H_{1/3} \eta_{3/1} + H_{1/2} \eta_{2/1} + H_{2/3} \eta_{3/2} + 1, \\
X_2 &= H_{1/3} \eta_{3/1} - H_{1/2} \eta_{2/1} + H_{2/3} \eta_{3/2} - 1, \\
X_3 &= H_{1/3} \eta_{3/1} - H_{1/2} \eta_{2/1} - H_{2/3} \eta_{3/2} + 1, \\
X_4 &= H_{1/3} \eta_{3/1} + H_{1/2} \eta_{2/1} - H_{2/3} \eta_{3/2} - 1,
\end{aligned} \tag{2}$$

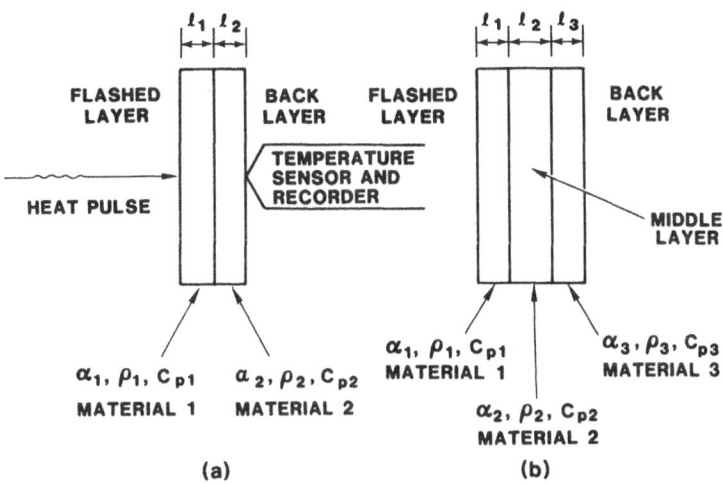

Fig. 1. Cross sections of layered pulse diffusivity samples.

$$\omega_1 = \eta_1/3 + \eta_2/3 + 1,$$

$$\omega_2 = \eta_1/3 + \eta_2/3 - 1,$$

$$\omega_3 = \eta_1/3 - \eta_2/3 + 1,$$ \hfill (3)

$$\omega_4 = \eta_1/3 - \eta_2/3 - 1,$$

The parameter T_∞ is related to the shape of laser input pulse and scales Equation (1) so that $0 \leqslant T/T_\infty \leqslant 1$. The T_∞ parameter is defined with the discussion of the expression for Q below. The γ_k in Equation (1) are the sequential positive real roots of the characteristic equation

$$X_1\sin(\omega_1\gamma) + X_2\sin(\omega_2\gamma) + X_3\sin(\omega_3\gamma) + X_4\sin(\omega_4\gamma) = 0 \qquad (4)$$

The $H_{i/j}$ and $\eta_{i/j}$ are defined as

$$H_{i/j} = H_i/H_j \qquad (5)$$

$$\eta_{i/j} = \eta_i/\eta_j \qquad (6)$$

where

$$H_j = a\rho_j c_{pj} \ell_j , \qquad (7)$$

$$\eta_j = (\ell_j{}^2/\alpha_j)^{1/2} \tag{8}$$

In the last expression, a is the cross sectional area of the sample, while α_j is the thermal diffusivity, ℓ_j is the thickness, ρ_j is the density, and c_{pj} is the specific heat at constant pressure of the jth sample layer.

The expression $Q(\gamma_k, \tau_p, \tau_3)$ in Equation (1) is related to the shape of the input heat pulse from the laser. For this analysis, the triangular pulse analysis of Heckman[3] was chosen. From Heckman's analysis

$$Q(\gamma_k, \tau_p, \tau_j) = \frac{2}{\beta}\left(\frac{1}{\gamma_k{}^4 \tau_p{}^2}\right)\left(e^{-\gamma_2{}^2\tau_j}\right)\left[1 + \frac{1}{(1-\beta)}\left\{\beta e^{\gamma_k{}^2\tau_p} - e^{\beta\gamma_k{}^2\tau_p}\right\}\right] \tag{9}$$

where, τ_j is the Fourier modulus, $\alpha_j t/\ell_j{}^2$, based on properties of the back layer of the sample, τ_p is the Fourier modulus based again on properties of the back layer for the pulse width of the base of the triangular pulse, and β is the fraction of the total pulse time before the peak of the triangle is reached. For the current analysis, values 0.28 for β and a pulse time of 1.35 ms were chosen as typical for the Sandia system.

For a triangular pulse shape, the temperature scaling factor, T_∞, for Equation (1) is [3]

$$T_\infty = \frac{F_0\, t_p}{2\,\rho_j\, c_{pj}\, \ell_j} \tag{10}$$

where F_0 is the magnitude of the pulse peak power in W/cm^2 and t_p is the the pulse width at the base of the triangle in seconds. The value of T_∞ is both system and sample dependent. For convenience in the present analysis, T_∞ is assumed to be unity. The value chosen for T_∞ does not affect the results given later in the paper.

The normalized temperature rise of a two-layer sample is obtained from Equation (1) by equating the properties for the second and third layers. The two-layer solution is presented in Reference 2.

While Lee's expressions are useful for the current analysis they are based on a one-dimensional analysis and do not contain the information to assess the impact of the following effects: non-uniformity of the laser beam, variation in the thickness across the layers, effect of temperature-dependent material

528

properties, and the lack of homogeneity in the material properties. Analysis of these effects would currently require the use of much less general techniques of numerical heat-transfer analysis.

One result which is not obvious from Lee's solutions is that the time versus temperature history measured at the back face of the sample is the same regardless of which sample face becomes the flashed face. This can be demonstrated rigorously for the two-layer case with the solutions for more general boundary conditions provided by Wirth and Rodin[4]. From numerical calculations with Equation (1) the same consistency of temperature history at the back face also appears true for the three-layer case of Equation (1) as long as the entire sample is simply rotated so that the opposite face is flashed. Rigorous proof of this effect for the three-layer case may also be possible with the multiple-layer solutions of Wirth and Rodin, but was not attempted for the current analysis.

SENSITIVITY ANALYSIS

Two different techniques have been used to assess the sensitivity of the calculated thermal diffusivity to the various parameters required for data reduction. The first method, based on analysis techniques described by Beck and Arnold[5], relies on the evaluation of sensitivity coefficients. For the second method, a data reduction program determines the effect on calculated diffusivity caused by varying one or more of the program input parameters.

Sensitivity coefficients, as defined by Beck and Arnold[5], are

$$S_i = \frac{\partial T}{\partial \beta_i} \tag{11}$$

where T is the temperature at the back sample face derived from expressions such as Equation (1) and the β_i are the problem parameters such as diffusivity, density-specific heat product, and thickness of the various layers. An alternative or normalized form of the parameters more suitable to sensitivity analysis is

$$S_i = \beta_i \frac{\partial T}{\partial \beta_i} \tag{12}$$

which may be viewed as the change δT in T given a small fractional change in β_i, i.e., $\delta T \cong S_i \, \delta\beta_i / \beta_i$. As shown in Figure 2, when the sensitivity coefficients are plotted versus time or Fourier modulus, the relative importance of various parameters to the temperature can be examined, and the time at which the temperature is most sensitive to the parameter may also be seen.

To determine the effect of other parameters on the measured diffusivity, an approximate technique was developed which uses ratios of the sensitivity coefficients. Since the sensitivity to most parameters peaks near the half rise time, $t_{1/2}$, (see Figure 2), the change in measured diffusivity relative to other parameters not contained in α can be approximated with the expression

$$\left.\frac{\partial \alpha}{\partial \beta_i}\right|_{t=t_{1/2}} \cong \left(\left.\frac{\partial T}{\partial \beta_i}\right|_{t=t_{1/2}}\right)\left(\left.\frac{\partial T}{\partial \alpha}\right|_{t=t_{1/2}}\right)^{-1} \tag{13}$$

and the resulting error in α approximated by

$$\frac{\Delta \alpha}{\alpha} = \frac{1}{\alpha} \quad \frac{\partial \alpha}{\partial \beta_1} \quad \Delta \beta_1 \tag{14}$$

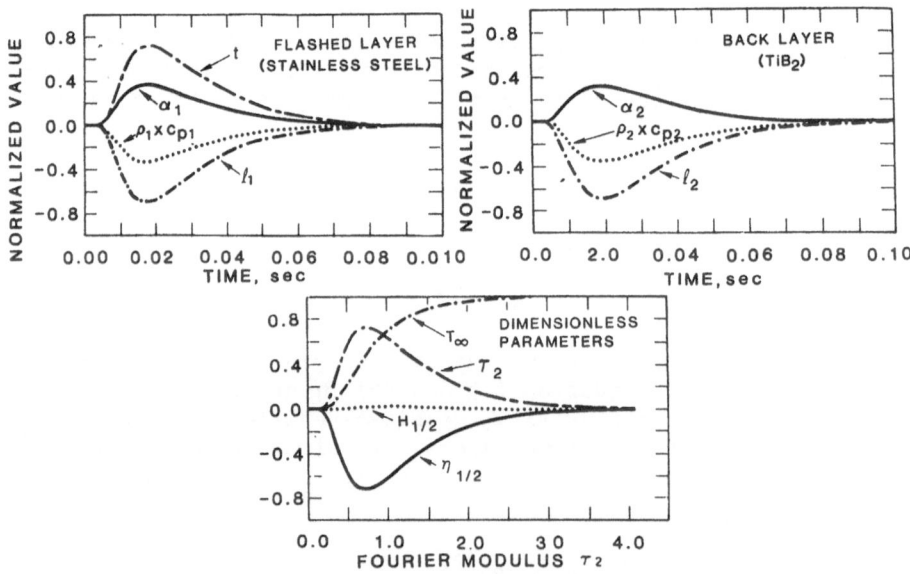

Fig. 2. Normalized sensitivity coefficients for simulated run with 0.24 cm thick TiB$_2$ coating on a stainless steel substrate 0.046 cm thick. Upper curves show sensitivity to dimensional parameters for flashed stainless steel and TiB$_2$ layers. Bottom curve shows sensitivity to dimensionless problem parameters. Material properties are the same as given with Table 1.

Table 1. Errors in Measured Diffusivity Resulting From Measurement Error in Other Parameters*

Measured Known Parameter	Assumed % Error in Known Parameter	Resulting % Error in Diffusivity, α_2, From Data Reduction Program	Resulting % Error in Diffusivity, α_2, Predicted From Sensitivity Coefficients
α_1	± 5	-6.8 $+8.3$	∓ 7.2
$(\rho c_p)_1$	± 5	$+5.3$ -4.7	± 5.2
$(\rho c_p)_2$	± 5	± 0.4	± 0.4
ℓ_1	± 1	$+2.2$ -2.1	± 2.1
ℓ_2	± 5.5	$+11.8$ -10.5	± 11.4
$t_{1/2}$	± 1.7	-3.6 $+3.7$	∓ 3.8

* For simulated case of TiB_2 coating on stainless steel substrate. For stainless steel assumed parameters are $\alpha_1 = 0.048$ cm^2/s, $(\rho c_p)_1 = 4.32$ J/cm^3-K, $\ell_1 = 0.0457$ cm. For TiB_2, $\alpha_2 = 0.023$ cm^2/s, $(\rho c_p)_2 = 2.56$ J/cm^3-K, $\ell_2 = 0.024$ cm.

As shown in Table 1, agreement between errors calculated from Equation (14) and errors estimated by varying the input to a data reduction program is good.

To facilitate the analysis, computer codes have been developed which numerically calculate and plot the normalized sensitivity coefficients for both the two- and three-layer cases. Simple difference formulas of the form

$$\frac{\partial T}{\partial \beta} \cong \frac{T(\beta + \delta\beta) - T(\beta)}{\delta\beta} \qquad (15)$$

are used to approximate the partial derivatives. Equations (1) and Lee's two-layer solution have also been used as the basis for data reduction codes which determine the diffusivity of the unknown material, given the necessary input parameters and the time to reach half the peak temperature at the back face.

Similarity of the shapes of the various curves for the sensitivity coefficients shown in Figure 2 indicates that the curves are linearly dependent, and curve-fitting techniques could not accurately determine separate values for thermal diffusivity and density-specific heat products of an unknown sample layer from pulse diffusivity data alone. If diffusivity and the density-specific heat product could be used as separate fit parameters in a nonlinear least-squares curve fit, for example, the necessity for separate experimental determination of densities and specific heats could be eliminated. A further discussion of the consequences of such linear dependence on parameter estimation methods is included in Beck and Arnold[5].

ANALYSIS FOR THE TWO-LAYER CASE

The figure and table provided as examples in the previous section are based results obtained by Roth with the pulse diffusivity apparatus at Sandia[6]. The sample consisted of a 0.024 cm thick plasma-sprayed coating of TiB_2 on a Type 304 stainless steel substrate 0.0457 cm thick. The stainless steel side of the sample was exposed to the laser pulse. The particular case chosen as an example represents results obtained at approximately 350°C.

Inspection of Figure 2 indicates that the overall error sensitivity of the layers is roughly equal. This is in contradiction to a simple diffusion-time analysis in which the parameters ℓ^2/α for each layer are compared. For the TiB_2 layer, ℓ^2/α is 0.025 s while for the stainless substrate, ℓ^2/α is 0.044 s, which would indicate that the properties of stainless steel would dominate the analysis, leading to large errors in determining the diffusivity of TiB_2. The calculated results demonstrate that this is not the case. Thus, while the diffusion time estimate may be useful as a rule of thumb, it does not provide an accurate quantitative assessment of probable errors.

The data reduction program for the two-layer case was also used to determine what would happen if all the errors in Table 1 were applied simultaneously. Total errors of + 35% and - 21% were determined by adjusting input parameters so that all changes in input values simultaneously increased or decreased the measured diffusivity. This bounding technique could be extended with statistical methods[5] to determine confidence limits on the measured diffusivity.

ANALYSIS FOR THE THREE-LAYER CASE

To examine the various effects which can occur with a three-layer sample, a hypothetical case has been considered for full-density TiC coatings of unknown diffusivity on both sides of a known substrate material. This simulates what might be obtained with new low-pressure-plasma-spray coating techniques[7] in which the plasma spray process is completed under vacuum. The full material density represents a best case from the plasma-spray view point, but a worst case for the accurate determination of the diffusivity of the coating. Coatings on both sides of the substrate were simulated to maximize the amount of TiC in the sample, because current plasma-spray techniques limit TiC coating thickness to approximately 0.025 cm.

In Figure 3a the same stainless steel substrate material considered for the TiB_2 case in the previous section is simulated. Because of the high diffusivity of the full density TiC (0.1 cm^2/s) and the relatively low diffusivity (0.04 cm^2/s) of the stainless steel, the transit time of the energy pulse is dominated by the substrate and the sensitivity to the TiC coatings is small. Thus, the high thermal diffusivity which makes TiC an attractive armor coating also makes determination of the diffusivity difficult.

In Figure 3b, the stainless steel substrate is replaced with a 0.025 cm copper substrate. Because of its higher diffusivity, 0.08 cm^2/s, and smaller thickness, one half of the stainless steel thickness, the sensitivity to the diffusivity of the TiC is improved. Note that the total sensitivity to the two TiC layers is twice that for the single layer value shown in Figure 3b. Even in this case, the sensitivity to the coating layers is marginal.

An interesting simulation result is shown in Figure 4. Here the same total thickness, 0.05 cm, of TiC has been applied to only one side of the 0.025 cm copper substrate with the result that the sensitivity to the TiC is substantially increased. This indicates that the measurement sensitivity is influenced by position as well as by total thickness. This effect can be qualitatively predicted by a simple diffusion time analysis, since the ℓ^2/α evaluated for the single TiC layer is twice the total diffusion time for two individual layers with half the original thickness. In this example the three-layer case worked against accurate determination of the diffusivity. If

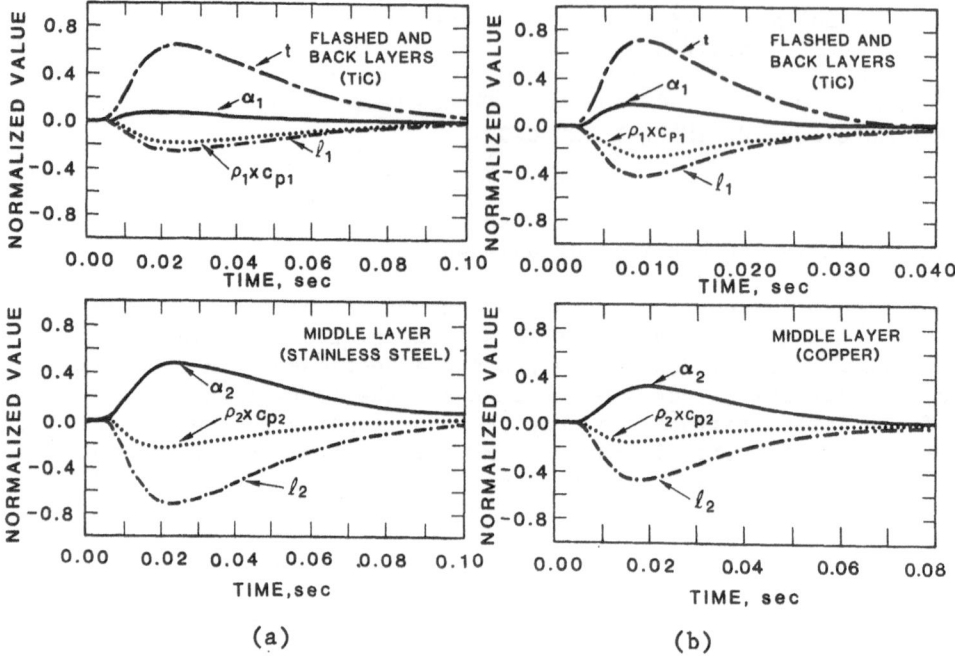

Fig. 3. Sensitivity coefficients for two different three-layer
samples with 0.025 cm thick TiC coatings of full density on both sides of substrate. Left figures represent sensitivities with 0.05 cm thick stainless steel
middle layer, while 0.025 cm thick copper middle layer
is assumed on the right. Assumed material properties
are as follows: for TiC, α = 0.095, ρc_p = 4.2; for
stainless steel, α = 0.048 ρc_p = 4.32; for copper
α = 0.083, ρc_p = 4.2.

the middle layer were the unknown layer, then this effect would
enhance the accuracy with which of the diffusivity can be determined.

Another effect noticed during analysis of three-layer samples is that the time-temperature history at the back face can
be changed when the actual sequence of layers is altered rather
than simply reversed, i.e., when the material of the middle
layer is made the outer layer. This means that when determining
the thermal diffusivity with three-layer samples, knowledge of
the actual sequence of materials is important to the data reduction process.

534

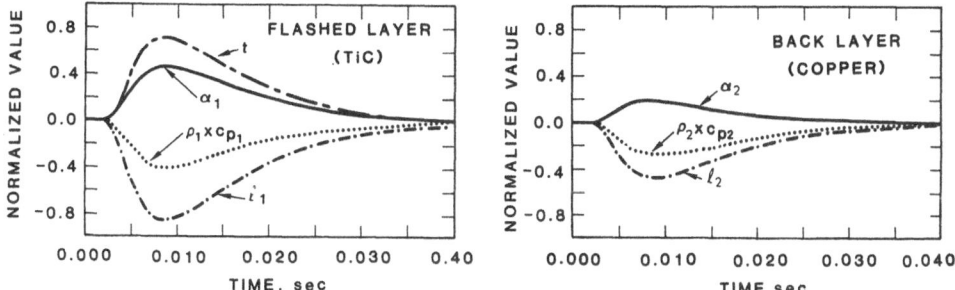

Fig. 4. Sensitivity coefficients for two-layer sample with same
materials and total thickness as Figure 3b. Note
increased sensitivity to TiC layer.

CONCLUSIONS

A methodology for sensitivity analysis of specific two- and
three-layer pulse diffusivity analysis has been demonstrated,
and some unusual effects noted. Further analysis is indicated
to provide a more general understanding of the effects in terms
of the dimensionless parameters present in the temperature expressions such as Equation (1). Inspection of plots of the
sensitivity coefficients, besides showing the relative importance
of the problem parameters, indicated that separate determination
of diffusivity and density-specific heat products could not be
made with pulse diffusivity data alone. Separate determination
of the density and specific heat of all sample layers is required.

In the cases presented, simple rules of thumb involving heat
diffusion times of the various layers, ℓ^2/α, did not adequately
predict accuracy of the diffusivity determination. This does
not mean that simple error-prediction correlations for accuracy
could not be developed, however.

For a given multi-layer sample, the temperature-versus-time
history at the back face appears to be invariant with regard to
which face which is exposed to the laser flash. For the two-
layer case, this can be rigorously demonstrated to be true. For
the three-layer case, specific calculations indicate this invariance may also be true, but a rigorous proof has not been
attempted.

For three-layer samples, rearranging the order of the materials in the layers has been observed to alter the temperature-versus-time history at the back face. This means that the correct order of layering must be used in data analysis programs.

When one layer of a two-layer sample is split and rearranged to form a three-layer sample, the sensitivity to the split material has been observed to be reduced, while the sensitivity to the material in the middle layer is enhanced.

REFERENCES

1. R. E. Taylor, Heat Pulse Thermal Diffusivity Measurements, High Temperature - High Pressure, 11: 43 (1979).

2. H. J. Lee, "Thermal Diffusivity in Layered and Dispersed Composites," Doctoral Thesis in Mechanical Engineering, Purdue University, 1975.

3. R. C. Heckman, Finite Pulse-Time and Heat-Loss Effects in Pulse Thermal Diffusivity Measurements, J. Appl. Phys., 44: 1455 (1973).

4. P. E. Wirth and E. Y. Rodin, A Unified Theory of Linear Diffusion in Laminated Media, in: "Advances in Heat Transfer, Vol. 15," J. P. Hartnett and R. F. Irvine, Jr. Eds., Academic Press, New York, 1982.

5. J. V. Beck and K. J. Arnold, "Parameter Estimation in Engineering and Science," John Wiley and Sons, New York (1977).

6. E. P. Roth, Measurement of Thermal Conductivity from High Temperature Pulse Diffusivity and Calorimetry Measurements, This Volume.

7. R. W. Smith, W. F. Schilling, and H. M. Fox, Low-Pressure Plasma Spray Coatings for Hot-Corrosion Resistance, Jour. of Eng. for Power, 103: 146 (1981).

HEAT PULSE THERMAL DIFFUSIVITY MEASUREMENTS-THERMAL PROPERTIES

TEMPERATURE DEPENDENCE AND NON-UNIFORMITY OF THE PULSE HEATING.

Alain Degiovanni, Gilbert Sinicki and Michel Laurent

I.N.S.A. - Laboratoire de Physique Industrielle

20, avenue Albert Einstein - 69621 Villeurbanne Cedex

ABSTRACT

In classical methods for measuring thermal diffusivity by pulse methods (laser, flash, electron beam), two hypotheses are generally assumed :
The thermal pulse is taken to be uniform over the front surface of the specimen. This paper deals with the cases where these hypotheses cannot be applied.
The heat transfer in the specimen obeys a system of linear equations; thermal characteristics and boundary conditions are constant.

Non-uniform pulse

Instead of doing a two-dimensional analysis of the heat transfer, we show that the use of an average temperature gives a simple solution which is the same as that which would be obtained if the heat pulse were uniform on the front face of the specimen. The temperature versus time curve which must be used to calculate the thermal diffusivity is the average temperature of the back face.
Practically, it is not necessary to know the energy distribution of the laser beam.

Nonlinear heat transfer

We investigate the influence of nonlinearities upon the measurements and we specify the range of validity of a previously proposed method for determining thermal diffusivity. We introduce three parameters of nonlinearity and study their influences upon the measured thermal diffusivities. We deduce the test temperature which must be used and the limitations of validity of the method. A majority of cases can be handled.

$$A_\lambda = \frac{Q}{(\rho c)_o \lambda_o \, e} \, (\frac{d\lambda}{dT})_{T_o}$$

$$A_{\rho c} = \frac{Q}{(\rho c)_o^2 \, e} \, (\frac{d(\rho c)}{dT})_{T_o}$$

$$B = \frac{Q}{(\rho c)_o e T_o}$$

c : specific heat
e : thickness
f(r): function
g(z) : function
h : heat transfer coefficient

$$H = \frac{4\sigma \, \varepsilon_t \, e T_o^3}{\lambda_o}$$

$$L = h \, e/\lambda$$

n,p : integers
P   : equation
Q   : pulse energy

$$\dot{q}^* = \frac{\dot{q} \, e^2}{Q\alpha_o}$$

$\dot{q}\,(t)$: flux corresponding to the flash energy
$\dot{q}_r(T)$: flux corresponding to heat losses
R   : radius of the sample
r   : coordinate
t   : time
T   : temperature

$t^* = \alpha t / e^2$

$\underline{T}^* = T\rho c\, e\,/\,Q$ or $(T-T_o)\rho c\, e/Q$

$\overline{T}$: average temperature

T_M : maximum temperature with or without heat losses

$t(1/3), t(1/2), t(2/3), t(5/6)$: defined on figure 2

z : coordinate

α : thermal diffusivity

β : coefficients corresponding to the temperature for which the actual diffusivity equals the calculated one.

ε_t : emmissivity

ε : small number

λ : thermal conductivity

λ^* : $\lambda\!\!\!/\, \lambda_o$

e : density

σ : Stephan constant

$\overline{\theta}_o$: $\dfrac{2}{R^2} \displaystyle\int_0^R f(r)\,dr$

$(\rho c)^* = \rho c / (\rho c)_o$

$A_{np}, A_1, A_2, A_3, A_4, F_p(r), G_m, V_{np}, u_n, \omega_p$ terms defined in the text.

Subscripts

e : at $z = e$

m : average value

o : at : $z = o$ or $t = o$ or $T = T_o$

R : at $r = R$

INTRODUCTION

A very simple method for determination of the diffusivity is one in which a sample in the form of a disk is brought to a steady temperature. A flash of thermal energy is supplied on the front surface of the sample and the transient temperature is observed on the back surface Fig.1.

This method has become very popular since it was first described by Parker, Jenkins, Buttler and Abbott[1].

In a previous paper[2] the thermal properties of the sample were considered to be constant even though the temperature on the front surface may greatly increase, and the flash of thermal energy was assumed to be uniform over the front surface of the sample although that is not always the case (for instance, for laser beams).

This paper deals with a way to allow for these difficulties or to determine the conditions under which they can be neglected.

NONUNIFORM FLASH OF ENERGY

When we take into account three heat transfer coefficients on the surfaces of the sample, we assume the thermal properties to be constant; and the pulse to be delivered instantaneously, the equations to be solved are :

$$\frac{1}{\alpha} \frac{\partial T}{\partial t} = \frac{\partial^2 T}{\partial r^2} + \frac{1}{r} \frac{\partial T}{\partial r} + \frac{\partial^2 T}{\partial z^2} \qquad (1)$$

$$\frac{\partial T}{\partial z} = \frac{h_o}{\lambda} T \ , \ z = 0 \qquad (2)$$

$$\frac{\partial T}{\partial z} = -\frac{h_e}{\lambda} T, \ z = e \qquad (3)$$

$$\frac{\partial T}{\partial r} = -\frac{h_R}{\lambda} T \ , \ r = R \qquad (4)$$

$$T = f(r).g(z), \ t = 0 \qquad (5)$$

with α : thermal diffusivity
λ : thermal conductivity
ρ : density
c : specific heat

and we obtain,

$$T(r,z,t) = \sum_n \sum_p A_{np}.F_p(r).G_n(z).\exp\left(-V_{np} \frac{\alpha t}{e^2}\right)$$

with

$$A_{np} = \frac{A_1.A_2}{A_3.A_4}$$

$$A_1 = \int_o^e g(z).G_n(z)\ dz$$

$$A_2 = \int_o^R rf(r).F_p(r)\ dr$$

$$A_3 = \int_o^e G_n^2(z)\ dz$$

$$A_4 = \int_o^R rF_p^2(r)\ dr$$

$$F_p(r) = J_o(\omega_p \frac{r}{R})$$

$$G_n(z) = \cos(u_n \frac{z}{e}) + \frac{h_o e}{\lambda u_n} \sin(u_n \frac{z}{e})$$

$$V_{np} = u_n^2 + \frac{e^2}{R^2}\omega_p^2$$

with u_n and ω_p given by :

$$(u_n^2 - \frac{h_o h_e e^2}{\lambda^2})\ tg\ u_n = \frac{(h_o + h_e)e}{\lambda} u_n$$

$$\omega_p.J_1(\omega_p) = \frac{h_R R}{\lambda} R.J_o(\omega_p)$$

where J_o and J_1 are Bessel functions of the first kind of order zero and one, respectively. When the heat pulse is uniform on the front surface, this solution becomes :

$$T(e,r_c,t^*) = 2 \sum_n \sum_p A_{np}.F_p(r_c).G_n(e).\exp(-V_{np}t^*)$$

with

$$g(z) = \frac{Q}{\varepsilon pc}\ ,\ 0 < z < \varepsilon$$

$$g(z) = 0,\quad \varepsilon < z < e$$

$$f(r) = 1,\quad 0 \leq r \leq R$$

$$A_{np} = \frac{u_n^2(u_n^2 + L_e^2)}{(u_b^2+L_o^2).(u_n^2+L_e^2)+(L_o+L_e)(u_n^2+L_o L_e)} \times \frac{2 L_R}{(L_R^2+\omega_p^2).J_o(\omega_p)}$$

$$T^* = T\ \rho ce/Q$$

$$t^* = \alpha t/e^2$$

$$L_o = h_o e/\lambda,\ L_R = h_R R/\lambda,\ L_e = h_e e/\lambda$$

From this result we previously demonstrated (2,3,4), the thermal diffusivity was given by :

$$\alpha = \frac{e^2}{t_{5/6}} \cdot P, \ast$$

with

$$P = 7.1793 \ (\frac{t(2/3)}{t(5/6)})^2 - 11.9554 \ \frac{t(2/3)}{t(5/6)} + 5.1365 \qquad (7)$$

or

$$P = 0.6148 \ (\frac{t(1/2)}{t(5/6)})^2 - 1.6382 \ \frac{t(1/2)}{t(5/6)} + 0.968 \qquad (8)$$

or

$$P = 1.0315 \ (\frac{t(1/3)}{t(5/6)})^2 - 1.8451 \ \frac{t(1/3)}{t(5/6)} + 0.8498 \qquad (9)$$

The terms $t(1/3)$; $t(2/3)$; $t(5/6)$ are defined on figure 2.

If the flash of energy is not uniform on the front surface of the sample, it is impossible to calculate the thermal diffusivity with these equations, and it seems necessary to know the function $f(r)$. But recording the dependence of energy on radius in the laser or electron beams makes the experiment more complicated. There is a way to suppress that difficulty :

If we multiply the equations (1) to (5) by $\frac{2}{R^2}$ rdr and if we integrate them from 0 to R, we obtain :

$$\frac{\partial^2 \bar{\bar{T}}}{\partial z^2} + \frac{2}{R^2} \int_o^R \frac{\partial}{\partial r} \ (r \ \frac{\partial T}{\partial r}) \ dr = \frac{1}{\alpha} \ \frac{\partial \bar{\bar{T}}}{\partial t}$$

$$\frac{\partial \bar{\bar{T}}}{\partial z} = - \frac{h_o}{\lambda} \ \bar{\bar{T}} \ , \ z = 0$$

$$\frac{\partial \bar{\bar{T}}}{\partial z} = \frac{h_e}{\lambda} \ . \bar{\bar{T}} \ , \ z = e$$

$$\frac{\partial \bar{\bar{T}}}{\partial r} = 0$$

$$\bar{\bar{T}} = \frac{2}{R^2} \int_o^R f(r) \ rdr.g(z), t = 0$$

with

$$\bar{\bar{T}} = \frac{2}{R^2} \int_o^R T \ rdr.$$

---

*P may be calculated from one of the equations 7,8 or 9; it represents the heat loss from the sample to the surrounding. From the experimental curve, we must find three values of P (Equations 7,8 and 9) within two percent. It is a way to check the shape of the experimental curve, and therefore the experimental procedure.

With the extra assumption $\lambda(\frac{\partial T}{\partial r}) = 0$ at $r = R$,

$$\frac{2}{R^2} \int_o^R \frac{\partial}{\partial R} (r \frac{\partial T}{\partial r}) \, dr = 0;$$

then we obtain

$$\frac{\partial^2 \bar{T}}{\partial z^2} = \frac{1}{\alpha} \frac{\partial \bar{T}}{\partial t}$$

$$\frac{\partial \bar{T}}{\partial z} = - \frac{h_o}{\lambda} \bar{T} \, , \quad z = 0$$

$$\frac{\partial \bar{T}}{\partial z} = \frac{h_e}{\lambda} \bar{T}, \quad z = e$$

$$\bar{T} = \bar{\theta}_o \, g(z) \quad \text{with} \quad \bar{\theta}_o = \frac{2}{R^2} \int_o^R f(r) \cdot r \, dr$$

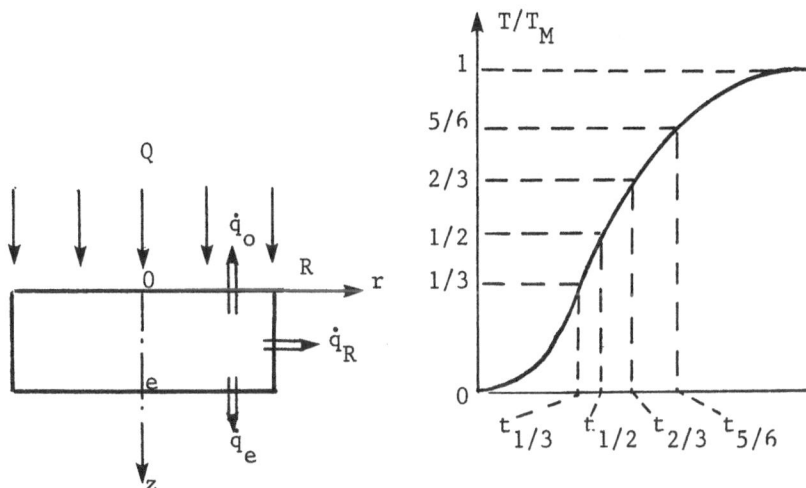

Fig. 1. Flash method on cylindrical sample

Fig. 2. Thermal diffusivity determination

From this new set of equations, one sees that if we measure the average temperature of the back surface of the sample, it does not matter if the flash of energy is not uniform on the front surface. Then, in order to determine the thermal diffusivity, we can use the equations for uniform pulse heating, equations (7) to (9). The radiative surface temperature measurements are suitable for measuring average temperature $\bar{T}$ (R,t).

In order to obtain this interesting result, we assumed ($\frac{\partial T}{\partial r} = 0$) no heat flux through the lateral surface of the sample.

But, by using the general solution (6) with lateral heat losses, corresponding to those of flash diffusivity measurements, we find no significant differences between the two average temperatures divided by the maximal average temperatures, with and without lateral heat losses.

At the beginning, to simplify the analysis, we have assumed the pulse to be delivered instantaneously; if the pulse duration is very long, we obtain the same result, by considering the long pulse as a succession of short pulses and adding the solutions corresponding to each short pulse.

Then for a nonuniform flash energy, it is sufficient to record the average temperature on the back surface to obtain the diffusivity. This "average temperature analysis" method can be used in other cases : for stacked disks having thermal contact resistance between them and nonuniform heating on the front surface and for cylindrical geometries.

By using this method for the study of the response of a calorimeter, we can show that the calorimeter response does not change for cylindrical cells if the heat is supplied in a small volume $2\pi r dr\, dz$ located at r and z in the cell, or if the heat is supplied all over the volume $h\, 2\pi r dr$ extending from the bottom to the top of the cell (h is the height of the cell).

THERMAL PROPERTIES VARIATIONS

We investigate how the flash energy acts upon the thermal diffusivity measurements with non-constant thermal properties. First we study the case of copper at low temperature and that of clay at room temperature. Then we study the influence of specific heat variations, thermal conductivy variations and radiative heat losses.

With non-constant thermal properties we obtain this set of equations :

$$\frac{\partial}{\partial z} \left(\left(\lambda(T)\, \frac{\partial T}{\partial z} \right) \right) = \rho c(T)\, \frac{\partial T}{\partial t}$$

$$T = T_o, \quad t = 0,$$

$$\lambda(T) \frac{\partial T}{\partial z} = - \dot{q}(t) + \dot{q}_r(T), \quad x = 0$$

$$\lambda(T) \frac{\partial T}{\partial z} = - \dot{q}_r(T), \quad x = e$$

Here we neglect lateral heat losses.
$\dot{q}(t)$ is the flux corresponding to the flash energy and, $\dot{q}_r(T)$ the heat losses.

With $z^* = \dfrac{z}{e}$; $t^* = \dfrac{\alpha_o}{e^2} t$; $T^* = \dfrac{T - T_o}{Q} (\rho c)_o e$

$$\lambda^* = \frac{\lambda}{\lambda_o} ; \quad (\rho c)^* = \frac{\rho c}{(\rho c)_o} ; \quad \dot{q}^* = \frac{\dot{q} e^2}{Q \alpha_o}$$

α_o : diffusivity at T_o

$(\rho c)_o$: specific heat by volume unit at T_o

λ_o : conductivity at T_o

$$Q = \int_o^\infty \dot{q}(t) dt$$

we obtain :

$$\frac{\partial}{\partial z^*} \left(\lambda^* \frac{\partial T^*}{\partial z^*} \right) = (\rho c)^* \frac{\partial T^*}{\partial t^*}$$

$$T = 0 \quad t^* = 0$$

$$\lambda^* \frac{\partial T^*}{\partial z^*} = - \dot{q}^*(t^*) + \dot{q}_r^*(T^*)$$

$$\lambda^* \frac{\partial T^*}{\partial z^*} = \dot{q}_r^*(T^*)$$

The nonlinear set of equations is solved by a method proposed by Zlamal[5] with a nonlinear point overrelaxation algorithm.
The flash energy is supplied during $t_f^* = 0.005$

Copper and clay measurements.

The thermal properties and other conditions are :

Copper :

$$\lambda^*(T^*) = 1/(T^* Q/94.2 + 1)$$

$$(\rho c)^*(T^*) = 2.54 - 1.54 \exp(-T^* Q/628)$$

$$T_o = 100K \ ; \ e = 2cm$$

$$\dot{q}_r^* = 0 \ ; \ 0.1 < Q < 50 \ J/cm^2$$

Clay :

$$\lambda^*(T^*) = \sqrt{1 + T^* Q/90}$$

$$(\rho c)^*(T^*) = 1$$

$$T_o = 300K \ ; \ e = 0.2cm$$

$$\dot{q}_r^* = 0 \ ; \ 1 < Q < 50 \ J/cm^2$$

Table 1 shows the thermal diffusivity divided by the thermal diffusivity at $T = T_o$ calculated from T^* (t^*) with equations (7) to (9) (α/α_o).

Table 1. Calculated diffusivity / α_o

	Copper				Clay		
Q in J/cm^2	0.1	1	10	50	1	10	50
Equation 7	0.97	0.96	0.86	0.60	0.98	1.03	1.26
Equation 8	0.98	0.97	0.87	0.62	0.99	1.04	1.23
Equation 9	0.99	0.98	0.88	0.63	1	1.05	1.22
$\frac{\alpha}{\alpha_o}$ at $T^*=1$	0.99	9.98	0.89	0.60	1.01	1.06	1.25
$T_M - T_o$	0.032	0.32	3.1	15.1	3.3	33.3	167

We can see that the thermal diffusivities coming from equations (7) to (9) are the same as the thermal diffusivity at $T^* = 1$. For the copper sample the temperature rise on the back surface is between 0.03 K to 16 K for a flash energy between 0.1 J/cm^2 to 50 J/cm^2, and the front-surface temperature rise is between 0.39 K and 210 K.
For the clay sample, the temperature rise on the back surface is between 3.3 K to 170 K for a flash energy between 1 J/cm^2 to 50 J/cm^2 and the front-surface temperature rise is between 170 K and 2200 K.

Specific heat variations and thermal conductivity variations.

By expanding $\rho c^*(T)$ and $\lambda^*(T)$ in a Taylor series we get :

$$(\rho c)^*(T^*) = 1 + \frac{QT^*}{(\rho c)_o^2 e} \left(\frac{d(\rho c)}{dT}\right)_{T_o^*} + \frac{1}{2} \left(\frac{QT^*}{e(\rho c)_o^2}\right)^2 \left(\frac{d^2(\rho c)}{dT^2}\right)_{T_o} + \ldots$$

$$\lambda^*(T^*) = 1 + \frac{QT^*}{(\rho c_o)\lambda_o e} \left(\frac{d\lambda}{dT}\right)_{T_o} + \frac{1}{2} \left(\frac{QT^*}{(\rho c)_o \lambda_o e}\right)^2 \left(\frac{d^2\lambda}{dT^2}\right)_{T_o} + \ldots$$

and by keeping the first term of the series, the heat transfer equation becomes :

$$\frac{\partial^2 T^*}{\partial z^{*2}} + A_\lambda \left[\left(\frac{\partial T^*}{\partial z^*}\right)^2 + T^* \frac{\partial^2 T^*}{\partial z^{*2}}\right] = \frac{\partial T^*}{\partial t^*} + A_{\rho c} T^* \frac{\partial T^*}{\partial t^*}$$

with

$$A_\lambda = \frac{Q}{(\rho c)_o \lambda_o e} \left(\frac{d\lambda}{dT}\right)_{T_o}$$

$$A_{\rho c} = \frac{Q}{(\rho c)_o^2 e} \left(\frac{d(\rho c)}{dT}\right)_{T_o}$$

The terms A_λ and $A_{\rho c}$ represent the temperature dependence of the thermal conductivity and specific heat, respectively.
For the radiative heat transfer, we obtain

$$\phi_r^* = \frac{H}{4B} \left((BT^* + 1)^4 - 1\right)$$

with

$$H = \frac{4\sigma \varepsilon_t e T_o^3}{\lambda_o}$$

$$B = \frac{Q}{(\rho c)_o e T_o}$$

with ε_t = emissivity.

In order to evaluate the values of $A_{\rho c}$, A_λ and B, we use the maximum temperature T_M reached by the sample with no heat loss :

$$A_{\rho c} = \frac{T_M - T_o}{(\rho c)_o} \left(\frac{d(\rho c)}{dT}\right)_{T_o}$$

$$A_\lambda = \frac{T_M - T_o}{\lambda_o} \left(\frac{d\lambda}{dT}\right)_{T_o}$$

$$B = \frac{T_M - T_o}{T_o} \; .$$

Then $A_{\rho c} < 0.5$, $A_\lambda < 0.5$ and $B < 0.1$, except for low temperature measurements ($T_o < 30$ K).

We consider three cases :

Case 1
$$\left| \begin{array}{l} \lambda^* = A_\lambda T^* + 1 \\ (\rho c)^* = 1 \\ \quad H = 0 \end{array} \right.$$

Case 2
$$\left| \begin{array}{l} \lambda^* = \dfrac{1}{A_\lambda T^* + 1} \\ (\rho c)^* = 1 \\ \quad H = 0 \end{array} \right.$$
(without the assumption of keeping only the first term of the Taylor series)

Case 3
$$\left| \begin{array}{l} \lambda^* = 1 \\ (\rho c)^* = A_{\rho c} T^* + 1 \\ \quad H = 0 \end{array} \right.$$

Table 2 shows α/α_o calculated with equations (7) to (9) and the diffusivity values α_M/α_o corresponding to $T^* = 1$ for $10^{-6} < A < 1$. when $A < 0.1$, the difference between α/α_o and α_M/α_o is smaller than two percent.

As W.A. Philips[6] did, we define the temperature T_e to be that for which the actual diffusivity equals the calculated one :

$$T_e = T_o + \beta(T_M - T_o) .$$

Table 2 Calculated diffusivity $/\alpha_o$

		A	10^{-6}	0.1	0.3	0.5	1
Case 1	α/α_o	Equation 9	0.987	1.085	1.239	1.382	1.702
		Equation 8	0.979	1.084	1.260	1.413	1.777
		Equation 7	0.989	1.090	1.269	1.419	1.800
	α_M/α_o		1	1.1	1.3	1.5	2
Case 2	α/α_o	Equation 9		0.895	0.768	0.677	0.553
		Equation 8		0.886	0.744	0.660	0.512
		Equation 7		0.883	0.729	0.632	0.451
	α_M/α_o			0.909	0.769	0.667	0.5
Case 3	α/α_o	Equation 9		0.920	0.842	0.783	0.722
		Equation 8		0.914	0.825	0.762	0.698
		Equation 7		0.912	0.819	0.755	0.672
	α_M/α_o			0.913	0.791	0.708	0.577

Table 3 Average value β_m and percentage deviation between α and the thermal diffusivity at T_e.

		β_m \\ A	0.1	0.3	0.5
Case 1	Equation 9	0.78	0.6	0.4	0.6
	Equation 8	0.84	0.0	0.6	0.5
	Equation 7	0.86	0.4	0.9	0.8
Case 2	Equation 9	0.99	1.6	0.4	1.2
	Equation 8	1.10	1.6	1.1	2.3
	Equation 7	1.20	1.1	0.8	1.1
Case 3	Equation 9	0.71	1.8	0.0	1.3
	Equation 8	0.80	1.6	0.0	1.3
	Equation 7	0.83	1.6	0.1	1.3

Table 3 shows the average value β_m and the percentage deviation from α for the thermal diffusivity at T_e. We do not find a general determination of β.

For a deviation less than one per cent we obtain :

$$\beta_m = \beta_o - 0.4 \, A$$

but β_o is different for each case

$$\beta_o = 0.97 \qquad \text{Case 1}$$
$$\beta_o = 1.20 \qquad \text{Case 2}$$
$$\beta_o = 0.9 \qquad \text{Case 3.}$$

Radiative heat transfer

Table 4 shows α/α_o for some value of H and B, it appears that the nonlinearity due to radiative heat transfer is not important and can be neglected.

Table 4. Calculated diffusivity $/\alpha_o$ for the case of radiative exchange.

H	B	Equation 9	Equation 8	Equation 7
0.02	$4 \ 10^{-3}$	0.99	0.99	1.02
0.02	$2 \ 10^{-2}$	0.99	0.99	1.015
2	$2 \ 10^{-3}$	1.01	1.00	0.995
2	$2 \ 10^{-2}$	1.04	0.99	0.995
2	10^{-1}	0.99	1.00	1.00

CONCLUSION

When a heat pulse is non-uniform over the front surface of the sample by using the average temperature of the back surface we can use the classical equations to determine the diffusivity; these equations coming from an analysis done with an uniform pulse.

The nonlinearity coming from radiative heat transfer is not important and can be neglected.

The nonlinearity coming from the conductivity and the specific heat variations with temperature may be neglected (error less than two per cent) if the term A of nonlinearity is smaller than 0.1, then the temperature corresponding to the diffusivity measurement is the maximum temperature of the recording. For A > 0.1 a specific study must be done for each case.

REFERENCES

1. W. J. Parker, R. H. Jenkins, C. P. Buttler, and G. L. Abbott, Flash Method of determining thermal diffusivity, Heat capacity and thermal conductivity, Journal of Applied Physics, Vol. 32, nb.9 (Sept. 1961).
2. H. Poncin avec la collaboration de A. Degiovanni, M. Laurent, and H. Lebodo, Thermocinétique impulsionnelle et mesure de la diffusivité thermique, Monographies du B.N.M., Editions Chiron (1978).
3. A. Degiovanni, Contribution à l'étude de la diffusivité thermique, Thèse de Doctorat d'Etat, n° d'ordre 75-19, Lyon (3 June 1975).
4. A. Degiovanni, Diffusivité et méthode flash, Revue générale de Thermique, nb.185 (May 1977).
5. M. Zlamal, A finite element solution of the non linear heat equation RAIRO, Analyse numérique, Vol. 2 (1980).
6. W. A. Philips, Non linear effects in heat pulse experiments, J. Appl. Phys. 51, 7 (1980).

THERMAL PROPERTIES OF SURFACE LAYERS USING

PULSE TRANSIENT HOT STRIP MEASUREMENTS

Silas E. Gustafsson

Department of Physics
Chalmers University of Technology
S-41296 Gothenburg, Sweden

ABSTRACT

A transient method for measuring thermal conductivity and thermal diffusivity has been developed for studies of insulating surface layers of thicknesses down to a few micrometers. The technique is based on the transient hot strip (THS) method but instead of a single transient heating of the metallic thin film (hot strip), a train of square pulses is passed through the thin film, the dimensions of which have been reduced to about 500 x 25 x 0.1 $(\mu m)^3$. The duration of the pulses has been varied from a few microseconds up to around 50 microseconds, which means that the depth of probing into the investigated silica glass sample varied from 4 to 10 μm. Because of the small probing depth and the size of the hot strip, this method makes it possible to study the thermal properties of vacuum deposited layers of insulators, which represent a group of solids hardly accessable for thermal investigations up to now.

INTRODUCTION

A transient method for measuring thermal conductivity and thermal diffusivity of solid surface layers with low electrical conductivity has recently been described {1}. The technique is based on the THS method {1,2,4} using an extended plane metallic strip (thin metal film) both as the heat source and the sensor of the temperature increase.

By reducing the strip size to the following dimensions: length 0.5 mm, breadth 25 μm and thickness 0.1 μm, and by exposing the

strip to a current pulse of very short duration it is in principle
possible to reduce the probing depth {5} to only a few micrometers.
The current pulse would in such a case have to be limited to a few
microseconds and the actual mapping of the voltage - versus - time
trace over this time period, while possible, would require quite
expensive electronic equipment. To reduce this problem of time
resolution we have used pulse non-linearity measurements, an elegant
technique developed and used for studies of thermal time constants
of non-linear resistors {6,7,8,9,10}. The technique is based on a
procedure by which a string of pulses is applied to an AC - coupled
circuit containing the hot strip actually acting as a non-linear
resistance.

If a circuit is constructed which contains a large blocking
capacitor in series with the hot strip its RC time constant will
be long compared to the pulse width and the strip will be exposed
to a train of square pulses.

Because of the AC coupling there will appear a base-line shift
of the instantaneous voltage over the strip. This is a consequence
of the fact that the current is changing direction twice during one
period. By working with duty cycles in the range of 0.02 to 0.05
the reverse current during the time between the positive pulses can
be made very small with the essential heating of the strip during
the positive excursions of the current.

Because of the non-linear part of the strip resistance, an
average voltage, recorded over a number of periods across the strip,
will develop and theory shows that it is inversely proportional to
the thermal conductivity of the substrate on top of which the thin
film resistor has been deposited.

The theoretical description of the heat dissipation from the
strip also shows that by making measurements at more than one pulse
frequency it is possible to derive information on the thermal diffusivity of the substrate.

THEORY

If an AC coupled network cosisting of linear components is
exposed to a train of square pulses, the time average of the output
voltage over a number of whole periods will be zero. However, if
we include a non-linear resistance in the circuit, the zero net
charge condition which prevails will give rise to a similar baseline shift, but the average output voltage will not be zero. The
average voltage will as a matter of fact adjust itself to a value
which reflects very directly the nonlinear part of the resistance.

The circuit used is shown in Fig. 1. The generator imposes the

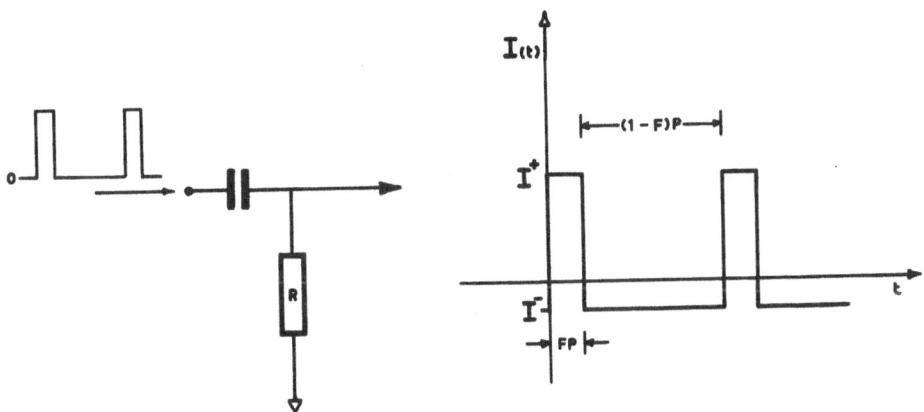

Figure 1: Electrical circuit and the pulse train.

pulse train on the THS probe through the large blocking capacitor. If the rectangular current pulses have a magnitude I and a duty cycle F, we have

$$I = I^+ + I^-, \tag{1}$$

where I^+ and I^- are the positive and negative excursions of the current, respectively.

From the zero net charge condition it is required that

$$F \, I^+ = (1 - F) \, I^- \tag{2}$$

As we will see below the most interesting arrangement is to use a small duty cycle, which means that the probe (hot strip) will be exposed to a strong current pulse during the first part of the period while the reverse current during the second part of the period will be rather small. The fact that there is a reverse current during the second part of the period will, however, mean that the temperature of the strip will not relax back to its initial temperature but will after each full period have experienced a small but finite temperature increase.

Considering a series of pulses from the generator, the average voltage, $\bar{U}_n$, over n periods may be expressed as

$$\bar{U}_n = (n \, P)^{-1} \int_o^{nP} u(t) \, dt, \tag{3}$$

where P is the pulse period and u(t) is the instantaneous voltage over the THS probe. Because of the complexity of $\bar{U}_n$, we may write

$$\bar{U}_n = n^{-1} \sum_{k=1}^{n} \bar{U}_k, \tag{4}$$

where $\bar{U}_k$ is the average voltage over the period number k, or more explicitly

$$\bar{U}_k = P^{-1} \{ I^+ \int_{(k-1)P}^{(k-1+F)P} R_k^+(t)dt - I^- \int_{(k-1+F)P}^{kP} R_k^-(t)dt \}, \tag{5}$$

where $R_k^+(t)$ and $R_k^-(t)$ are the time-dependent resistances of the strip during the positive and negative excursions of the current, respectively. These resistances are directly dependent on the two different heat dissipations, Q^+ and Q^-, which with good accuracy can be given by

$$Q^+ = R_o (I^+)^2 \tag{6}$$

and

$$Q^- = R_o (I^-)^2. \tag{7}$$

From Eq. (2) we have

$$Q^-/Q^+ = r^2, \tag{8}$$

where R_o is the initial resistance of the hot strip and

$$r = F/(1 - F). \tag{9}$$

If we assume that a long, resistive strip of negligible heat capacity on the surface of a solid substrate is electrically heated so that the output of power, Q, is constant, the time-dependent resistance, R(t), can be given {4} as

$$R(t) = R_o (1 + \frac{\alpha\, Q}{16\, \pi\, d^2 h\, (\Lambda_1 \Lambda_2)^{1/2}} \int_{t_1}^{t_2} (t - t')^{-1}\, dt' \; x$$

$$x \int_{-d}^{+d} dy \int_{-d}^{+d} dy'\, \exp \{ - \frac{(y - y')^2}{4\, \kappa_2 (t - t')} \}), \tag{10}$$

where α is the temperature coefficient of the strip resistivity (TCR), 2d is the width and 2h is the length of the strip. It is here assumed that the substrate is an anisotropic material with the principal conductivities Λ_1, Λ_2 and Λ_3 along the directions of the x-, y- and z-axis, respectively. The orientation of the axes is such that the z-axis extends along the length of the strip and the x-axis is perpendicular to the surface of the strip and thus also perpendicular to the surface of the substrate. The thermal diffusivities are defined by $\kappa_i = \Lambda_i/\rho c$ (i = 1,2,3) where ρc is the heat capacity of the substrate per unit volume. Eq. (10)

is general in the sense that it is assumed that the output of power in the strip only lasts from the time t_1 to t_2.

It is now possible to derive an analytical expression of $\bar{U}_n$ even if the derivation becomes quite long. To demonstrate how it is carried out one may calculate the average voltage over the first period $\bar{U}_1$ since this calculation actually contains the main features of the derivation of $\bar{U}_k$.

For the first part of period number one, defined by

$$0 \leq t \leq FP,$$

one gets from Eq. (10)

$$R_1^+(t) = R_o\{ 1 + Q^+B \int_o^t \Psi(t,t')\, dt' \}, \qquad (11)$$

where

$$B = \alpha(16\pi d^2 h)^{-1}(\Lambda_1 \Lambda_2)^{-1/2}. \qquad (12)$$

For the second part of period number one, defined by

$$FP \leq t \leq P,$$

one can write

$$R_1^-(t) = R_o\{ 1 + Q^+B \int_o^{FP} \Psi(t,t')\, dt' +$$

$$+ Q^-B \int_{FP}^t \Psi(t,t')\, dt' \}, \qquad (13)$$

where the following definition has been used:

$$\Psi(t,t') = (t-t')^{-1} \int_{-d}^{+d} dy \int_{-d}^{+d} dy'\, \exp\{ -\frac{(y-y')^2}{4\kappa_2(t-t')} \}. \qquad (14)$$

After evaluating the integrals Eqs (11) and (13) can be written accordingly:

$$R_1^+(t) = R_o\{ 1 + 8d^2\pi^{1/2}BQ^+\, f(\tau) \} \qquad (15)$$

and

$$R_1^-(t) = R_o(1 + 8d^2\pi^{1/2}BQ^+\{ f(\tau) - f[(\tau^2 - \tau_{FP}^2)^{1/2}] \} +$$

$$+ 8d^2\pi^{1/2}BQ^-\, f[(\tau^2 - \tau_{FP}^2)^{1/2}]), \qquad (16)$$

where

$$\tau = (t/\theta)^{1/2}, \qquad (17)$$

$$\tau_x = (x /\theta)^{1/2}, \qquad (18)$$

and
$$\theta = d^2/\kappa_2. \tag{19}$$

The function $f(\tau)$ is defined elsewhere {4}.
Finally the average voltage over the first period $\bar{U}_1$, using Eq (5), can be given as

$$\bar{U}_1 = C\, \sigma^{-1}\{ (1 + r) J_{F\sigma} - rJ_\sigma + r(1 - r^2)J_{(1-F)\sigma} \}, \tag{20}$$

where
$$C = \alpha(1-F)^3 U_{in}^3 (24\, \pi\, h\, R_o)^{-1}(\Lambda_1\Lambda_2)^{-1/2}. \tag{21}$$

U_{in} is the total pulse height given by

$$U_{in} = R_o(I^+ + I^-) \quad \text{and} \quad I^+ = U_{in}(1-F)/R_o . \tag{22}$$

We have further that
$$\sigma = P/\theta \qquad \text{and} \tag{23}$$

$$J_x = 8\, \pi^{1/2} \tau_x^3 \operatorname{erf} \tau_x^{-1} - \tau_x^2 \exp(-\tau_x^{-2}) -$$
$$- 3\tau_x^4\{1 - \exp(-\tau_x^{-2})\} - (1 + 6\tau_x^2)\operatorname{Ei}(-\tau_x^{-2}). \tag{24}$$

Applying the same procedure as outlined above one can derive expressions for $\bar{U}_2$, $\bar{U}_3$, $\bar{U}_4$ and so on; and during these derivations it will become clear that the number of terms containing the J-function will increase with three in going from one period to the next. It is also easy to derive a recursion formula for the average voltage over period number k, and the final expression may be given in the following form

$$\bar{U}_k = C\, \sigma^{-1}\{(1+r)(1-r^2)\Big[\sum_{i=0}^{k-2}\{J_{(i+F)\sigma} + J_{(i+1-F)\sigma}\} -$$

$$- 2\sum_{i=0}^{k-3} J_{(i+1)\sigma}\Big] - \{1 + (1+2r)(1-r^2)\}J_{(k-1)\sigma} +$$

$$+ (1+r)J_{(k-1+F)\sigma} + r(1-r^2)J_{(k-F)\sigma} - rJ_{k\sigma} \}, \tag{25}$$

where $k=1,2,\cdots,n$. It should be noted that whenever the upper limit of the summation in this equation is less than zero the sum should be considered identically zero.

Experimentally one is measuring the average voltage, $\bar{U}_n$, defined by Eq (4) over a certain number of periods. This voltage is then a function of several variables; a fact which may be expressed in the following way

$$\bar{U}_n = C\, \Phi(n,F,\sigma). \tag{26}$$

Experimentally it is observed that $\bar{U}_n$ tends to stabilize and become very little dependent on the number of periods beyond, say, twenty periods, which is easily verified by Eq (25).

Experiments have been performed using low power pulse generators {1}. In order to arrange for maximum power transfer to the hot strip in such measurements, the THS probe should have a resistance as near the internal resistance of the pulse generator as possible. For such arrangements the final expression of the average voltage (from Eqs (21), (25) and (26)) will be slightly modified according to the theory given in reference {12}.

DERIVATIONS FROM THE THERMAL MODEL

In the derivation of the thermal model for the heat dissipation from the hot strip and the temperature increase thereof it has been assumed that a) the heat loss through radiation is negligible, b) the THS probe is a part of an infinitely long strip and c) the heat capacity of the strip is zero.

Radiation: The temperature increase in a pulse THS measurement will only be slightly higher than in a normal THS experiment which means that the theory developed in references {2} and {5} applies. For work at high temperatures an obvious way of reducing the radiation losses in a single-shot transient recording is to reduce the total time of the experiment and thus reduce the temperature increase. In a pulse THS experiment one would similarly have to consider reducing the pulse width.

Electrical contacts: The influence of the end contacts on the temperature of the strip has been estimated in earlier work {2,5} with the recommendation to choose $h/d = 20$. Because of the slightly higher temperature increase caused by a train of pulses as compared to the increase during one pulse, the end contacts will have a somewhat larger influence. However, the contributions to the average voltage from the periods with higher values of n is very limited and for high-precision work it would be easy to increase the length of the strip considerably beyond the 0.5 mm used in the studies reported in reference {1}, and avoid any influence on the experimental results.

Heat capacity of the strip: A basic assumption in the theory given above is that the heat source or the hot strip does not have any heat capacity. The influence from the heat capacity of a strip with a certain thickness has been estimated {5} for transient experiments where one is following the voltage variation due to the heating from one single current pulse.

Because of the interest in possibilities of reducing the probing depth (Δ_p) defined {1} by

$$\Delta_p = 2 (\kappa_1 FP)^{1/2} \tag{27}$$

a more detailed estimation of the influence on a pulse THS measurement would seem appropriate. Assuming that the hot strip has a certain heat capacity ($\rho_v c_v$) per unit mass, we can use the following solution recently developed {11} for an infinitely wide strip:

$$R_p(t) = R_0\{ 1 + \alpha v_p(\tau_p) \}, \tag{28}$$

where

$$v_p(\tau_p) = \nu\gamma Q\pi^{-1/2}(dh\Lambda_1)^{-1} g(\tau_p), \tag{29}$$

$$g(\tau_p) = \tau_p - 0.5 \; \pi^{1/2}(1 - \exp \tau_p^2 \; \text{erfc} \; \tau_p), \tag{30}$$

$$\tau_p = (t/\theta_p)^{1/2} \quad \text{and} \quad \theta_p = (4\nu^2 \gamma^2/ \kappa_1)^{1/2}, \tag{31}$$

where 2ν is the thickness of the strip and γ is the ratio of the heat capacity of the strip and that of the substrate. This solution assumes a constant output of heat and can only be used as long as one can consider the heat flow from the strip to be unidirectional into the sample. After a certain period of time the solution given by Eq (10) will be taking over and be representing the resistance variation of the hot strip. That is, when the influence from the heat capacity of the strip will be negligible and the heat flow also in the y-direction into the substrate has become important. Eq (10) can be written in the form

$$R(t) = R_o\{ 1 + \alpha \; v(\tau)\}, \quad \text{where} \tag{32}$$

$$v(\tau) = Q \; (2h)^{-1}(\pi\Lambda_1\Lambda_2)^{-1/2} \; f(\tau). \tag{33}$$

An approximate form of the $f(\tau)$-function has been given {4} as

$$f(\tau) = \tau - (4\pi)^{-1/2} \tau^2 \tag{34}$$

valid for τ-values less than 1. Applying these two solutions to the same strip, one can show that the plane source solution gives a linear increase of the resistance versus time at the very beginning of the current pulse, while the strip source solution (THS solution) Eq (33) gives a $t^{1/2}$ dependence at the beginning. After a certain time the plane source solution (TPS solution) will over-shoot the THS solution, provided the ratio of the thickness to the width of the strip has a reasonably small value. The time, t_c, for the cross-over of the two solutions, can be calculated by equating the right hand sides of Eqs (28) and (32) and replacing t by t_c which gives the relation

$$1 - \mu^2(\theta_p/\theta)^{1/2} \pi^{-1} - \exp \mu^2 \; \text{erfc} \; \mu = 0 \tag{35}$$

$$\mu = (t_c/\theta_p)^{1/2}. \tag{36}$$

It is now possible for a given value of θ_p/θ to solve this equation and get the cross-over time in terms of θ. When deriving Eq (35) the approximate form of the $f(\tau)$-function, Eq (34), has been used Since θ_p/θ is rarely larger than 0.01 one can use an approximate solution of Eq (35)

$$\mu = (\pi^2\theta/\theta_p)^{1/4} \quad \text{and} \quad t_c = \pi(\theta\theta_p)^{1/2} \tag{37}$$

In order to estimate the influence on the average resistance increase during a single pulse, the length of which is estimated to θ seconds, one may calculate the ratio

$$R_\theta = \int_o^{t_c}\{ v(\tau) - v_p(\tau_p)\} \; dt \; / \; \int_o^\theta v(\tau) \; dt \tag{38}$$

560

This equation gives the error one should expect when using the THS solution instead of the TPS solution in the time interval $0 \le t \le t_c$. The final expression becomes quite complex but for values of θ_p^-/θ^c less than, say, 0.01, one can use the simple expression

$$R_\theta = 3(\theta_p/\theta) \quad \text{or} \quad R_\theta = 12v^2\gamma^2/d^2 \tag{39}$$

as a good estimate of the limits of validity of the THS solution.

In routine experiments it is rather easy to work with a thickness of the strip around 40 nm, which would mean that the minimum strip width can be estimated to 2 μm assuming R_θ to be less than 0.01. This strip width is also close to what one can easily achieve using standard deposition techniques.

DISCUSSION

The work performed so far with the pulse THS technique indicates that it is possible to measure the thermal properties of insulating surface layers down to a thickness of some 5 μm with rather good accuracy, provided the experiments are designed carefully. As shown above there are limitations on how small a thickness can be probed by this technique and a practical limit might at the moment be set at a thickness of the hot strip of 40 nm giving a strip width of 2 μm. The probing depth, Δ_p, defined by Eq (27) may be rewritten as

$$\Delta_p = 2d(\kappa_1/\kappa_2)^{1/2} \tag{40}$$

assuming the pulse width, FP, to be equal to the characteristic time θ. A strip width of 2 μm would according to this relation also give a probing depth of the same size. This means that a practical lower limit of the probing depth for the pulse THS method may be estimated at about 2 μm. In order to further reduce the probing depth one would have to take into consideration also the heat capacity of the hot strip and then specifically use the TPS solution represented by Eq (28) in deriving the average voltage. This work is presently in progress.

An area of particular interest for applying the pulse THS method would seem to be studies of thermal properties of insulating surface layers produced by vacuum deposition processes. It may also be noted that it has not been possible to detect any influence from a possible temperature difference between the outer surface of the insulating sample and the inner surface of the hot strip (deposited thin metal film) in spite of the fact that the probing depth was only a few micrometers.

ACKNOWLEDGEMENTS

This work was financially supported by the Swedish Natural Science Research Council, by "Sven och Dagmar Saléns Stiftelse" and by "Ollie och Elof Ericssons Stiftelse för vetenskaplig forskning", which is gratefully acknowledged.

REFERENCES

1. Gustafsson, S.E., Chohan, M.A., Ahmed, K. and Maqsood, A., "Thermal Properties of Thin Insulating Layers Using Pulse Transient Hot Strip Measurements", J. Appl. Phys. 55, (1984).

2. Gustafsson, S.E., Karawacki, E. and Khan, M.N., "Transient Hot Strip Method for Simultaneously Measuring Thermal Conductivity and Thermal Diffusivity of Solids and Fluids", J. Phys. D: Appl. Phys. 12, 1411 (1979).

3. Gustafsson, S.E., Karawacki, E. and Khan, M.N.,"Determination of the Thermal Conductivity Tensor and the Heat Capacity of Insulating Solids with the Transient Hot Strip Method", J. Appl. Phys. 52, 2596 (1981).

4. Gustafsson, S.E., Ahmed, K., Hamdani, A.J. and Maqsood, A., "Transient Hot Strip Method for Measuring Thermal Conductivity and Specific Heat of Solids and Fluids: Second Order Theory and Approximations for Short Times", J. Appl. Phys. 53, 6064 (1982).

5. Gustafsson, S.E., "Transient Hot Strip Method for Measuring Thermal Conductivity and Specific Heat of Non-conducting Solids and Liquids", Proc. Eighth Symp. Thermophys. Prop. Vol. II. Ed. Sengers, J.V., The American Soc. Mechanical Engineers, 345 East 47th Street, New York, N.Y. 10017.

6. Rosenthal, L.A., "Electrothermal Measurements of Bridgewires Used in Electroexplosive Devices", IEEE Trans. Instrum. Meas. IM-12, 17 (1963).

7. Rosenthal, L.A., "Method and Apparatus for the Measurement of Electrothermal Nonlinearity", Rev. Sci. Instrum. 43, 1575 (1972).

8. Rosenthal, L.A., "Heat Capacity Measurement Method for Bridgewires", Rev. Sci. Instrum. 45, 1523 (1974).

9. English, A.T., Miller, G.L., Robinson, D.A.H., Dodd, L.V. and Chynoweth, T., "Pulse Nonlinearity Measurements on Thin Conducting Films", J. Appl. Phys. 49, 717 (1978).

10. Hebard, A.F. and Steverson, W.M., "Thermal Time Constants of Thin-film Resistors Using Pulse Nonlinearity Measurements", J. Appl. Phys. 49, 5250 (1978).

11. Wolter, J., v. d. Doel, R., Weiss, K. and Wouters, M.C.H.M.,
    "Measurement of Transient Temperatures of Thin Metal Films",
    Thin Solid Films 74, 281 (1980).

12. Gustafsson, S.E., Karawacki, E. and Chohan, M.A., "Circuit
    Design for Transient Measurements of Electrical Properties
    of Thin Metal Films and Thermal Properties of Insulating
    Solids or Liquids", Rev. Sci. Instrum. 55, (1984).

SURFACE TEMPERATURE MEASUREMENTS WITH INTRINSIC CONTACTING THERMO-
COUPLES - STEADY ERRORS AND TRANSIENT EFFECTS - ANALYTICAL AND
EXPERIMENTAL STUDY

Annie Gery, Alain Degiovannni, and Michel Laurent

I.N.S.A. - Laboratoire de Physique Industrielle
20, avenue Albert Einstein - 69621 Villeurbanne Cédex

ABSTRACT

When thermoelectric elements are individually pressed on a
solid, the thermocouple junction is made by the solid itself or by
a thin deposited conducting layer. The temperature field is modified
by this intrinsic thermocouple from which the measurement is taken.

A theoretical analysis for steady-state conditions shows the
influence of the main parameters : thermal conductivity and shape
of the thermoelectric elements, thermal conductivity of the solid,
thermal conductivity and thickness of the deposited layer, contact
conditions and heat transfer with the surroundings.

From this analysis, an error term is computed ; in this expression appear thermal resistance terms corresponding to the macro and
micro constrictions of the heat flow near the surface and to thermoelectric elements. The measured temperature "inside" the contact
is determined from an analytical study of the thermal contact resistance (microconstrictions) which take into account the deposited
layer.

The experimental part of the study in steady state conditions
corroborates the theoretical analysis. Measurements were done for
semiconducting materials (Bi_2Te_3), and for many solids, their thermal conductivity ranges from 400W/m.K (copper) to 0.2W/m.K (perspex)
To change the thermal contact resistance, the load applied on the
thermoelectric element and the pressure of the environmental atmosphere are set up at some values.

In transient conditions, the theoretical analysis deals with a

heat pulse absorbed by the specimen. The response of the thermocouples is matched with a temperature variation versus time curve
of the surface when there is no element on the surface. It appears
that the thermal behaviour is not easy to understand; the many parameters : specimen thickness and thermal properties, thermal contact
resistance, geometry and thermal properties of thermoelectric maté-
rials act in different ways and it is quite difficult to separate
their effects. However, we can calculate the response for every
configuration.

The main parts of the apparatus for the transient experimental
study are a flash device (the light is driven from the flash tube
to the specimen by optical fibers), a photomultiplier to obtain
the energy versus time curve and a transient recorder to pick up
the surface temperature and photomultiplier output.

The experimental results show the importance of the contact
conditions. With Bi_2Te_3 elements the delay times of the response
does not exceed 100µs. Practically, this study demonstrates the
quality of surface temperature measurement with intrinsic thermocouples using Bi_2Te_3 for steady and transient conditions.

PRELIMINARY STUDY OF THE THERMAL DIFFUSIVITY OF

SOURCE SPRAYED MATERIALS BY THE FLASH LASER METHOD

Lech Pawlowski*, Christian Martin,
Pierre Chagnon, and Ahmed Mahlia

Céramiques Nouvelles, L.A. 320 C.N.R.S.
Equipe Thermodynamique et Plasma
123 avenue A. Thomas F87060 Limoges Cedex

ABSTRACT

We have measured the thermal characteristics of plasma-sprayed coatings for thermal barriers. These coatings should be able to resist the thermal stresses caused by the presence of thermal gradients and the difference of thermal dilatation coefficients between the materials of the coating and of the substrate.

We have applied the plasma spraying process to multicoating systems and optimized the number and thicknesses of the coatings in order to reduce or eliminate the stresses. At this time approximately 60% of the thermal barriers are realized by the plasma spraying of zirconia with different stabilisers. We would like to find the thermal characteristics of the thermal barriers by measuring the diffusivity, in order to optimize the plasma spraying parameters.

We measure the thermal diffusivity by the flash method, from ambient temperature up to 1600 K and a pressure of 10^{-6} Torr. The flash is sent by a neodymium-glass Laser (energy pulse up to 100 J and pulse duration of about 450 μs) The samples have a diameter of about 1 cm and thickness up to several millimeters. The signal is detected by an I.R. detector and after amplification is sent to a digital oscilloscope which transfers the data to a minicomputer for calculating the diffusivity. For testing the apparatus we have

---

*Permanent address: Institute of Inorganic Chemistry and Metallurgy of rare elements, Technical University of Wroclaw, 50-370 Wroclaw, W.Wyspianskiego 27, Poland.

measured copper samples having good accordance with known diffusivity values.

The theoretical solution of the heat diffusion equation is:

$$T(L,t) = \frac{Q}{\rho\,C\,L} \left(1 + 2 \sum_{n=1}^{\infty} (-1)^n \exp\left(\left(-\frac{n^2\pi^2}{L^2}\right)\alpha t\right)\right)$$

where T is the temperature of the sample at the time t and position L; Q is the density of the energy given by the flash; ρ, C and α are respectively: density, specific heat and thermal diffusivity. This result was obtained by the supposition that the sample is adiabatic and that the energy of the flash is uniformly deposited over the surface of the sample. We would like to take into account that the sample is not adiabatic and also the spatial distribution of the flash. For this reason we have measured the emissivity of the materials.

For zirconia we have also made a study of the transparency of this material for small thickness and high temperature. We have done a preliminary study of the influence of the transparency on measurements of the temperature of the back face of the sample.

SESSION S

Invited Paper

SPEAKER

G. W. Milton
Cornell University
Ithaca, NY.

SESSION CHAIRMAN

R. U. Acton
Sandia National Laboratories
Albuquerque, NM.

THERMAL CONDUCTION IN COMPOSITES

G. W. Milton

Baker Laboratories
Cornell University
Ithaca, N. Y. 14853

K. Golden

Courant Institute
New York University
New York, N. Y. 10012

ABSTRACT

We consider the effective thermal conductivity of two-component
isotropic composites and review bounds obtained through analytic
continuation of the effective conductivity as a function of the
component conductivities. The connection between this conductivity
function and Stieltjes functions is emphasized. Many of the well-
known bounds on the effective thermal conductivity correspond to
bounds on Stieltjes functions and these bounds, in turn, are
closely related to Padé approximants.

INTRODUCTION

The emphasis of this paper is different from that of the
conference presentation. Here we review analytic properties of the
effective thermal conductivity, $\bar{\lambda}_0$, of composites, rather than
proving that the well-known effective-medium approximation [1,2]
is exact for a specific class of model composites. These model
composites, which have a self-similar or fractal-like character,
are described elsewhere [3].

The effective conductivity of a statistically isotropic and
homogeneous two-component composite is defined as the constant of

proportionality, $\bar{\lambda}_0$, in the equation

$$\{\underset{\sim}{J}_T(\underset{\sim}{x})\}_\Omega = -\bar{\lambda}_0 \ \{\nabla T(\underset{\sim}{x})\}_\Omega \tag{1}$$

relating the average of the heat current $\underset{\sim}{J}_T(\underset{\sim}{x})$ to the average of
the temperature gradient $\nabla T(\underset{\sim}{x})$ in the composite. The brackets are
used to denote averages, in this case an average over a large cubic
volume, Ω, of the composite. The fields $\underset{\sim}{J}_T(\underset{\sim}{x})$ and $T(\underset{\sim}{x})$ are often
calculated by solving the microscopic equations of thermal conduc-
tivity:

$$\underset{\sim}{J}_T(\underset{\sim}{x}) = -\lambda(\underset{\sim}{x})\nabla T(\underset{\sim}{x}), \qquad \nabla \cdot \underset{\sim}{J}_T(\underset{\sim}{x}) = 0, \tag{2}$$

where the local conductivity $\lambda(\underset{\sim}{x})$ takes two values: λ_1 in component
1 and λ_2 in component 2. Our work is based on these equations. To
justify their use we make four simplifying assumptions. First, we
suppose the components are in good thermal contact so that there
is no temperature discontinuity across interfaces. Second, the
temperature must be sufficiently high to ensure that the mean free
path of phonons or free electrons is much smaller than the characteristic size of inhomogeneities. Third, the temperature must be
low enough to neglect heat transport due to radiation and convection.
Last, we suppose the cross-coupling between different fields is
negligible. (This is reasonable when the thermopowers and thermo-
elastic coefficients are small.)

Our overall aim is to reproduce known bounds on the effective
conductivity that have practical applications and then provide an
elegant generalization of them. We start by reviewing the well-
known connection between thermal conductivity in composites and
other transport processes. We then discuss a perturbation solution
for $\bar{\lambda}_0$, which is useful when the structure is arbitrary and
$\delta_\lambda \equiv \lambda_1 - \lambda_2$ is small. Finally, following Bergman [4,5] and others
[6-11] analytic properties of $\bar{\lambda}_0(\lambda_1, \lambda_2)$ are studied and used to
obtain bounds on the effective conductivity. We draw attention to
the connection between these bounds and the bounds on Stieltjes
functions derived by Baker [12,13].

RELATED EFFECTIVE CONSTANTS

Electrical conductivity in a composite is described by the
equations

$$\underset{\sim}{J}_E(\underset{\sim}{x}) = - \sigma(\underset{\sim}{x})\nabla V(\underset{\sim}{x}), \qquad \nabla \cdot \underset{\sim}{J}_E(\underset{\sim}{x}) = 0, \tag{3}$$

where $J_E(\underset{\sim}{x})$ is the electrical current, $V(\underset{\sim}{x})$ the electric potential
and $\sigma(\underset{\sim}{x})$ is the local electrical conductivity taking values σ_1 in

component 1 and σ_2 in component 2. The similarity between (2) and (3) implies that the effective electrical conductivity $\bar{\sigma}_0(\sigma_1, \sigma_2)$ is precisely the same function of σ_1 and σ_2 as $\bar{\lambda}_0(\lambda_1, \lambda_2)$ is of λ_1 and λ_2, for a given composite with fixed structure. Similarly the effective dielectric constant $\bar{\varepsilon}_0(\varepsilon_1, \varepsilon_2)$ is the same function of ε_1 and ε_2. (See the remarks of J. D. Patterson in these proceedings.) Hence if the ratios σ_1/λ_1 and σ_2/λ_2 are equal, then $\bar{\sigma}_0/\bar{\lambda}_0$ also shares the same value. This implies that if the Wiedemann-Franz law [14] applies to both component materials, then it works for the composite as well. The similarity does not extend to the effective elastic constants of composites, which relate average stress to average strain fields. Nevertheless, simple correlations between the thermal conductivity and elastic properties of composites have been established [3].

A PERTURBATION SOLUTION

One approach to estimating $\bar{\lambda}_0$ using a perturbation scheme is due to Brown [15], who deduced formal expressions for the coefficients, y_m, in the series expansion,

$$\bar{\lambda}_0 = \{\lambda\}_0 + \delta_\lambda \sum_{m=1}^{\infty} y_m (\delta_\lambda / \{\lambda\}_0)^m, \qquad (4)$$

where

$$\delta_\lambda \equiv \lambda_1 - \lambda_2, \quad \{\lambda\}_0 \equiv f_1\lambda_1 + f_2\lambda_2, \qquad (5)$$

in which f_1 and $f_2 = 1 - f_1$ are the volume fractions occupied by components 1 and 2. Brown related y_m to a multiple integral over an $(m + 1)$ - point correlation function which gives the probability that all $m + 1$ points lie in component 1 (or 2). In the special case $m = 1$, this integral is independent of the structure of the composite and Brown proves $y_1 = - f_1f_2/d$, where d is the dimensionality of the composite. The next important contribution to $\bar{\lambda}_0$ comes from the term involving y_2. This coefficient can be expressed [15,16] in the form

$$y_2 = f_1f_2[f_2 + (d-1)\zeta_1 - df_1]/d^2 = -f_1f_2[f_1+(d-1)\zeta_2-df_2]/d^2, (6)$$

where, for a three-dimensional composite, we define

$$\zeta_1 \equiv 1 - \zeta_2 = \frac{9}{2f_1f_2} \int_0^\infty dr \int_0^\infty ds \int_{-1}^{+1} du \, \frac{f_{111}(r,s,u)}{rs} \, P_2(u), \qquad (7)$$

573

in which $P_2(u)$ is a Legendre polynomial and $f_{111}(r,s,u)$ is the probability that all three vertices of a given triangle lie in component 1; the triangle having sides r, s and included angle $\cos^{-1}u$. This fundamental geometric parameter ζ_1 has been calculated for a variety of composites, including cell materials [17] (described below), regular arrays of spheres [18], and systems of both penetrable spheres [19] and hard spheres [20] randomly inserted in a matrix. The simplest results are for the cell materials, constructed by dividing space into cells and then flipping the same weighted coin in each cell: heads, which occurs with probability f_1, means the cell is to be filled with component 1; tails means it is to be filled with component 2. Miller [17], who devised these materials, proves $\zeta_1 = f_1$ for spherical cells, $\zeta_1 = f_2$ for platelike cells, and $\zeta_1 = (f_2 + 3f_1)/4$ for needlelike cells. The results have been generalised to spheroidal cells of arbitrary eccentricity [18]. For cells to be considered truly platelike, the aspect ratio of the equatorial axis to the longitudinal axis must exceed 100. For needle-like cells, this aspect ratio must be smaller than 1/10.

In the following section, the coefficients y_m in the expansion (4) are incorporated in bounds on $\bar{\lambda}_0$. These bounds often give reliable estimates for the effective conductivity. Note that the series expansion is expressible in various equivalent forms through a change of variables. For instance, $1/\bar{\lambda}_0$ can be expressed as a series in terms of

$$\delta_{1/\lambda} \equiv 1/\lambda_1 - 1/\lambda_2, \quad \{1/\lambda\}_0 \equiv f_1/\lambda_1 + f_2/\lambda_2, \tag{8}$$

having the same basic form as (4), but with different coefficients.

ANALYTIC PROPERTIES AND BOUNDS

The function $\bar{\lambda}_0(\lambda_1, \lambda_2)$ has some very beautiful properties first recognized by Bergman [4] and rigorously proved by Golden and Papanicolaou [10]. Specifically the function can be expressed in the form

$$\bar{\lambda}_0(\lambda_1, \lambda_2) = a_0\lambda_1 + b_0\lambda_2 + \lambda_1 g_0(\lambda_1/\lambda_2), \tag{9}$$

where a_0 and b_0 are real-valued satisfying

$$a_0 \geq 0, \qquad b_0 \geq 0, \qquad a_0 + b_0 \leq 1, \tag{10}$$

and $g_0(z)$ has the integral representation

$$g_0(z) = \int_0^\infty \frac{d\phi_0(u)}{1 + zu}, \tag{11}$$

in which $\phi_0(u)$ is a bounded non-decreasing function normalized to ensure

$$\bar{\lambda}_0(1, 1) = 1. \tag{12}$$

The representation (11) is, in fact, the defining equation for a _Stieltjes function_. The properties of Stieltjes functions have been extensively studied in the mathematics literature; Baker [13] gives an excellent review. Evidently from (11), $g_0(z)$ is an analytic function of z with singularities restricted to the negative real z axis. These singularities include simple poles with positive residues, essential singularities and branch cuts. For example, Bergman [21] and McPhedran and McKenzie [22] consider regular arrays of identical spheres (and cylinders) of conductivity λ_1 in a matrix of conductivity λ_2. Their numerical work indicates that for these composites $g_0(z)$ has an essential singularity at $z = -1$ which forms the accumulation point of a set of poles located along the negative real axis. In any composite with some degree of randomness, the poles are likely to get smeared out and form a branch cut [4]. Indeed, Dykhne [23] proves $\sqrt{\lambda_1\lambda_2}$ is the _exact_ effective conductivity of any two-dimensional, possibly random, composite that is symmetric in the two-components, like a chequerboard. This implies $g_0(z) = 1/\sqrt{z}$, with a branch cut along the entire negative real axis.

Qualitative features of the function $\bar{\lambda}_0(\lambda_1, \lambda_2)$ can sometimes be determined from measurements of the effective dielectric constant taken over those frequencies where the wavelength greatly exceeds the characteristic size of inhomogeneities. (At higher frequencies the radiation is scattered.) The dielectric constants ε_1, ε_2 and $\bar{\varepsilon}_0$ are then generally complex with imaginary parts that govern the absorption of incident radiation. For example, in the cermet Ag - SiO_2 the silica has a real and relatively constant dielectric dielectric constant, ε_2, whereas the dielectric constant, ε_1, of silver varies with frequency along a trajectory in the upper half of the complex plane. This trajectory passes very close to the negative real ε_1 axis, i.e., near where the singularities of $g_0(z)$ are expected to have the greatest influence. The measurements of $\bar{\varepsilon}_0$ with $f_1 = 0.39$, taken by Abeles and Gittleman [24] exhibit a pronounced _absorption peak_ at a wavelength of about 0.45 microns. This resonant absorption, not present in either pure silver or pure silica indicates the presence of a pole (or short branch cut singularity) in $g_0(z)$. Physically it is due to the same sort of resonance occurring in capacitor-inductor networks: the silica acts as a capacitor and the silver serves as an inductor.

Although the analytic properties of $\bar{\lambda}_0(\lambda_1, \lambda_2)$ have been known since the pioneering work of Bergman [4], the connection with Stieltjes functions has been recognized only recently. In fact, many of the bounds on $\bar{\lambda}_0$ deduced from first principles by Bergman [4,5], Felderhof [8] and one of us (GWM) [6,7] could have been directly obtained from the bounds on Stieltjes functions derived by Baker [12]. These bounds incorporate the coefficients in the series expansion (4) up to any given order and are closely related to Padé approximants [13]. To see how they arise, we follow the approach of Baker [12], also adopted by others [11,42]. For simplicity, let us suppose $\lambda_1 > \lambda_2$. First note that the inequality

$$\frac{1}{z} \int_0^\infty \frac{d\phi_0(u)}{1 + u} \leq g_0(z) \leq \int_0^\infty \frac{d\phi_0(u)}{1 + u}, \quad \text{for all } z > 1, \tag{13}$$

which follows from (11), can be combined with (9), (10) and (12) to establish the elementary bound,

$$\lambda_2 \leq \bar{\lambda}_0 \leq \lambda_1. \tag{14}$$

Now the invariance properties of Stieltjes functions imply that if $\bar{\lambda}_0(\lambda_1, \lambda_2)$ has the series expansion (4) with $y_1 = -f_1 f_2/d$, then, recalling the definitions (5) and (8),

$$\bar{\lambda}_1(\lambda_1, \lambda_2) \equiv (d - 1)^{-1} \lambda_1 \lambda_2 (\bar{\lambda}_0 \{1/\lambda\}_0 - 1)/(\{\lambda\}_0 - \bar{\lambda}_0) \tag{15}$$

is a function of the same character as $\bar{\lambda}_0(\lambda_1, \lambda_2)$, i.e. it can be expressed in the form

$$\bar{\lambda}_1(\lambda_1, \lambda_2) = a_1 \lambda_1 + b_1 \lambda_2 + \lambda_1 g_1(\lambda_1/\lambda_2), \tag{16}$$

where $g_1(z)$ is a Stieltjes function, and a_1 and b_1 are non-negative with sum $a_1 + b_1 \leq 1$. The normalization constant $(d-1)^{-1}$ in (15) serves to ensure $\bar{\lambda}_1(1, 1) = 1$. Since all the terms in (16) contributing to $\bar{\lambda}_1$ are positive when $\lambda_1, \lambda_2 > 0$, it follows that $\bar{\lambda}_1 > 0$ and through (15) this implies

$$\{1/\lambda\}_0^{-1} \leq \bar{\lambda}_0 \leq \{\lambda\}_0. \tag{17}$$

These bounds, established by Wiener in 1912 using an entirely different method [25], were the best available for 50 years. Hashin and Shtrikman [26] made the next major advance. They discovered new variational principles (incorporating trial "polarization" fields) which yielded an improved set of bounds. By analogy with (14) we have

$$\lambda_2 \leq \bar{\lambda}_1 \leq \lambda_1 \tag{18}$$

which in conjunction with (15) yields bounds on $\bar{\lambda}_0$ that are in fact equivalent to the bounds of Hashin and Shtrikman. They prove there exist composites [26] which attain these bounds. The lower bound in (18) is attained when the composite is an assemblage of

spheres of component 1 each coated with a shell of component 2 such
that the components are in the same proportion, $f_1 : f_2$, in every
coated sphere. In order to fill all space the coated spheres must
have a variety of sizes, ranging to the infinitesimal. The upper
bound corresponds to a similarly constructed material where the
roles of the components are interchanged. Hence (18) are the best
possible bounds that incorporate f_1 (or f_2) and no other information
about the isotropic composite. For a comparison with experiment,
see Corson [27] and DeVera and Strieder [28]. Recently, Tartar and
Murat [29] and Lurie and Cherkaev [30] have generalized these bounds
to anisotropic composites, using the method of compensated compact-
ness. Their bounds are attained when the composite is an aggregate
of coated ellipsoids that are aligned and fill all space.

Note that (4) and (6) when substituted in (15), imply

$$\bar{\lambda}_1(\lambda_1, \lambda_2) = \{\lambda\}_1 + 0(\delta_\lambda^2) = \{1/\lambda\}_1^{-1} + 0(\delta_\lambda^2), \qquad (19)$$

where

$$\{\lambda\}_1 \equiv \zeta_1 \lambda_1 + \zeta_2 \lambda_2, \quad \{1/\lambda\}_1 \equiv \zeta_1/\lambda_1 + \zeta_2/\lambda_2. \qquad (20)$$

Hence the geometric parameters ζ_1 and $\zeta_2 = 1 - \zeta_1$, defined by (7),
now play a similar role to the volume fractions f_1 and $f_2 = 1 - f_1$.
By analogy with (17) we deduce the bounds

$$\{1/\lambda\}_1^{-1} \leq \bar{\lambda}_1 \leq \{\lambda\}_1. \qquad (21)$$

which are identical with bounds due to Beran [31] when expressed in
terms of $\bar{\lambda}_0$. [Beran's bounds have been considerably simplified by
Torquato and Stell [32] and Milton [16]: the connection with (21)
becomes evident only after this simplification]. Few experimental
tests of these bounds have been made owing to the difficulty in
measuring the three point correlation function $f_{111}(r,s,u)$ needed
to evaluate ζ_1 via (7) [27]. If, however, the elastic properties
of the composite are known, then this provides an alternative method
for estimating ζ_1. The bounds on the effective bulk and shear moduli
of composites deduced by Beran and Molyneux [33], McCoy [34] and
Milton and Phan-Thien [35], in fact, incorporate ζ_1 [16].

An entire sequence of bounds can be constructed by generalizing
the preceding arguments [11,12]. We introduce a hierarchy of con-
ductivity functions $\bar{\lambda}_j(\lambda_1, \lambda_2)$ defined for $j = 1, 2, \ldots$, in
terms of $\bar{\lambda}_0(\lambda_1, \lambda_2)$ through the recursion relation

$$\bar{\lambda}_{j+1}(\lambda_1, \lambda_2) \equiv n_j \lambda_1 \lambda_2 (\bar{\lambda}_j \{1/\lambda\}_j - 1)/(\{\lambda\}_j - \bar{\lambda}_j), \qquad (22)$$

where the normalization constant $n_j > 0$ is chosen to ensure $\bar{\lambda}_{j+1}(1, 1) = 1$ and the averages

$$\{\lambda\}_j \equiv w_{1,j}\lambda_1 + w_{2,j}\lambda_2, \quad \{1/\lambda_j\} \equiv w_{1,j}/\lambda_1 + w_{2,j}/\lambda_2, \qquad (23)$$

incorporate non-negative weights $w_{1,j}$, $w_{2,j}$ defined by

$$w_{1,j} \equiv 1 - w_{2,j} \equiv \left.\frac{\partial \bar{\lambda}_j(\lambda_1,1)}{\partial \lambda_1}\right|_{\lambda_1=1}. \qquad (24)$$

The normalization constants n_j and weights $w_{1,j}$ can be expressed, using (22), in terms of the coefficients y_m in the series expansion (4). They respectively depend on $2j + 2$ and $2j + 1$ point correlation functions. From (4), (15) and (20) we have

$$n_0 = (d - 1)^{-1}, \quad w_{1,0} = f_1, \quad w_{1,1} = \zeta_1. \qquad (25)$$

In the special case where $\bar{\lambda}_0(\lambda_1, \lambda_2)$ is a rational function of λ_1 and λ_2, the hierarchy terminates when $\bar{\lambda}_j = \{\lambda\}_j$ or $\{1/\lambda\}_j^{-1}$: beyond this the conductivity functions are clearly not defined.

As j increases, the inequalities

$$\lambda_2 \leq \bar{\lambda}_j \leq \lambda_1, \qquad \{1/\lambda\}_j^{-1} \leq \bar{\lambda}_j \leq \{\lambda\}_j \qquad (26)$$

imply successively tighter bounds on $\bar{\lambda}_0$ which include progressively more information about the composite. The nested sequence of bounds generated by the first pair of inequalities are known as even-order bounds, since they depend on correlation functions up to an even order, $2j$. The second pair of inequalities generate a nested sequence of odd-order bounds.

Common [36] and Baker [12] obtain the corresponding sets of bounds on Stieltjes functions. The problem was first suggested by Common who deduced a hierarchy of inequalities, later sharpened (and extended to complex λ_1/λ_2) by Baker. Beran [31] and Kröner [37] describe how odd-order and even-order bounds up to arbitrarily high order can be deduced from variational principles. Their bounds coincide with the bounds on $\bar{\lambda}_0$ implied by (26), which in turn are equivalent to the bounds in Refs. 7, 9 and 11 [38]. This generalizes the work of Bergman [4], who established that the Wiener [25] and Hashin-Shtrikman [26] bounds follow form the analytic properties of

$\bar{\lambda}_0(\lambda_1, \lambda_2)$; a conclusion reached in the above analysis.

Beran [31] first raised the question of whether the bounds converge as $j \to \infty$. The answer is yes, provided λ_1/λ_2 is not zero, infinite, or real and negative. Indeed, as Baker establishes in his monograph [13], the moment problem is determinate for a Stieltjes series with a non-zero radius of convergence. McPhedran and Milton [18] calculate series expansions for the conductivity of regular arrays of nearly touching spheres of conductivity λ_1 in a matrix having conductivity λ_2. They find the bounds on $\bar{\lambda}_0$ converge rapidly as j increases for conductivity ratios λ_1/λ_2 up to 10,000. In fact, the width of these bounds is negligible when $j \geq 7$. In this sense, the normalization constants n_j and the weights $w_{1,j}$ characterize $\bar{\lambda}_0(\lambda_1, \lambda_2)$. To calculate them is in general a tedious and difficult task, requiring knowledge of high-order correlation functions which are not usually available. Two exceptions are worth noting. First, for symmetric materials we have $\bar{\lambda}_0(\lambda_1, \lambda_2) = \bar{\lambda}_0(\lambda_2, \lambda_1)$. This implies $w_{1,j} = w_{2,j} = \frac{1}{2}$ for all j. Second, Keller's identity [39],

$$\bar{\lambda}_0(1/\lambda_1, 1/\lambda_2) = 1/\bar{\lambda}_0(\lambda_1, \lambda_2), \qquad (27)$$

which holds for any two-dimensional (isotropic) composite implies $n_j = 1$ for all j. When $d > 2$, (27) takes the form of an inequality, derived by Schulgasser [40], and the normalization constants n_j for $j \geq 1$ all depend on the structure of the composite. The effective medium approximation [1,2] for an aggregate of spheres corresponds to a choice of parameters $n_j = (d - 1)^{-1}$ and $w_{1,j} = f_1$ for all j[41].

The above analysis extends to anisotropic composites. For these composites, the effective thermal conductivity is represented by a d-dimensional symmetric matrix $\bar{\lambda}_0$ with non-negative eigenvalues $\lambda_{0,k}$, $k = 1, 2, \ldots, d$. For simplicity, let us suppose the structure of the composite remains unchanged under spatial reflection, which ensures that the eigenvectors (or principal axes) of $\bar{\lambda}_0$ do not rotate as the ratio λ_1/λ_2 varies. Then any eigenvalue, as a function of λ_1 and λ_2, is expressible in the familiar form [4,6,10],

$$\bar{\lambda}_{0,k}(\lambda_1, \lambda_2) = a_{0,k}\lambda_1 + b_{0,k}\lambda_2 + \lambda_1 g_{0,k}(\lambda_1/\lambda_2) \qquad (28)$$

where the positive constants $a_{0,k}$ and $b_{0,k}$ have sum at most 1 and $g_{0,k}(z)$ is a Stieltjes function. Furthermore $\lambda_{0,k}(\lambda_1, \lambda_2)$ has a series expansion in δ_λ and $\{\lambda\}_0$, of the form (4) with new coefficients $y_{m,k}$($m = 1, 2, \ldots \infty$). The leading coefficients are correlated through the identity [6]

$$\sum_{k=1}^{d} y_{1,k} = - f_1 f_2. \qquad (29)$$

In conjunction with the representation (28) this identity implies bounds on $\tilde{\lambda}_0$ that are precisely equivalent to the ones obtained by Tartar and Murat [29] and Lurie and Cherkaev [30]. To see this it is necessary to introduce functions $\bar{\lambda}_{1,k}(\lambda_1, \lambda_2)$ each related to $\tilde{\lambda}_{0,k}(\lambda_1, \lambda_2)$ through an equation analogous to (22).

One would hope that similar considerations apply to the effective thermal conductivity of multicomponent composites and polycrystalline materials. Despite significant progress [4,7,11], this remains an outstanding problem deserving more attention.

ACKNOWLEDGMENTS

We thank G. A. Baker, Jr., and J. G. Berryman for pointing out the connection between the bounds on the thermal conductivity and Padé approximants. We have also profited from many stimulating discussions with M. E. Fisher and G. Papanicolaou. The award of a Hertz Foundation Fellowship to K. G., and the award of a Sydney University Travelling Scholarship and Sage Fellowship from Cornell University to G.W.M. are gratefully acknowledged. The National Science Foundation provided additional support, in part through the Materials Science Center at Cornell University.

REFERENCES

1. D.A.G. Bruggeman, Berechnung verschiedener physikalischer Konstanten von heterogenen Substanzen, Annalen der Physik 24:636 (1935).
2. R. Landauer, Electrical conductivity in inhomogeneous media, in: "Electrical Transport and Optical Properties of Inhomogeneous Media," J.C. Garland and D.B. Tanner, eds., American Institute of Physics, New York (1978).
3. G.W. Milton, Correlation of the electromagnetic and elastic properties of composites and microgeometries corresponding with effective medium approximations, in: "Physics and Chemistry of Porous Media," D.L. Johnson and P.N. Sen, eds., American Institute of Physics, New York (1984).
4. D.J. Bergman, The dielectric constant of a composite material-a problem in classical physics, Phys. Rep. 43:377 (1978).
5. D.J. Bergman, Rigorous bounds for the complex dielectric constant of a two-component composite, Ann. Phys. 138:78 (1982).
6. G.W. Milton, Bounds on the complex permittivity of a two-component composite, J. Appl. Phys. 52:5236 (1981).
7. G.W. Milton, Bounds on the transport and optical properties of a two-component composite material, J. Appl. Phys. 52:5294 (1981).

8. B.U. Felderhof, Bounds for the effective dielectric constant of disordered two-phase materials, J. Phys. C 15:1731 (1982).

9. B.U. Felderhof, Bounds for the complex dielectric constant of a two-phase composite, submitted for publication.

10. K. Golden and G. Papanicolaou, Bounds for effective parameters of heterogeneous media by analytic continuation, Commun. Math. Phys. 90:473 (1983).

11. K. Golden, Bounds for the effective parameter of multicomponent media by analytic continuation, Ph.D. thesis, Courant Institute of Mathematical Sciences, New York University (1984).

12. G.A. Baker, Jr., Best error bounds for Padé approximants to convergent series of Stieltjes, J. Math. Phys. 10:814 (1969).

13. G.A. Baker, Jr., "Essentials of Padé Approximants," Academic Press, New York (1975).

14. N.W. Ashcroft and N.D. Mermin, "Solid State Physics," Saunders College, Philadelphia (1976).

15. W.F. Brown, Solid Mixture Permittivities, J. Chem. Phys. 23:1514 (1955).

16. G.W. Milton, Bounds on the electromagnetic, elastic and other properties of two-component composites, Phys. Rev. Lett. 46:542 (1981).

17. M.N. Miller, Bounds for effective electrical, thermal and magnetic properties of heterogeneous materials, J. Math. Phys. 10:1988 (1969).

18. R.C. McPhedran and G.W. Milton, Bounds and exact theories for the transport properties of inhomogeneous media, Appl. Phys. A 26:207 (1981).

19. S. Torquato and G. Stell, Microstructure of two-phase random media. III. The n-point matrix probability functions for fully penetrable spheres, J. Chem. Phys. 79:1505 (1983).

20. B.U. Felderhof, Bounds for the effective dielectric constant of a suspension of uniform spheres, J. Phys. C. 15:3953 (1982).

21. D.J. Bergman, The dielectric constant of a simple cubic array of identical spheres, J. Phys. C. 12:4947 (1979).

22. R.C. McPhedran and D.R. McKenzie, Electrostatic and optical resonances of arrays of cylinders, Appl. Phys. 23:223 (1980).

23. A.M. Dykhne, Conductivity of a two-dimensional two-phase system, Soviet Physics JETP, 32:63 (1971).

24. B. Abeles and J.I. Gittleman, Composite materials films: optical properties and applications, Appl. Optics 15:2328 (1976).

25. O. Wiener, "Abhandlungen der Mathematisch-Physischen Klasse der Königlichen Sächsischen Gesellschaft der Wissenschaften," 32:509 (1912).

26. Z. Hashin and S. Shtrikman, A variational approach to the theory of the effective magnetic permeability of multiphase materials, J. Appl. Phys. 10:3125 (1962).

27. P. B. Corson, Correlation functions for predicting properties of heterogeneous materials. IV. Effective thermal conductivity of two-phase solids, J. Appl. Phys. 45:3180 (1974).

28. A.L. De Vera, Jr., and W. Strieder, Upper and lower bounds on the thermal conductivity of a random, two-phase material, J. Phys. Chem. 81:1783 (1977).

29. L. Tartar and F. Murat, private communication (1981).

30. K.A. Lurie and A.V. Cherkaev, Exact estimates of conductivity of composites, Proc. Roy. Soc. to appear; see also Dokl. Akad. Nauk. 264:1128 (1982).

31. M. Beran, Use of a variational approach to determine bounds for the effective permittivity of a random medium, Nuovo Cimento 38:771 (1965).

32. S. Torquato and G. Stell, "Microscopic Approach to Transport in Two-Phase Random Media," (CEAS Report #352, 1980).

33. M. Beran and J. Molyneux, Use of classical variational principles to determine bounds for the effective bulk modulus in heterogeneous media, Q. Appl. Math. 24:107 (1966).

34. J.J. McCoy, On the displacement field in an elastic medium with random variations of material properties, in: "Recent Advances in Engineering Science," Pergamon Press, New York, 5:235 (1970).

35. G.W. Milton and N. Phan-Thien, New bounds on the effective elastic moduli of two-component materials, Proc. Roy. Soc. Lond. A380:305 (1982).

36. A.K. Common, Padé approximants and bounds to series of Stieltjes, J. Math. Phys. 9:32 (1968).

37. E. Kröner, Bounds for the effective elastic moduli of disordered materials, J. Mech. Phys. Solids 25:137 (1977).

38. G.W. Milton, A comparison of two methods for deriving bounds on the effective conductivity of composites, in: "Macroscopic Properties of Disordered Media," R. Burridge, S. Childress and G. Papanicolaou, eds., Springer Verlag, New York (1982).

39. J.B. Keller, A theorem on the conductivity of a composite medium, J. Math. Phys 5:548 (1964).

40. K. Schulgasser, On the phase interchange relationship for composite materials, J. Math. Phys. 17:378 (1976).

41. J.G. Berryman, private communication (1983).

42. Y. Kantor and D. J. Bergman, Improved rigorous bounds on the effective elastic moduli of a composite material, J. Mech. Phys. Solids, 32:41 (1984).

SESSION T

Theory and Numerical Analysis

SESSION CHAIRMAN

R. U. Acton
Sandia National Laboratories
Albuquerque, NM.

MATHEMATICAL MODELS FOR

EFFECTIVE THERMAL CONDUCTIVITY

Lis Marcussen

Department of Chemical Engineering
The Technical University of Denmark
Building 229, DK-2800 Lyngby

ABSTRACT

Different mathematical models for effective thermal conductivity are described and discussed in view of experimental results for porous materials.

Models which apply the porosity of the solid as the only structural information are found not to be satisfactory. The degree of continuity in the phases is identified as an important structural parameter. Consequently, such a parameter is included in the models.

Model I   : A solid lattice is assumed.

Model II   : No specific structure of the solid is assumed, but a distribution in space is anticipated for the porosity.

Model III : An extended use of the EMA-model (Effective Medium Approximation for calculation of dielectric constants) to predict effective thermal conductivity leads to a model which involves a factor interpreted as a measure of the degree of continuity in the solid phase.

It is concluded that models II and III are the most promising models.

INTRODUCTION
(See notation section for an explanation of symbols)

Models for λ_e of two-phase porous materials applying porosity as the only structural parameter are found not to be satisfactory, since prediction of λ_e on the basis of porosity and thermal conductivity of the two phases can only limit λ_e to a domain. The simple

585

fundamental limits of this domain are found by considering the thermal resistances to be in series or in parallel:

Upper limit (resistances in parallel): $\lambda_e = \varepsilon \cdot \lambda_g + (1-\varepsilon) \cdot \lambda_s$ (1)

Lower limit (resistances in series): $\lambda_e = [\varepsilon/\lambda_g + (1-\varepsilon)/\lambda_s]^{-1}$ (2)

These bounds can be restricted by including more structural information, e.g. by assuming statistically homogeneous materials, a random distribution of the two phases, isotropy, etc., but the domain is still too broad to be useful for prediction of λ_e.

However, by studying the bounds of λ_e for structures where the solid phase is continuous and the gas phase discrete (upper limit) or where the gas phase is continuous and the solid phase discrete (lower limit) it is realized that the degree of continuity in the phases is an important structural parameter.

Consequently, some models with structural parameters which may be regarded as a measure of the degree of continuity in the solid phase are described and compared with experimental results.

MODEL I

This model is based on the analogy with electrical networks. The unit cell shown in fig. 1 is suggested by Dul'nev and Novikov [1] who solved the equations for resistances in parallel.

This unit cell represents for $\xi = 0$ one phase dispersed in a matrix of the other phase, while $\xi = 1$ reduces the model to the cubic lattice contained in Luikov's model [2] for A = 0.

The unit cell in fig. 1 is not symmetrical with respect to the two phases. Consequently, the way the phases are arranged in the unit cell is important. In the following formulas the solid is represented by a cube with sides of length ℓ and the bars with side lengths $\xi\ell$, $\xi\ell$ and $(1-\ell)$. Formulas for the opposite choice may easily be constructed.

By inspection of fig. 1 it is realized that

$\varepsilon = 1-\ell^3 - 3 \cdot (\xi\ell)^2 \cdot (1-\ell)$ (3)   and $0 \le \xi \le 1$ (4)

ξ is chosen as the free-structure parameter since it has a well-defined interval of definition according to equation (4). ℓ can be found from (3) when ξ and ε are known.

The application of electrical network theory, i.e. using the analogy between the integrated Fourier law and Ohm's law is not a unique method. Consequently, two expressions for λ_e are developed for resistances in parallel and in series, respectively:

586

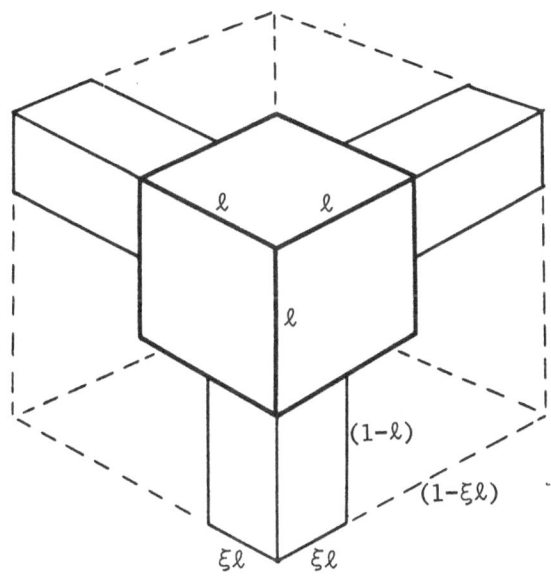

Fig.1. Unit cell of model I.

$$\lambda_{par.} =$$

$$\lambda_g \cdot [1-2\xi\ell(1-\ell)-\ell^2]+\lambda_s \cdot (\xi\ell)^2+\lambda_g \cdot \lambda_s \cdot [\frac{2\xi\ell(1-\ell)}{\lambda_s(1-\xi\ell)+\lambda_g\xi\ell} + \frac{\ell^2-(\xi\ell)^2}{\lambda_s(1-\ell)+\lambda_g\ell}] \quad (5)$$

$$\lambda_{ser.} =$$

$$\frac{1}{\dfrac{\xi\ell}{\lambda_s+(\lambda_g-\lambda_s)[1-\ell^2-2\xi\ell(1-\ell)]} + \dfrac{\ell-\xi\ell}{\lambda_s+(\lambda_g-\lambda_s)(1-\ell^2)} + \dfrac{1-\ell}{\lambda_s+(\lambda_g-\lambda_s)[1-(\xi\ell)^2]}}$$

$$(6)$$

Equations (5) and (6) give the bounds of λ_e for the assumed structure.

For $\xi = 1$, i.e. for the simple cubic structure, good results are obtained by calculating λ_e from the geometric mean of $\lambda_{par.}$ and $\lambda_{ser.}$:

$$\lambda_e = \lambda_{par.}^{\varepsilon} \cdot \lambda_{ser.}^{(1-\varepsilon)} \quad (7)$$

It is assumed that (7) can also be used for $\xi \neq 1$.

In vacuum ($\lambda_g = 0$) (5) and (6) reduce to nonzero expressions.

MODEL II

Model I has the advantage that it depicts the porous material in a simple way. It is, however, quite restrictive to assume a uniform lattice. Instead it can be assumed that the porous region is distributed in space. The distribution function may be chosen in different ways.

As an illustration, the porosity distribution of model I is shown in figs. 2 and 3 for resistances in parallel and in series.

Model I implies a distribution of porosity in steps. For a statistically homogeneous and random structure it might be more reasonable to assume a continuous distribution of porosity in space corresponding to a lattice with an infinite number of steps.

Figs. 4 and 5 show an equal distribution of porosity, i.e. all porosities occupy equal volume fractions. The abscissa in these diagrams is the distribution function $F(\varepsilon)$. The area under the curves is equal to ε.

Resistances in parallel (fig.4)

The porosities $\varepsilon = 0$ and $\varepsilon = 1$ are allotted the volume fractions χ and ψ. χ is chosen to be nonzero. Consequently, in vacuum the condition $\lambda_{par.} \neq 0$ is fulfilled.

The relationship between ε, χ and ψ is found from the area under the distribution curve:

$$\varepsilon = \psi \cdot 1 + \tfrac{1}{2} \cdot (1-\chi-\psi) = \tfrac{1}{2} \cdot (1-\chi+\psi) \tag{8}$$

$\chi = 1+\psi-2\varepsilon$ is inserted into $0 < \chi < 1 \Rightarrow 2\varepsilon-1 < \psi < 2\varepsilon \cdot$ $\tag{9}$

$$\text{and } 0 < \chi + \psi < 1 \Rightarrow \varepsilon - \tfrac{1}{2} < \psi < \varepsilon \cdot \tag{10}$$

From (8), (9) and (10): $\max (0, 2\varepsilon-1) < \psi < \varepsilon \cdot$ $\tag{11}$

The result for $\lambda_{par.}$ is:

$$\lambda_{par.} = \psi \cdot \lambda_g + (1+\psi-2\varepsilon) \cdot \lambda_s + \left[\frac{\lambda_g \cdot \lambda_s \cdot 2 \cdot (\varepsilon-\psi)}{\lambda_s - \lambda_g} \right] \ell n \left(\frac{\lambda_s}{\lambda_g} \right) \tag{12}$$

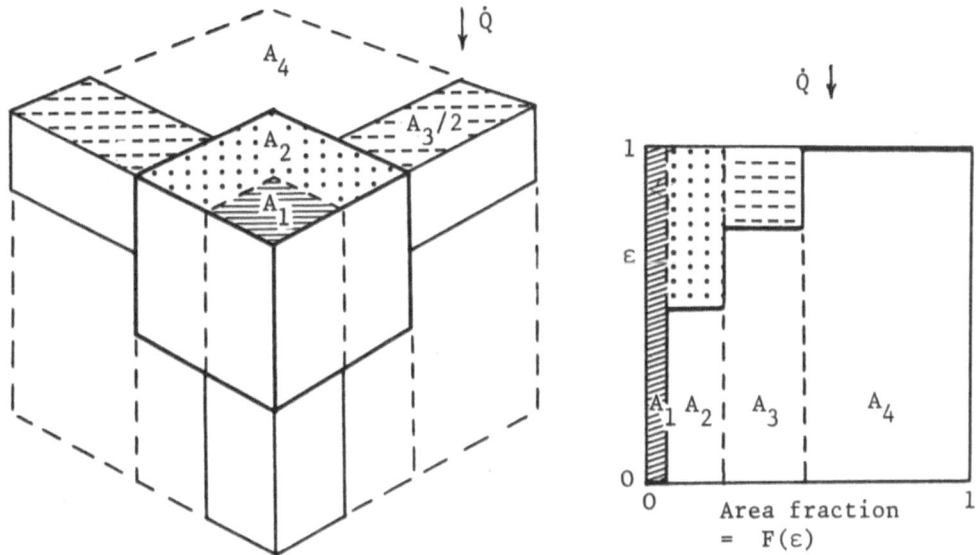

Fig. 2. Porosity distribution of model I. Resistances to heat
conduction in parallel.

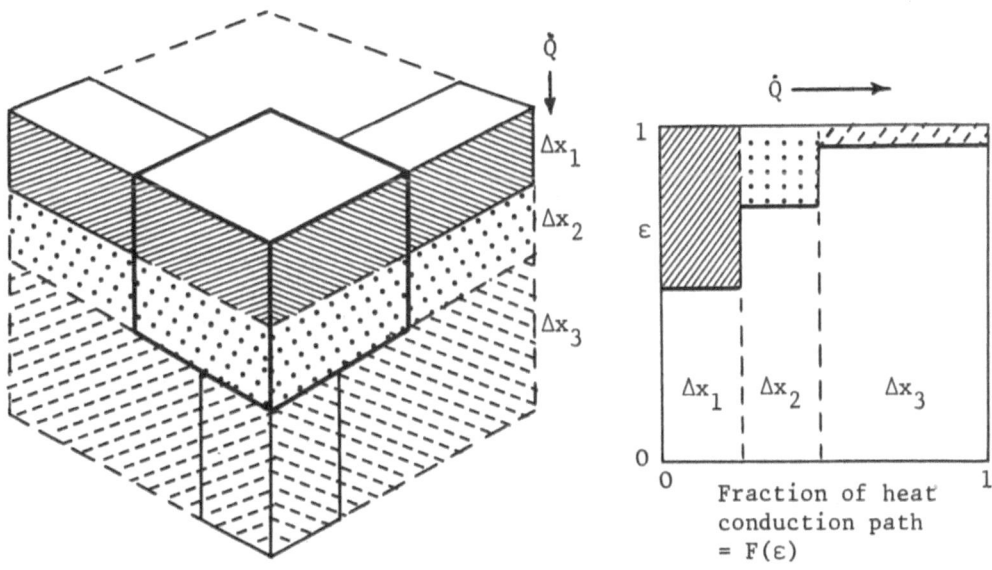

Fig. 3. Porosity distrivution of model I. Resistances to heat
conduction in series.

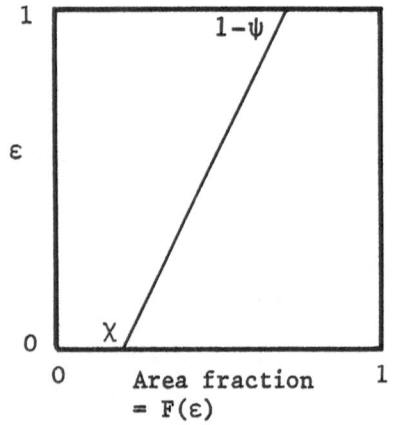

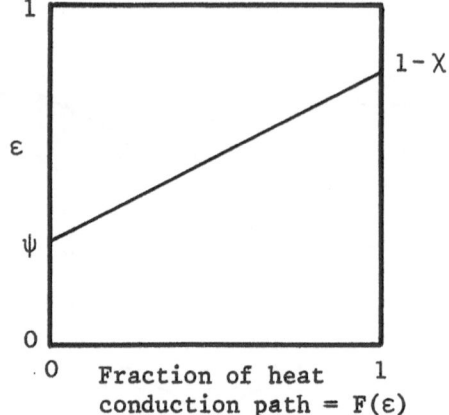

Fig 4: Resistances in paralle.   Fig. 5: Resistances in series.

Figs. 4 and 5. Perosity distribution of Model II. In the examples shown here, all porosities occupy equal volume fractions.

Resistances in series (fig. 5)

With the distribution function shown in fig. 5, no fraction of the heat conduction path is allotted the porosity $\varepsilon = 1$. Consequently, the effective thermal conductivity in vacuum is nonzero.

The result for $\lambda_{ser.}$ is:

$$\lambda_{ser.} = 2 \cdot (\lambda_g - \lambda_s) \cdot (\varepsilon - \psi) \cdot \left[\ell n \left(\frac{(\lambda_g - \lambda_s) \cdot (2\varepsilon - \psi) + \lambda_s}{(\lambda_g - \lambda_s) \cdot \psi + \lambda_s} \right) \right]^{-1} \tag{13}$$

with ψ bounded as given in equation (11).

MODEL III

Model I and II are based on the integrated Fourier law (Ohm's law) whereas model III is derived from the fundamental field equation using Maxwell's field theory. The model is derived by Bruggeman [3] and Niesel [4]. The derivation which originally was performed for dielectric constants of composite media uses the Effective Medium Approximation (EMA).

The pores are approximated by ellipsoids (closed pores) dispersed in a solid matrix. By analogy with the result for dielectric constants [4]:

$$\lambda_e = \lambda_m \cdot \left[1 + \delta_g \left[\frac{\lambda_g - \lambda_m}{\lambda_m} \right] \cdot \sum_{i=a}^{c} \frac{\lambda_{local} \cdot \cos^2 \alpha_i}{\lambda_{local} + (\lambda_g - \lambda_{local}) \cdot F_i} \right] \quad , \tag{14}$$

590

where a, b and c are the ellipsoid axes and α_i are the angles between the temperature gradient ∇T and the axes.

F_i is an analogue to the depolarization factor, depending on the geometry of the ellipsoids:

$$F_i = \int\limits_0^\infty \frac{ds}{(i^2+s)\cdot\sqrt{(a^2+s)\cdot(b^2+s)\cdot(c^2+s)}} \qquad (15)$$

$$\sum_{i=a}^c F_i = 1 \qquad (16) \qquad \text{and} \qquad \sum_{i=a}^c \cos^2\alpha_i = 1 \qquad (17)$$

(14) is valid for small values of δ_g when the ellipsoids are oriented randomly or when all ellipsoids are oriented either parallel to or at right angles to $\langle\nabla T\rangle$.

Unsymmetrical materials

A material with thermal conductivity λ is constructed by dispersion of a volume v_g of ellipsoids with thermal conductivity λ_g in a volume v_s of material with thermal conductivity λ_s. The material is assumed to be quasihomogeneous($\lambda_m = \lambda = \lambda_{local}$) which means that further dispersion of an infinitesimal volume of ellipsoids dv_g leads to a thermal conductivity $\lambda + d\lambda$ which is calculated by insertion of $\delta_g = dv_g/(v_g + v_s)$ in (14):

$$d\lambda = (\lambda + d\lambda)-\lambda = \frac{dv_g}{v_g+v_s} \cdot (\lambda_g-\lambda)\cdot \sum_{i=a}^c \frac{\lambda\cos^2\alpha_i}{\lambda+(\lambda_g-\lambda)\cdot F_i} \cdot \qquad (18)$$

Introduction of $v_g+v_s = V-v$ and using separation of variables yields:

$$\int\limits_\varepsilon^o \frac{-d(v/V)}{1-v/V} = \int\limits_{\lambda_s}^{\lambda_e} \left(\sum_{i=a}^c \frac{\lambda\cdot(\lambda_g-\lambda)\cdot\cos^2\alpha_i}{\lambda+(\lambda_g-\lambda)\cdot F_i}\right)^{-1} \cdot d\lambda \qquad . \qquad (19)$$

Eq. (19) may be integrated [5] for rotational ellipsoids (using $F_b = F_c = F \Rightarrow F_a = 1 - 2F$ and $\cos^2\alpha_b+\cos^2\alpha_c = 1-\cos^2\alpha$, where $\alpha = \alpha_a$).

By inserting different values for F and $\cos^2\alpha$, very different structures may be represented. Some examples are given in table 1 with the solutions for special geometries given by Niesel [4].

Symmetrical materials

Equation (19) describes an unsymmetrical material: One phase is added to the other in such a way that it forms isolated particles, pores or droplets which are approximated by ellipsoids. This leads

Table 1. Solutions to the unsymmetrical model III (One phase dispersed in the other).

Symmetry of dispersed phase	F	Orientation	$\cos^2\alpha$	Solution to (19)
Spheres	$\frac{1}{3}$	Random	$\frac{1}{3}$	$1-\varepsilon = \dfrac{\lambda_g-\lambda_e}{\lambda_g-\lambda_s} \cdot \left(\dfrac{\lambda_s}{\lambda_e}\right)^{1/3}$
Plates	0	Random	$\frac{1}{3}$	$\lambda_e = \lambda_g \cdot \left[\dfrac{3\lambda_s+2\varepsilon\cdot(\lambda_g-\lambda_s)}{3\lambda_g-\varepsilon\cdot(\lambda_g-\lambda_s)}\right]$
		Axis of rotation at right angle to ∇T	0	$\lambda_e = \varepsilon\cdot\lambda_g+(1-\varepsilon)\cdot\lambda_s$ cfr. equation (1)
		Axis of rotation parallel to ∇T	1	$\lambda_e = \left(\dfrac{\varepsilon}{\lambda_g} + \dfrac{(1-\varepsilon)}{\lambda_s}\right)^{-1}$ cfr. equation (2)
Cylinders	$\frac{1}{2}$	Random	$\frac{1}{3}$	$1-\varepsilon = \dfrac{\lambda_g-\lambda_e}{\lambda_g-\lambda_s} \cdot \left(\dfrac{\lambda_g+5\lambda_s}{\lambda_g+5\lambda_e}\right)^{\frac{2}{5}}$

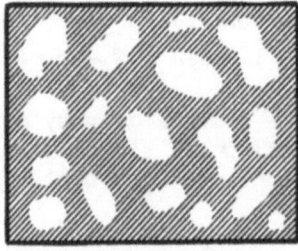

Fig.6a. Unsymmetrical material.

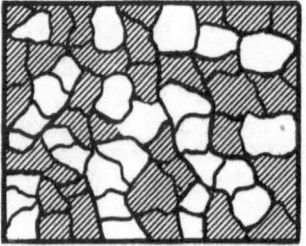

Fig. 6b. Symmetrical material.

to an unsymmetric distribution of the phases.

In many two-phase materials the phases are symmetrically distributed as shown in fig. 6b and should consequently be treated equal in the derivation of λ_e. Bruggeman and Niesel's way of doing this is, for the heat transfer analogue:

An infinitesimal volume fraction δ_g' of an initially homogeneous material with thermal conductivity λ_e is substituted by a material with thermal conductivity λ_g. This is assumed to result in a quasihomogeneous material with a slightly different thermal conductivity λ_e which is calculated from (14):

$$\lambda_e' = \lambda_e \cdot \left[1 + \delta_g' \cdot \left[\frac{\lambda_g - \lambda_e}{\lambda_e} \right] \cdot \sum_{i=a}^{c} \frac{\lambda_e \cos^2 \alpha_{g,i}}{\lambda_e + (\lambda_g - \lambda_e) \cdot F_{g,i}} \right] \tag{20}$$

Subsequently, another infinitesimal volume fraction δ_s' is substituted by a material with thermal conductivity λ_s in such a way that the thermal conductivity of the resulting quasihomogeneous material is brought back to λ_e:

$$\lambda_e = \lambda_e \cdot \left[1 + \delta_g' \cdot \left[\frac{\lambda_g - \lambda_e}{\lambda_e} \right] \cdot \sum_{i=a}^{c} \frac{\lambda_e \cos^2 \alpha_{g,i}}{\lambda_e + (\lambda_g - \lambda_e) \cdot F_{g,i}} \right] \cdot$$

$$\cdot \left[1 + \delta_s' \cdot \left[\frac{\lambda_s - \lambda_e'}{\lambda_e'} \right] \cdot \sum_{i=a}^{c} \frac{\lambda_e' \cos^2 \alpha_{s,i}}{\lambda_e' + (\lambda_s - \lambda_e') \cdot F_{s,i}} \right] \tag{21}$$

Since δ_g' and δ_s' are infinitesimal, $\lambda_e \cong \lambda_e'$ and second order terms $(\delta' \cdot \Sigma)_g \cdot (\delta' \cdot \Sigma)_s \ll (\delta' \cdot \Sigma)_{g \text{ or } s}$:

$$\delta_g' \cdot \sum_{i=a}^{c} \frac{(\lambda_g - \lambda_e) \cos^2 \alpha_{g,i}}{\lambda_e + (\lambda_g - \lambda_e) F_{g,i}} + \delta_s' \cdot \sum_{i=a}^{c} \frac{(\lambda_s - \lambda_e) \cos^2 \alpha_{s,i}}{\lambda_e + (\lambda_s - \lambda_e) F_{s,i}} = 0 \tag{22}$$

Now a quasihomogeneous material with thermal conductivity λ_e is constructed and the entire procedure can be repeated until a two-phase material with volume fractions δ_g and δ_s has been formed.

It follows from (22) that the final result is

$$\delta_g \cdot \sum_{i=a}^{c} \frac{(\lambda_g - \lambda_e) \cos^2 \alpha_{g,i}}{\lambda_e + (\lambda_g - \lambda_e) F_{g,i}} + \delta_s \cdot \sum_{i=a}^{c} \frac{(\lambda_s - \lambda_e) \cos^2 \alpha_{s,i}}{\lambda_e + (\lambda_s - \lambda_e) F_{s,i}} = 0 \tag{23}$$

The crucial approximation in this derivation is the assumption of quasihomogeneity which means that the thermal conductivity λ_{local} in the surroundings of each particle is postulated to be equal to the effective thermal conductivity λ_e of the material. Deviations (between λ_{local} and λ_e and between ∇T_{local} and $\langle \nabla T \rangle$) arising from interactions between neighboring ellipsoids are neglected in this way.

The model is capable of predicting a critical percolation concentration δ_s^*, i.e. the value of δ_s in a two-phase mixture with $\lambda_s \neq 0$ and $\lambda_g = 0$, at which λ_e just becomes zero because the conducting particles at $\delta_s = \delta_s^*$ transform from a continuous network to isolated particles.

The predicted value of δ_s^* for spheres is 0.33 while the correct value is about 0.15. This is caused by the fact that the EMA theory neglects the interactions which arise from the tendency of the spheres to form chains. The error is, however, only significant near δ_s^*. Consequently the model can be considered as a reasonable approximation when $\delta_s > \delta_s^*$ or $(\lambda_g, \lambda_s) \neq 0$.

The model predicts $\delta_s^* = 0$ for cylinders, i.e. a continuous network of conducting material down to zero concentration, and consequently a formula for λ_e with $\lambda_e \neq 0$ in vacuum for all $\varepsilon < 1$.

In principle, F can be determined by microscopy using computerized image-processing techniques.

The value of F determined in this way does not necessarily lead to the best agreement between experiments and model predictions, especially not for structures with many restrictions in the solid phase, i.e., for structures with a porosity close to the critical percolation threshold.

Consequently, F is kept as a structure parameter making the model more suited for describing materials with many restrictions in the solid phase; cfr. the parameters ξ and ψ in model I and II.

Random orientation of the ellipsoids ($\cos^2\alpha_{g,i} = \cos^2\alpha_{s,i} = 1/3$) and identical form of the two phases is assumed:

$$F \equiv F_{g,b} = F_{g,c} = F_{s,b} = F_{s,c} = (1-F_{g,a})/2 = (1-F_{s,a})/2 \qquad (24)$$

F is then a parameter describing how "particle-like" the solid structure is.

Eq. (24) is inserted in (23):

$$\varepsilon \cdot (\lambda_g - \lambda_e) \cdot \left[\frac{1}{\lambda_e + (\lambda_g - \lambda_e) \cdot (1-2F)} + \frac{2}{\lambda_e + (\lambda_g - \lambda_e) \cdot F} \right]$$

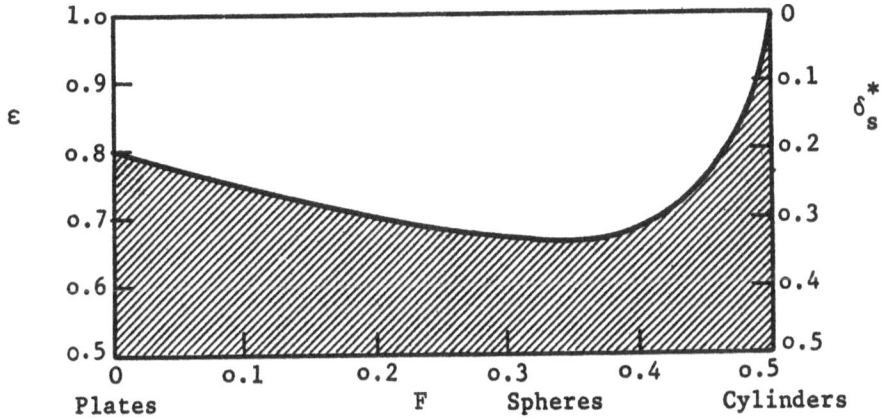

Fig. 7. Curve: Critical percolation concentration.
Hatched area: Allowed domain of (F,ε). (Hatched area $0<\varepsilon<0.5$ not shown).

$$+(1-\varepsilon)\cdot(\lambda_s-\lambda_e)\cdot\left[\frac{1}{\lambda_e+(\lambda_s-\lambda_e)\cdot(1-2F)}+\frac{2}{\lambda_e+(\lambda_s-\lambda_e)\cdot F}\right]=0 \qquad (25)$$

If the solid phase forms a skeleton, F is only allowed to vary within a limited domain, given by $\lambda_e = \lambda_g = 0$ in equation (25). Inserting the critical percolation concentration $\delta_s^* = 1-\varepsilon^*$ leads to:

$$6F^2 - F - 9\cdot F\cdot\delta_s^* + 5\cdot\delta_s^* -1 = 0 \qquad (26)$$

$\delta_s^*(F)$ and the allowed domain (hatched area) of (F,ε) is shown in fig. 7.

EXPERIMENTS

The effective thermal conductivity of three types of refractory brick was measured by means of a probe, in which heat was generated at constant rate.

The probe temperature T_p was recorded as a function of time t. λ_e was determined by comparing the experimental values of $\Delta T_p(t)$ with the solution to Fourier's equation, for a boundary condition which considers a thermal resistance between the probe surface and the surrounding brick.

Experiments were performed in vacuum and at atmospheric pressure with nitrogen or helium in the pores.

595

Table 2. Comparison of models and experiments.

Brick type	Model I Eqs. (5),(6),(7); Model II Eqs. (12),(13),(7); Model III Equation (25)	Model fitted to experiments at vacuum and at atmospheric pressure with Nitrogen in the pores					Model fitted to experiments at vacuum and at atmospheric pressure with Helium in the pores				
		ξ	ψ	F	$\frac{\lambda_{s,\text{model III}}-\lambda_s}{\lambda_{s,\text{model III}}}\cdot100\%$	$\frac{\lambda_{e,\text{exp}}-\lambda_{e,\text{model}}}{\lambda_{e,\text{exp}}}\cdot100\%$ for helium	ξ	ψ	F	$\frac{\lambda_{s,\text{model III}}-\lambda_s}{\lambda_{s,\text{model III}}}\cdot100\%$	$\frac{\lambda_{e,\text{exp}}-\lambda_{e,\text{model}}}{\lambda_{e,\text{exp}}}$ for nitrogen
A	MODEL I					Not realistic	> 1				Model fitting impossible
	MODEL II		0.64		(31%)	) −7%		0.66		(45%)	+ 7%
	MODEL III			>0.5 / If F ≡ 0.5	(0%)	Model fitting impossible / −17%			>0.5 / If F ≡ 0.5	(0%)	Model fitting impossible
B	MODEL I					Not realistic	0.32			−60%	+ 11%
	MODEL II		0.22		−12%	− 8%		0.24		− 1%	+ 4%
	MODEL III			0.12	0%	− 7%			0.19 / 0.38	0% / 0%	+ 5% / + 5%
C	MODEL I					Not realistic	0.15			−368%	+ 15%
	MODEL II		0.27		−30%	−11%		0.28		−10%	+ 5%
	MODEL III			0.11 / 0.36	+ 7% / − 7%	− 6% / − 8%			0.15 / 0.40	+ 2% / − 2%	+ 4% / + 5%

Values of λ_e are in the range 0.049 – 0.67 W/(m°C).
Values of ε are in the range 0.5 – 0.8.

CONCLUSIONS

Predicted and measured values of λ_e are compared in table 1. It is concluded that

Model I   is unsuited for the porous material investigated here.

Model II   can be fitted to all bricks and pore fillings investiga-
           ted. The deviations between predicted and measured values
           are between -11% and +6%. The predicted λ_e-values are com-
           parable to those calculated by model III, although the
           deviations between predicted and measured λ_e are somewhat
           greater for model II, but still quite satisfactory when
           the complexity of the porous structure and the simplicity
           of the model is considered.

Model III cannot be fitted to brick type A but gives predictions of
           λ_e within -8% to + 5% for types B and C.
           Inspection of microphotographs of the bricks, gives an
           idea of the reason for F > 0.5 in the case of brick A and
           lower values of F for brick B and C:
           Brick A looks like a sponge with the solid forming a con-
           tinuous network which should result in $\xi \simeq 1$ in model I
           and $F \simeq 0.5$ in model III, (cfr. fig 7) while bricks B and
           C show a more coarse and granular structure (many restric-
           tions in the solid phase) which in accordance with the ex-
           perimental results and model derivations should corres-
           pond to $\xi < 1$ and $F < 0.5$.

It is noticed that model II and III lead to nearly the same value of λ_s for each type of brick.

In summary, models II and III are the most promising models.
Model III might be preferred because it is based on the fundamental
field equation while model II is based on the integrated field equation. Model III fails, however, on brick A, whereas model II is capable of describing all the materials investigated.

NOTATION

a,b,c = ellipsoid axes (m)

F   = structure parameter in model III

$F(\varepsilon)$ = distribution function in model II

ℓ = side length of cube in model I (shown in fig. 1), (m)

t   = time (s)

T   = temperature (K)

T_p = probe surface temperature (K)

Greek symbols

α = angle between ellipsoid axis and mean temperature gradient

δ = volume fraction

ε = porosity (m^3gas/(m^3gas + m^3 solid)), or volume fraction of a material in a two-phase mixture

λ = thermal conductivity (W/(mK))

λ_e = effective thermal conductivity (W/(mK))

ξ = dimensionless parameter representing restrictions in the lattice of model I (shown in fig. 1)

χ = parameter in model II (shown in figs. 4 and 5)

ψ = parameter in model II (figs. 4 and 5)

Subscripts

a,b,c = ellipsoid axes

g   = gas phase (or material 1 in a two-phase mixture)

local = value in the nearest surroundings of an ellipsoid (model III)

m   = matrix

par. = resistances to heat conduction in parallel

s   = solid phase (or material 2 in a two-phase mixture)

ser. = heat conduction resistances in series

Other symbols

*   = value at the percolation point

< > = mean value

LITERATURE

1. Dul'nev, G.N., and Novikov, V.V. "Conductivity of Nonuniform Systems," J.Eng.Phys. 36(5), 601 (1979).

2. Luikov, A.V., Shashkov, A.G., Vasiliev, L.L., and Fraiman, Yu.E. "Thermal Conductivity of Porous Systems", Int.J.Heat Mass Trans. 11, 117 (1968).

3. Bruggeman, D.A.G. "Berechnung verschiedene physikalischer Konstanten von Heterogenen Substanzen," Annalen der Physic 5. Folge 24, 636 (1935).

4. Niesel, W. "Die Dielektrizitätskonstanten heterogener Mischkörper aus isotropen und anisotropen Substanzen", Annalen der Physik 6. Folge 10, 336 (1952).

5. Reynolds, J.A. and Hough, J.M. Phys.Soc.Proc. B 70, 769 (1957).

FINITE ELEMENT ANALYSIS OF UNGUARDED HOT-PLATE THERMAL
CONDUCTIVITY APPARATUSES

E. Ashworth and T. Ashworth

Mining Engineering and Physics Departments
South Dakota School of Mines and Technology
Rapid City, SD 57701-3995

ABSTRACT

By utilizing a simple thermal conductivity measuring apparatus
with well-defined geometry and by having the ability to model the
system in detail, accurate thermal conductivity values can be
determined. We have used unguarded hot-plate systems with
components of highly contrasting thermal conductivity. This
arrangement provides well-defined boundary conditions for the
modeling. Finite element analysis has been used as the modeling
tool for two such apparatuses. This method was chosen because of
its ability to easily model complex material systems. An outline of
the finite element method together with a comparison with the finite
difference method is given in the paper. By adjusting the thermal
parameters of the model to obtain a match between the calculated and
measured temperature values, the thermal conductivity of either or
both samples can be determined. It has been possible with this
method to show the variation in thermal conductivities of two
supposedly similar samples of gneiss.

INTRODUCTION

Numerical techniques for solving complex problems have
developed along with improved computers. Finite difference
techniques have been used traditionally for heat flow studies and
other diffusion type problems. Finite element analysis, on the
other hand, has been used traditionally to solve structural problems
where stress and displacements are calculated. Both methods have
many advocates. This paper will show the use of finite element
analysis in a heat flow problem, specifically modeling heat flow in
two unguarded hot-plate thermal conductivity apparatuses.

599

FINITE DIFFERENCE METHOD

The finite difference technique was developed along with numerical techniques for solving differential equations. Therefore, the user starts with the differential equation, in this case the diffusion equation

$$\nabla^2 T = \frac{\rho\,C_p}{k}\frac{\partial T}{\partial t} \tag{1}$$

where
∇^2 = Laplacian Operator
T = temperature
t = time
ρ = density of the material
C_p = specific heat at constant pressure for the material
k = thermal conductivity of the material.

Equation (1) is converted into a finite difference equation with the assumptions :

$$\frac{\partial T}{\partial x} \simeq \frac{T_i^n - T_{i-1}^n}{\Delta x} \qquad \frac{\partial^2 T}{\partial x^2} \simeq \frac{T_{i+1}^n - 2T_i^n + T_{i-1}^n}{(\Delta x)^2} \qquad \frac{\partial T}{\partial t} \simeq \frac{T_i^{n+1} - T_i^n}{\Delta t} \tag{2}$$

where
T_i^n = the temperature at node i in time interval n
Δx = the x-distance between the nodes
Δt = the time interval. Equations similar to (2) are used for y and z or for alternative coordinate systems.

Assumptions (2), substituted in equation (1), give the basic finite difference equation, which for one-dimensional heat flow in the x-direction, is

$$\frac{T_{i+1}^n - 2T_i^n + T_{i-1}^n}{(\Delta x)^2} = \frac{\rho\,C_p}{k}\left(\frac{T_i^{n+1} - T_i^n}{\Delta t}\right) \tag{3}$$

An iterative equation is obtained by rearranging terms so that nodal temperatures at time t, T_i^{n+1}, are written as a function of nodal temperatures at the previous time, $t-\Delta t$.

By specifying the temperatures at all nodes at some time, say t = 0, the temperatures at all nodes at any later time can be calculated. Additional complexities can be added, as desired:
Fixed-temperature boundary conditions can be added by requiring T_j^n to be a fixed value for all n and all nodes j located on the boundary. Other boundary conditions can be similarly included.
Any boundary having a complex geometry has to be specially considered by numerical approximation to already specified nodes or by modification of nodal positions to match the boundary. See, for example, Figure 1(a).
Nodes are usually equally spaced. If temperature variation is expected to be significantly non-linear then the size of Δx can be varied within the body. The iteration process has to then include a non-constant $x_i - x_{i-1}$.

600

Different materials can be accommodated by using material properties ρ_i, C_{pi}, and k_i, that depend on the location of node i.

The appropriate equations are easily iterated for a specified time or until convergence to a required accuracy is reached for a steady-state solution. Convergence can be achieved more quickly by using the technique called successive over-relaxation, in which the current value T_i^n is modified by weighting it with the previous value to give $T_i^{n'}$

$$T_i^{n'} = \alpha \, T_i^n + (1 - \alpha) \, T_i^{n-1} \tag{4}$$

α is called the over-relaxation parameter and is usually between 1 and 2. The $T_i^{n'}$ is then used in the next iteration.

FINITE ELEMENT ANALYSIS

The finite element method grew out of the matrix method used for studying forces and displacements in structures. Elements were beams and the nodes were the joints between the beams. The basic method was expanded to include a continuum divided into "elements" connected at specified points called nodes. Application of the basic equations of equilibrium (minimum potential energy) leads to a system of constraining equations with the displacements at each node as the unknown variables.

For heat flow by conduction, the differential equation (1) is modified to an equivalent integral equation, through the use of a "functional". In the case of steady-state heat flow, with convection boundary conditions on any surface S of the body, the functional F is

$$F = \iiint_V \tfrac{1}{2} k \left[\left(\frac{\partial T}{\partial x}\right)^2 + \left(\frac{\partial T}{\partial y}\right)^2 + \left(\frac{\partial T}{\partial z}\right)^2 \right] dV + \iint_S \tfrac{1}{2} h (T - T_0)^2 dS \tag{5}$$

where   V = volume of the continuum
         S = surface of the continuum
         h = coefficient of convective heat transfer across S
         T_0 = air temperature outside S.

The minimum of this functional is found with respect to the nodal temperature values, as is done in finding the minimum potential energy in a stress-strain problem. A system of equations results which can be written in matrix form:

$$[K]\{T\} = \{Q\} \tag{6}$$

where   [K] = conductivity matrix of the finite element mesh
         {T} = unknown nodal temperatures
and   {Q} = equivalent nodal heat flow.

The thermal conductivity matrix [K] is calculated from information about the

      a) type of elements in the mesh
      b) location of the nodes
      c) material types and their properties
      d) interconnection of the elements
      e) boundary conditions.

For example

$$[K] = \sum_e \iiint_{V^e} [B]^{e^T} [k]^e [B]^e \, dV^e \tag{7}$$

where   e = element number
        V^e = volume of element e
        $[B]^e$ = shape matrix of element e
        $[k]^e$ = thermal conductivity matrix of element e
and $[\ ]^T$ = matrix transpose.

In the case of two-dimensional triangular elements, the integrand in Equation (7) is a constant so that [K] can be calculated without numerical integration. For more complicated elements, numerical integration has to be performed.

The unknown nodal temperatures are found by matrix inversion and multiplication using Equation (6). For a significant problem, [K] will be a very large matrix. However, it is symmetrical, "heavy" diagonal and banded. These facts allow special techniques to be used to calculate the inverse $[K]^{-1}$ and thus {T}.

In order to calculate $[k]^e$, the temperature variation within one element is assumed to be polynomial; a linear variation is assumed for two-dimensional triangular elements. Once the type of element is chosen, the continuum is covered by elements in such a way that smaller elements are used where the temperature gradient is expected to be the largest, and larger elements where the temperature gradient is expected to be linear. This allows more accurate values of temperature to be calculated while economizing on computer storage and calculation time. An appropriate finite element mesh is shown in Figure 1(b) as a direct comparison with a similar finite difference mesh in Figure 1(a).

In some senses, a finite element mesh is less abstract than a finite difference mesh. For example, the heat source or sources can now be assumed to be within a certain element instead of at a node and the boundaries can be better modeled. It is possible for each element to be made of a different material. In addition, anisotropic thermal conductivities can be incorporated easily into the model. The angle of anisotropy can be varied with each element thus making this type of analysis ideal for studying heat flow in materials having complex structures such as geological materials, concretes or composites.

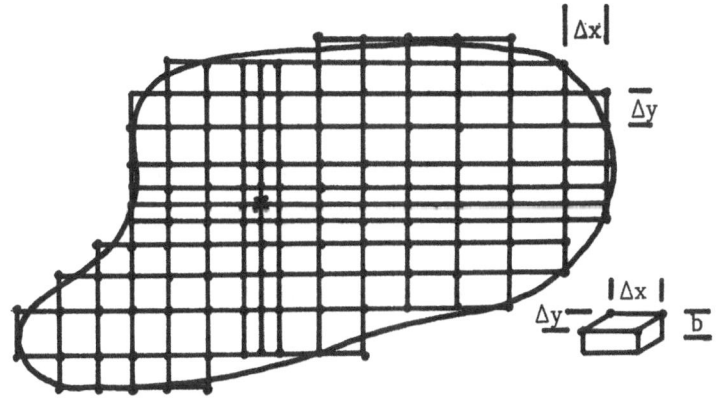

| x   Heat Source (Node)

(a) Finite Difference Mesh

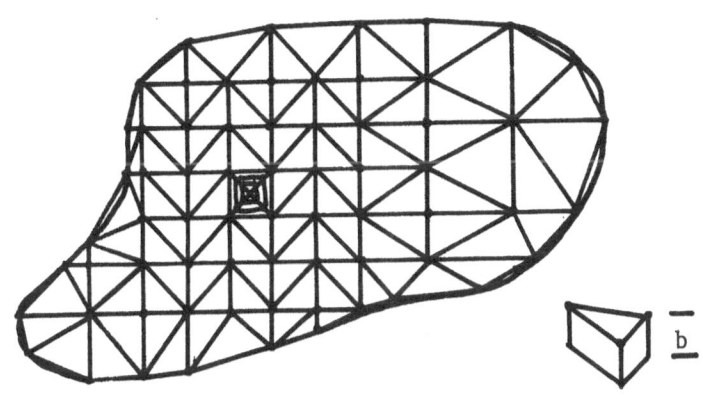

Heat Generating Material

(b) Finite Element Mesh

FIGURE 1   Example of Two Types of Meshes

No iteration is needed for steady-state conditions. For time-dependent studies an iteration procedure needs to be added in which Equation (6) would include terms in $\{T\}^n$ and $\{T\}^{n-1}$. In this type of study, a finite difference technique could be considered slightly advantageous, as an iteration process is automatically built into the technique. The finite element method would need both a matrix inversion and an iteration process. In the case when the thermal properties depend significantly on temperature the matrix would need to be calculated and inverted in each iteration.

Parametric studies can be accomplished by both techniques but it is believed that changing the parameters of the model for parts of a material is easier to do using finite element analysis.

Table 1 summarizes the main advantages and disadvantages of the two methods. The authors have used both techniques but have not made a direct comparison for the studies presented here. General comparisons have been published by others; for example Betts, et al [1] discuss both techniques in detail. Besides, the characteristics of the systems presented here fall into the advantage column of the finite element method.

SPECIFIC COMPUTER CODE

The computer code, HEATRAN, used in this project came originally from the University of California at Berkeley in 1966. The algorithm of the code is explained by Wilson and Nickell [2].

The program is unsophisticated and easy to use. While more advanced codes exist, it was considered inappropriate to go beyond a

TABLE 1

Comparison of Finite Element Method and Finite Difference Method.

	ADVANTAGES	DISADVANTAGES
FEM	No iteration for steady-state Matches complex boundaries Easily models multimaterials Complex material laws included Handles various boundary types	Requires large matrix inversion Error analysis difficult Time studies need iteration Numerical integration needed for complex elements

FD	Finite difference mesh easy
	Time studies easily added to
	  iteration process
	Upper bounds on errors can
	  be calculated
		Approximates complex
		  boundaries
		Long time for convergence
		Multimaterials and complex
		  laws difficult

simple heat conduction program with linear temperature variation in triangular elements. It solves two-dimensional problems. The assumption of cylindrical symmetry allows one to calculate the variations in temperature along the r and z directions. Alternatively, the assumption of planar symmetry allows one to calculate the temperature variations in the x and the y directions. Additional symmetry in the problem can be used to reduce the number of elements required to model the situation, and hence the time of computation, with no loss in accuracy.

Two types of boundary conditions are allowed. These are fixed temperatures (Dirichlet condition) or convective boundary conditions (Neumann condition). The required material properties are thermal conductivity, specific heat and density. The latter two are only required for time studies when the diffusivity of the material is calculated. Heat generation in an element is included also as a material property, so that a heater or other heat source can be modeled. Modifications (Ashworth [3]) to the code allow for materials with anisotropic thermal conductivity to be studied.

Partial mesh generation is performed by the computer program allowing for easier input of data. The user draws an appropriate mesh using quadrilaterals and triangles. The quadrilaterals are subdivided into triangles by the program.

UNGUARDED HOT-PLATE APPARATUS #1

This unguarded system is described in detail in Murdock [4]. Basically it was designed to be used in conjunction with an existing hydraulic rock-testing machine so that measurements of thermal conductivity of geological samples under uniaxial pressure could be made. The steel rams of the testing machine act as two heat sinks so that the thermal conductivity stack consists of a ram, one sample disc, a copper heater disc, another sample disc and the second ram. Thermistors were chosen to measure temperatures and these were embedded in copper discs at either side of the samples. The supplied heat passes through both samples and heat losses at the edges of the stack are minimized by surrounding the stack by fibreglass and rigid styrofoam.

The apparatus can be used in either of two modes:
a) both samples are of the same material with an average conductivity calculated from the amount of heat supplied, estimated heat losses and temperature differences across the samples;
or, b) one sample is a standard material whose thermal conductivity is known and similar to that of the unknown sample. From the amount of heat supplied, heat flow across the standard sample, estimated heat losses and the temperature difference across the unknown sample, its thermal conductivity can be calculated.

The samples are discs about 5.75 inches (14.6 cm) in diameter and about 1 inch (2.54 cm) in thickness. In most experiments, the temperature gradient across the samples was small (1.2°F/in or 25 K/m), in order to match commonly occurring geological temperature gradients. In this case guarding, by supplying an external temperature gradient similar to that found in the stack, was considered to be an unnecessary sophistication. Heat losses, as indicated previously, were minimized by using insulation. These heat losses were estimated by using measurements on two standard samples in the apparatus.

In addition a complete finite element study was conducted. Finite element modeling was ideally suited as the system involves different materials, steady state, heat source and two different boundary conditions (fixed temperature at the rams, convective heat transfer at the outer circumference of the styrofoam). Axisymmetry was assumed to allow use of two-dimensional analysis. Triangular elements were assumed so that numerical integration was not needed. The finite element mesh used was given in detail by Ashworth [3].

As an example of results obtained, data for the experiment using two samples of Black Hills granite from the Harney Peak area are presented. These experimental results were presented by Alexander [5]. The parameters used in the model are listed in Table 2 with the modeled temperatures in Table 3.

Temperature values in the model show that the temperature variations throughout the samples, for a given radius, are linear in all cases studied. There was a small change in the temperature radially, the largest being 0.03°C. Heat losses were found to be less than 3% in the worst case and compared well with those estimated from laboratory measurements. The actual temperatures measured by Alexander [5] were very close to those found in the finite element analysis. The temperature difference across sample one was measured at 3.18°C and modeled at 3.26°C, while the values for sample two were 3.87°C and 3.76°C respectively. Since one sample has a positive difference between measured and modeled values and the other sample a negative difference, this indicates that the two samples did not have the same conductivity value as assumed in the model. The finite element model was thus adjusted to give two different values of thermal conductivity for the granite samples and the values adjusted until a closer match of temperatures was obtained. The two conductivities were 2.3 W/mK and 2.0 W/mK respectively.

Absolute temperature measurements have a probable certainty of only 0.1°C. In each measurement sequence the stack was allowed to equilibriate under zero heating rate. In this way the calibration of the thermistors was frequently compared so that the temperature

TABLE 2

Material Parameters used in the Finite Element Model

Material #	Material Description	Thermal Conductivity (W/mK)	Heat Generation /unit volume (W/m^3)
1	Styrofoam (insulation) *	0.001 [0.100]	0.0
2	Steel (rams)	16.76	0.0
3	Copper (thermistor discs)	40.22	0.0
4	Copper (heater disc)	40.22	1.103 x 10^5
5	Granite (samples) +	2.17	0.0

* Insulation thermal conductivity was not known so three values were used, the best and the worst [ ] cases are reported here.
+ Average value obtained in the actual experiment.

TABLE 3

Modeled Temperature Values (°C) as a Function of Location and Conductivity of the Insulation for Apparatus #1.

Node #	Radius (m) 0.00	0.03	0.06 Stack	0.07	0.072	0.076 Styrofoam
241	21.55 [21.50]	21.55 [21.50]	21.55 [21.50]	21.55 [21.50]	21.55 [21.50]	21.61 [21.55]
Thermistor Plate						
227	21.58 [21.53]	21.58 [21.52]	21.58 [21.52]	21.58 [21.52]	21.58 [21.52]	21.77 [21.71]
Sample 1						
181	24.90 [24.78]	24.88 [24.75]	24.87 [24.72]	24.87 [24.71]	24.87 [24.71]	23.94 [23.80]
Heater						
158	24.90 [24.77]	24.88 [24.75]	24.88 [24.72]	24.88 [24.71]	24.88 [24.71]	24.06 [23.92]
Thermistor Plate						
149	24.86 [24.74]	24.84 [24.71]	24.84 [24.69]	24.84 [24.68]	24.84 [24.67]	23.92 [23.78]
Sample 2						
80	21.04 [21.01]	21.04 [21.01]	21.04 [21.01]	21.04 [21.01]	21.04 [21.01]	21.29 [21.24]
Thermistor Plate						
66	21.02 [20.99]	21.02 [20.99]	21.02 [20.99]	21.02 [20.99]	21.02 [20.99]	21.12 [21.08]

differences could be measured, and modeled, to ±.01°C. However the actual location of the isotherms in the vicinity of the thermistors is not known accurately and so the confidence in the modeled conductivities is only quoted to one decimal place while the average value from measurements is quoted to two.

UNGUARDED HOT-PLATE APPARATUS #2

This unguarded system is described in detail in Ashworth [6]. It is a one-sample stack with a heater at one side of the sample and a heat sink connected to a constant temperature bath at the other side. The system is surrounded by insulation. Originally the system was used in a drift mode in which the heater was turned on and the temperature of the sample allowed to rise. A record of temperature differences was obtained. The heater was turned off and the temperature of the sample allowed to fall. The temperature history was again measured. From these results an estimate of heat loss and the thermal conductivity of the sample were calculated. Also the system can be used in the usual steady-state mode; however the heat losses from the back side of the heater need to be estimated accurately. These heat loss corrections were confirmed by finite element analysis to be less than 2% of total heat input. Other possible errors were studied by finite element analysis. The conclusions were:

    a) the location of the thermocouple radially was unimportant

    b) non-uniformity of the heater was insignificant

    c) the highest error was caused by non-uniformity of the sample which would give in the worst case an error in calculated thermal conductivity of 1.2%.

A new study is being undertaken to use finite element analysis to investigate the heat flow in the vicinity of thermistor/ thermocouple wells in the sample and/or copper discs. Also contact resistance is being modeled by a thin layer of material whose thermal conductivity can be varied to represent thermal resistance.

DISCUSSION

It has been shown that finite element modeling can be used to study heat flow in two unguarded hot-plate thermal conductivity systems. It is ideally suited when the apparatuses are used in the steady-state mode. The boundary conditions are well defined and the parameters in the multimaterial finite element model can be adjusted until a good match is obtained between the measured and modeled temperatures. Thus heat losses and other errors can be investigated and an accurate value for the thermal conductivity of the sample obtained.

This allows an alternative method to expensive, highly sophisticated systems in which heat losses are reduced to an absolute minimum: Build a simple system in which the boundary conditions are well defined, add additional thermocouples to measure temperature profiles and use finite element analysis to model the system. The simpler the system the more benefit is obtained since it is easier to model with unique parameters.

REFERENCES

[1] Betts, P.L., Haslam, J.C. and Lidder, J.S., "Comparisons of Four Computer Programs for Two-dimensional Convection in Closed Cells", Numerical Methods in Thermal Problems, R.W. Lewis and K. Morgan, Eds., Proc. First Int. Conf., Swansea, pp 243-252 (1979).

[2] Wilson, E.L., and Nickell, R.E., "Application of the Finite Element Method to Heat Conduction Analysis", Nuclear Engineering and Design 4, North Holland Publishing Co., pp 276-286 (1966).

[3] Ashworth, E., "The Applications of Finite Element Analysis to Thermal Conductivity Measurements", M.S. Thesis, South Dakota School of Mines and Technology (1983).

[4] Murdock, R.A., "Determination of Thermal Conductivity of Naturally Occurring Materials under Varying States of Stress", M.S. Thesis, South Dakota School of Mines and Technology (1979).

[5] Alexander, T.M., "Investigation of Thermal Conductivity of Natural Materials", M.S. Thesis, South Dakota School of Mines and Technology (1981).

[6] Ashworth, T., Lacey, W.C., and Ashworth E., "Drift Measurement Techniques Applied to Poor Conductors", ASTM Special Technical Publications 660, R.P. Tye, Ed., pp 426-436 (1978).

SESSION U

Composites

SESSION CHAIRMAN

D. L. McElroy
Oak Ridge National Laboratory
Oak Ridge, TN.

THERMAL DIFFUSIVITY IN SITU MEASUREMENTS

OF CARBON/CARBON COMPOSITE REINFORCEMENTS

Agnés Juc-Bouhali, Renée Pujola and Daniel Balageas

Division de Thermophysique
Office National d'Etudes et de Recherches Aérospatiales
(ONERA)
BP 72 92322 Châtillon Cédex (France)

SUMMARY

The diffusivity of directional reinforced composites (3-D C/C) was measured using the flash method. Problems related to the heat losses and the pulse duration were considered. The variation in the diffusivity determined at various moments during the heating confirms the numerical simulations: instead of being a nearly constant quantity as it is true for a homogeneous medium, the apparent diffusivity for these materials is a monotonic function that decreases with time and depends on the sample thickness.

An original method is used which changes nothing in the usual measurement techniques but permits to find the homogenized diffusivity for the whole composite, the in situ axial diffusivity of the parallel reinforcement with controlled heat flux, and the diffusivity of an equivalent matrix representing the actual matrix with the transverse reinforcements. This method, using an optical measurement of the mean temperature on the rear surface of the sample, is compared by numerical simulation with the method determining the diffusivity of the reinforcement from the temperature-time history by a point measurement on the reinforcement alone. Results of this comparison show that the intrinsic precision of the proposed method is better.

The reinforcement axial diffusivity of several 3-D C/C materials was found to vary from 2.6×10^{-4} to 3.1×10^{-4} $m^2 s^{-1}$, values three times greater than the diffusivities of the equivalent homogeneous materials, which can be determined on very thick samples or calculated from steady-state measurements of the thermal conductivity and specific heat.

INTRODUCTION

Of the many composite materials now coming into more general use, those made with highly anisotropic fibers embedded in preferential directions in a homogeneous matrix have a thermal behavior that raises difficulties in modeling and characterization[1]. Of these directional reinforced composites (DRC), the carbon/carbon materials are particularly important.

When determining the thermal properties of such materials, experimentors[2,3,4] consider them to be thermally homogeneous media. They use essentially the flash method[5], and their data are not always satisfactory. It is for this reason that a theoretical analysis of the transient thermal behavior of DRC's was sought[1,6]. It was possible to identify the governing parameters. These theoretical results need experimental confirmation.

Following a brief statement of the flash method and its application to DRC's, a description of the facility and validation of the method by measurements on Armco iron is given. Experimental data obtained with various carbon/carbon tridirectional composites (3-D C/C) are presented. The stated apparent variations of diffusivity are comparable to what was obtained by numerical simulations. The influence of sample thickness and the fact that, when this thickness becomes large, the apparent diffusivity approaches the homogenized diffusivity, are discussed. Afterwards, a very recently proposed method[7] is recalled which permits determination of the axial in-situ diffusivity of the parallel reinforcement with controlled heat flux from the temperature-time history. This method is applied to the present 3-D C/C composites. Finally, the diffusivity of the equivalent matrix representing the actual matrix with the transverse reinforcements is estimated.

FLASH MEASUREMENT OF THE DIFFUSIVITY - APPLICATION TO DIRECTIONAL REINFORCED COMPOSITES (DRC's)

The temperature T of the rear face of a homogeneous sample of thickness L, submitted to a Dirac function heat flux, may be used to determine at each instant t an apparent diffusivity $\alpha(t)$. Let $T/T_{lim} = f(t)$ be the nondimensionalized rear face temperature, T_{lim} being the final temperature and the initial temperature being zero. Then:

$$\alpha(t) = \frac{L^2}{t} \mathcal{F}^{-1}[f(t)] \qquad (1)$$

where $\mathcal{F}^{-1}$ is the inverse function of the exact analytical solution for application of a Dirac function heat flux to a homogeneous sample without loss:

$$\mathcal{F}\left(\frac{\alpha t}{L^2}\right) = 1 + 2\sum_{n=1}^{\infty} (-1)^n \exp\left(-\pi^2 n^2 \frac{\alpha t}{L^2}\right) = \frac{T}{T_{lim}} \qquad (2)$$

In the case of a homogeneous medium, the function $\alpha(t)$ has a constant value (the actual diffusivity of the medium).

The method may be applied to DRC's. We shall always consider here the restricted case of controlled heat flux parallel to one of the directions of reinforcement. Furthermore we suppose that for DRC's $\alpha(t)$ is obtained using the mean temperature of the rear face $\bar{T}$.

The measurement conditions (ambient temperature, low global temperature rise, fast variation) lead to low heat losses. We use the correlation of Balageas[8]:

$$\log[\log(\frac{T_{lim}}{T_{max}})] = -\frac{2}{3}(\frac{t_{max}}{t_{1/2}}) + 1.13 \tag{3}$$

to find the adiabatic limit temperature T_{lim} that the sample would have reached without losses, from the maximum temperature reached T_{mx}. In this expressin $t_{1/2}$ is the time value corresponding to half rise of temperature. This correction remains small under the prevailing test conditions: from 1 to 8 %. Eq. (3) is used for DRC's after verifying by numerical simulations that it applies to such media as long as the losses remain small (Biot number below 0.05), which was true in these experiments. On the other hand, the duration t_p of the laser pulse (approximatively 400 µs) is non-negligible compared with the characteristic thermal time L^2/α of the samples. Because of this, the measured diffusivity is underestimated at every instant[10]. The effect of the pulse shape and duration on the estimation of the diffusivity, in a homogeneous loss-free sample, was studied with emphasis on the low values of T/T_{lim} at the beginning of the experiment[11]. The correction proposed by Azumi and Takahashi[12] to correct the Parker diffusivity $\alpha_{1/2}$ for this effect was used here to correct the apparent diffusivity function $\alpha(t)$. This correction consists of taking the pulse energy center of gravity as the time origin rather than the start of the pulse. The time shift thus introduced is expressed by:

$$t_g = \frac{\int_0^{t_p} t \cdot P(t)\, dt}{\int_0^{t_p} P(t)\, dt} \tag{4}$$

where $\alpha(t)$ is the shape of the pulse, i.e. proportional to the power density during the pulse. The diffusivity is then given by:

$$\alpha(t) = \frac{L^2}{(t - t_g)} \mathcal{F}^{-1}[f(t - t_g)] \tag{5}$$

This new relation (5) was used after verifying by numerical simulations that the correction was also valid for the DRC's.

ONERA FLASH THERMAL DIFFUSIVITY SETUP

A system has been developed to measure thermal diffusivity at ambient temperature by the flash method. The temperature is measured

with a HgCdTe photovoltaic cell operating mainly between 6 and 11 μm.
The mode-locked ruby laser outputs approximatively 1 J, leading to a
temperature rise in the sample of a few degrees. A ZnSe lens is used
to focus the image of the center part of the sample rear face on the
300x300 μm sensing surface of the cell. The observed sample area is
some six mm in diameter. The amplifier system passband is greater than
10 kHz.   The signal is digitized and stores in an on-line mini-
computer, at a rate of 100 kHz.

Validation of the setup and identification method was achieved by
preliminary measurements on Armco iron. The specimens were thin enough
to demonstrate the effect of the pulse duration. The use of the
proposed correction makes the diffusivity values nearly independent of
the time or of the thickness, thus demonstrating the good quality of
the measurements. The mean value of the diffusivity that can be
concluded from the tests is 2.07×10^{-4} m^2 s^{-1} which matches that
recommended by Ho and Powell[13] for this material at 300 K.

MEASUREMENTS ON CARBON/CARBON COMPOSITES

The flash method was applied as described previously to various
samples of two 3-D C/C composites. Table 1 summarizes the properties
of these composites. Fig. 1a shows the convention adopted for these
geometric characteristics.

For the 3-D C/C #1, 8 mm dia. cylindrical specimens of various
thicknesses were used, the laser flux being parallel to the Z axis.
For the 3-D C/C #3, some samples were used with laser flux parallel to
the Z axis (3-D C/C #3 //) and others parallel to the X or Y axes (3-D
C/C #3⊥). The X and Y axes are equivalent since the reinforcements
are the same in both directions. Fig. 1b indicates the respective
directions of the reinforcements and of the laser flux for the 3-D C/C
#3. The data characterizing the specimens, in particular the specific
contact surface $\sigma = L\Sigma$, are given in Table 1. This specific contact
surface is dimensionless. It characterizes the geometry of the
composite specimen and governs the heat transfers in the DRC, in
conjunction with the specific contact thermal resistance $\rho = k_M R\Sigma$ (R:
thermal contact resistance per contact area unit, k_M: matrix
conductivity). All of these quantities are defined by Balageas and
Luc[1,6]. In particular, the parameter Σ is a specific contact surface
between the matrix and the reinforcement, per volume unit of composite.
It depends on the shape of the straight section, on the volume
fraction τ and on the space period ϖ of the reinforcement. It is $4\sqrt{\tau_z}$
L/ϖ_x for the 3-D C/C #1 and #3 // and is $6\sqrt{\tau_x} L/\varpi_y$ or $3\sqrt{\tau_x}L/\varpi_z$ for the
3-D C/C #3⊥.

The apparent diffusivity does not approach a quasi constant value
throughout the heating process as it does in a homogeneous material:
rather, it is a monotonically decreasing function of time that depends

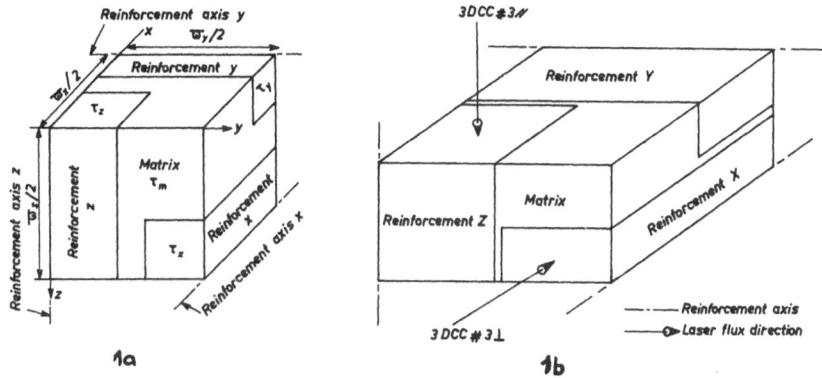

Fig. 1. Volume element of a 3-D composite: Definition of the
geometric parameters (Fig. 1a) and relative configuration
of the reinforcements and laser flux in measurements on
the 3-D C/C #3 (fig. 1b).

Table 1. Characteristics of the tested materials and specimens

Material	Material characteristics				Sample characteristics	
	Space periods (mm)	Reinfor. volume content	Bulk density (kg m^{-3})	estimated specific heat (J kg^{-1} m^{-3})	sample thickness L, (mm)	$\sigma = L\Sigma$
3D C/C #1	$\varpi_x = 1.6$ $\varpi_y = 1.6$ $\varpi_z = 1.6$	$\tau_x = 0.12$ $\tau_y = 0.12$ $\tau_z = 0.30$	1930 1930 1930	730 730 730	1.60 2.94 4.82	2.2 4.0 6.6
3D C/C #3	$\varpi_x = 1.6$ $\varpi_y = 1.6$ $\varpi_z = 0.8$	$\tau_x = 0.22$ $\tau_y = 0.22$ $\tau_z = 0.25$	1930 1930 1930	730 730 730	3D C/C #3 // 1.62 3.21 4.81 6.42 8.20	2.0 4.0 6.0 8.0 10.2
					3D C/C #3 $\perp$ 3.21 4.80 6.42	5.6 8.4 11.2

on the test sample thickness. In Fig.2a, its evolution is presented versus the mean temperature of the rear face of the sample in the case of 3-D C/C #3 //. The thicker the specimen, the lower the slope and the higher the final diffusivity. The specific contact surface of the thickest sample is not large enough for it to behave like a homogeneous medium having a constant apparent diffusivity. This does not contradict measurements of R.E. Taylor[4] on fine-weave C/C composites leading to the conclusion that these composites have a constant apparent diffusivity matching the values calculated from direct measurements of thermal conductivity and specific heat. In effect, for values of $L\Sigma$ greater than 10 (probable in Taylor's measurements) it is very difficult to see any significant variation in the apparent diffusivity for T/T_{lim} greater than 0.2. The present data may be compared to previous numerical simulations results[1,6,14] showing that the apparent diffusivity varies uniformly with time, starting at the diffusivity α_R of the most conductive phase (the reinforcement) at the beginning of the experiment, and ending at a value between the diffusivity $\bar{\alpha}$ of the homogeneous equivalent medium, whose expression is:

$$\bar{\alpha} = [\tau k_R + (1-\tau) k_{\bar{M}}] / [\tau c_R + (1-\tau) c_M] \tag{6}$$

and the diffusivity $\alpha_{\bar{M}}$ of the equivalent matrix representing the actual matrix and the transverse reinforcements. Fig. 2b gives the

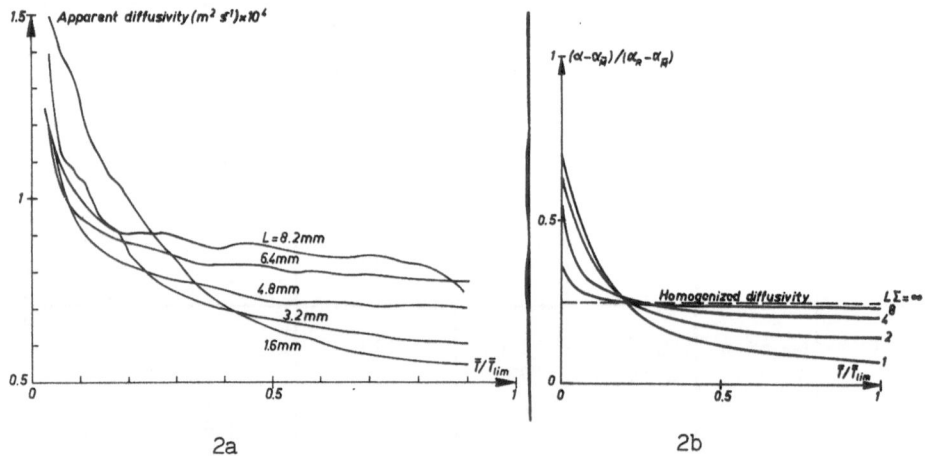

2a   2b

Fig. 2.   Variation of the apparent diffusivity with temperature of the rear face of the specimen: a) experimental results from specimens of 3-D C/C #3 // of various thicknesses ; b) numerical simulation results showing the influence of the specific contact surface $L\Sigma$ for a DRC characterized by : $\tau = 0.25$, $k_{R/M} = 5$, $c_{R/M} = 1$, $\rho = 0.2$.

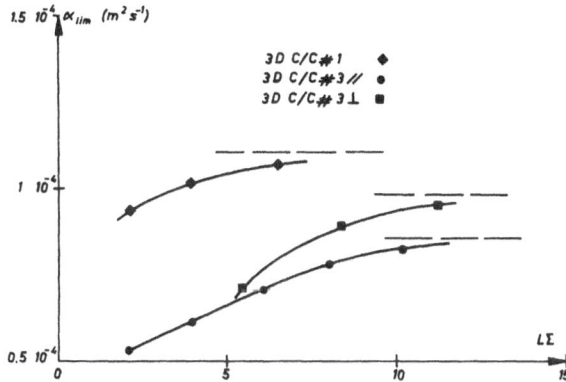

Fig. 3. Apparent diffusivity of the C/C materials as measured at the end of the experiment, with T/T_{lim} approaching unity: effect of the thickness of specimens. Determination by extrapolation of the diffusivity $\bar{\alpha}$ of the equivalent homogeneous medium.

apparent diffusivity variation from the numerical simulation of a DRC having characteristics similar to those of the composites studied experimentally. The pronounced influence of the specific contact surface $L\Sigma$, and thus of the specimen thickness L, is quite analogous to what is found experimentally. Fig. 2b shows that Parker's diffusivity $\alpha_{\frac{1}{2}}$ (T/T_{lim} =0.5) is lower than the homogenized diffusivity $\bar{\alpha}$. This probably explains the fact reported by R. Taylor[3] and Procter from their experiments and those of others[15, 16] , that the thermal conductivities of C/C composites deduced from flash diffusivities by Parker's method are generally lower than the conductivities measured in steady-state regime.

Furthermore, it is shown in Fig. 2b, that the final diffusivity values α_{lim} - that is those obtained for T/T_{lim} tending toward unity - approach the diffusivity of the equivalent homogeneous medium $\bar{\alpha}$ as $L\Sigma$ increases. Thus, we can get $\bar{\alpha}$ by extrapolating the curve $\alpha_{lim}(L\Sigma)$. This method of estimating the homogenized diffusivity was applied to the experimental curves (see Fig. 2a) producing the $\alpha_{lim}(L\Sigma)$ laws given in Fig. 3. The following asymptotic values can easily be determined to $\pm$ 5 %: $\bar{\alpha}$ = 1.1 x$10^{-4} m^2 s^{-1}$ for the 3-D C/C #1, $\bar{\alpha}$ = 0.8 x$10^{-4} m^2 s^{-1}$ for the 3-D C/C #3 // and $\bar{\alpha}$ = 0.97 x$10^{-4} m^2 s^{-1}$ for the 3-D C/C #3⊥ .

IN SITU MEASUREMENT OF THE REINFORCEMENT AXIAL DIFFUSIVITY α_R

Measurement problems

The Fig. 2b shows that if we want to find $\bar{\alpha}$, we must make observation as late as possible in the experiment and on the thickest possible specimens. On the other hand, if we want to determine α_R, this must be done as early as possible and on the thinnest possible specimens. The simplest way of determining α_R may consist of extrapolating the apparent diffusivity curves $\alpha(t)$, calculated from the mean temperature of the rear face, toward the time origin. However, this extrapolation is risky because the slope of the curve is increasing, tending practically toward infinity at the origin. We are led to devise a new way of interpreting the temperature-time histories. If the reinforcement and the matrix were not thermally coupled, α_R could be found simply from the variation in the mean temperature of the reinforcement alone, $\bar{T}_R$, by:

$$\alpha_R = \frac{L^2}{t}\, \mathcal{F}^{-1}\, [\,\frac{\bar{T}_R}{\bar{T}_{Rlim}}\,(t)\,] \tag{7}$$

In practice, the two phases are always thermally coupled. This disturbs $\bar{T}_R$ and its final value $\bar{T}_{Rlim}$. However, the closer we come to the start of the heating process the less time there is for matrix-reinforcement heat transfer, and thus the closer the diffusivity calculated from $\bar{T}_R$ is to α_R. Although it is not always easy to measure $\bar{T}_R$ directly, this alternative was chosen by R.E. Taylor[17]. Taylor measures $\bar{T}_R$ with a very fine thermocouple embedded in a fiber yarn reinforcement in the rear face, parallel to the flux. Aside from the fact that this temperature is not normalized satisfactorily in the referenced study, it should be noted that this measurement is neither easy nor reliable and that the way the thermocouple averages the temperature is difficult to estimate. Another approach consists of still measuring the mean temperature of the composite $\bar{T}/\bar{T}_{lim}$, as is done for a homogeneous material, but then calculating from this the normalized mean temperature $\bar{T}_R/\bar{T}_{Rlim}$, which can be used to identify α_R. This method, proposed by Balageas[7], has been used here.

Measurement method [7]

At the start of the experiment, the heating of the rear face is essentially due to the heating of the reinforcement, since the reinforcement has the greater diffusivity. The mean temperature T of the rear face is thus roughly:

$$\bar{T} \underset{t \to 0}{\simeq} \tau\, \bar{T}_R \tag{8}$$

and, considering the difference in the volume specific heat between the matrix and the reinforcement, the normalized temperature $\bar{T}_R/\bar{T}_{Rlim}$ can be expressed for times close to the origin by:

$$\bar{T}_R/\bar{T}_{Rlim} \underset{t \to 0}{\simeq} \psi\, \bar{T}/\bar{T}_{lim} \tag{9}$$

where: $\psi = \tau^{-1}\,[\tau + (1-\tau)/c_{R/M}]^{-1}$. In case where the ratio $c_{R/M}$ of the volume specific heats is unity (approximately verified for the C/C materials), ψ can be reduced to: $\psi = 1/\tau$. From (7) and (9) we get:

$$\alpha_R = \lim_{t \to 0} \left[\frac{L^2}{t} \mathcal{F}^{-1}(\psi \frac{\overline{T}}{\overline{T}_{lim}}(t)) \right] = \lim_{t \to 0} [\alpha^*] \tag{10}$$

Thus the diffusivity α^* obtained from the temperature-time history and from the a priori evaluation of ψ tends toward α_R as $t \to 0$. Fig. 4 compares the shape of this $\alpha^*(\overline{T}/\overline{T}_{lim})$ curve with the corresponding apparent diffusivity law $\alpha(\overline{T}/\overline{T}_{lim})$. These are numerical simulation data. Either of these curves can be extrapolated to the origin to find a value for α_R, but the conditions for extrapolating α^* are more attractive because the curve is nearly horizontal. The diffusivity curve calculated from the temperature of the reinforcement $\overline{T}_R/\overline{T}_{Rlim}$ is also given. This law too can be extrapolated to find α_R under conditions analogous to the α^* law. However α_R is approached more closely if we use the α^* law.

The curve α^* is again given in the Fig. 5a for various specimen thicknesses. The thicker the specimen the lower the heat transfer resistance between the matrix and the reinforcement, and the more pronounced is the minimum α^*_{min} of the curve, below the value α_R we are looking for. The maximum error in estimating α_R occurs at the minimum of the curve. Numerical simulations show that the quantity α^*_{min} varies almost linearly with $L\Sigma$ when $L\Sigma$ is small. It tends toward αR as $L\Sigma \to 0$. αR can thus be extrapolated from measurements made on several specimens. The linear extrapolation is independent of Σ, which is the same in all specimens. Only the slope is dependent on Σ, but not the ordinate at the origin. Therefore, we can plot the value α^*_{min} again the thickness L alone, which can be measured with a good precision (a few %).

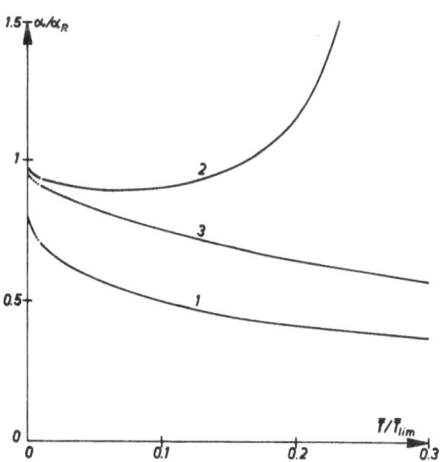

Fig. 4. Apparent diffusivity from numerical simulation for the DRC characterized by: $\tau = 0,25$, $k_{R/M} = 5$, $c_{R/M} = 1$, $\rho = 0.2$, $L\Sigma = 2$.
curve 1: $\alpha = L^2 t^{-1} \mathcal{F}^{-1}(\overline{T}/\overline{T}_{lim})$; curve 2: $\alpha^* = L^2 t^{-1} \mathcal{F}^{-1}(\psi \overline{T}/\overline{T}_{lim})$
curve 3: $\alpha = L^2 t^{-1} \mathcal{F}^{-1}(\overline{T}_R/\overline{T}_{Rlim})$.

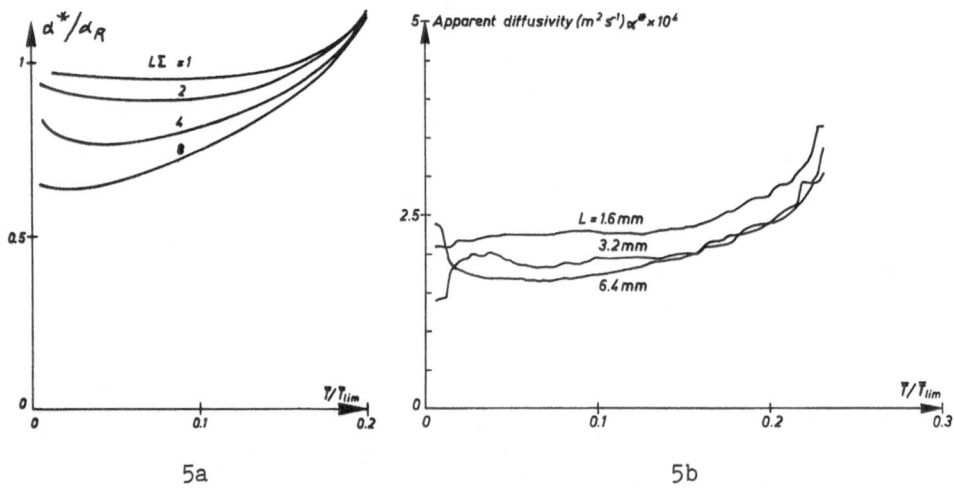

5a   5b

Fig. 5.   Apparent diffusivity laws α^*. a) Numerical simulation results for the same DRC as Fig. 4. Influence of LΣ. b) Experimental data for 3-D C/C #3 //.

Experimental results for 3-D C/C composites

      The method was used to interpret the temperature-time histories of the 3-D C/C. Once corrected for the effects of losses and pulse duration, the laws $\overline{T}/\overline{T}_{lim}$ (t) were multiplied by the factor $\psi = 1/\tau$ ($c_{R/M}$ assumed to be unity from the chemical nature of the two phases). The corresponding apparent diffusivity laws $\alpha^*(\overline{T}/\overline{T}_{lim})$ are given in Fig. 5b for the 3-D C/C #3 //. The shape of the curves is comparable with the simulation results of Fig. 5a. The minimum values of α^* are plotted against LΣ in the Fig. 6. This gives an extrapolation line intersecting with the L =0 axis at the reinforcement axial diffusivity we are seeking, i.e.: $\alpha_R = 3.0 \times 10^{-4}$ m^2 s^{-1} for 3-D C/C #1, $\alpha_R = 2.6 \times 10^{-4}$ m^2 s^{-1} for 3-D C/C # // and $\alpha_R = 3.1 \times 10^{-4}$ m^2 s^{-1} for 3-D C/C #3$\perp$. Note that the diffusivity of the reinforcement is in a ratio of three with respect to the homogenized diffusivity $\overline{\alpha}$.

ESTIMATION OF THE EQUIVALENT MATRIX DIFFUSIVITY $\alpha_{\overline{M}}$

      A multidirectional DRC with controlled heat flux parallel to one of the reinforcement directions can be likened to a unidirectional DRC having the same reinforcement parallel to the flux, but with a matrix equivalent to the actual matrix homogenized with the transverse reinforcements[1,6]. Let $\alpha_{\overline{M}}$ be the diffusivity of this equivalent matrix. Knowing $\overline{\alpha}$ and α_R, $\alpha_{\overline{M}}$ can be estimated. If $c_{R/M}$ is unity, Eq. (6) reduces to:

$$\overline{\alpha} = \tau \alpha_R + (1-\tau)\alpha_{\overline{M}} \qquad (11)$$

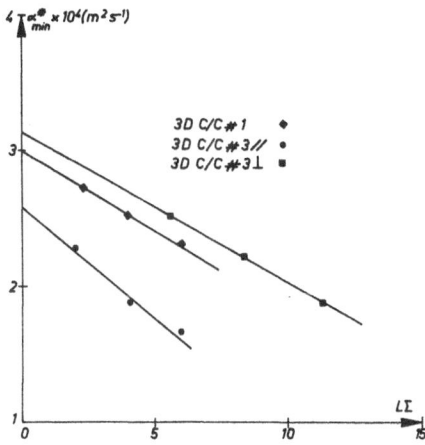

Fig. 6. Minimum diffusivity α^*_{min} of the 3-D C/C composites: effect of the specimen thickness. Extrapolated diffusivity α_R.

from which we get the expression for $\alpha_{\bar{M}}$: $\alpha_{\bar{M}} = (\bar{\alpha} - \tau\alpha_R)/(1 - \tau)$. The formula yields the following values for the three composites: $\alpha_{\bar{M}} = 2.9$ $\times 10^{-5}$ m^2 s^{-1} for 3-D C/C #1, $\alpha_{\bar{M}} = 2.7 \times 10^{-5}$ m^2 s^{-1} for 3-D C/C #3 / and $\alpha_{\bar{M}} = 3.6 \times 10^{-5}$ m^2 s^{-1} for 3-D C/C #3$\perp$.

CONCLUSION

The present study has shown the possibility of determining the diffusivities of the homogenized composite, of the reinforcement parallel to a given flux direction and of the matrix equivalent to the actual matrix homogenized with the transverse reinforcements. Numerical simulation shows that the results obtained are weakly influenced by the thermal contact resistance characterizing the matrix-reinforcement thermal coupling. The measurements form a consistent set of data for carbon/carbon composites (see Table 2).

Table 2. 3-D C/C composites: Thermal properties deduced from the temperature-times histories.

Material	axe $\parallel$ φ	Homogenized properties		Reinforcement		Equivalent Matrix	
		$\bar{\alpha}$ (m^2 s^{-1}) $\times 10^{-4}$	$\bar{k}$ (W m^{-1} K^{-1})	α_R (m^2 s^{-1}) $\times 10^{-4}$	k_R (W m^{-1} K^{-1})	$\alpha_{\bar{M}}$ (m^2 s^{-1}) $\times 10^{-4}$	k_M (W m^{-1} K^{-1})
3-D C/C #1	z	1.1	150	3.0	420	0.29	41
3-D C/C #3 $\parallel$	z	0.85	120	2.6	360	0.27	37
3-D C/C #3 $\perp$	x,y	0.97	130	3.1	430	0.36	50

For this type of material the diffusivity ratio between the reinforcement and the matrix is on the order of ten and the axial diffusivity of the reinforcement reaches very high values which are probably a result of the material processing involving temperature cycles up to 2700 ° C.

References

1. D.L. Balageas and A.M. Luc, Nonstationary thermal behavior of directional reinforced composites. Limit of application of thermal property homogenization, AIAA Paper 83-1471 (1983)
2. W. Fritz, W. Huttner, K. Maier and R. Brandt, Thermophysical properties of carbon-fibre/carbon matrix composites at high temperatures, Rev. Int. Hts Temp. Refr. 16:350 (1979).
3. R. Taylor and R. Procter, Measurement of thermal diffusivity of carbon fibre carbon composites from 300-3000 K, Univ. Manchester/UMIST(U.K.), Report AFOSR-TR-81-0638 (1981).
4. R.E. Taylor, H. Groot and R. Shoemaker, Thermophysical properties of fine weave C/C composites, AIAA 81-1103 (1981).
5. W.J. Parker, R.J. Jenkins, C.P. Butler and G. Abbott, Flash method of determining thermal diffusivity, heat capacity and thermal conductivity, J. Appl. Phys. 32:9:1679 (1961)
6. A.M. Luc and D. Balageas, Comportement thermique des composites à renforcement orienté soumis à des flux impulsionnels, to be published in High-Temp. High Pressure (1984).
7. D.L. Balageas, Détermination par méthode flash des propriétés thermiques des constituants d'un composite à renforcement orienté, idem.
8. D.L. Balageas, Nouvelle méthode d'interprétation des thermogrammes pour la détermination de la diffusivité par méthode impulsionnelle, Rev. Phys. Appl.17:227 (1982).
9. M. Assouline, Private communication.
10. R.E. Taylor and L.M. Clark III, Finite pulse time effects in flash diffusivity method, High Temp. High Pres.6:65 (1974)
11. N. Coulon, Private communication.
12. T. Azumi and Y. Takahashi, Novel finite pulse-width correction in flash thermal diffusivity measurement, Rev. Sc. Instr. 52:9:1411 (1981).
13. C.Y. Ho, and R. Powell, Thermal conductivity of the elements: a comprehensive review, J. Phys. Chem. Ref. Data 3, suppl. 1(1974).
14. A.M. Luc and D.L. Balageas, Le problème de la définition de la mesure de la diffusivité thermique des matériaux composites à renforcement, ONERA TP 1981-30 (1981).
15. H.J. Lee and R.E. Taylor, Thermophysical properties of carbon/graphite fibres and MOD-3 fibre reinforced graphite, Carbon 13 (1975).
16. M.L. Minges AFML-TR-74-96 (1975).
17. R.E. Taylor, Thermal diffusivity of composites, TPRL 253 (1981).

ON THE INTERPRETATION OF OFF-AXIS THERMAL DIFFUSIVITY DATA

FOR COARSE-WEAVE COMPOSITES*

M. S. Deshpande

Center for Information and Numerical Data Analysis
and Synthesis (CINDAS), Purdue University
West Lafayette, Indiana

ABSTRACT

An off-axis test for the determination of thermal diffusivity along the principal axes of coarse-weave composites was suggested recently due to its potential in reducing percentage rise-fraction effects. In the present paper an evaluation of off-axis testing is presented based on the available data suggesting that more careful study is needed to conclude that off-axis testing is a more reliable means of obtaining on-axis thermal diffusivity values. It is also suggested that the off-axis sample cut at 45° direction may not always result in the 45° off-axis thermal diffusivity values.

INTRODUCTION

New material development, especially the capability of tailoring materials to specific requirements, is a path being increasingly taken to satisfy the needs of product developers and designers. Requirements such as efficient use of energy, retention of good mechanical strength at higher temperatures along with low density are some of the criteria which have been invoked in designing these new materials. The broad class of composite materials, composed of fiber reinforcement embedded in a more-or-less homogeneous matrix, can be cited as an example of these new materials.

---

*This publication is approved for release by the Materials Laboratory of the Air Force Wright Aeronautical Laboratories as per ASD-83-2167 on October 4, 1983.

Composites are becoming increasingly important in the new technological applications. One sub-group of these composites is known as carbon/carbon materials. Their applications have been surveyed by Fitzer and Heym [1] and have been emphasized by Schmidt [2]. Some of the important applications are their use in rocket nozzles, re-entry heat shields, aeroplane disc brakes, prostheses, etc. Naslain et al. [3] have pointed out drawbacks of these materials, which restrict their usage for several potential energy production related applications. They have also suggested a few ways to overcome them. For instance, they suggested that poor oxidation resistance could be overcome by replacing, partly or totally, the carbon matrix by refractory materials, inert towards carbon and having good oxidation resistance, such as silicon carbide (see Christin et al. [4]). The C-C/SiC composites, produced by Naslain et al. [3] by chemical vapor infiltration (CVI), were found to exhibit higher mechanical strength than related C-C composites and their oxidation resistance is excellent up to 1500°C (at least when they are well processed). Due to better compatibility of TiC with carbon at higher temperatures and better thermal stability of TiC, the use of C/TiC hybrid matrix has been suggested by Aggour et al. [5]. A higher thermal stability, equivalent mechanical strength, but somewhat low air-oxidation resistance, could be expected for C-C/TiC composites as compared to the C-C/SiC related materials. Hence, the C-C/TiC composites are potential candidates for high-temperature structures. In fact, TiC is among the refractory materials that could be used as a first wall in nuclear fusion reactors (see Whitley et al. [6]).

Thermal property behavior along with strength behavior is of central importance in using these composites. In fact, the measurement of the thermal conductivity or diffusivity is of special concern, since the data are typically used to (a) ensure adequate structural properties at a given temperature, (b) assess thermal shock resistance, and (c) maintain allowable substrate temperatures. More specifically and to the point of this discussion, the thermal conductivity or diffusivity measurements are often required in an 'off-axis' direction to the fiber bundles. This work deals with the interpretation of results of such measurements obtained by the present flash diffusivity technique in order to understand the data produced and the measurement technique.

COMPOSITE CONSTRUCTION

The composite materials may be viewed as fiber bundles embedded in a more-or-less homogeneous matrix. A three-directional reinforced composite is composed of fibers such as thermally stable graphite yarn bundles oriented along each of the principal axes and a matrix within the yarn bundles and interstitials. The matrix may be composed, for instance, of an organic resin, a metallic

626

composition, or a carbonaceous composition. The various composite constructions and methods for their manufacture are well described in the literature, and the interested reader should refer to discussions such as Latchman et al. [7] and Fitzer and Huttner [8]. The composite structure utilized in the present study has a typical cylindrical configuration and contained a carbon matrix derived by pyrolysis of a pitch precursor. A schematic of the cylindrical construction and the geometrical directions are shown in Fig. 1. Additional details about the material are given by Taylor et al. [9], who have provided the results of measurements done in an off-axis direction for the present analysis.

OFF-AXIS TESTING

In a flash diffusivity experiment a small disc is exposed to an intense heat flux for a very short time and the temperature-time history is subsequently recorded on the specimen back face. In the testing of composite materials, where materials are designed to have specified fiber fractions, the back face is not heated uniformly, because the fiber bundles represent preferred paths for heat flow. Taylor and Groot [12] have pointed out the problems that stem from this nonuniform heating. The problems related to sample-size requirements received attention from Deshpande et al. [10], who established diameter requirements in relation to viewing spot size from a purely geometrical approach. An analysis by Jortner [11] quantified requirements of sample thickness. Taylor and Groot [12] pointed out the problem of establishing a unique thermal diffusivity value at different time-rise fractions. They have characterized this by calculating the thermal diffusivity values at 20% and 80% rise times, and half-rise times.

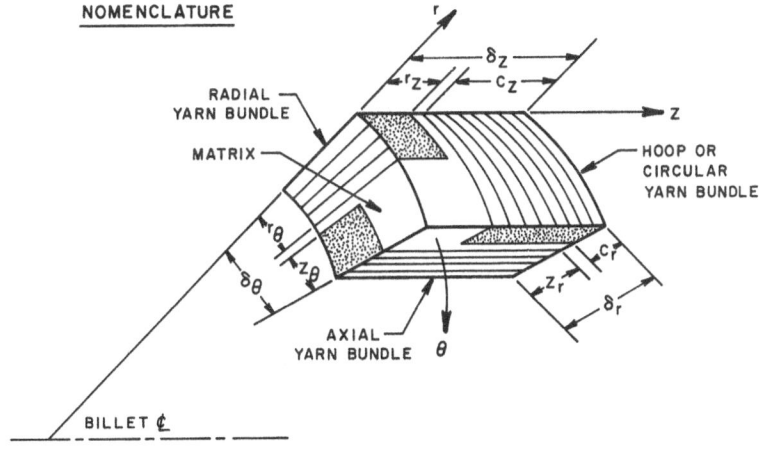

Figure 1. Nomenclature for 3-D Cylindrical Unit Cell.

627

Recently, Jortner [13] suggested an off-axis test as a means of reducing the percent rise-fraction effect in measuring thermal diffusivity by the flash technique. An initial attempt for such measurements was made by Taylor et al. [9]. The potential of the off-axis test as a reliable means to reduce percent rise-fraction effects is now completely proven for these composites (see Jortner [13]). However, the proposition that an off-axis thermal diffusivity value can produce reliable on-axis thermal diffusivity data needs further verification.

For the reduction of the off-axis data to the on-axis values, Taylor et al. [9] used the following equation

$$\alpha_{nn} = \sum_{i=1}^{3} \cos^2 \theta_i \alpha_i. \qquad (1)$$

Here, α_{nn} is the thermal diffusivity in the direction considered, θ_i is the angle made by this direction with the ith principal direction, and α_i is the thermal diffusivity along the ith principal direction. For the directional cosines ℓ, m, and n respectively, eq. (1) reduces to

$$\alpha_{nn} = \ell^2 \alpha_1 + m^2 \alpha_2 + n^2 \alpha_3. \qquad (2)$$

Taylor et al. [9] considered thermal diffusivity measurements along principal directions and also along the direction which lies in a plane formed by the two principal axes and which forms equal angles with them. Then, the directions considered for testing represent $n = 0$ and $\ell = m = \cos 45^\circ$. Thus, eq. (2) can be written as

$$\alpha_{nn} = \cos^2 45^\circ \, \alpha_1 + \cos^2 45^\circ \, \alpha_2 = \frac{1}{2} (\alpha_1 + \alpha_2). \qquad (3)$$

It is important to note that eq. (1) assumes that the thermal diffusivity measurement is carried out in a direction normal to isotherms (see Carslaw and Jaeger [14], sec. 1.20). The isotherms in the sample during the transient measurement are not fully known at present. However, due to the difference in effective thermal diffusivities in the orthogonal directions, it is clear that nonuniform heating of the back-face will result. Moreover, the isotherms are not planar during the transient. Therefore, the assumption $n = 0$, $\ell = m = \cos 45^\circ = 1/\sqrt{2}$ used in reducing eq. (1) to eq. (3) is not precise and does not represent the physical reality. In the flash technique the front-face of the sample is exposed to a constant heat flux for a very short time and the temperature-time history is recorded at the back-face. This suggests that one may be measuring thermal diffusivity along the direction of net heat-flux during such a transient experiment rather than along the normal to isotherms. For heterogeneous (anisotropic) materials the direction of heat flow is not normal to isotherms (see Carslaw and Jaeger [14], sec. 1.17 and 1.20) and the relation derived for thermal diffusivity in the heat flux direction is given by

$$\frac{1}{\alpha_{mm}} = \sum_{i=1}^{3} \frac{\cos^2 \theta_i}{\alpha_i}, \tag{4}$$

where α_{mm} is thermal diffusivity in the heat flux direction. θ_i is the angle made by heat flux with i^{th} principal direction, and α_i is thermal diffusivity in i^{th} principal direction. For the composites considered, eq. (4) becomes

$$\frac{1}{\alpha_{mm}} = \frac{\ell^2}{\alpha_1} + \frac{m^2}{\alpha_2} + \frac{n^2 (n=0)}{\alpha_3}, \tag{5}$$

and n = 0. However, the correct values for ℓ and m are not known since the net heat flux direction is not known. Since the effective thermal diffusivity in adjacent principal directions is different, the net heat flux direction need not coincide with the direction in which the sample is cut.

Thus, identifying the measured α_{flash} with a proper direction becomes an important aspect and should receive attention. A simple approximate approach is presented here to analyze the available data of Taylor et al. [9] and to verify the applicability of relations (1) and (4) for conversion of the off-axis thermal diffusivity values to the on-axis thermal diffusivity values.

ANALYSIS OF NET HEAT-FLOW DIRECTION DURING THE TRANSIENT

The intent here is to design a tool for an approximate qualitative understanding of the net heat-flux direction for the case studied by Taylor et al. [9]. For the directions tested, at least one directional cosine is zero, and the other two principal directions make angles of 45° with the sample axis. A schematic of the sample setting for the off-axis flash test is shown in Fig. 2, which shows the front and back faces of the sample, the sample axis, and the η and ξ directions that make 45° angles with the sample axis. Let us assume that the axial reinforcement always runs along the η direction and the radial bundles run along the ξ in the axial-radial testing, with the circumferential bundles replacing the radials in the axial-circumferential testing. When a flash of thermal energy is provided on the front face at time t = 0, the heat flows primarily along the bundles, because they are the preferred paths. Thus,

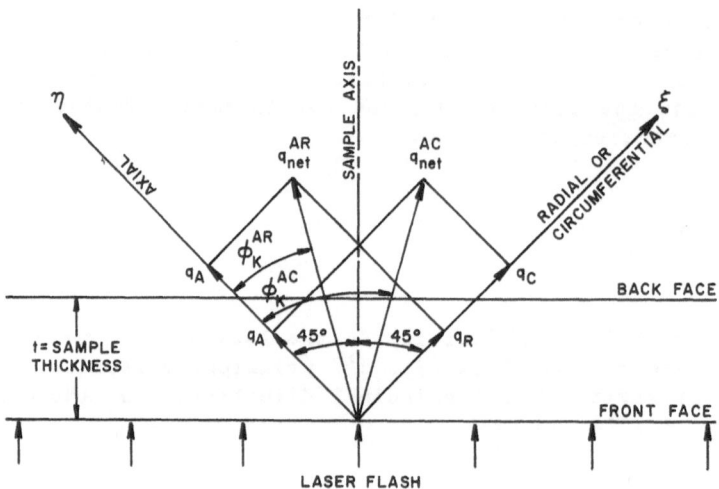

Figure 2. Details of Sample Heating in Flash Diffusivity Testing.

the instantaneous net heat-flux in the sample during the transient is controlled by the amount of heat piped through the bundles. If ϕ_K is the angle made by the instantaneous net heat flux with the η direction (axial principal direction), then

$$\phi_K^{AR} = \tan^{-1} \frac{q_R}{q_A}, \text{ and } \phi_K^{AC} = \tan^{-1} \frac{q_C}{q_A} \tag{6}$$

where q_R, q_C, and q_A are effective instantaneous heat-fluxes in the radial, circumferential, and axial directions, respectively, and ϕ_K^{AR} and ϕ_K^{AC} are angles made by the net heat flux with the η axis in the axial-radial and axial-circumferential cases, respectively.

This expression under simplifying engineering approximations as discussed by Deshpande [15] reduces to

$$\phi_K^{AR}(T,t) = \tan^{-1} \frac{\alpha_R(T,t)}{\alpha_A(T,t)}, \text{ and } \phi_K^{AC} = \tan^{-1} \frac{\alpha_C(T,t)}{\alpha_A(T,t)}. \tag{7}$$

Equation (7) is only approximate and would merely represent the qualitative behavior of net instantaneous heat-flow direction. This is clear from careful examination of eq. (7), as time $t \to 0$ and $t \to \infty$. Deshpande [15] has discussed these limiting cases to conclude that the use of these computed values in converting the

on-axis to the off-axis thermal diffusivity values are justified
only when the effective thermal diffusivity concept is valid and the
reasonably accurate values are known (i.e., for 80% rise time to
steady state). For samples cut in a direction other than 45° in rθ
and rz planes, this approach may not work, as the approximations may
fail.

RESULTS AND DISCUSSION

In order to apply the approximate analysis, the thermal diffusivity data obtained by Taylor et al. [9] were smoothed by third
order polynomial fits in order to obtain values at regular temperature intervals. The results of these efforts can be found in
Deshpande [15]. The net heat-flux direction was analyzed and the
corresponding results for the axial-radial and axial-circumferential
cases are presented in Fig. 3. It is observed in Fig. 3 that ϕ_K
changes with time in a specific pattern. From 100°C to 700°C the ϕ_K
values decrease at each percent time-rise-fraction and approach a
unique ϕ_K-steady state value (100% rise time); but for T > 700°C,
the ϕ_K values show a reversal in nature. The ϕ_K observed during the
transient for 900°C and 1000°C are drastically higher than those at

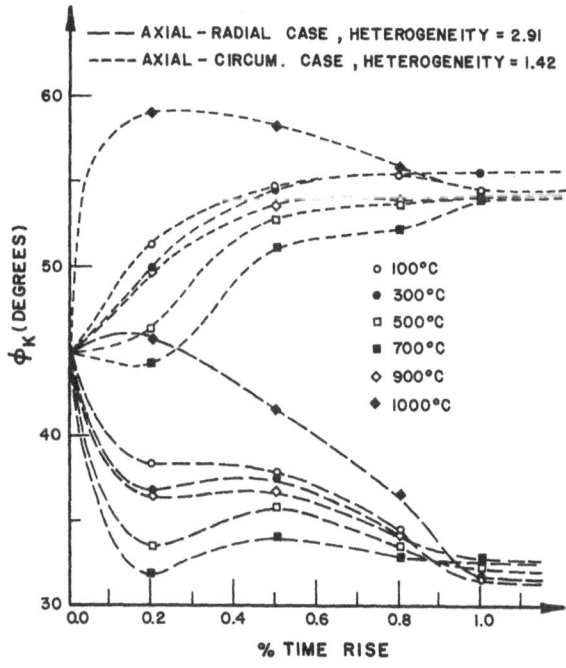

Figure 3. ϕ_K as a Function of Time.

700°C, and they also approach the same unique steady-state value. This pattern was observed for both the axial-radial and the axial-circumferential cases, which represent two entirely different heterogeneity cases. Here, the heterogeneity of the material being tested is defined as the ratio of fiber fractions in the η direction to that in the ξ direction (i.e., heterogeneity = f,axial/f,radial or f,axial/f,circumferential). The pattern is essentially independent of heterogeneity for the sample tested. To quantify and demonstrate this pattern the $\phi_{K-effective}$ values for a given temperature were computed, where the $\phi_{K-effective}$ is defined from its history as,

$$\phi_{K-effective}(T) = \frac{1}{\tau} \int_0^\tau \phi_K(T,t) \, dt$$

$$= \int_0^1 \phi_K(T,t/\tau) \, dt/\tau. \qquad (8)$$

Results of applying eq. (8) are given in Fig. 4, which shows a minima at 700°C for both cases studied. It also indicates that the variations in $\phi_{K-effective}$ with respect to temperature (T) for the two different heterogeneities are identical relative to each other.

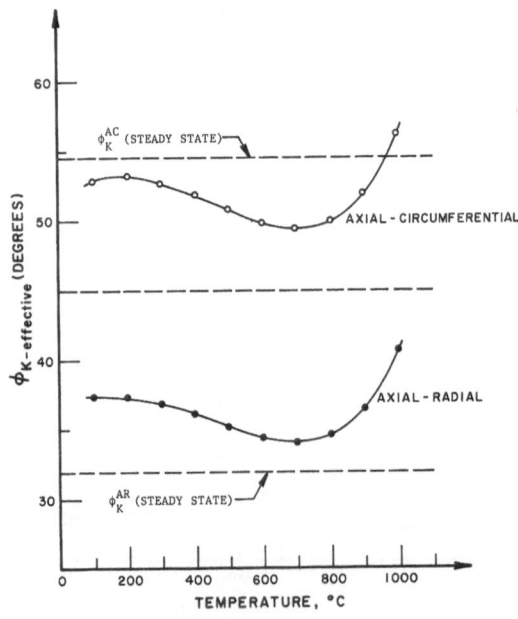

Figure 4. Effective Net Heat Flow Direction During Flash Diffusivity Testing as a Function of Temperature.

Therefore, the direction of net heat flux is decided by the specific choice of direction in which the sample is cut rather than the heterogeneity of the cases considered. Figure 4 also reveals that the sample is cut at a 45° angle and is expected to achieve a totally different ϕ_K-steady state angle (which depends on fiber fraction along adjacent orthogonal directions), and that the sample heating during transient is occurring at angles totally different than these angles. Thus, clearly the experiment does not obey assumptions of the relation given by eqs. (1) or (4), which are derived for one unique direction, whereas the α_{flash} is obtained from a temperature-time history, which refers to the set of net-heat flux directions given in Fig. 3.

We will now use this set of directions computed for different temperatures (T) and different rise-time fractions (t/τ) to calculate the thermal diffusivity by a relation given by eqs. (1) and (4) for further evaluation of the applicability of these equations as an engineering approximation. Results of these calculations are presented here for the steady-state thermal diffusivity values and 80% time-rise values (almost steady state case) in Table 1. Table 1 presents a comparison of the measured thermal diffusivity with the calculated thermal diffusivity values for the axial-radial case. The measured thermal diffusivity is compared with the values computed by eq. (1) for a 45° angle and the values calculated by eq. (4) for a ϕ_K-steady state angle. Such comparison is done for both steady state and 80% rise-fraction values. A similar comparison for the axial-circumferential case was also done. Such comparison for steady-state values does not give conclusive insight to validate eqs. (1) or (4) mainly because determination of the steady-state values are subject to personal bias.

The comparisons carried out for the 80% rise-time fraction thermal diffusivity values reveal many important facts. First, it is noted that for T > 300°C the thermal diffusivity calculated both by eq. (1) using 45° as an angle (α_{nn}-45°) and using eq. (4) with ϕ_K-steady state as an angle (α_{mm}-ϕ_K) are almost same. Secondly, for T < 300°C, the α_{nn}-45° (obtained from eq. 1) agree better with the measured values than α_{mm}-ϕ_K. Thus, it is impossible to conclude that any one equation is better than the other.

Evaluation of eq. (1) was also carried out by calculating the deviation of net heat-flux from the sample axis and noting the difference in the measured and the calculated α_{nn}-45° values of the thermal diffusivity. These calculations are presented in Fig. 5, which reveals that, as net heat-flux deviates more from the sample axis, the agreement in measured diffusivity and calculated α_{nn}-45° is better for higher heterogeneity. For smaller heterogeneity, however, the reverse is true. These observations discourage the use of eq. (1) as a better engineering approximation for data reduction and also indicate that the off-axis thermal diffusivity values are dependent on the net heat-flux direction.

Table 1. Calculations and Comparison: Axial-Radial Case

Heterogeneity = f,axial/f,radial = 2.91
Steady state off-axis (45°) cut thermal diffusivity values

Temp. °C	α_{nn}* 45° cm²/sec	α_{nn}* 32° cm²/sec	α_{mm}** 32° cm²/sec	$\alpha_{measured}$+ steady state cm²/sec
100	0.461	0.509	0.486	0.478
200	0.337	0.366	0.31	0.313
300	0.275	0.298	0.286	0.262
400	0.233	0.254	0.244	0.222
500	0.205	0.225	0.215	0.195
600	0.186	0.204	0.195	0.174
700	0.171	0.187	0.179	0.157
800	0.159	0.175	0.167	0.147
900	0.150	0.165	0.158	0.136
1000	0.140	0.156	0.149	0.126

REMARKS: It can be seen that $\alpha_{measured}$ given for steady state values are closer to α_{mm} than α_{nn}. Since steady state values are obtained by extrapolation, we need to repeat this exercise for 80% rise values to see correct comparison. Again 80% values are near 'steady state,' thus the conclusions of such comparison should be valid.

Thermal diffusivity comparison: 80% time rise values
Axial-radial case (cm²/sec)

Temp. °C	α_{nn}* 45° 80% time rise values	α_{mm}** 32° 80% time rise values	$\alpha_{measured}$+ (smoothed) 80% time rise values	α_{mm}** 45° 80% time rise values
100	0.470	0.486	0.481	0.453
200	0.375	0.385	0.320	0.359
300	0.299	0.310	0.237	0.288
400	0.248	0.257	0.206	0.238
500	0.214	0.223	0.202	0.205
600	0.195	0.201	0.199	0.185
700	0.181	0.180	0.171	0.172

+Values from Taylor et al. [9].
*Calculations are done using eq. (1).
**Calculations are done using eq. (4).

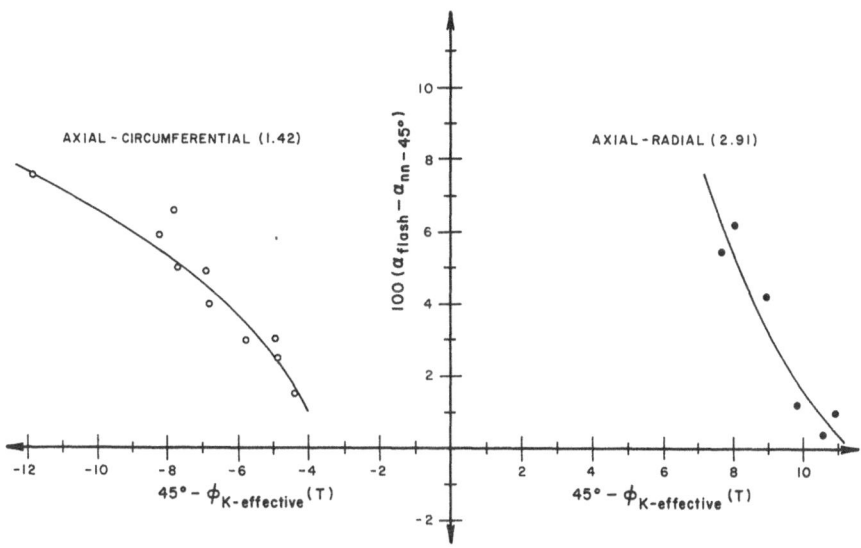

Figure 5. Difference Versus Deviation of Net Heat Flux.

Thus, we have a twofold insight into off-axis testing. First, we have realized that use of eq. (1) is not always justified. The difference versus deviation plot (Fig. 5) shows that the heterogeneity of the composite plays a significant role in controlling agreement between measured and calculated α_{nn}. Secondly, in application of eq. (4), it was found that ϕ_K is changing during the transient and is therefore not unique, whereas eqs. (1) and (4) assume a unique direction. Therefore, the applicability of eqs. (1) and (4) may not always be justified. One of them could be a better engineering approximation, but at present the available data are simply not adequate to decide this.

CONCLUSIONS

We present a summary of the qualitative conclusions that can be drawn from the above analysis, which are considered tentative based on a necessarily approximate model. The availability of more data could modify our insight in the problems involved.

1. In the interpretation of data on off-axis thermal diffusivity data from flash-diffusivity experiments, one should be very careful in attributing the resulting thermal diffusivity value to a particular off-axis direction. It is now shown that for the off-axis flash diffusivity experiment on coarse-weave composite constructions the nonuniform heating of the back face is an important factor that decides the

net heat-flux direction. The net heat flow is not along the direction in which the sample is cut. Thus, a sample cut at a 45° angle ($\ell = m = 1/\sqrt{2}$, n = 0) need not be characterizing the heat-flow property in that direction.

2. Effective heat flow during the transient was calculated approximately for the material under evaluation. These calculations reveal that the off-axis testing incorporates a typical pattern in sample heating, irrespective of its heterogeneity. This poses a serious problem in associating obtained thermal diffusivity with a proper heat-flow direction. Thus, the interpretation of the on-axis data from the off-axis thermal diffusivity data without knowing a proper ϕ_K is a difficult task.

3. The relations given by eqs. (1) and (4) assume a unique direction in deriving these relations. The approximate analysis reveals that the flash technique neither obeys conditions assumed in deriving eqs. (1) and (4) nor does it measure diffusivity for a unique direction, but rather uses a set of directions in measuring thermal diffusivity. It is proposed that one of these relations might be a better engineering approximation. However, the question as to which is better needs further examination.

4. In view of the ability of the off-axis testing to reduce rise-time effects, the off-axis testing should be carried out as routine measurement along with the on-axis measurements, so that an exercise to check self consistency can be carried out. This will be useful to answer questions raised here and will help designers to use this data with greater confidence.

ACKNOWLEDGMENTS

The author acknowledges support by the Materials Laboratory of Air Force Wright Aeronautical Laboratories and the Defense Technical Information Center of Defense Logistics Agency.

The author also acknowledges helpful discussions held with Professor R. E. Taylor, Professor H. M. James, and Dr. R. H. Bogaard. Their suggestions and comments were of great help in the preparation of this manuscript. Special thanks are extended to Professor R. E. Taylor for providing data and for the keen interest and encouragement received. Finally, thanks for patience and skillful typing are extended to Ms. Diane Coffing.

REFERENCES

[1] Fitzer, E. and Heym, M., 'High Temperature Mechanical Proper-
    ties of Carbon and Graphite,' High Temp.-High Pressures, 10,
    29-66 (1978).

[2] Schmidt, D., 'Carbon/Carbon - The Light Weight Composite for
    Hot Spots,' Mech. Eng., 91(1), 62-3 (1980).

[3] Naslain, R., Rossignol, J.Y., Hagenmuller, P., Christin, F.,
    Heraud, L., and Choury, J.J., 'Synthesis and Properties of New
    Composite Materials for High Temperature Applications Based on
    Carbon Fibers and C-SiC or C-TiC Hybrid Matrices,' Rev. Chim.
    Min., 18, 544-64 (1981).

[4] Christin, F., Naslain, R., and Bernard, C., in Proceedings of
    CVD-VII (Sedwick T.O. and Lydin, H., Editors), The Electro-
    chemical Society, Princeton, 499-514 (1979).

[5] Aggour, L., Fitzer, E., and Schlichting, J.J., in Proceedings
    of CVD-V (Blocher, J.M., et al., Editors), The Electrochemical
    Society, Princeton, 600-10 (1975).

[6] Whitley, J.B., Mullendour, A.W., and Langley, R.A., Thin Solid
    Films, 73, 81-90 (1980).

[7] Latchman, W.L., Crawford, J.A., and McAllister, L.E., 'Multi-
    directionally Reinforced Carbon-Carbon Composites,' Proc. 2nd
    Int. Natl. Conf. Composite Materials, Toronto, Canada, AIME,
    1302-19 (1978).

[8] Fitzer, E., and Huttner, W., 'Structure and Strength of Carbon/
    Carbon Composites,' Review Article, J. Phys. D: Appl. Phys.,
    14, 347-71 (1981).

[9] Taylor, R.E., Jortner, J., and Groot, H., 'Thermal Diffusion in
    Carbon/Carbon Composites,' TPRL 256, a special report for AFOSR
    Grant F 49620-81-K-0011 (June 1982).

[10] Deshpande, M.S., Bogaard, R.H., and Taylor, R.E., 'Variances in
     the Measurement of Thermal Diffusivity on Coarse Weave C-C
     Composites and the Dependence Upon Fiber Fraction,' Int. J.
     Thermophys., 2(4), 357-70 (1981).

[11] Jortner, J., 'Analysis of Transient Temperature Response in a
     Carbon-Carbon Composite During Flash Method Thermal Diffusivity
     Tests,' Science Applications, Inc., Irvine, CA, 92715 (30 Jan.
     1981) (also appeared as appendix to 12).

[12] Taylor, R.E. and Groot, H., 'Thermophysical Properties of
     Carbon-Carbon Composites,' TPRL 244, a special report for AFOSR
     Grant 77-3280 (May 1981).

[13] Jortner, J., 'On the Use of Off-Axis Testing to Characterize
     the Thermal Diffusivity of Orthogonally Reinforced Carbon-
     Carbon Composites,' Science Applications, Inc., Irvine, CA,
     92715 (Feb. 1982) (also appeared as appendix to 9).

[14] Carslaw, H.S. and Jaeger, J.C., Conduction of Heat in Solids,
     Second Ed., Oxford at Clarendon Press (Nov. 1959).

[15] Deshpande, M.S., 'On Interpretation of Off-Axis Thermal Diffu-
     sivity Data for Coarse Weave Composites,' CINDAS Special In-
     House Rept., in preparation (1984).

ON THE INFLUENCE OF IRRADIATION ON THERMAL CONDUCTIVITY OF POLYETHYLENE

P. Sukiennik

Politechnika Krakowska
Instytut Fizyki ul Podchorazych 1
Pl-30-084 Krakow, Poland

Measurements have been carried out on the influence of irradiation-induced crosslinks on the thermal conductivity of polyethylene under conditions of constant temperature during irradiation. On the basis of additional tests of thermal treating of the irradiated sample, including microcalorimetry, the author suggests that irradiation-produced crosslinks may influence the thermal conductivity of polyethylene only indirectly, by influencing the crystallinity.

AN INVESTIGATION OF THERMAL CONTACT RESISTANCE

IN THERMAL CONDUCTIVITY MEASUREMENTS OF A THIN NYLON SAMPLE

M. Satter and T. Ashworth

Physics Department
South Dakota School of Mines and Technology
Rapid City, South Dakota 57701

ABSTRACT

An unguarded plate apparatus was constructed to investigate the effects of contact resistance on measurement accuracy of thermal conductivity of rock samples using a minimum of sample preparation. Measurements were made on a well-characterized Comco nylon sample [1]. Interfacial resistance between copper thermometer plates and the nylon sample was clearly evident; using heat fluxes of 10 to 260 Wm^{-2}, apparent thermal conductivity values were respectively 12.0% and 10.3% lower than the values obtained in a linear flow adiabatic apparatus [2]. The inferred contact resistance was apparently slightly dependent upon the heat flux, decreasing from 0.0034 m^2KW^{-1} at 10 Wm^{-2} to 0.0023 m^2KW^{-1} at 260 Wm^{-2}. All measurements were made with an applied interfacial pressure of 6400 Pa. Conduction enhancement materials (grease, paste and a thin plastic film) were used. The Omegatherm paste [3] and silicone vacuum grease [4] reduced the apparent contact resistance only when large heat fluxes were used. The thin plastic film (polyethylene) did not reduce the apparent contact resistance. Thus, with conductivity measurements of low thermal conductors in the range near 2 $Wm^{-1}K^{-1}$ (typical of rocks) in this type of apparatus, contact resistance must be taken into account.

INTRODUCTION

The continuing depletion of the world's oil reserves has provided impetus to develop alternate energy resources. Viable energy alternatives include solar, wind, coal, geothermal and nuclear energy. However, further development of geothermal and nuclear energy may depend on additional research relating to heat transfer in geological materials (rocks). To illustrate, the focal point in a feasibility study on the possible utilization of geothermal energy is the ratio of the extractable heat/cost of extraction (joules/dollar). Thus, accurate thermal conductivity values of the hot rocks in a geothermal energy system are required to determine the amount of heat energy that can be obtained. In a complete nuclear energy system, conductivity values are needed to evaluate rock formations as possible nuclear waste repositories. Also, a deep mine ventilation engineer requires rock conductivities to calculate the heat transfer into the drifts and shafts. Rock thermal conductivities are usually in the range of 0.5 to 5.0 $Wm^{-1}K^{-1}$ [5]. The extractable heat can be estimated by inputting the representative temperatures and thermophysical properties into a finite element (or a difference) model of the geothermal system.

OVERALL OBJECTIVE

The overall objective is to develop an apparatus and a technique to determine to adequate accuracy (better than $\pm$ 5%) the thermal conductivity (λ) of rocks with minimal sample preparation. When achieved, a rock having undergone only standard preparation (coring, slicing and polishing) will be placed in the conductivity apparatus and a measurement taken.

These requirements were dictated by desired ease of measurement and the difficulty of grooving many samples without risk or significant disturbance to the surface; also cementing a thermocouple in place can modify the thermal properties of a porous sample in the vicinity of the thermocouple attachment. These requirements in turn dictated that temperatures should be determined by thermometers mounted in the plates which sandwich the sample; preferably the thermometers could be mounted in such a way that they have physical contact with the sample. Thus, a measurement system with the basic configuration shown in Figure 1 must be employed. Besides corrections for heat losses for this type of apparatus one must also consider the possible effects of thermal contact resistance (TCR). Usually with poor conductors, contact resistance is not a serious problem, but in previous measurements on rocks it appeared to be significant. This study determined the effective contact resistance in an apparatus of the type shown in Figure 1, and the influences of a grease, a paste or a soft plastic film applied to the sample surfaces.

As heat flows through the plate-sample-plate stack there is a resistance to flow at each interface. This resistance arises because the bodies touch each other at a limited number of points. The actual area of contact is a small fraction of the projected area. In a well-insulated system the heat flow will be linear within the

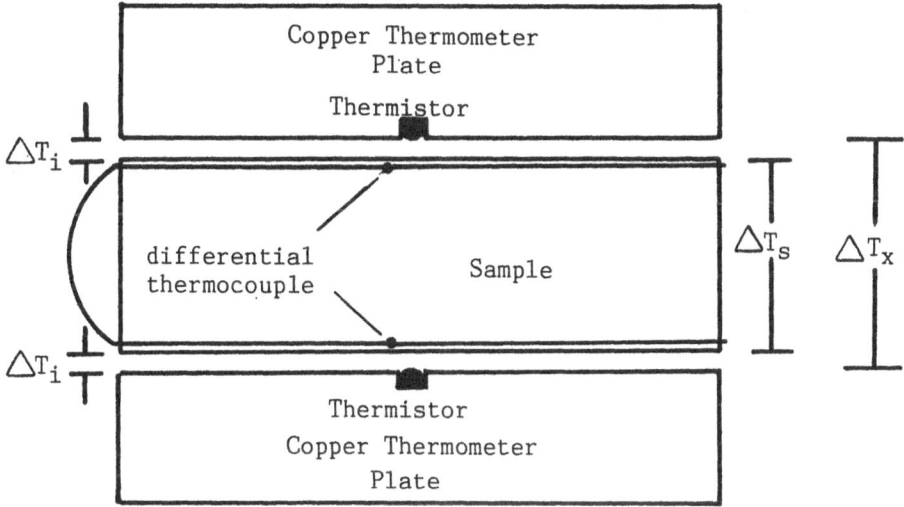

ΔT_x = Apparent temperature drop across from thermometer plates

ΔT_s = True temperature drop across sample from thermocouple

ΔT_i = Interfacial temperature drop

Figure 1. Scheme of Temperature Measurement in Stack.

bodies, but at each interface this flow becomes three-dimensional as the heat flow lines are pinched in at contact points. This constriction of heat flow, together with any resistance to energy transfer between the materials, creates a drop in temperature across each interface. If the temperature measurements are taken on the plate surfaces, the interfacial temperature drops are added to the temperature drop across the sample, and an error will result in determining the λ of the sample. Thus, assuming linear heat flow and that the temperature drop across each interface is the same, the apparent thermal conductivity (λapp) will be

$$\lambda\text{app} = \frac{\dot{q}\Delta x}{\Delta T_x} = \frac{\dot{q}\Delta x}{\Delta T_s + 2\Delta T_i} \tag{1}$$

where ΔT_i is the drop in temperature across one interface, ΔT_x is the apparent drop across the sample, ΔT_s is the drop in temperature across the sample (as illustrated in Figure 1) of thickness Δx, and $\dot{q}$ is the heat flux through the projected area A. The error in the conductivity measurement ($\Delta k/k$) is proportional to the ratio $\Delta T_i/\Delta T_s$ for small values of the latter ratio.

Thermal contact resistance for a single interface (TCR) in Km^2/W is defined as

$$TCR = \frac{(\Delta T_i)}{\dot{q}}$$

The average temperature drop T_i across one interface equals

$$\frac{\Delta T_x - \Delta T_s}{2}$$

Under identical conditions the error introduced by contact resistance will be much larger for a high-λ material than for one with low-λ. For the former the ratio of interfacial temperature drop, $\Delta T_i/\Delta T_s$ will be large compared to that of the latter. Thus, under identical contact conditions, the experimental accuracy obtainable for highly conductive materials is worse than for low conducting materials. Medvedev and Savicheva [6] report that for a specimen of thickness Δx = 20 mm, with an additional layer of Vaseline lubrication of thickness $\Delta x'$ = 0.01 mm, to enchance the thermal contact, the percentage error in determining the λ of glass, bismuth and silver by the two temperature-time interval method is as follows: for glass $\Delta \lambda/\lambda$ = 0.3%, for bismuth $\Delta \lambda/\lambda$ = 9.0%, and for silver $\Delta \lambda/\lambda$ = 64%.

The TCR may be reduced by increasing the actual contact area; this can be done by applying various materials (greases, soft metals) at the interface. TCR is also reduced by increasing the compressive load (pressure) on the stack.

THERMAL CONDUCTIVITY MEASUREMENTS ON ROCKS

Rocks have fairly low thermal conductivity (0.5 - 5 $Wm^{-1}K$) and due to their composition and formation usually only thin samples may be obtained. Cavities, chips, cracks and foliation in rocks require that the height-to-diameter ratio for many samples be kept low. Most rocks are porous and may contain an appreciable amount of moisture. The apparent conductivity of a 'wet' rock is significantly increased due to moisture migration as reported by Lacey [7]. Moisture migration can be minimized by establishing only small temperature gradients across the sample. Even if dry rocks are studied, only small temperature gradients should be employed, to match in-situ conditions. Thus, in a typical λ measurement of a rock, a small temperature gradient is established across a thin sample (T_s small). Under these conditions the TCR may be large and an appropriate correction necessary to avoid an inaccurate determination of λ.

APPARATUS AND METHOD

A thermal conductivity apparatus was constructed and a schematic drawing of the apparatus is shown in Figure 2. The design used was similar to that suggested in ASTM C-177 [8] but the system was unguarded. A thin, right circular disk of sample material was sandwiched between thermometer plates. This geometry requires only minimal sample preparation, as required. The thermometer plates were composed of copper disks containing stable Unichip (41U12401) thermistors manufactured by Victory Engineering Corp. [9]. The thermistors measured the apparent surface temperatures of the specimen and so determined the apparent drop in temperature ΔT_x across the specimen. All the thermistors were calibrated against a standard platinum thermometer (Rosemount 162D) with calibration traceable to the IPTS-68. A Hewlett Packard 3455A Digital Multimeter [10] was used for all

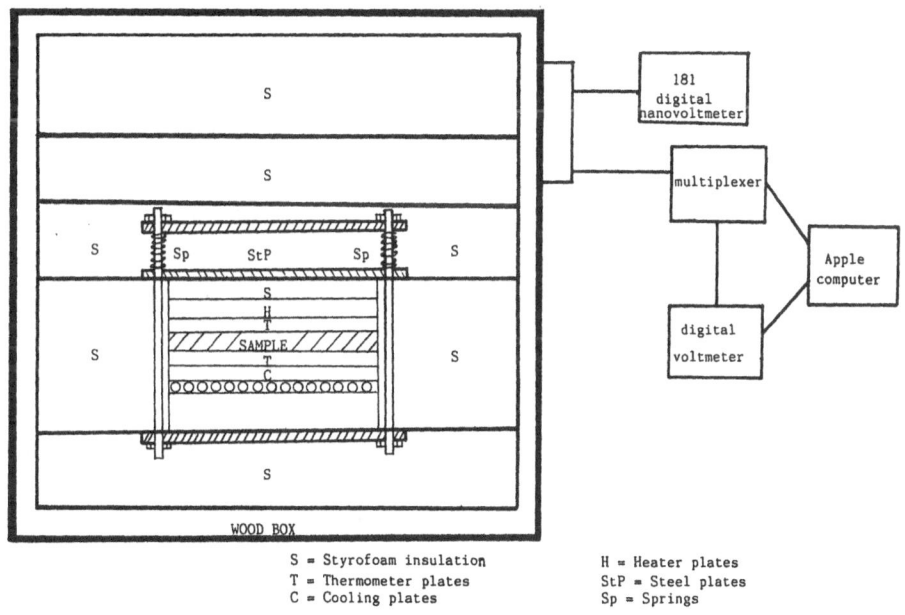

S = Styrofoam insulation   H = Heater plates
T = Thermometer plates   StP = Steel plates
C = Cooling plates   Sp = Springs

Figure 2

resistance measurements. With proper implantation, the resistance of the thermistors is independent of the interfacial pressure. This was verified by experiment. Thermistors were implanted at the center and edge of each thermometer plate to verify the condition of linear heat flow. With the plates in contact (no sample), the central thermistors registered a temperature discrepancy of 0.02°K at room temperature (300°K).

Heating and cooling plates created the heat flux. The power dissipated by the heater plate was the product of the current through and the potential drop across the heater element. The stack was placed within a pressure-plate device, spring loaded to enhance the thermal contact at all the interfaces. The assembly was contained in a box filled with insulation. Heat losses from the stack were determined by modeling the apparatus as a thermal circuit. In such a circuit, heat flow across thermal resistors creates temperature drops. Resistance to heat flow out the top of the stack was experimentally measured. The resistance to heat flow out the side of the stack was estimated. The total heat loss was obtained from the measured temperature differences across the thermal resistors. Variations in room temperature were the limiting factor in evaluating these losses; it is estimated that an error of no greater than 0.0015 W was present in the heat loss determination 0.15 with the lowest heat flux (corresponds to an error of 5%) and 0.15 W (error of 5%) with the highest flux.

In this investigation a sample of Comco nylon [1] was used. This material was chosen because its thermal conductivity is known over a wide range of temperatures [2,11,12,13], and its value (about 0.37 $Wm^{-1}K^{-1}$) is close to that of many rocks. Use of a material with a well defined, known conductivity was advantageous because it allowed several checks on the performance of the system. Also nylon is easy to machine and does not absorb moisture. The sample used was 0.1016 m (4") in diameter and 0.1335 m (1/2") thick, giving an aspect ratio (D/ΔX) of 8:1. Surfaces were machined and estimated flat to within 25 mm. A square groove just deep enough to accommodate a fine (0.15 mm diameter) copper-constantan differential thermocouple was cut across both faces of the nylon. The thermocouple was fixed in place and its leads thermally tempered with a small amount of General Electric 7031 Adhesive and Insulating Varnish [14]. This thermocouple, monitored by a Keithley 181 Digital Nanovoltmeter [15], allowed determination of the true temperature difference ΔT_s across the specimen. With the specimen enclosed in an isothermal chamber, the apparent voltage output of the thermocouple averaged 0.44 microvolts (0.01°K) with a fluctuation less than 0.05 microvolts. In combination with the measurement of ΔT_x, the thermocouple (provided ΔT_s) allowed determination of the interfacial temperature drop ΔT_i, and hence the thermal contact resistance.

An absolute thermocouple (copper-constantan) was also imbedded on the upper surface of the nylon sample (hot side closest to heater plate). The temperature of an insulated copper plate (average room temperature) was used as the reference for this thermocouple; the junction plate was attached to the outside of the wooden box housing the apparatus. The plate provided an isothermal surface whose temperature could be conveniently monitored and it served as a junction device providing connections between the temperature sensors in the stack and the external monitoring equipment. The junction plate was constructed by cementing threaded copper junction strips onto a heavy (1/4") copper plate with an epoxy-alundum mixture. Good thermal contact and electrical isolation were both achieved. Plate temperature was monitored with an Omega Linear Response Component, #44202 [16] manufactured by Omega Engineering, Inc. The combination of the significant thermal mass of the junction plate and the thermal insulation placed around it resulted in a stable junction temperature.

RESULTS

All data were compiled with an applied force of 147N, which produced an interfacial pressure of 6400 Pa. Measurements were first taken with direct copper-nylon interfaces. The results, expressed as apparent thermal conductivities calculated using ΔT_x and ΔT_s, are shown in Figure 3. Corrections for heat losses have been made in these data. With $\Delta T_s \leq 1$ K (mean sample temperature of $\approx$ 298 K), all the values obtained with ΔT_s are in good agreement (within 3%) of the value 0.365 $Wm^{-1}K^{-1}$ at 298°K obtained in a linear flow adiabatic apparatus [2]. These values show some scatter due to variations in room temperature which influenced spurious emf's and heat losses. For $\Delta T_s > 1K$, $\Delta T_s = 5K$ and $T_s \cong 10K$, the values obtained using ΔT_s are within 1% of those reported in reference 2.

Values of thermal conductivity calculated from ΔT_x clearly illustrate the influence of interfacial resistance. With the smallest temperature difference across the sample (0.34°K), the conductivity is 129% below the value obtained in reference 2, and with a 10°K difference the value is below the value obtained in reference 2; $\Delta T_s \cong 0.34$°K corresponds to a heat flux of 10 Wm^{-2}, and $\Delta T_s = 10$°K corresponds to a heat flux of 260 Wm^{-2}.

These data, translated into contact resistance, are shown in Figure 4; a slight dependence on heat flux (gradient) is evident. Also shown in Figure 4 is the influence on contact resistance of a very thin layer (~10μm) of silicone vacuum grease [4] manufactured by Dow-Corning Co. and a layer (0.2 mm) of Omegatherm 201 High Thermal Conductivity Paste [3] applied (separately) to the interfaces. While both materials reduce the interfacial resistance, there is still a large apparent contact resistance with the lower heat fluxes. Using this data for Omegatherm Paste at a ΔT_s of 10K, the apparent thermal

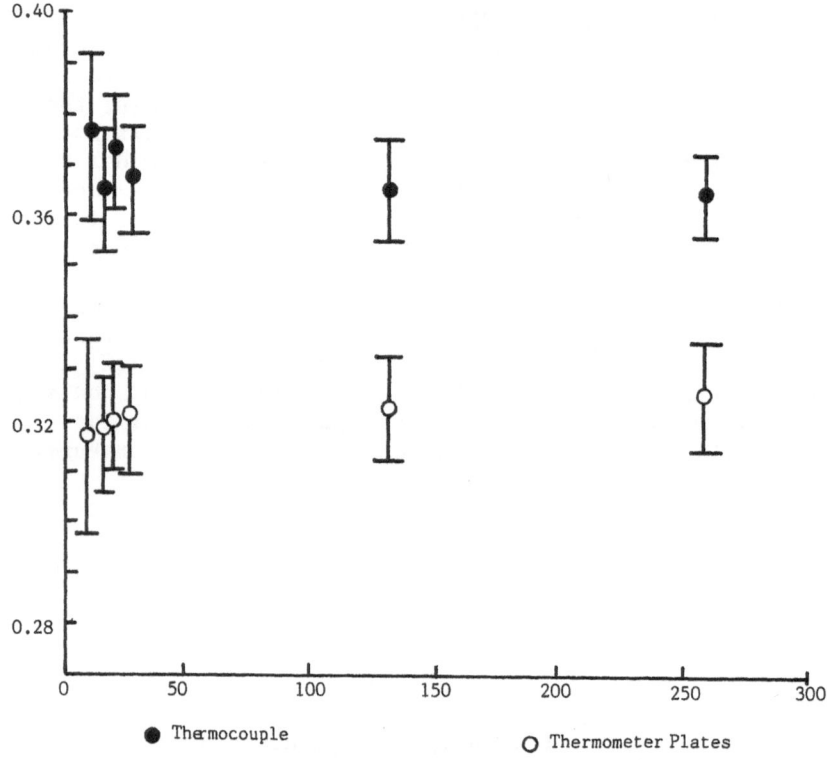

Figure 3. Apparent Thermal Conductivity Versus Heat Fluz.

conductivity calculated from ΔT_x is still 8% below the most probable value. The contact resistance for interfaces to which silicone grease or Omegatherm paste has been applied is apparently dependent to a significant degree upon heat flux. Figure 4 also shows the effect of placing a thin (1 mil) polyethylene film between the copper and nylon surfaces; the plastic film increased the apparent contact resistance.

CONCLUSIONS

In determining conductivity of low thermal conductors employing a system in which thermometers are embedded in plates which sandwich the specimen, corrections must be made for the interfacial resistance.

It is possible that some of the pressure dependence of thermal conductivity in rocks reported by Murdock [17] and Clark [18] may

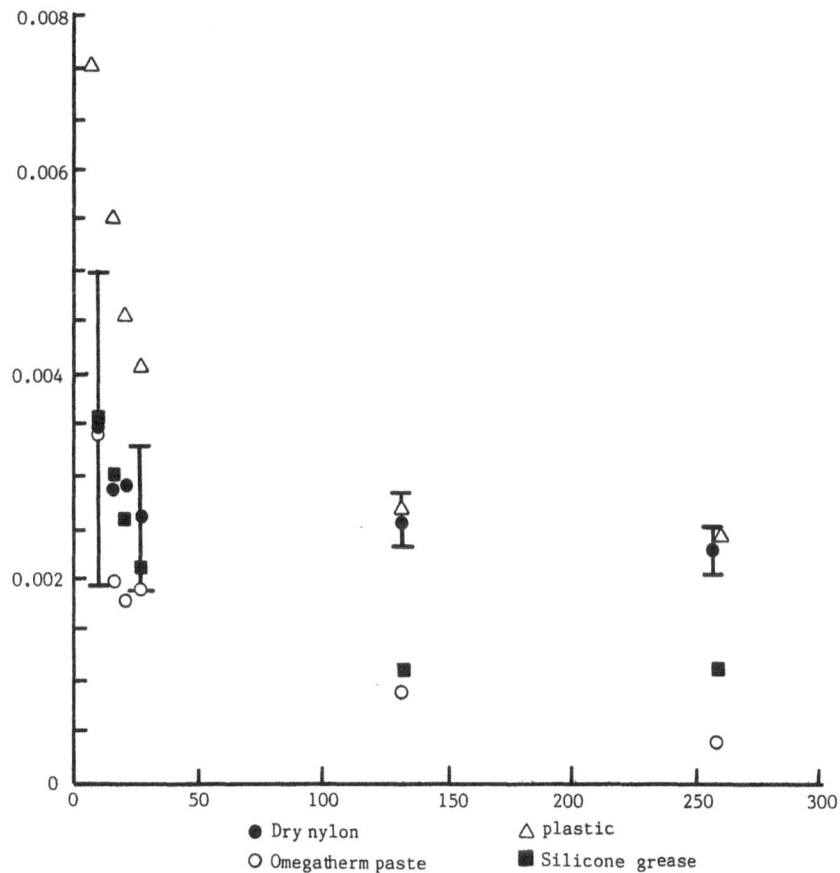

Figure 4. Thermal Contact Resistance Versus Heat Fluz.

be due to diminishing interfacial resistance as more load is applied.
We plan to extend this investigation to actual rock specimens and to
determine the effect of interfacial pressure.

REFERENCES

1. Comco Nylon, Boulder Plastics Inc., Boulder, Colorado.
2. Ashworth, T., Johnson, L.R., Hsiung, C.Y., and Kreitman, M.M.
       (1973), "Use of the Linear Heat Flow for Poor Conductors and
       its Application to the Thermal Conductivity of Nylon," Cryo-
       genics, Vol 13, 43p.
3. Omegatherm 201 High Thermal Conductivity Paste, Omega Engineer-
       ing, Inc., Stamford, Connecticut.
4. Silicone Vacuum Grease, Dow Corning Corporation, Midland,
       Michigan.

5. Morgan, M.T. and West, G.A. (1980), "Thermal Conductivity of the Rocks in the Bureau of Mines Standard Rock Suite," Oak Ridge National Laboratory, Report TM-7052, 54p.

6. Medvedev, N.N. and Savicheva, Z.M. (1975), "Determination of Thermal Conductivity," Inzhenerno-Fizicheskii Zhurnal, Vol 29,pp 326-331.

7. Lacey, W.G. (1975), "A Low Temperature Gradient Method of Measuring the Thermal Conductivity of Microconcrete," M.S. Thesis, South Dakota School of Mines and Technology.

8. 1974 Annual Book of ASTM Standards, Part 18, Standard C177-71, American Society for Testing and Materials, Philadelphia, 1974.

9. Victory Engineering Corporation, Victory Road, Springfield, New Jersey.

10. Digital Voltmeter Model 3455A, Hewlett Packard Company, P.O. Box 301, Loveland, Colorado.

11. Berman, R., Foster, E.L., and Rosenberg, H.M. (1955), "The Thermal Conductivity of Some Technical Materials at Low Temperature," British Journal of Applied Physics, Vol 6, pp 181-189.

12. Kolouch, R.J. and Brown, R.G. (1968), "Thermal Conductivities of Polyethylene and Nylon from 1.2° to 20°K," Journal of Applied Physics, Vol 39, pp 3999-4002.

13. Kreitman, M.M. and Ashworth, T. (1972), "Thermal Conductivity of Nylon," presented to the Twelfth International Thermal Conductivity Conference, Birmingham, Alabama.

14. Adhesive and Insulating Varnish, No. 7031, General Electric Company, Schenectady, New York.

15. Digital Nanovoltmeter Model 181, Keithley Instruments, Inc., Cleveland, Ohio.

16. Omega Linear Response Component, #44202, Omega Engineering, Inc., Stamford, Connecticut.

17. Murdock, R.A. (1979), "Determination of Thermal Conductivity of Naturally Occurring Materials under Varying States of Stress," M.S. Thesis, South Dakota School of Mines and Technology.

18. S.P. Clark, Jr., "Thermal Conductivity," Handbook of Physical Constants, The Geological Society of America, Inc., New York 1966.

SESSION V

Geological Materials

SESSION CHAIRMAN

A. E. Beck
University of Western Ontario
Ontario, Canada

HEAT TRANSPORT IN GEOLOGIC MEDIA:

WHAT WE CAN LEARN IN THE LABORATORY

M. L. Linvill and R. O. Pohl

Laboratory of Atomic and Solid State Physics
Cornell University
Ithaca, New York 14853

ABSTRACT

Based on a study of a variety of minerals of the feldspar and the pyroxene groups, it is suggested that lamellar structures with a spacing of the order of tens of Å are responsible for the low thermal conductivity observed in these minerals above 100 K.

INTRODUCTION

Much of our understanding of the thermal conductivity of solids of practical importance has been derived from investigations of carefully prepared specimens with controlled lattice disorder. In the example to be discussed in this paper, however, involving the thermal conductivity of natural minerals, we have been unable to draw on this fund of knowledge; the thermal conductivity in these minerals appears to be determined by a phonon scattering process which has not yet been observed in solids prepared under controlled conditions in the laboratory. This paper is a progress report on our efforts to elucidate this scattering process.

Through the pioneering work by Birch and Clark [1], it has been demonstrated that rocks and minerals have widely differing thermal conductivities at room temperature and above. While the conductivity λ of natural rock salt at 300 K can be almost as high as that of single crystal NaCl prepared carefully in the laboratory, ($\lambda = 5.3 \times 10^{-2}$ and 8.0×10^{-2} W cm^{-1} K^{-1}, respectively [1,2]), certain natural single crystal plagioclase feldspars (mixtures of albite (Ab, NaAlSi$_3$O$_8$) and anorthite (An, CaAl$_2$Si$_2$O$_8$) have conductivities close

653

to that of vitreous silica ($\lambda = 1.4 \times 10^{-2}$ W cm^{-1} K^{-1}). This indicates a phonon mean free path $\bar{\ell}$, defined through the gas kinetic equation $\lambda = (1/3)C_v$ v $\bar{\ell}$, of the order of the intermolecular spacing ($\bar{\ell} \sim 4$ Å) [1]. Physical or chemical lattice defects which are likely to be present in natural rocks and minerals, are known to scatter phonons and will thus lower the thermal conductivity [3]. However, an average phonon mean free path of the order of the intermolecular spacing, which means that even the phonons with the shortest wavelengths travel, on average, less than one wavelength before being scattered, requires a scattering strength which to date has been observed only in fully amorphous solids, and not in only partly disordered crystalline solids. The occurrence of such a low thermal conductivity in the crystalline plagioclase feldspars is thus an indication of a scattering process which has not been observed in crystalline solids prepared under controlled conditions in the laboratory and which we want to understand.

Birch and Clark [1] had also shown that the observed thermal conductivity λ_{tot} of aggregate igneous rocks was determined by the conductivity of their constituent minerals according to the formula

$$\frac{1}{\lambda_{tot}} = X_A \frac{1}{\lambda_A} + X_B \frac{1}{\lambda_B} + \dots \, , \tag{1}$$

where the letters A, B... refer to the different constituent minerals, and X to their volume fraction. Eq. (1) is generally valid if heat travels through a random array of tightly bonded grains of the conductivities λ_A, λ_B, etc. Good agreement for quartz-bearing and for gabbroic rocks was achieved using fairly uniformly low thermal conductivities for all the feldspars, and high ones for quartzite, the pyroxenes, and the olivines. This remarkable observation, which has since been verified in other investigations (see, e.g. ref. [4]), indicated that the wide spectrum of conductivities observed in the major constituents of the earth's crust, i.e. the igneous rocks, is determined predominantly by the concentration of the feldspars they contained; chemical impurities or other lattice defects play a relatively minor role, in contrast to what the solid state physicist expects on the basis of studies performed on perfect and disordered crystalline solids prepared in the laboratory.

These two observations by Birch and Clark [1], i.e. that feldspars can have thermal conductivities as low as amorphous solids, and that it is their presence in the igneous rocks, and not random lattice defects, which determines the conductivities of these rocks, and thus the heat flow in the earth's crust, formed the starting points for this investigation. In this paper, we will report on a systematic study of the thermal conductivity of several feldspar minerals, and we will also present preliminary data on pyroxenes.

EXPERIMENTAL METHOD AND SAMPLES USED

Below 100 K, the thermal conductivity was measured with the steady state gradient method, using several cryostats of standard design. Above 100 K, where radiation losses reduce the accuracy of this technique, a modified Angstrom technique was used to determine the thermal diffusivity [5]. Thermal conductivity was calculated using published values of the specific heat.

The samples used in this study were the following:

α-quartz and calcite: Natural single crystals, purchased from Wards Mineral Establishment. Heat flow along c-axis.

Albite: A low albite, single crystal specimen from Madagascar; given to us by Dr. H. Pentinghaus from the University of Munster, W. Germany.

Microcline: Polycrystalline, from Keystone, SD, Wards.

Red Albite: Heavily altered, polycrystalline, from Bancroft, Ontario, Wards.

Peristerite: Polycrystalline, Ab rich plagioclase feldspar, sample from Valhalla, Westchester County, NY; Smithsonian Museum No. 116810.

Labradorite: Single crystal from Lake St. John, Quebec; Ab_{33} $An_{65.5}$ Or_2, Wards. (Or stands for orthoclase, $KAlSi_3O_8$).

Sm. Labradorite, calcic: ("Sm." indicates it was obtained from the Smithsonian Museum No. 135512-1); Sagebrush Flat, Warner Valley, Lake Co., OR.

Glassy Labradorite: A glass with the composition Ab_{50} An_{50}, prepared by Dr. H. Pentinghaus.

Silica: Commercial amorphous SiO_2.

Augite: A mineral of the pyroxene group ($Ca-Mg-Fe-SiO_3$). This sample obtained from the Smithsonian Museum (Sample No. 120049, from Rozier, Gorges du Tarn, Tarn, France).

Although some of the mineral samples were single crystals, their crystallographic orientation has not yet been determined. However, because of the low symmetry of these minerals we expect little anisotropy in the thermal conductivity.

EXPERIMENTAL RESULTS AND DISCUSSION

The data obtained on the different feldspar samples and on the amorphous Ab_{50} An_{50} are shown in Fig. 1; the data for quartz, calcite, and silica are shown for comparison. Although all feldspars have conductivities considerably smaller than quartz and calcite, in agreement with the earlier observation [1], our data reveal a remarkably wide spread between the different feldspars. The laboradorites

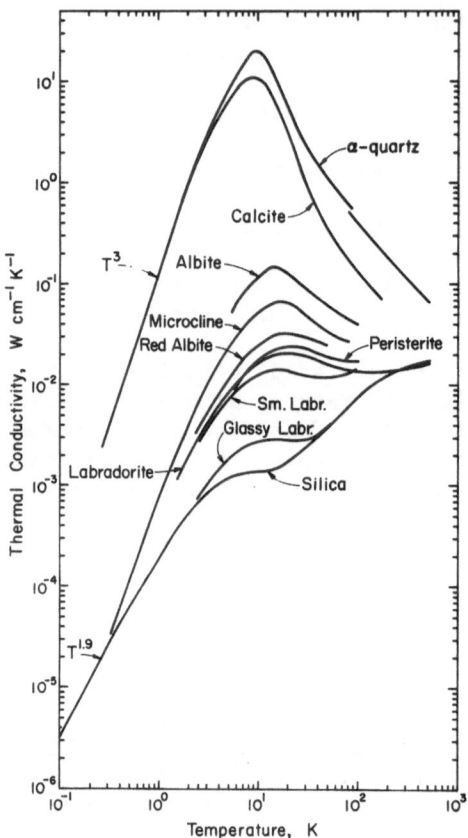

Fig. 1: Thermal conductivity of several natural feldspars are described in the text. For comparison, data on highly perfect single crystal SiO_2 and $CaCO_3$, and on samples of fully amorphous silica and of fully amorphous $Ab_{50}An_{50}$ are also shown.

seem to have the lowest thermal conductivity. Near room temperature, it approaches that of the amorphous $Ab_{50}An_{50}$, which in turn resembles that of all amorphous solids (the conductivity of silica is shown as an example). Note that the thermal conductivity of the labradorite also increases with increasing T at high temperatures, which is also typical for amorphous solids. Albite and microcline, by comparison, have higher conductivities, and also show the temperature dependence characteristic for crystals, i.e. a decrease of λ with increasing T.

We begin our search for the possible origin of the <u>high</u> temperature phonon scattering by inspecting the <u>low</u> temperature data.

Below 1 K, the thermal conductivity of all rocks and minerals approaches a T^3 power law, indicative of a phonon wavelength independent scattering process [7]. This phonon mean free path $\bar{l}$ can vary from less than 1 µm to hundreds of µm, depending on the sample. Grain boundaries can give rise to such a wavelength independent phonon scattering. It has been found, however, that in most cases studied, the experimental $\bar{l}$ was very much smaller than the grain sizes observed in the samples [7,8]. A close correlation, however, was discovered between the low temperature mean free path $\bar{l}$ and the average spacing of intragranular lamellar structures common to many rocks and minerals; these structures can have different origins. In marbles and in feldspars, for example, they can arise from twinning; and in pyroxenes, from chemical exsolution. In a single crystal of labradorite, finally, $\bar{l}$ correlated with the spacing of long platelets of precipitated impurities (ilmenite). Phonon scattering by colloidal precipitates has been observed before, e.g. in diamond [9] and in alkali halides [10]. The role played by twin boundaries has also been demonstrated in plastically deformed single crystals of bismuth [11] and in calcite [12]. Thus, the importance of the intragranular lamellae for the heat transport in minerals and rocks near 1 K is well established, and the phonon scattering observed in Fig. 1 in this temperature range is likely to be caused by such lamellar structures, too. However, how can we explain the phonon scattering above 100 K? As mentioned before, the average mean free path in labradorite near room temperature is of the order of 10 Å; hence, lamellar structures with spacings of 10^4 Å or more or, for that matter, grain boundaries with even larger spacings cannot possibly be important. It does turn out, however, that in the plagioclases very closely spaced lamellae do exist. These lamellae arise from the fact that a mixture of $NaAlSi_3O_8$ and $CaAl_2Si_2O_8$ cannot form an ordered lattice. Hence, as the melt solidifies, some exsolution occurs, leading to lamellar structures with a large spacing of the order of 10^3 Å [13]. Within these lamellae, however, a further structure has been observed, which has been explained as a variation of the order within the Al and Si sublattices and a variation of the Na^+ and Ca^{++} distribution. This order modulation has been shown to have a period of approximately 50 Å [13]. Thus, regions in which the Al and Si ions are ordered will alternate with regions in which they are disordered; their length scale is 50 Å/4 ≈ 10 Å.

While we can offer no direct evidence at this time that this order-modulated structure can lead to the strong phonon scattering observed in the plagioclase feldspars, we can offer the following four arguments as supporting evidence:

Firstly, as has been reviewed above, a variety of lamellar interfaces can be very efficient phonon scatterers. It is therefore not unlikely that the variation of the ordering of the Al and Si ions, together with the variation of the ordering of the Na and Ca ions, will lead to similar phonon scattering (perhaps through changes in the force constants between the order-modulated regions).

Secondly, this order modulation is absent in the alkali feldspars albite and microcline. Their conductivity should, therefore, be higher, as has been observed (Fig. 1). The reason why their conductivities differ somewhat among each other may be explained through small amounts of anorthite present in these samples.

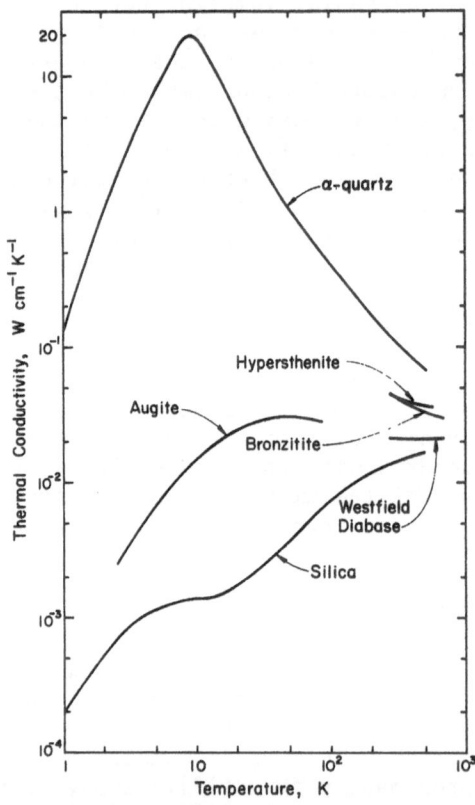

Fig. 2: Thermal conductivity of minerals of the pyroxene group (Ca-Mg-Fe-SiO$_3$). The augite sample is described in the text. The Westfield diabase consists of 67% labradorite and 33% augite, while the hypersthenite and bronzitite are predominantly Mg$_2$Si$_2$O$_6$. The latter three sets of data after ref [1].

Thirdly, electron-microscopic studies have shown that a fine structure of similar spacing can also occur in other silicates. In the chain silicate pyroxene, calcium-rich augite and calcium-poor pigeonite phases form by exsolution during cooling of the magma. A second stage exsolution can occur at lower temperatures, and finally a microstructure on the scale of 50 - 100 Å has been observed in the electron microscope [14]. Fig. 2 shows preliminary measurements of the thermal conductivity of an augite sample. Its conductivity maximum is low, and the conductivity extrapolates towards that of the Westfield diabase (a mixture of 66 volume % of $Ab_{46}An_{54}$ (labradorite) and 32% augite, measured by Birch and Clark [1]). By comparison, hypersthenite and bronzitite, rocks consisting of $Mg_xFe_{1-x}Si_2O_3$, a pyroxene very low in calcium, have a noticeably higher thermal conductivity (very close to that of quartzite (polycrystalline quartz)) [1]. Conceivably, in the absence of calcium, no exsolution microstructure can occur, and therefore the phonon scattering is less.

Fourthly, lamellar structures should be ineffective as scatterers for phonons of a wavelength larger than their spacing. We note that in the plagioclase feldspars, the conductivity increases as the temperature decreases to 20 K. The wavelength of the phonons carrying most of the heat in these solids is approximately 50 Å at 10 K. Hence one would expect that lamellar interfaces separated by 10 Å would become less effective scatterers in this temperature range.

SUMMARY

In summary, there appears to be a good amount of (admittedly circumstantial) evidence that a lamellar microstructure present in plagioclase feldspars is responsible for the glass-like thermal conductivity observed in these minerals at high temperatures. More work is obviously needed. Regardless what the source of the low conductivity is in these minerals, however, it would be of particular interest, from a fundamental as well as from a practical point of view, to learn how to produce similarly strong phonon scattering also in man-made materials.

ACKNOWLEDGEMENT

This research was supported by the US National Science Foundation, Grant No. DMR-8207079.

REFERENCES

[1]. Birch, F. and Clark, H., The thermal conductivity of rocks and its dependence upon temperature and composition, Am. Jour. Sci. 238 529-612, (1940).

[2]. Caldwell, R. F. and Klein, M. V., Experimental and theoretical study of phonon scattering from single point defects in sodium chloride, Phys. Rev. 158, 851 (1967).

[3]. For a recent review, see Berman, R., "Thermal Conduction in Solids", Clarendon Press, Oxford, 1976.

[4]. Sibbitt, W. L., Dodson, J. G., and Tester, J. W., Thermal conductivity of rocks associated with energy extraction from hot dry rock geothermal systems, in XV. ITCC Proceedings, Plenum, NY, 1978, p. 399.

[5]. Vandersande, J. W., and Pohl, R. O., Simple apparatus for the measurement of thermal diffusivity between 80 - 500 K using the modified Angström method, Rev. Sci. Instrum. 51, 1694 (1980).

[6]. Horai, K. I., "Thermal Conductivity of Rock-Forming Minerals", J. Geophys. Res. 76, 1278 (1971).

[7]. Vandersande, J. W., and Pohl, R. O., Negligible effect of grain boundaries on the thermal conductivity of rocks, Geophys. Res. Lett. 9, 820 (1982).

[8]. Pohl, R. O., and Vandersande, J. W., Thermal conductivity of nuclear waste forms and geologic media, in Mat. Res. Soc. Symp. Proc. Vol. 15 (1983), Elsevier Science Publishing Co., Inc., p. 711.

[9]. Vandersande, J. W., Phonon scattering by nitrogen aggregates in intermediate type natural diamonds, in "Phonon Scattering in Condensed Matter", H. J. Maris, Plenum Press, N. Y. (1980), p. 247.

[10]. Worlock, J. M., Thermal conductivity in sodium chloride containing silver colloids, Phys. Rev. 147, 636 (1966).

[11]. Matsuo, T. and Suzuki, H., Effect of plastic deformation on the thermal conductivity of bismuth crystals, J. Phys. Soc. Japan 41, 1692 (1976), and 43, 1974 (1977).

[12]. Vandersande, J. W., Chopra, P. N., and Pohl, R. O., Phonon scattering by twin planes, in "Phonon Scattering in Condensed Matter", Springer Verlag, Berlin, W. Eisenmenger, et al., eds., 1984, p. 182.

[13]. Putnis, A. and McConnell, J. D. C., "Principles of Mineral Behavior", Elsevier, New York, 1980, p. 246 ff.

[14]. Champness, P. E., and Lormer, G. W., Exsolution in silicates, in "Electron Microscopy in Mineralogy", Springer Verlag, Berlin, 1976, H.-R. Wenk, ed., p. 182.

THERMAL CONDUCTIVITY AND THERMAL DIFFUSIVITY MEASUREMENTS OF SALT ROCKS BY DIFFERENT METHODS

Johannes Kopietz and Wolfgang Neumann

Bundesanstalt für Geowissenschaften und Rohstoffe
Hannover, Federal Republic of Germany

Österreichisches Forschungszentrum Seibersdorf
Wien, Austria

ABSTRACT

Thermal conductivity measurements are being performed on a large number of drill core samples from the boreholes for exploration of the Gorleben salt dome for nuclear waste disposal. A laboratory instrument for rapid thermal conductivity determination based on the familiar heat-flow-meter method (ASTM C518) is used for this purpose. Thermal conductivity data for selected samples from these measurements have been compared with measurements on the same samples by a comparative method under steady state longitudinal heat flow conditions. In addition, thermal diffusivity measurements were made. The comparison of the different methods was made with respect to accuracy, reproducibilty and possible measurement errors. Special emphasis was given to the thermal contact resistance within the experimental set-up.

INTRODUCTION

The Gorleben salt dome is a prospective site for the disposal of heat-generating radioactive waste in the Federal Republic of Germany. Thermal conductivity measurements are being made by the Bundesanstalt für Geowissenschaften und Rohstoffe (BGR) on samples from drill cores from deep boreholes in this salt dome (1), to obtain site-specific thermal conductivity data representative of the different types of salt rock occurring with varying impurities. In the course of the geothermal exploration it is necessary to make measurements on a large number of specimens. This calls for a method that consumes as little time as possible, even if it is associated with an inevitable loss of accuracy. This disadvantage can be compensated for by the statistical treatment of the large amount of data. The use of mean values makes it possible not only to improve the accuracy of a single measurement, but also to average the inhomogeneities in the mineral composition of the rock layers, as well as the inhomogeneities of the sample itself. Nevertheless, the reliability of the data, especially with respect to systematic instrument errors, should be verified by tests with different methods. The samples which have been used in such tests can be considered to have well-defined thermal properties and may serve as a standard for other tests. A commercial laboratory instrument based on the familiar heat-flow-meter method (ASTM C518)* is used for the rapid thermal conductivity measurements. The comparison measurements have been carried out by the Österreichisches Forschungszentrum Seibersdorf (FZS) on a commercial instrument using the comparative method under steady-state, longitudinal heat-flow conditions**. In addition, the extent to which the laser-flash method is applicable to thermal diffusivity measurements on salt rock specimens was tested.

MEASURING METHODS

The basic principle of the heat flow meter method, as well as the experimental device used, are described in Ref. (2). A detailed description of the comparative method used in these experiments is in Ref. (3). Schematic diagrams of these methods are shown in Fig. 1. In both methods, the determination of thermal conductivity is based on the following equation:

$$\lambda = \frac{V}{A} \cdot \frac{\Delta X}{\Delta Ts}$$

where

λ = thermal conductivity
V = heat flow
A = cross section area of the specimen
$\frac{\Delta Ts}{\Delta X}$ = temperature gradient in the specimen

---

* Dynatech C-matic
** Dynatech TCFCM – N20

This equation is valid provided a uniaxial steady-state heat flow exists. The significant difference between the heat-flow and the comparative method lies in the way V and $\Delta T/\Delta X$ are determined.

Modified heat-flow-meter method

- determination of V: The heat flow across the sample is determined with a heat flow transducer, which is located between heat source and specimen. This heat flow transducer has to be calibrated by measuring the temperature drop across a specimen of known conductivity and geometrical dimension. This reference specimen replaces the sample during the calibration procedure.

- determination of $\Delta Ts/\Delta X$: The temperature difference between top and bottom of the specimen is obtained by measuring the temperature in the lower and upper plates or with thermocouples mounted in grooves in the top and bottom surfaces of the samples. Both methods were used for all of the samples in the present tests.

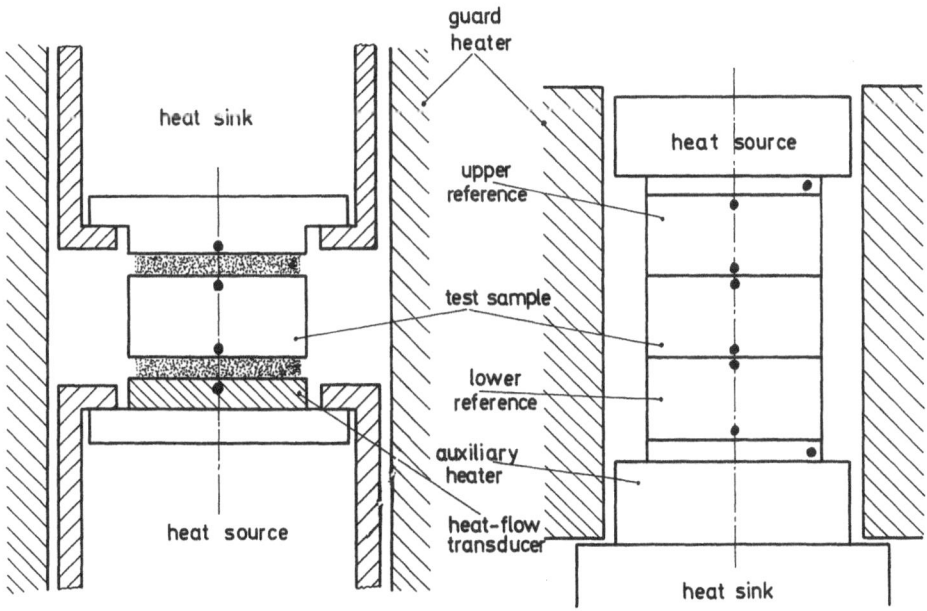

Figure 1. Schematic Diagram of Measuring Methods

Comparative method

- Determination of V: The actual heat flow need not be measured, if one assumes that any radial heat loss is negligible. This assumption is checked at every test temperature. In fact, the temperature gradients across the upper and lower reference are measured, which makes it easy to calculate real heat flow.

- determination of $\Delta Ts/\Delta X$: The temperatures on the top and bottom surfaces of the specimen are measured by thermocouples fixed in grooves on these surfaces.

It should be pointed out that the dimensions of the specimen for the tests in both apparatus are the same, i.e. cylinders 50 mm in diameter and 20 mm thick. The experimental set-up permits special attention to be paid to the thermal contact resistance.

Laser flash method

The laser flash method for thermal diffusivity measurement is also well-known (4). It is based on a transient method completely different from the above-described methods. The disk-shaped sample is only 1 mm thick with a diameter of 19 mm. The thermal diffusivity is derived from the temperature increase on the surface of this disc due to the absorption of the energy of a laser pulse incident on the opposite side. The measurement time is only about a few seconds.

RESULTS AND DISCUSSION

The results of the two steady-state methods are in good agreement with each other, usually better than 5 %. This agreement could not be expected, for although the same specimens were used for these rather similar methods, there were some important differences in the procedures, including replacement of the thermocouples fixed in the specimens.

Full use of the rapid measurement capability using the heat-flow-meter apparatus is possible only if the thermocouples are not mounted on the specimen. In this case, the temperature difference ΔTp measured between the upper and lower plates has to be corrected for the interfacial temperature drops, which are assumed to be proportional to the heat flow V through the sample according to

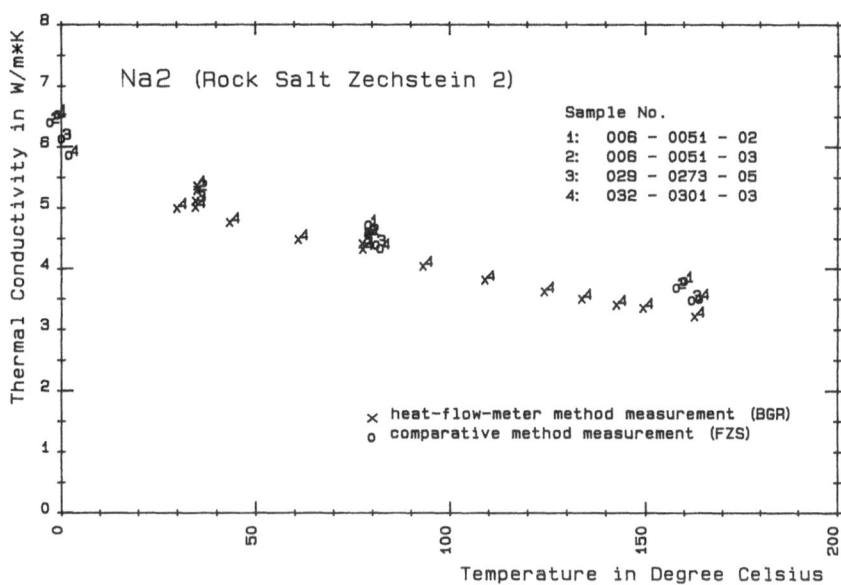

Figure 2.   Comparison of Thermal Conductivity Measurements
on Rock Salt of Zechstein 2

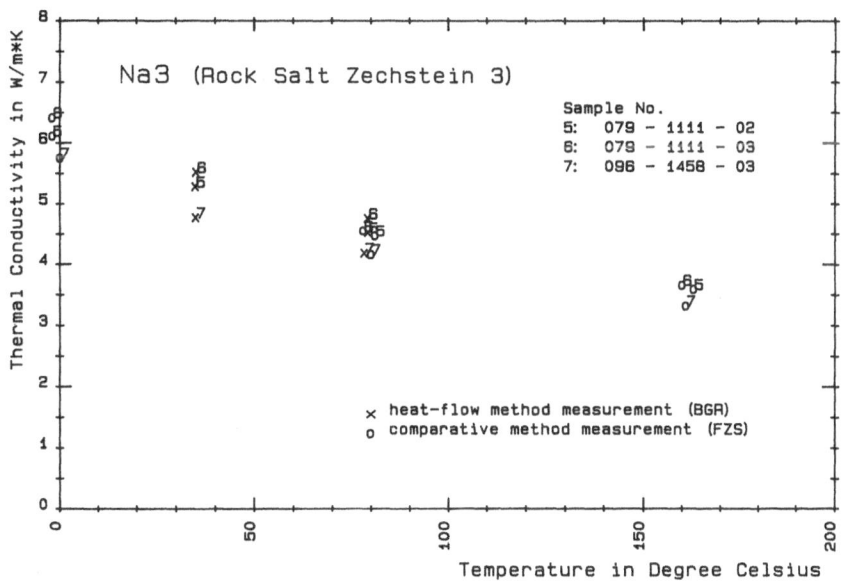

Figure 3.   Comparison of Thermal Conductivity Measurements
on Rock Salt of Zechstein 3

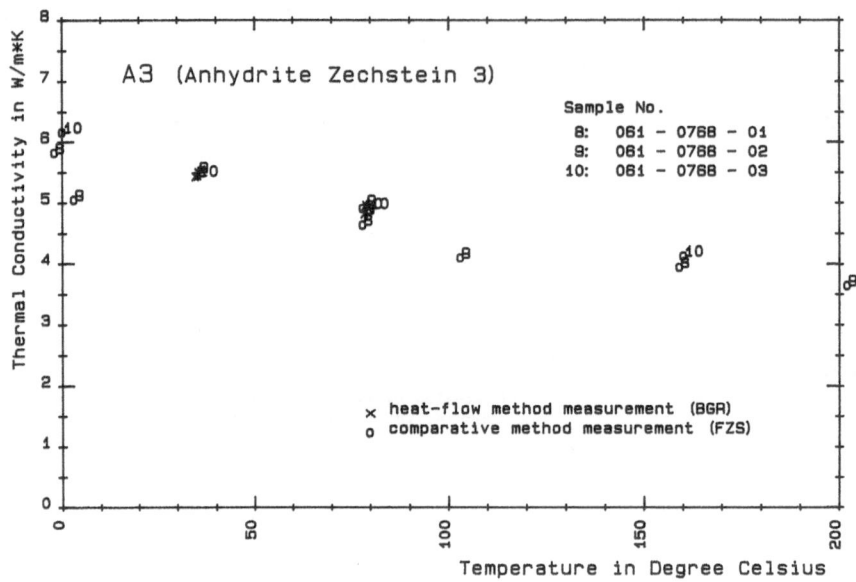

Figure 4. Comparison of Thermal Conductivity Measurements on Anhydrite of Zechstein 3

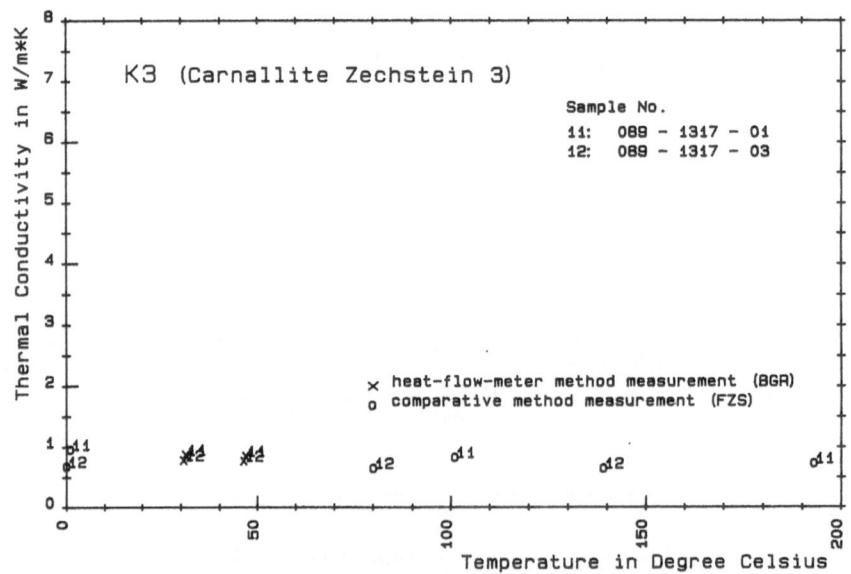

Figure 5. Comparison of Thermal Conductivity Measurements on Carnallite of Zechstein 3

the following equation (2):

$$\Delta T_s = \Delta T_p - a \cdot V$$

The error in thermal conductivities derived from measurements of plate temperature using this equation, together with the formula for thermal conductivity given above, is given by the following expression:

$$\frac{\delta\lambda}{\lambda} = \frac{\Delta T_p - \Delta T_s}{\Delta T_s} \cdot \frac{\delta a}{a}$$

According to this equation, the deviation in the coefficient of surface heat transfer a yields an error in λ less than 1 % for most of the values for thermal conductivity derived from plate temperature and heat flow measurements on the same sample, as can be seen from the data in Table 1. Considering all of the measurements listed in this Table the error in λ is, for the most part, less than about 5 %. In the most unfavorable case, the interfacial temperature drop amounts to 58 % of the actual temperature drop across the sample; even together with the 18 % maximum deviation for a from its total average (sample 029-0273-05 on Table 2), that extreme high interfacial temperature drop results in an error for λ of only about 10 %.

The comparative method makes it possible to check the required conditions, i.e. uni-directional steady-state heat flow at every measurement temperature. In the experiments described here, slight differences between the heat flow in the upper and lower reference samples were observed, but a significant influence on the results could not be detected (Table 3). Nevertheless, the heat-flow conditions may not be neglected, because the first experiments, which showed significant heat flow differences, lead to significantly lower thermal conductivity values due to radial heat losses. In addition to heat flow, the thermal contact between reference and specimen and, in particular between thermocouples and samples has to be considered carefully. As illustrated in Table 4, tests on the same sample with different thermal contacts resulted in different values, although in both experiments the heat flow differences were similar. The maximum error due to unavoidable temperature drops was more than 30 %. For temperature drops of about 1 $^{\circ}$C, the results were satisfactorily reproducible and in good agreement with the results obtained by the heat-flow-meter method.

To prepare thin slices of rock salt samples is very

Table 1.   Results from Heat-Flow-Meter Method

Sample	T (°C)	ΔTp (K)	ΔTs (K)	Tp-ΔTs (K)	$\dfrac{\Delta Tp-\Delta Ts}{Ts}$ (K)	V (mV)	a (K/mV)	(W/m·K)
Na 2								
006-0051-02	79	28.7	20.4	8.39	0.41	28.9	0.35	4.59
	79	28.8	20.4	8.46	0.42	24.0	0.35	4.60
	35	10.7	6.8	3.95	0.58	10.1	0.39	5.36
006-0051-03	80	28.9	20.7	8.17	0.39	24.1	0.34	4.55
	79	28.6	20.6	8.07	0.39	24.0	0.34	4.56
	35	10.6	6.8	3.80	0.56	10.0	0.38	5.29
029-0273-05	77	29.8	23.3	6.53	0.28	23.9	0.27	4.41
	78	29.8	23.4	6.46	0.28	23.9	0.27	4.41
	35	11.0	7.7	3.26	0.42	9.9	0.33	5.11
032-0301-03	78	30.4	23.6	6.72	0.28	23.4	0.29	4.27
	78	30.3	23.6	6.67	0.28	23.4	0.28	4.28
	35	11.0	7.8	3.27	0.42	9.5	0.34	4.88
Na 3								
079-1111-02	79	28.8	20.5	8.28	0.40	23.6	0.38	4.51
	79	28.6	20.5	8.12	0.40	23.7	0.34	4.51
	79	28.6	20.4	8.20	0.40	23.7	0.35	4.54
	35	10.9	6.9	4.00	0.58	10.1	0.40	5.28
079-1111-03	79	28.2	20.0	8.21	0.41	24.4	0.34	4.76
	80	28.5	20.3	8.20	0.40	24.6	0.33	4.75
	35	10.5	6.6	3.86	0.58	10.1	0.38	5.52
096-1458-03	79	29.7	21.9	7.89	0.36	23.3	0.34	4.17
	79	29.6	21.8	7.79	0.36	23.4	0.33	4.18
	35	11.2	7.4	3.78	0.51	9.8	0.39	4.77
A 3								
061-0768-01	78	27.7	20.3	7.46	0.37	25.0	0.30	4.81
	79	27.8	20.4	7.47	0.37	25.2	0.30	5.43
	35	10.4	6.9	3.52	0.51	10.4	0.34	4.83
061-0768-02	79	27.5	19.4	8.16	0.42	24.3	0.34	4.93
	79	27.9	19.6	8.28	0.42	24.8	0.33	4.98
	79	28.4	19.6	8.75	0.45	24.9	0.35	4.97
	79	28.2	19.5	8.66	0.44	24.6	0.35	4.95
	35	10.5	6.7	3.81	0.57	10.1	0.38	5.50
061-0768-03	79	26.9	19.4	7.54	0.39	23.4	0.31	4.89
	79	29.9	19.4	7.56	0.39	24.2	0.31	4.92
	35	10.1	6.6	3.50	0.53	9.9	0.35	5.44

Table 2

Average Coefficients of Surface Heat Transfer to be Used for Correction of Plate Temperature Measurements.

Sample No.	a (mV/K)	
	35 °C	79 °C
Na2		
006-0051-02	0.39	0.35
006-0051-03	0.38	0.34
029-0273-05	0.33	0.27
032-0301-03	0.34	0.29
average	0.36	0.31
Na 3		
079-1111-02	0.40	0.35
079-1111-03	0.38	0.34
096-1458-03	0.39	0.33
average	0.39	0.34
A 3		
061-0768-01	0.34	0.34
061-0768-02	0.38	0.34
061-0768-03	0.35	0.31
average	0.36	0.33
total average	0.37	0.33

Table 3

Results from Comparative Method.

Sample		Temperature (°C)	Thermal conductivity (W/m·K)			Heat Flow Differences (%)		
			av	max	min	V_{av}	V_{max}	V_{min}
Na2		0	6.23	5.86	6.53	6.3	12.8	1.2
		80	4.50	4.72	4.33	6.1	12.0	3.1
		160	3.61	3.79	3.48	4.9	8.8	0
Na3		0	6.09	6.41	5.75	4.7	7.7	1.9
		80	4.40	4.55	4.16	4.0	5.7	2.7
		160	3.50	3.66	3.32	2.1	3.2	1.5
A3		0	5.68	6.16	5.05	9.2	16.6	3.7
	*	80	4.78	4.91	4.64	4.6	7.0	2.1
	*	160	4.04	4.13	3.94	1.7	3.1	0.3
K3	*	0	0.82	0.96	0.68	12.6	18.0	7.2

* Average value of two measurements only

Table 4.   Thermal Conductivities Measured under Different Thermal Contact

Sample No.	good thermal contact				poor thermal contact			
	Temperature	Thermal Conduct.	Heat Flow Difference	Average Temp. Drop	Temperature	Thermal Conduct.	Heat Flow Difference	Average Temp. Drop
	(°C)	(W/m·K)	(%)	(°C)	(°C)	(W/m·K)	(%)	(°C)
006-0051-02	- 1	6.53	- 6.1	1.5	2	4.98	9.0	6.0
	79	4.72	- 6.1	1.5	102	3.78	5.7	5.7
	160	3.79	- 2.1	1.7	202	3.06	- 4.1	4.5
006-0051-03	- 3	6.40	- 4.9	0.9	- 1	6.36	6.0	7.3
	79	4.57	- 3.2	1.3	79	4.61	4.0	8.0
	158	3.68	0	1.0	162	3.67	2.0	6.4
029-0273-05	0	6.13	- 1.2	1.1	2	4.10	- 9.3	6.8
	81	4.39	3.1	1.4	101	3.19	- 9.6	6.7
	162	3.48	8.5	1.4	202	2.64	12.5	5.5
032-8301-03					2	5.86	- 12.8	9.7
					82	4.33	- 12.0	10.1
					164	3.50	- 8.8	7.4

Table 5: Properties of anhydrite (A3) at room temperature.

Thermal Diffusivity α (cm2/s)	Specific Heat c (cal/g·K)	Density ρ (g/cm3)	Thermal Conductivity	
			derived λ (W/m·K)	measured
0.036	0.182	2.891	8.09	5.7

difficult and sometimes impossible. Therefore, only one specimen of anhydrite could be tested in the laser flash apparatus. For comparison with the other results, the thermal conductivity λ was calculated from the measured thermal diffusivity α, and specific heat c and density ρ using the formula

$$\lambda = \alpha \cdot \rho \cdot c$$

CONCLUSIONS

- Thermal conductivity measurements on selected samples of salt rock using a comparative-method instrument confirmed the results obtained with an instrument for rapid thermal conductivity measurements based on the heat-flow method. For the most part, the results agree to within 5 %.

- in special cases the very rapid laser-flash method may be suitable for thermal diffusivity measurements on rock samples. Besides the homogeneity requirements for the samples, it should be possible to easily prepare thin discs of the specimens so that the method is sufficiently efficient.

REFERENCES

(1)   Kopietz, J., Thermal conductivity measurements related to the geothermal exploration of the Gorleben salt dome, Proc. 18th ITCC, these Proceedings.
(2)   Comou, K.G., Tye, R.P., A laboratory instrument for rapid determination of thermal conductivities in the range 0.4 to 5 W m-1 K-1, High Temperatures - High Pressures 13:695 (1981).

(3)   Francl, J., Kingery, W.D., Apparatus for deter-
        mining thermal conductivity by a comparative
        method, J. Am. Ceram. Soc. 37:80 (1954).
(4)   Parker, J., et al., Flash method of determining
        thermal diffusivity, heat capacity, and thermal
        conductivity, J. Appl. Phys. 32:1679 (1960).

THERMAL CONDUCTIVITY MEASUREMENTS IN RELATION TO

THE GEOTHERMAL EXPLORATION OF THE GORLEBEN SALT DOME

Johannes Kopietz

Bundesanstalt für Geowissenschaften
und Rohstoffe (BGR)
Hannover, Federal Republic of Germany

ABSTRACT

The investigations of the suitability of the Gorleben
salt dome for permanent storage of heat-generating radioactive waste include thermal conductivity measurements on
core samples from shallow and deep boreholes drilled for
site exploration. The thermal conductivity data, together
with temperature data from the exploration boreholes, serve
as basic input for model calculations of heat dissipation
and thermomechanical processes associated with high-level
waste disposal in a salt dome. In addition, the geothermal data are relevant to the hydrogeological and geological
exploration of the salt dome area. Results of thermal conductivity measurements on rock salt and associated structures are presented in this paper. Thermal conductivity
data obtained from the laboratory measurements on the core
material are compared with high-precision temperature gradient logs from the exploration boreholes.

INTRODUCTION

An extensive investigation program is being carried
out on the Gorleben salt dome in nothern Germany to judge
its suitability as a radioactive waste disposal site, as
well as for site-specific repository design. There are
about 300 hydrogeological exploration boreholes and four

deep exploration boreholes drilled into the flanks of the salt dome. Two boreholes have been drilled in the central part to explore the area where shafts are to be sunk from which drifts are to be driven for further exploration. The investigation program includes taking of drill cores (over the total length of up to 2000 m in the deep boreholes) and all available methods of geophysical borehole measurements. The main objective of the geothermal exploration of the Gorleben salt dome and the rock above and around it is to obtain the temperature and thermal conductivity data required as basic input for model calculations of the heat dissipation process that would be associated with the permanent disposal of high-level radioactive waste in the salt dome. These data are also necessary for determinations of natural heat flow in the salt dome area. In addition, thermal conductivity measurements, together with borehole temperature measurements, are an aid to geological exploration and are used to separate conductive and convective heat transfer in order to identify groundwater flows.

The particular suitability of rock salt as host rock for the disposal of heat-generating waste is largely due to its relatively high thermal conductivity. Of the accessory constituents in the North German salt domes, carnallitite is of particular interest due to its low thermal conductivity compared to that of rock salt, thus hindering the dissipation of heat in the salt formation. Since the mineral carnallite at elevated temperatures is thermally unstable and decomposes with the release of its water of hydration, it must be certain that the temperature of the carnallitite layers in the salt dome will not rise as high as that at which the dehydration takes place. Hence -- depending on the thickness and location of carnallitite layers within the salt dome -- the relatively low thermal conductivity of carnallitite might be of relevance for the siting and designing of the repository mine.

Thermal conductivities published for different types of rock salt show a great variability. Hence, in the past, much consideration was given to the question of the applicability of laboratory data to in-situ conditions. This question is of particular interest in geoscientific investigations of material properties, which may be significantly different in a particular rock sample from those of a large, undisturbed rock mass in a deep geological formation. Comprehensive thermal conductivity measurements -- both in the laboratory and in situ -- have been performed in earlier investigations on Asse rock salt. From these measurements, it could be concluded that for rock salt the sample fabric is of only minor influence on thermal conductivity. However, the impurities contained in the salt formations (e.g. carnallite, kieserite, and clay) can have a significant influence on thermal conductivity, as can be concluded from the present measurements on Gorleben salt rocks. In further experiments therefore, it will be attempted to determine a general relationship between thermal conductivity and mineral composition.

MEASUREMENT TECHNIQUE

Due to the inhomogeneity of the rocks, measurements must be made on a rather large number of samples in order to obtain representative thermal conductivity data of the internal structures of salt formations. Thus, equipment for the laboratory measurements was selected on the basis of its capability of handling a large number of samples. The unavoidable loss of precision relative to more time consuming methods was accepted, because under the given conditions, the accuracy of the individual measurement is less important than having a large amount of data for statistical treatment. A commercial instrument (Dynatech C-matic) was chosen, which is based on the heat-flow-meter method. The adjustable heat flow through the sample is measured with a heat-flow transducer, together with the temperature gradient in the sample. The apparatus is calibrated on a reference sample for each set of test conditions. The precision of the instrument is about 5 %. The method has been described in detail by Coumou & Tye (1). An estimate of the accuracy and reliability of the measurements reported on here will be described in another paper presented at this meeting (2).

SAMPLE DESCRIPTION

The measurements were made on samples 5 cm in diameter and 2 cm long. All of the samples were taken from drill cores from the exploration borehole Go 1003 in the Gorleben salt dome. Pieces about 10 cm long were selected from the drill cores with the advice of a geologist as being representative of the main layers of the salt dome; three samples were cut from each of these cores pieces. A macroscopic description of the samples is given in Table 1. A chemical analysis has been carried out on some of these samples, the results of which are given in Table 2.

Table 1. Macroscopic Description of Samples from the Main Rock Layers in the Gorleben Salt Dome

---

STASSFURT ROCK SALT Na2

Grey medium-grained halite rock. It contains fragments of rock salt crystals, some of which are coarser than the sample. These crystal fragments are cloudy due to extremely fine inclusions that are usually brine and/or gas, rarely anhydrite.

LEINE ROCK SALT Na3

Fine to medium-grained crystalline, white to light brown, very pure halite rock with traces of anhydrite.

HAUPTANHYDRIT A3

Fine-grained, blue-grey anhydrite rock. The anhydrite contains hair-fine, brown, clayey, magnetic flaser. It also contains clusters of brick-red carnallite, usually several mm in diameter.

STASSFURT POTASH SEAM K2; RONNENBERG POTASH SEAM K3

Medium-grained, sometimes fine-grained crystalline carnallitite consisting of halite and carnallite. The carnallite is for the most part coarser grained than the halite. The carnallite is generally colored red by tiny hematite particles. The halite is usually colorless. It sometimes contains small anhydrite particles and a noticeable amount of fine-grained kieserite.

STASSFURT POTASH SEAM K2 / KIESERITIC TRANSITION BEDS Na2K

A fine- to medium-grained crystalline carnallitite. The original color of the rock was probably light grey to light brown. The carnallite is colored red by very small hematite particles in some places. In addition to halite, the rock contains some anhydrite and fine-grained crystalline kieserite nodules about 1 cm in diameter.

Table 2. Mineral Composition of Rock Samples

| Sample No. | Weight Percent (%) | | | |
	Rock Salt	Anhydrite	Carnallite	Kieserite
070-0930-01	67.9	1.12	27.4	3.26
070-0930-02	67.7	0.72	27.4	3.63
070-0930-03	82.9	0.82	11.8	2.88
098-1490-01	18.0	0.55	80.0	<0.03
098-1490-02	20.8	0.59	76.5	<0.03
098-1490-03	33.9	0.62	64.2	0.03
039-0370-01	61.8	0.65	32.7	3.47
039-0370-02	32.8	0.38	51.1	14.6
039-0370-03	53.6	0.58	18.3	25.5

RESULTS

Thermal conductivities compiled for different types of rock salt from a number of publications reveal a range of variation from about 4 to 10 W/m·K at temperatures near 0 °C (3,4,5). The highest values are those of pure, single crystals grown artificially from solution or melt. From a comparative study of elastic wave velocities and thermal conductivity in rock salt the question arose, whether the thermal conductivity in an undisturbed salt deposit could be assumed to be as high as that of very pure single crystals (6). The results of previous thermal conductivity measurements of the BGR (3) on samples of Asse rock salt with very different fabric vary only by about 20 % in the temperature range of 30 - 350 °C and are in good agreement with the thermal conductivity values given by Birch & Clark (7). Fig. 1 shows as an example a comparison of measurements on two samples with extremely different fabric, one of which is an almost completely transparent, natural single crystal and the other one a polycrystal aggregate, full of microcracks. From these unexpected results it was concluded that for rock salt the fabric of the samples and even microfracturing in the sample resulting from the taking of the drill cores and cutting the specimens is only of minor influence on the thermal conductivity and that for this reason it might be permissible to apply the conductivities measured in the laboratory to in-situ conditions. This conclusion is strengthened by the results of

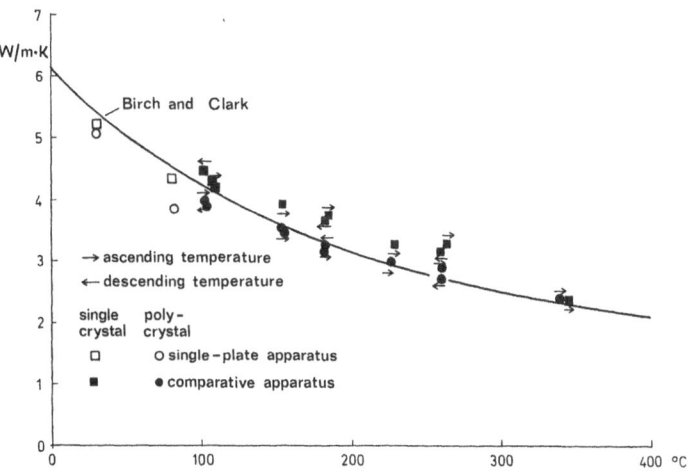

Figure 1. Thermal Conductivity of Rock Salt Samples
With Extremely Different Fabric

677

in-situ thermal conductivity measurements and the results of model calculations for underground in-situ heating experiments in salt formations (8). It is also confirmed by the results of earlier investigations by Bridgman (9) and more recent ones by Durham (10) on the pressure dependence of the thermal conductivity of rock salt, which proved to be negligible in the range that is relevant to nuclear waste disposal. The results of thermal conductivity measurements on samples of pure Stassfurt and Leine rock salt taken from the Gorleben salt dome, and also those on samples of anhydrite, differ only insignificantly from the measurements on Asse rock salt as shown in Fig. 2.

Considering the values to be recommended for the thermal and thermomechanical model calculations for the prospective Gorleben salt dome repository, it did not appear, on the basis of these results, to be appropriate for the present to propagate values different from the thermal conductivity data for rock salt given by Birch & Clark, which have already been extensively used as input data for model calculations related to the underground disposal of heat-generating waste. But it must be emphasized that these recommended values are, in general, only applicable to structures of pure rock salt or pure anhydrite.

In order to describe the temperature-dependent thermal conductivity an equation of the following form (4) is recommended:

$$\lambda = \frac{\lambda_0}{1 + CT} \quad W/m\cdot K$$

with
λ_0 = 6.1 W/m·K (thermal conductivity at 0 °C)
C = 0.0045/°C
T = temperature (°C)

According to the available experimental results, the application of this empirical formula is confined to the temperature range from 0 to 400 °C for rock salt and from 0 to about 200 °C for anhydrite. The thermal conductivity measurements on pure anhydrite samples yield constant or slightly rising values for temperatures above 160 °C. This effect must be examined more closely; it may possibly be due to thermally unstable carnallite contained in the anhydrite (see below).

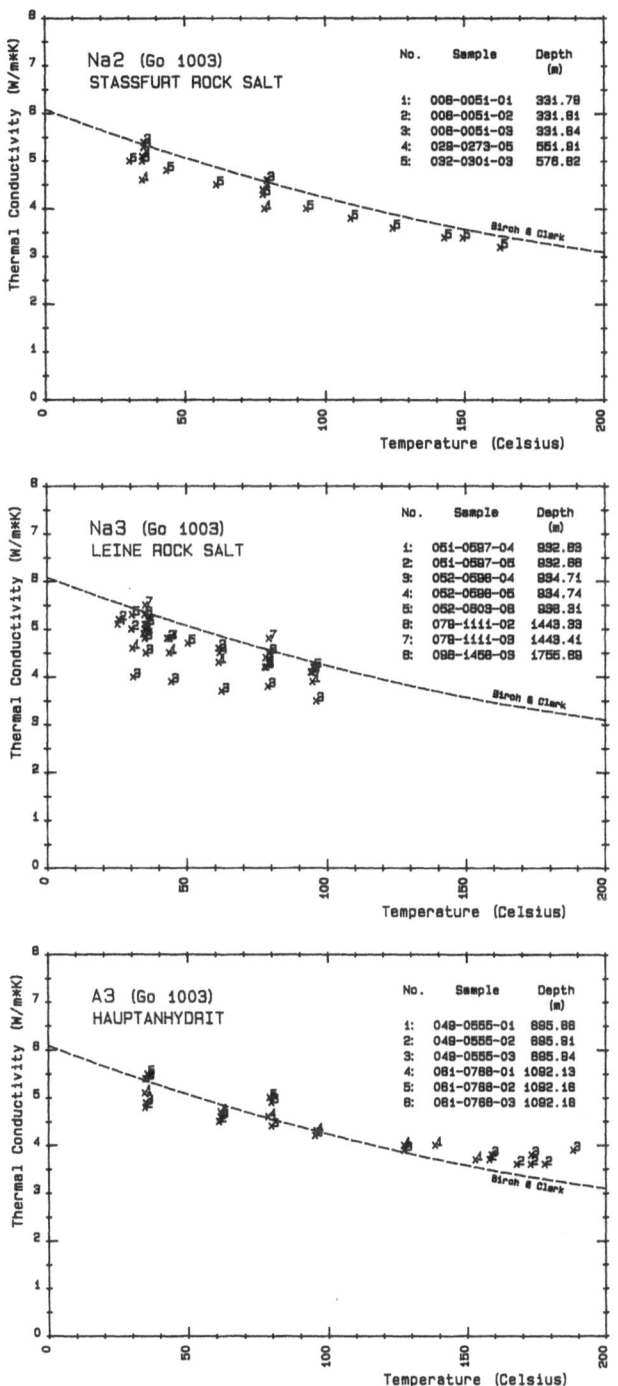

Figure 2. Thermal Conductivity of Pure Gorleben
Rock Salt and Anhydrite

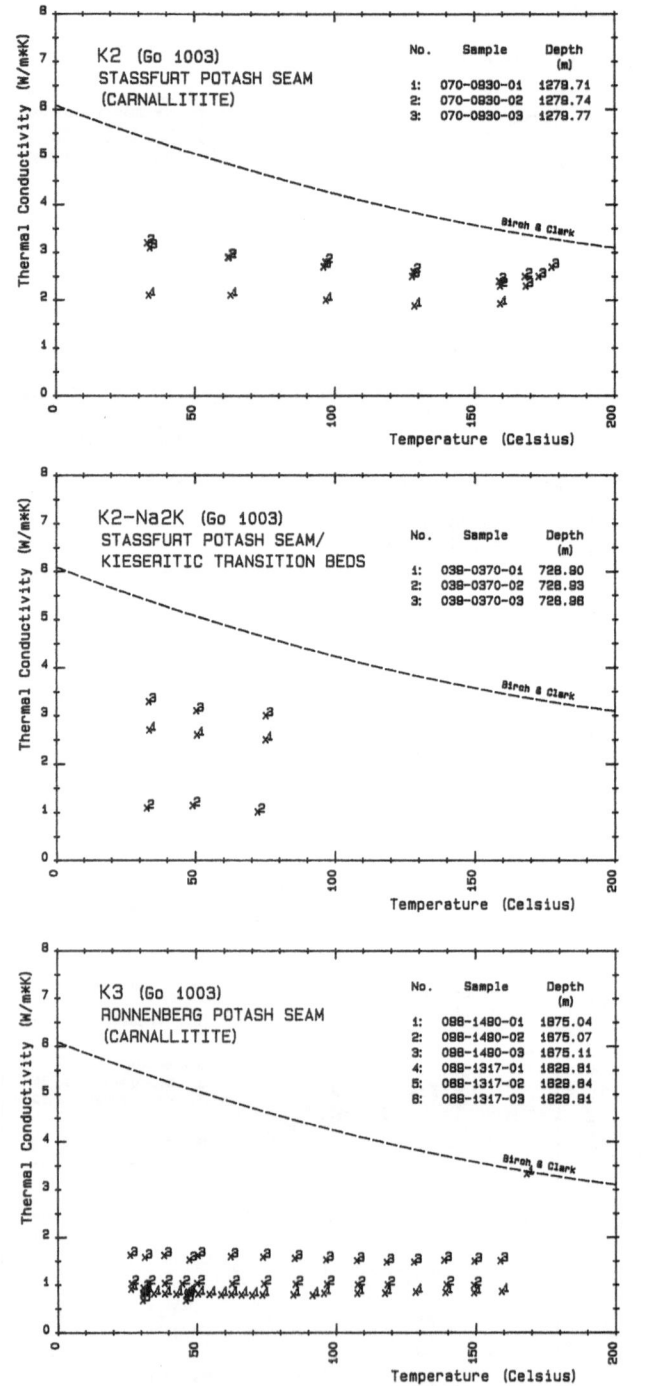

Figure 3. Thermal Conductivity of Gorleben
           Salt Dome Impurities

Extremely low values were obtained for the thermal conductivity of carnallitite samples from the Go 1003 borehole (Fig. 3). Most of them are below 1 W/m·K (with a minimum value of 0.67 W/m·K). The thermal conductivities measured for carnallitite samples are nearly independent of temperature. For temperatures above 160 °C, carnallite decomposes with the release of water of hydration (the higher reading in this temperature range arises from the thermal energy released by the dehydration process; it is only an apparently higher thermal conductivity).

The thermal conductivity values obtained for samples from the Stassfurt potash seam, especially those from the kieseritic transition beds, vary considerably. This is due primarily to differences in the mineral composition (see Table 1). The dependence of the thermal conductivity on the composition will be analyzed in further investigations.

Table 3 compares the thermal conductivity values for rock salt from this study at two temperatures with those of Birch and Clark and those of other investigations in the USA related to radioactive waste disposal.

Table 3. Comparison of Thermal Conductivities Measured on Salt Rocks from Different Locations

| | | Thermal Conductivity (W/m∗K) | |
		35 °C	75 °C
BGR[+)]	Gorleben Leine rock salt	5.1 (4.6 – 5.4)	4.4 (4.0 – 4.6)
	Gorleben Stassfurt rock salt	5.0 (4.5 – 5.5)	4.3 (3.8 – 4.5)
	Gorleben anhydrite	5.2 (4.8 – 5.5)	4.6 (4.4 – 5.0)
	Asse rock salt	4.8 (4.1 – 5.3)	3.9 (3.3 – 4.1)
Birch & Clark	NaCl single crystals	5.3	4.6
Yang[++)]	NaCl single crystals	6.0	5.4
Durham & Abey	Avery Island rock salt	6.2	5.5
Morgan	Avery Island rock salt	4.0	3.4
Sweet & McCreight	Southwestern New Mexico rock salt	5.9	4.9

[+)] average and minimum – maximum of measured values
[++)] recommended values generated by compilation of published experimental data

BOREHOLE TEMPERATURE LOGGING

For the geothermal exploration of the Gorleben salt
dome area borehole temperature measurements with an
overall accuracy of about 0.01 °C and 0.001 °C resolution
are being performed (11). The equipment for continuous
borehole logging in depth intervals between 1 and 10 cm
set up for this purpose basically consists of a high-reso-
lution digital multimeter (HP 3456A) for 4-wire resistance
measurement of a 200-ohm Platinum resistance temperature
detector. This instrument is controlled by a desktop computer with a tape cartridge drive for recording the data
(HP 9825) and a plotter (HP 7225A) for monitoring the
measurements, which are triggered in preselected depth
intervals from a depth signal generator continuously
driven by the cable winch of the logging device.

A computerized evaluation of temperature measurements
performed in the Go 1003 exploration borehole, including
the results of the stratigraphic evaluation based on the
drill cores, is shown in Fig. 4. The temperature and temperature gradient logs shown are obtained by sliding averaging (linear regression) over 200 temperature readings at
resp. 20 m intervals. The peaks in the temperature gradient, which is inversely proportional to thermal conductivity, show a remarkable correlation with the carnallitite
layers. The shifts in the temperature logs illustrate the
impeding effect on heat dissipation resulting from the low
thermal conductivity of the carnallite compared to that of
rock salt. The temperature logs are also useful to locate
flows of brine, hydrocarbons, or gas into a borehole or in
its surroundings. The temperature anomaly between 400 and
450 m shown on Fig. 4 for example arises from intrusion of
brine into the borehole, which was encountered during
drilling.

Temperature and temperature gradient logs obtained
from the hydrogeological exploration boreholes show the
very varied temperature regime typical for salt dome
areas. A correlation of these logs with the corresponding
stratigraphic logs are used to identify the effect of convective heat transfer by groundwater flows. From the
gathered temperature and thermal conductivity data the
flow velocities of the groundwater are determined on the
base of the differential equations for conductive and convective heat transfer according to Bredehoeft and
Papadopoulos (12).

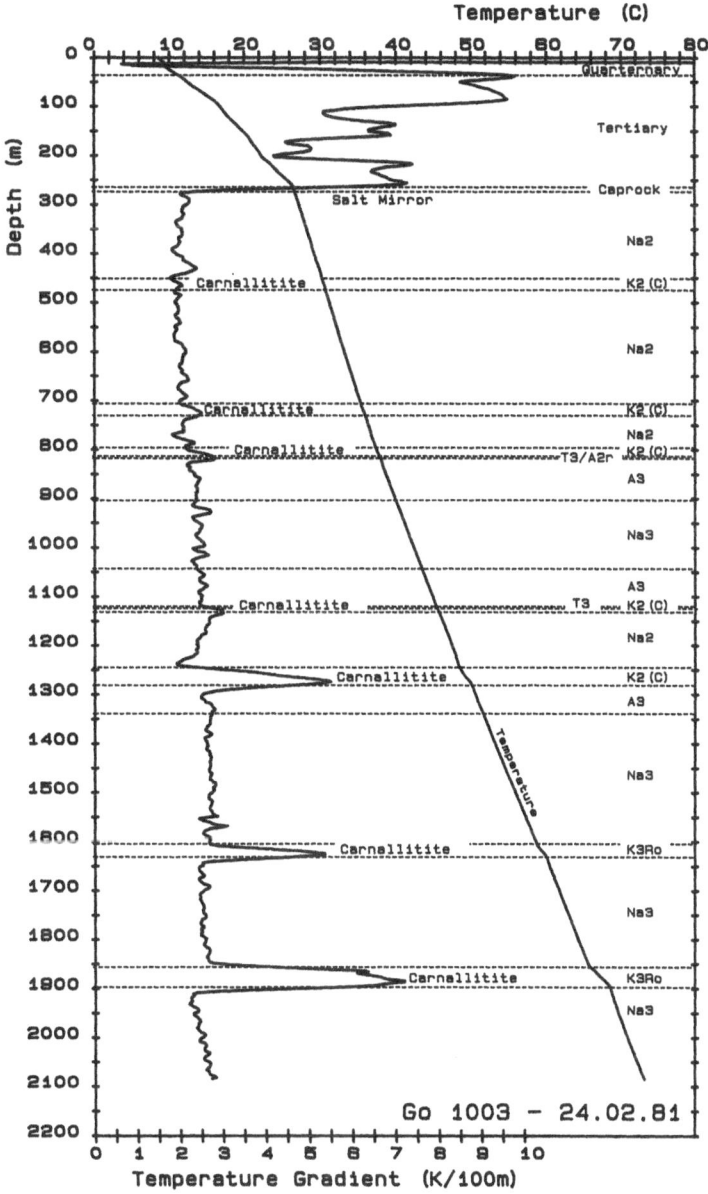

Figure 4. Geothermal Logs from a Gorleben Salt Dome
Exploration Borehole

CONCLUSIONS

- For Stassfurt and Leine rock salt, as well as for anhydrite (Hauptanhydrit), the results of this thermal conductivity study available up to now agree rather well with those of Birch & Clark. Thus, it does not seem appropriate, at present, to propagate recommended values for thermal and thermomechanical model calculations for the prospective Gorleben salt dome repository different from the thermal conductivity data for rock salt given by Birch & Clark.

- Very low thermal conductivity values were obtained for carnallitite samples. The lowest values of about 0.7 W/m·K were as much as eight times lower than the values for pure rock salt. The thermal conductivity of carnallitite was nearly independent of temperature up to a temperature of 160 °C, at which water of hydration is released.

- The thermal conductivities obtained for samples from the contact zone rock salt/carnallitite varied considerably. The dependence of thermal conductivity on the mineral composition of salt rocks must be studied in further studies to be used in a future, more refined model of the salt dome.

ACKNOWLEDGEMENTS

This work was commissioned by the Federal Ministry of the Interior and the Federal Ministry of Research and Technology (BMFT). Dr. R. Fischbeck from the Engineering Geology Division of the BGR provided the macroscopic sample descriptions. The chemical analysis of the samples was done by the Geochemistry and Mineralogy Division of the BGR.

REFERENCES

(1) Coumou, K.G., Tye, R.P., A Laboratory Instrument for Rapid Determination of Thermal Conductivities in the Range 0.4 to 5 W/m·K, in: "High Temperatures - High Pressures", 13:695 (1981).
(2) Kopietz, J., Neumann, W., Thermal Conductivity Measurements on Salt Rocks by Different Methods, in: "Proc. 18th ITCC", these Proceedings.

(3)   Kopietz, J,, Jung, R., Geothermal In-situ Experiments
        in the Asse Salt Mine, in: "Proc. of the Seminar
        on In-situ Heating Experiments in Geological Forma-
        tions", Ludvika-Stripa, NEC/OECD Paris (1978).
(4)   Cermak, V., Rybach, L., Thermal properties, in: Landolt-
        Börnstein, Physical properties of rock, vol. 1,
        Heidelberg-New York (1982).
(5)   Yang, J.M., Thermophysical Properties, in: "NBS
        Monograph Physical Properties Data for Rock Salt",
        Washington (1981).
(6)   Giesel, W., Elastic Wave Velocities and Thermal Conduc-
        tivity in Rock Salt, in: "Z. Geophys.", 33:9
        (1967).
(7)   Birch, F., Clark, H., The Thermal Conductivity of Rocks
        and its Dependence upon Temperature and Composition,
        in: "Am. J. Sc.", 238(I):529 (1940).
(8)   Liedtke, L., Kopietz, J., Thermomechanical calculations
        related to thermally induced rock loosening, Com-
        puters & Structures, 17:5-6, (1983).
(9)   Bridgman, P.W., The Thermal Conductivity and
        Compressibility of Several Rocks under High
        Pressure, in: "Am. Jour. Sc.", 7:81 (1924).
(10) Durham, W.B., Abey, A.E., Thermal Properties of Avery
        Island Salt to 573 K and 50 MPa Confining Pressure,
        in: "Proc. of the 17th ITCC", Gaithersburg (1981).
(11) BGR report 90035/KWA 2060 0 (1981) − to be
        published.
(12) BGR report 93344/KWA 2060 0 (1983).
(13) Sweet, J.N., McCreight, J.H., Thermal Conductivity of
        Rock Salt and Other Geologic Materials from the Site
        of the Proposed Waste Isolation Plant, in: "proc. of
        the 16th ITCC", Chicago (1981).
(14) Acton, R.U., Thermal Conductivity of S. E. New Mexico
        Rock Salt and Anhydrite, in: "Proc. of 15th ITCC",
        (1978).

THERMAL CONDUCTIVITY OF PERMIAN BASIN

BEDDED SALT AT ELEVATED PRESSURE

W. B. Durham, C. O. Boro, and J. M. Beiriger

University of California
Lawrence Livermore National Laboratory
Livermore, California 94550

ABSTRACT

Measurements of thermal conductivity were made on five core samples of bedded rock salt from the Permian Basin in Texas. The sample size was 100 mm in diameter by 250 mm in length. Measurements were conducted under confining pressures ranging from 3.8 to 31.0 MPa and temperatures from room temperature to 473 K. Conductivity showed no dependence on confining pressure, but showed a monotonic, negative temperature dependence. Four of the five samples showed conductivities clustered in a range of 5.6 ± 0.5 W/m·K at room temperature, falling to 3.6 ± 0.3 W/m·K at 473 K. These values are approximately 20% below the values for pure halite, reflecting perhaps the 5-20% non-halite component of the samples. The fifth sample showed a conductivity vs. temperature dependence much like that of halite.

INTRODUCTION

Rock salt formations are among the leading contenders as sites for underground nuclear waste repositories. Interest in the viability of rock salt waste repositories has clearly passed form general overview to focused site-specific attention. Among sites under consideration in the U.S. are the Permian Basin bedded salt formations in the Texas panhandle. Before a respository is constructed in these formations, it is necessary to predict with the best possible accuracy how the repository will respond to the heat from the nuclear waste load. Such predictions depend on an accurate knowledge of the pertinent physical properties of the repository medium under in situ physical and chemical conditions.

We report here measurements of one such physical property, thermal conductivity, made on five core samples taken from nearby locations in the Permian Basin bedded rock salt. Conductivity, which characterizes heat energy transport in the steady-state situation, and thermal diffusivity, which characterizes transport in the transient situation, must be known in order to predict temperature profiles in the future repositories.

SAMPLE AND STANDARD REFERENCE MATERIAL

Sample material was cored from the Cycle 4 and Cycle 5 bedded salt formations in the Permian Basin in Deaf Smith County, Texas panhandle. Five salt samples were measured (Table 1): two horizons were sampled in each of the G. Friemel #1 and Detten #1 wells, with the lower horizon in the Detten #1 well sampled twice.

Petrographic analyses are available for all horizons except the lower one at Detten #1 (Dixon, 1982; Fukui, 1982) and are summarized in Table 1. The material varies between 78 and 94% halite with the balance being primarily a mix of clay minerals and anhydrite. The grain size of the halite ranges from medium (< 10 mm) to coarse (∿ 50 mm).

A reference standard was fabricated from Pyroceram Code 9606, a microcrystalline ceramic material manufactured by Corning Glass Works (CGW), Corning, NY. Material for our reference standard was created during the slabbing operation at the end of the single 1982 production run at CGW.

Pyroceram 9606 is a particularly uniform and chemically stable material which was under consideration by the National Bureau of Standards nearly 20 years ago as a standard reference material for thermal conductivity and diffusivity (Flynn et al.,1964).

Table 1: Sample Material

			----Comp (%)----			
Sample	Well	Depth(ft)	Halite	Clay	Anhydrite	z^c
TP7	Dtn[a]	2454-55	78-85	6-12	8-9	5
TP8	G Frml[b]	2523-24	93-94	2-5	1-3	10
TP9	G Frml	2308-09	89	2	8	1.5
TP10	Dtn	2654-55	analysis not available			~180
TP11	Dtn	2655-56				~180

[a]Well Detten #1
[b]Well G. Friemel #1
[c]Distance to nearest petrographic analysis (ft)

Pyroceram 9606 is essentially 100% dense, i.e., it has zero porosity, so its physical properties would not be expected to exhibit any extrinsic (crack- and pore-related) pressure dependence. Since Pyroceram 9606 has a bulk modulus in the range of 80 GPa, it should not show any measurable intrinsic pressure effects until well outside the 0-30 MPa pressure range covered in this study. The independence of thermal diffusivity and pressure between 0.1 and 200 MPa was confirmed by Mirkovich et al. (1982).

SAMPLE AND STANDARD PREPARATION

Salt cores were machined to thick-walled cylinders 254 mm long with inner and outer diameters of 21 and 102 mm, respectively. The cyclinders were encapsulated in metallic jackets as shown in Fig. 1 to exclude the high pressure confining medium (argon gas) from the pores and cracks of the salt in order to retain the validity of the simulation of lithostatic pressure.

Six thermocouples were placed at six logarithmically spaced radii in the central radial plane of a sample (Fig. 1). The thermocouples were distributed azimuthally to minimize thermal shadowing effects of one upon the other. Sheathed, 1.5-mm diameter Type J (iron-constantan) thermocouples were used. They were

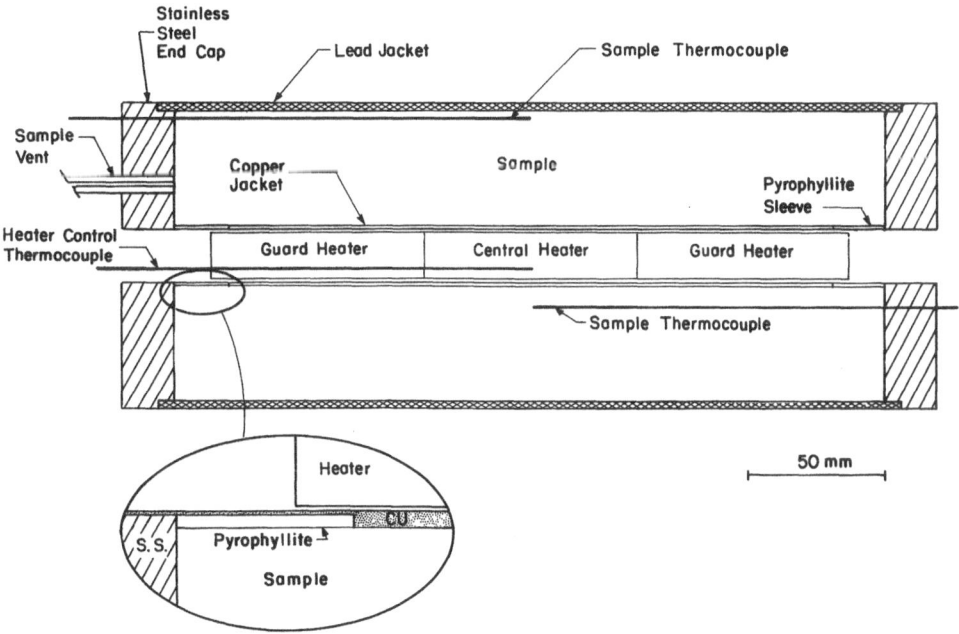

Fig. 1. Scale drawing of the sample assembly. Only two of the six sample thermocouples are shown.

introduced axially through the ends of the sample assembly, three
from each end, in close-fitting holes drilled in the salt.
Thermocouple sheaths were brazed to the sample end caps to retain
the pressure seal around the sample. Also introduced through one
end cap was a sample vent of high pressure tubing which ultimately
passed through the pressure vessel (see below) to the outside. The
vent maintained pore pressure at 0.1 MPa and also served as a leak
detecting device.

The Pyroceram 9606 reference standard was received from Corning
Glass Works in its final form, lacking only the six thermocouple
holes, which were cored in our laboratory prior to jacketing. The
standard was fabricated from six discs of approximately equal
thickness which were cemented together using Corning Code 7574
devitrifying solder glass. The joints between discs spanned less
than 0.25 mm and were barely visible. The manufacturer could not
guarantee that the joints were free of air pockets, but since the
joints lay in the plane normal to the axis and heat flow was
primarily radial, no problem was anticipated. A numerical
simulation of the experiment demonstrated that altering the
conductance of joints from infinite to zero has an effect that is
barely perceptible. The standard was jacketed in precisely the
manner of the salt samples.

MEASUREMENT TECHNIQUE

The measurement technique is a refined version of that
described by Abey et al. (1982) and Durham and Abey (1983). The
sample assembly (Fig. 1) resides inside an externally heated
pressure vessel wherein pressurized argon gas and power supplied to
the external heaters provide the hydrostatic pressure (P) and
temperature (T) to simulate appropriate ambient conditions in the
earth. The vessel has a design range to P = 200 MPa (simulating
burial depths of 6 to 8 km) and T = 773 K. A temperature gradient
in the sample, needed to measure thermal conductivity, is provided
by the three low-power internal heaters shown in Fig. 1. The
gradient is measured by the six sample thermocouples. Power to
each of the three core heaters is also measured. Voltage taps for
the power measurements are taken at the point of emergence of the
heater leads from the axial hole in the sample. A host of
thermocouples within and around the sample assembly and pressure
vessel are used for diagnostic purposes and to control the internal
and external heaters.

Thermal conductivity (λ) is measured by the infinite line
source method wherein $T(r)$ is related to the conductivity and power
per unit length of the line source (q) by:

$$T(r) = \frac{q}{2\pi\lambda} \log_e r \qquad (1)$$

The measurement technique has been refined to allow determination of λ in situations where (1) is not strictly applicable. The refinement was necessitated by the extreme experimental difficulty in identifying the conditions under which (1) does strictly apply. The interested reader can find details of the refinement in Durham et al. (1984).

The sequence of (P,T) conditions under which the measurements were made is shown schematically in Fig. 2. Utility and a desire to minimize thermally-induced damage were factors in choosing the sequence: rocks generally suffer permanent damage by micro-fracturing as temperature increases at low pressures (rock salt, because of its nearly isotropic character, may be an important exception), hence lower temperatures and higher pressures were explored first. All pressures were sampled in sequence at a given temperature simply because the system can change pressure much more rapidly than it can change temperature. Note that heating was always done at 31 MPa, the highest pressure used. The identical path was followed for all five salt samples. Neither the effects of cycling nor the effects of following a different path in (P,T)

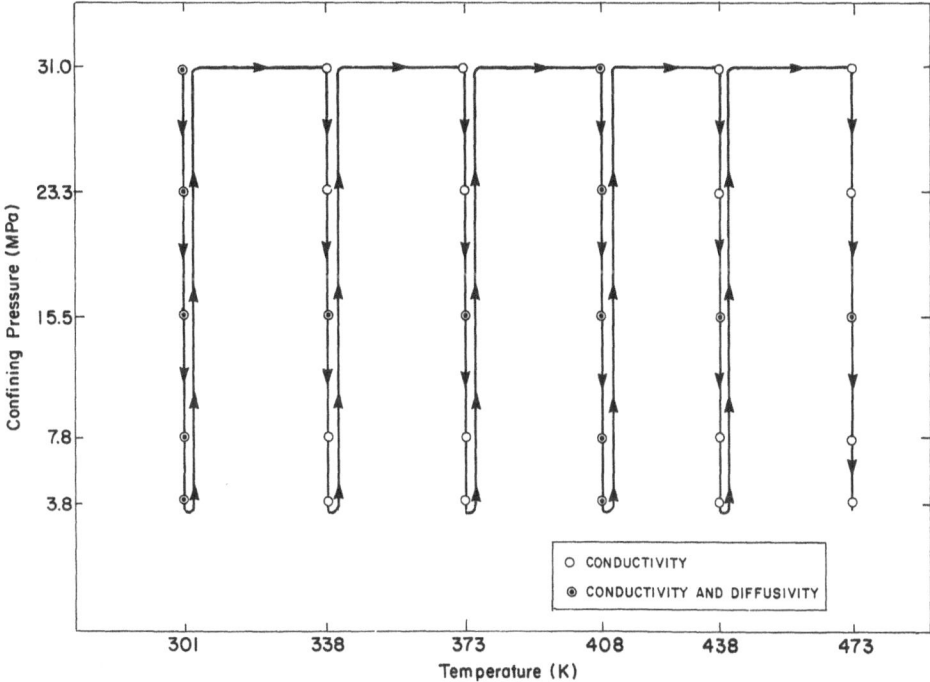

Fig. 2. Path followed in pressure-temperature space for the measurements. Note that conductivity was measured at each stopping point, diffusivity only at some of the stopping points.

space were investigated. The path followed for the reference standard was the same, except a second excursion to 338 K was added following the measurement at 408 K, and the measurements at room temperature were done last.

RESULTS

Measurements of conductivity were made at all six temperatures at all five pressures (Fig. 2) on samples TP8-TP11 and on the reference standard. Measurements at 7.8 and 23.3 MPa were eliminated from the test matrix of TP7. The final results are plotted in Fig. 3.

Approximately half a million individual data (times, temperatures, pressures, currents, voltages) were gathered, of which approximately 60% fell into the category of diagnostic and control. The remaining 200,000 data actually used to determine conductivity are still prohibitively numerous to include here. All data remain stored on magnetic disc, in duplicate, and are easily accessible upon request.

DISCUSSION

Thermal conductivity of the five rock salt samples decreases monotonically with increasing temperature. Any pressure effect on the data must be exceedingly small. There is no obvious order to the five symbols (i.e. pressures) between any of the temperatures in any of the runs in Fig. 3, but the point has not been checked quantitatively. If the measurement reproducibility is taken as ± 0.25 W/m•K as estimated below, then any pressure effect on λ is easily less than ± 0.1 W/m•K over the 30 MPa range measured here. Compared run to run in Fig. 4, the conductivities of samples TP7, TP8, TP9, and TP11 are indistinguishable over the entire (P,T) range. The fifth sample, TP10, which would be expected to closely resemble TP11 on the basis of Table 1, has a distinctly higher conductivity than the rest of the samples. Excepting TP10, the rock salt measured here is less conductive, by 0.5 to 1.0 W/m•K, than pure halite (Yang, 1981) but still considerably more conductive than most crystalline rocks (see, for instance, Touloukian and Ho, 1981). The temperature sensitivity of the conductivity of TP10 is strong and at T < 350 K TP10 actually shows a higher conductivity than pure halite.

Most of the relationships shown in Fig. 4 are reasonable. It is likely that the non-halite components of the rock salt samples measured here have a lower conductivity than halite (the conductivity of anhydrite is reported in the Touloukian and Ho (1981) compendium as 5.1 ± 0.6 W/m•K; values are not available for clay minerals) so the conductivity of the bulk rock should be somewhat lower than that of pure halite, depending on morphological

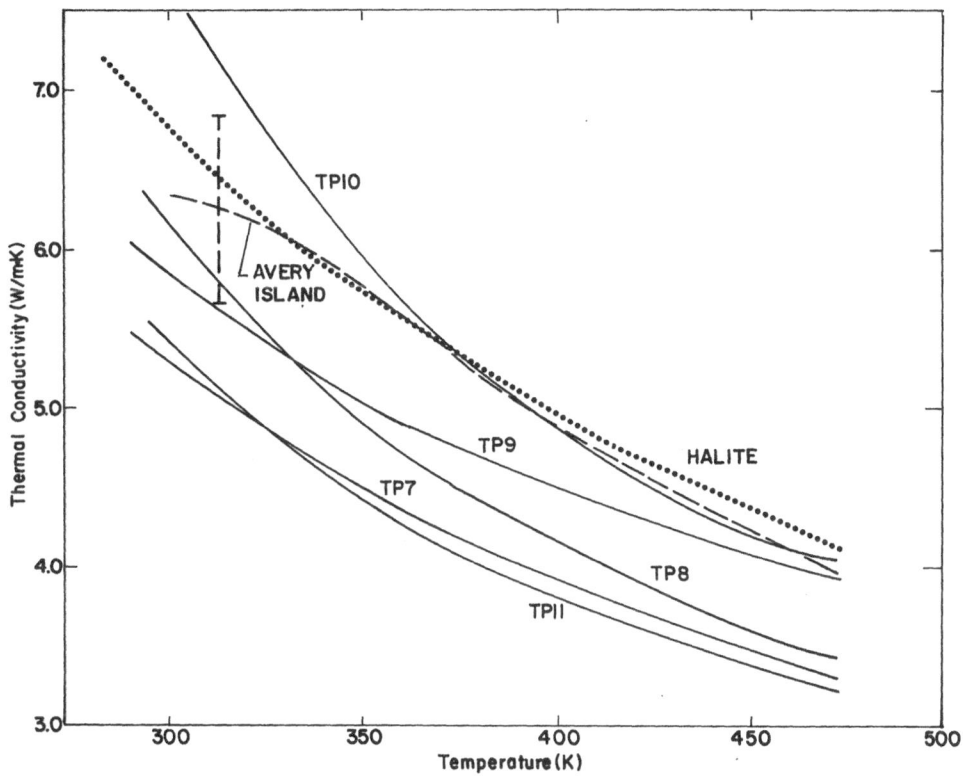

Fig. 4. Summary of thermal conductivity measurements for salt. The solid lines are estimated fits to the the curves in Fig. 3. The error bar shows the scatter in the Avery Island data, which are taken from Durham et al. (1981).

relationships between the different phases (see, for instance Walsh and Decker, 1966). Acton (1978) measured the thermal conductivity of a number of rock salts with > 50% halite content and found conductivities at room temperature increasing from 3 to 8 W/m·K with increasing halite concentration. Note that the conductivity of a nearly pure rock salt from Avery Island, Louisiana, is essentially identical to that of halite (Durham et al., 1981). The conductivity of TP10, therefore, is anomalous. While a petrographic examination of the sample has not been made, the "normal" results from a sample immediately adjacent (TP11) suggest nothing abnormal about the composition of rock at this depth. Measurement error is a possibility, although there is no clue in either the Pyroceram run or in the other salt runs as to its cause. Note that TP10 and TP11 were run immediately before and immediately after, respectively, the Pyroceram run.

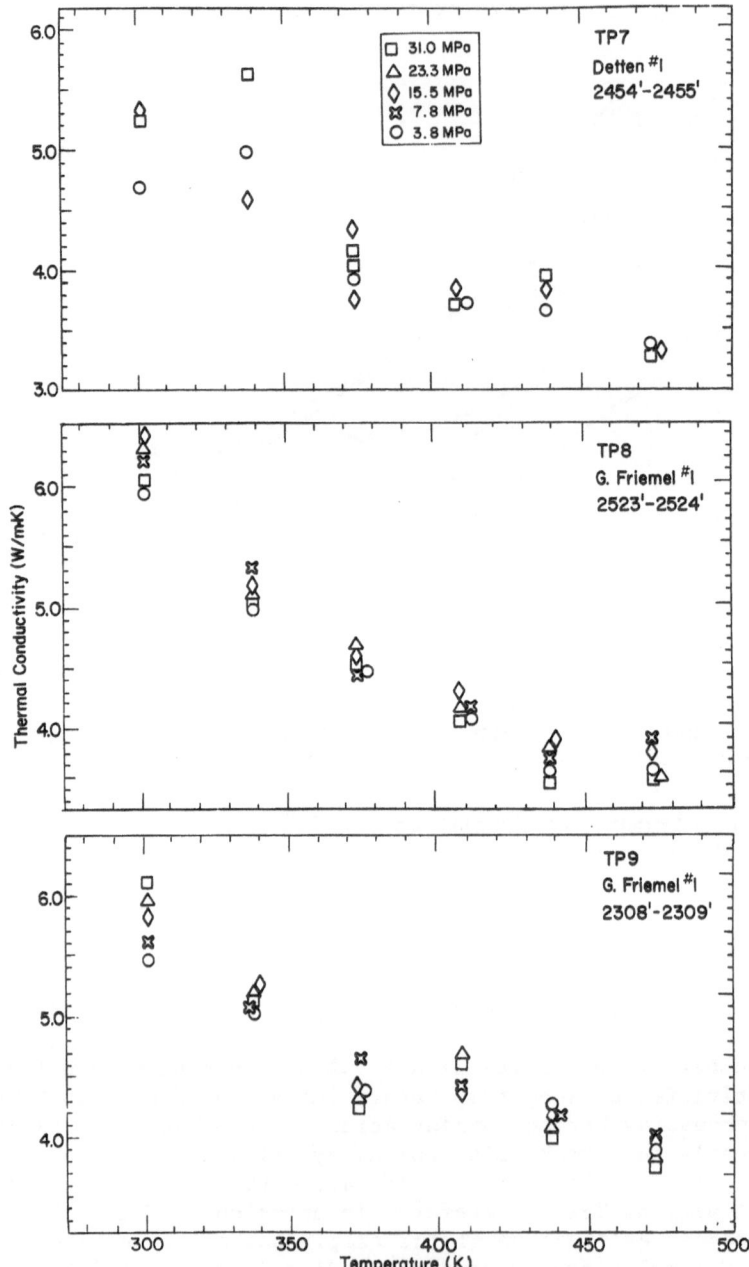

Fig. 3. Results of thermal conductivity vs. temperature and
pressure for five rock salt samples and the reference
standard. The curve through the data for the Pyroceram
9606 is taken from Touloukian et al. (1970).

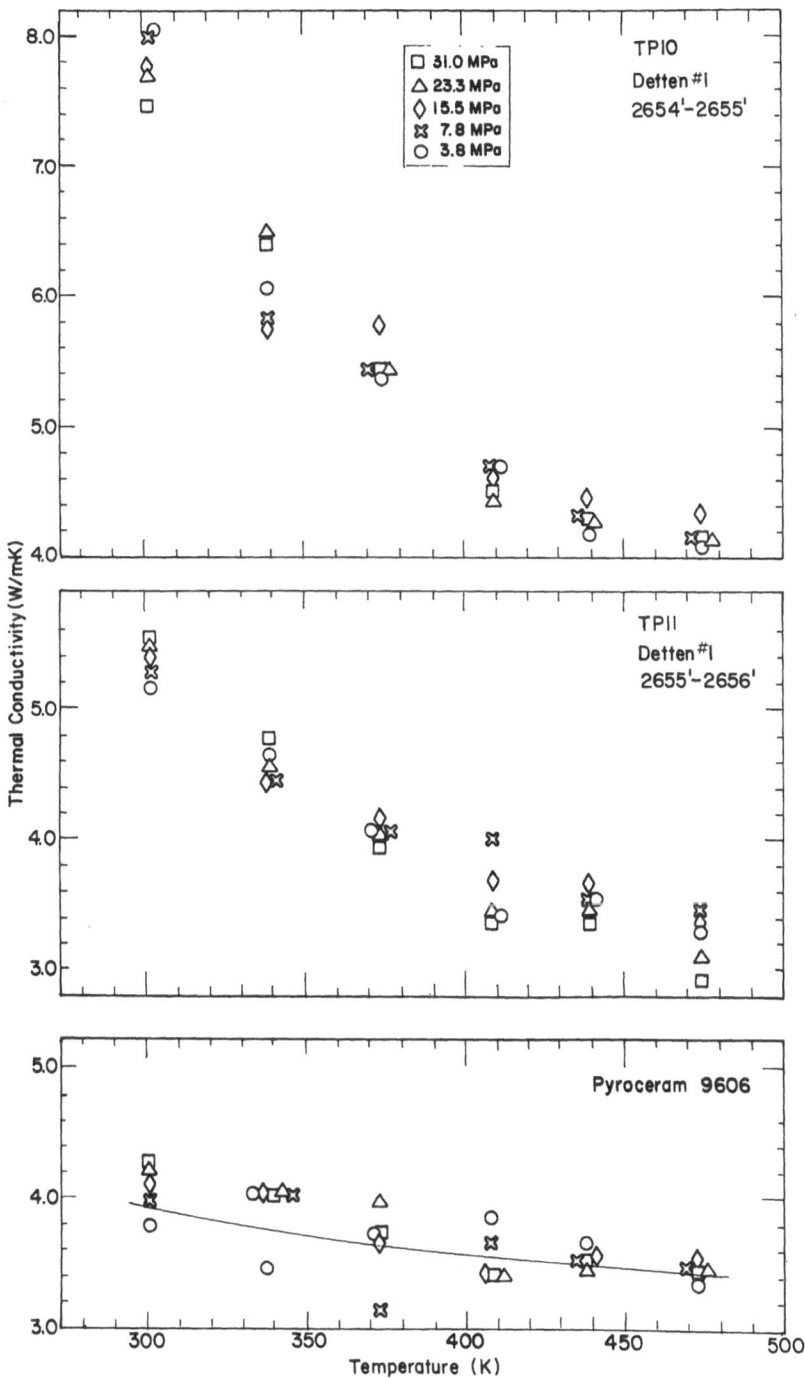

Fig. 3. (Cont.)

The lack of a pressure effect in our measurements is also reasonable, but would not be reasonable for almost any other crystalline rock. Confining pressure in the range up to 200 MPa, which is applicable to the upper crust, typically has a pronounced effect on many rock physical properties such as strength, transport properties, and elastic moduli. Most of these effects can be related to the effect which pressure has on the microfractures which pervade most rocks: pressure acts to hold closed existing cracks and to prevent new ones from being created. In contrast to most other rocks, rock salt is composed of a single mineral, halite, whose symmetry is cubic. Elastic mismatches under a change in lithostatic load, and thermal expansion mismatches under uniform heating, two important causes of microfractures in other rocks, are less important in rock salt. Durham et al. (1981) also found Avery Island rock salt to have thermal properties which were independent of pressure. It should be pointed out that the pressure effects discussed here pertaining to the upper crust are extrinsic effects, to be contrasted with intrinsic effects seen at higher pressures. Above 1 GPa the halite lattice itself becomes sufficiently compressed that increases in the thermal transport properties are readily detectable. (See, for instance Bridgman, 1952.)

Rock salt has another atypical physical property which pertains to the absence of a pressure effect in our measurements: a relatively low plastic flow strength (Carter and Heard, 1970; Carter and Hansen, 1980). Fractures which do form in rock salt through, say, the application of a non-uniform load or non-uniform temperature change, can be closed permanently by plastic deformation under modest pressures even at room temperature (ductility increases as temperature increases). Sutherland and Cave (1980) and Wawersik and Hannum (1980) found that permanent reductions in room pressure porosity occurred in rock salt following pressurization to between 10 to 30 MPa. This effect is very important to laboratory testing of core samples. A coring operation applies non-uniform stresses to the core and therefore may damage (i.e. put fractures and microfractures in) the core. That core when tested in the laboratory may show differing physical properties at low (e.g. 0.1 MPa) pressures depending on whether or not it had been pressurized after coring. Note in particular that our sample preparation treatment involved an initial pressurization at 10 to 12 MPa. Further, note that our test matrix began at P = 31 MPa (Fig. 2). In the case of Avery Island salt, thermal conductivity measurements on core at 0.1 MPa are substantially lower if the core has not been pressurized (Morgan, 1979) than if it has been pressurized (Durham et al., 1981).

CONCLUSIONS

Measurements of thermal conductivity (λ) have been made on five 100-mm diameter by 250-mm length rock salt cores from the

Detten #1 and G. Friemel #1 wells in the Permian Basin Cycle 4 and
Cycle 5 formations in Deaf Smith County, Texas. Measurement
temperatures (T) covered 300<T<473 K and pressures (P) covered
3.8<P<30 MPa. We found the following:
    1. λ does not exhibit a dependence upon P beyond the
measurement resolution ± 0.25 W/m·K. Comparing groups of
measurements of λ indicates the variation of λ over the range
of P used in the experiments is < ± 0.10 W/m·K.
    2. λ exhibits a monotonic, negative temperature dependence
in all five samples tested. Values were generally lower than for
pure halite at any T, perhaps reflecting the 5-20% non-halite
component of the rocks. For four of the five samples, λ fell
from 5.6 ± 0.5 W/m·K at room temperature to 3.6 ± 0.3 W/m·K
at 473 K, i.e. approximately 20% below the curve for pure halite.
The fifth sample showed λ = 7.7 W/m·K (± 0.25 W/m·K) at
room temperature, falling to 4.25 ± 0.25 W/m·K at 473 K.

ACKNOWLEDGMENTS

    Work performed under the auspices of the U.S. Department of
Energy by Lawrence Livermore National Laboratory under Contract
W-7405-Eng-48.

REFERENCES

Abey, A. E., W. B. Durham, D. A. Trimmer, and L. L. Dibley,
        1982, An apparatus for determining the thermal properties
        of large geological samples at pressures to 0.2 GPa and at
        and temperatures to 750K, Rev. Sci. Instrum., 53: 876-879.
Acton, R. U., 1978, Thermal conductivity of S.E. New Mexico
        rock salt and anhydrite, Thermal Conductivity 15
        (Proceedings of the 15th International Conference on
        Thermal Conductivity), V. V. Mirkovich, ed., Plenum Press,
        New York, 263-276.
Bridgman, P. W., 1952, "The Physics of High Pressure," G. Bill
        and Sons, London, 320-329.
Carter, N. L. and H. C. Heard, 1970, Temperature and rate
        dependent deformation of halite, Am J. Sci., 269: 193-249.
Carter, N. L. and F. D. Hansen, 1980, "Mechanical Behavior of
        Avery Island Halite: A Preliminary Analysis," ONWI-100,
        Prepared for Office of Nuclear Waste Isolation, Battelle
        Memorial Institute, Columbus, OH.
Dixon, M. L., 1982, "Petrographic Report: Insoluble Residue
        Analysis Permian Cycle 4 Salt Well G. Friemel #1 Palo Duro
        Basin, Texas," prepared for Battelle Memorial Institute and
        the U.S. Department of Energy National Waste Terminal
        Storage Program, Columbus, OH.
Durham, W. B., A. E. Abey, and D. A. Trimmer, 1981, "Thermal
        Properties of Avery Island Rock Salt to 573 K and 50 MPa
        Confining Pressure," UCRL-53128, University of California,
        Lawrence Livermore National Laboratory, Livermore, CA.

Durham, W. B. and A. E. Abey, 1983, Thermal conductivity and diffusivity of climax stock quartz monzonite at high pressure and temperature, Thermal Conductivity 17, (Proceedings of the 17th ITCC), J. Hust, ed., Plenum Press, New York, 459-468.

Durham, W. B., C. O. Boro, J. M. Beiriger, and D. N. Montan, 1984, "Thermal Conductivity of Permian Basin Bedded Salt at Elevated Pressure and Temperature," UCRL-53476, University of California, Lawrence Livermore National Laboratory, Livermore, CA.

Flynn, D. R., H. E. Robinson, and I. L. Martz, 1964, Present status of Pyroceram code 9606 as a thermal conductivity reference standard, Proceedings of the 4th Conference on Thermal Conductivity (13-16 October 1964, San Francisco, CA), U.S. Naval Radiological Defense Laboratory, San Francisco, CA, I-F-1 to I-F-27.

Fukui, L. M., 1982, "Petrographic Report: Insoluble Residue Analysis Permian Cycle 5 Salt G. Friemel #1 and Detten #1 Wells Palo Duro Basin, Texas," prepared for Battelle and the U.S. Department of Energy National Waste Terminal Storage Program, Columbus, OH.

Mirkovich, V. V., W. B. Durham, and H. C. Heard, 1983, Measurement of thermal diffusivity of rocks at high pressure, Proceedings of the 8th European Conference on Thermophysical Properties (Baden-Baden, FRG, 1982), 255-264.

Morgan, M. T., 1979, "Thermal Conductivity of Rock Salt from Louisiana Salt Domes," ORNL/TM-6809, Oak Ridge National Laboratory, Oak Ridge, TN.

Sutherland, H. J. and S. P. Cave, 1980, Argon gas permeability of New Mexico rock salt under hydrostatic compression, Int. J. Rock Mech. Min. Sci., 17:281-288.

Touloukian, Y. S., R. W. Powell, C. Y. Ho, and P. G. Klemens, 1970, "Thermal Conductivity of Non-Metallic Solids," Thermophysical Properties of Matter, Vol. 2, IFI/Plenum, New York, 942.

Touloukian, Y. S. and C. Y. Ho., eds., 1981, "Physical Properties of Rocks and Minerals," McGraw-Hill/CINDAS Data Series on Material Properties, Vol. II-2, McGraw-Hill, New York, Chpt. 12.

Walsh, J. B. and E. R. Decker, 1966, Effect of pressure and saturating fluid on the thermal conductivity of compact rock, J. Geophys. Res., 71:3053-3061.

Wawersik, W. and D. W. Hannum, 1980, Mechanical behavior of New Mexico rock salt in triaxial compression up to 200°C, J. Geophys. Res., 85:891-900.

Yang, J. M., 1981, Thermophysical properties, in: "Handbook of Rock Salt Properties Data," L. H. Gevantman, ed., National Bureau of Standards Monograph 167, U. S. Government Printing Office, Washington, D.C., Chpt. 4.

EFFECTIVE THERMAL DIFFUSIVITY OF POWDERED COAL

Clifford J. Cremers

Department of Mechanical Engineering
University of Kentucky
Lexington, KY 40506-0046

ABSTRACT

   The line heat-source technique was used for the direct determination of the effective thermal diffusivity for two samples of
powdered coal at two different particle size distributions. Measurements were made on Western Kentucky No. 9 and No. 11 coals at
three different densities corresponding to light, moderate and
dense packing. Particle size distributions were established by
using samples that would pass through 10 and 200 mesh Tyler screens.
The mean value of the thermal diffusivity for these was 1.29×10^{-7}
m^2/s. Values for the dense packing were about 10% higher than the
average and about 10% lower for the loose packing. Coal type was
not found to be a significant parameter but there was a slight
effect of particle size with the finer particles not showing as
strong a dependence on packing as did the coarser particles.

INTRODUCTION

   Coal is an increasingly important source of thermal energy for
the generation of electrical power and process steam. Also, as
stocks of natural gas and crude oil become depleted and their prices
rise, coal will become an important feed stock for the chemical industry, either for use directly or after gasification or liquifac-
tion to provide a cleaner hydrocarbon. In all of these applications
coal is frequently used in the pulverized state. Consequently it is
important for the proper design of coal handling and conversion
devices that the thermophysical properties be well established.

   Much is already known about the thermophysical properties of
solid coal. For instance, Tye, Desjarlais and Singer [1] recently

699

published a paper on the thermophysical properties of Pittsburgh seam coal over a range of temperatures between 50 and 850°C. Also, results have begun to appear recently for powdered coal. Vargha-Butler et al. [2] recently measured the specific heat of coal powder by differential scanning calorimetry. Further, Miura et al. [3] recently published results of measurements on the effective thermal conductivity of packed coal during carbonization. The focus here was on conditions similar to those in coking ovens. To the author's knowledge there have been no published results on the direct determination of the thermal diffusivity of powdered coal.

The diffusion of thermal energy in a particulate medium is a complicated process involving combined modes of heat transfer. For instance, it has been shown [4] that the radiative contribution to energy transport in a porous medium at room temperature and under vacuum is of the same order as the conductive transport. Further, in the same paper, it is shown that the thermal conductivity of a fine particulate medium increases by a factor on the order of 100 when the ambient pressure increases from 10^{-3} torr to one atmosphere. These measurements were made on ultrafine powdered rock samples, the lunar fines, but they do illustrate the complexity of the problem. As a consequence, one may define only effective values of the thermal transport properties. Also, because of the complex nature of the problem, the uncertainties in property measurement are large and so one is on weak ground in calculating any such property based on measurements of another property. This is frequently done for the thermal diffusivity using the thermal conductivity as the measured quantity. For these reasons it is desirable to make direct measurements.

The present paper presents some measurements on two substantially different coal samples for two screen sizes of pulverized coal and for three different densities each. Measurements are made using a line heat-source technique which is particularly well suited for use with pulverized materials.

MATHEMATICAL MODEL

A method was proposed by Jaeger [5] in 1959 for the direct determination of both the thermal conductivity and thermal diffusivity of a sample using the line heat-source technique. In that paper it was suggested that the method is particularly well suited for use with powdered materials although the author instead modified it for illustrative measurements on rock slabs.

The working equation for the line heat-source technique is:

$$\frac{\partial^2 T}{\partial r^2} = \frac{1}{\alpha} \frac{\partial T}{\partial t} \; ; \quad T(r,0) = 0 \; , \; q(0,t) = \begin{cases} 0, \; t \leq 0 \\ q, \; t > 0. \end{cases} \tag{1}$$

Here T is the temperature rise in a medium of infinite extent at a distance r from an infinitely long line heat-source of constant strength q switched on at zero time. Equation (1) has the solution [6]

$$T(r,t) = \frac{q}{4\pi\lambda} E_1 \left(\frac{r^2}{4\alpha t}\right)$$

where   (2)

$$E_1 \left(\frac{r^2}{4\alpha t}\right) = \int_{\frac{r^2}{4\alpha t}}^{\infty} \frac{e^{-z}}{z} dz.$$

E_1 is the first order exponential integral and is tabulated [7].

The thermal conductivity can be eliminated from the above by taking the ratio of temperature rises at a given point r for two different times t_2 and t_1. This gives

$$\frac{T(r,t_2)}{T(r,t_1)} = \frac{E_1 \left(\frac{r^2}{4\alpha t_2}\right)}{E_1 \left(\frac{r^2}{4\alpha t_1}\right)}$$ (3)

The behavior of the exponential integral E as a function of its argument is shown in Figure 1. Now let $t_1 = nt_0$, $t_2 = 2nt_0$, $x = \frac{r^2}{8n\alpha t_0}$ where t_0 is a convenient reference time. It follows that

$$\frac{T(r,2nt_0)}{T(r,nt_0)} = \frac{E_1(x)}{E_1(2x)}$$ (4)

Equation (4) is the working equation for Jaeger's technique. One records the temperature rise with time at position r after the heat source is switched on, chooses a convenient base time t_0 and forms the ratio of temperatures shown in equation (4) for a number of different multiples of t_0. Then, the argument x can be determined from a plot of the ratio $E_1(x)/E_1(2x)$ as shown in Figure 2. The thermal diffusivity is subsequently obtained from x using the measured separation distance r.

The model described above is commonly used for the direct determination of the thermal conductance of particulate media and so affords one the opportunity of using the same set of data for the determination of this property in addition to the thermal diffusivity. This gives a means of checking the consistency of the results as there are thermal conductivity data in the literature with which one can compare. To do this the exponential integral in equation (2) is

701

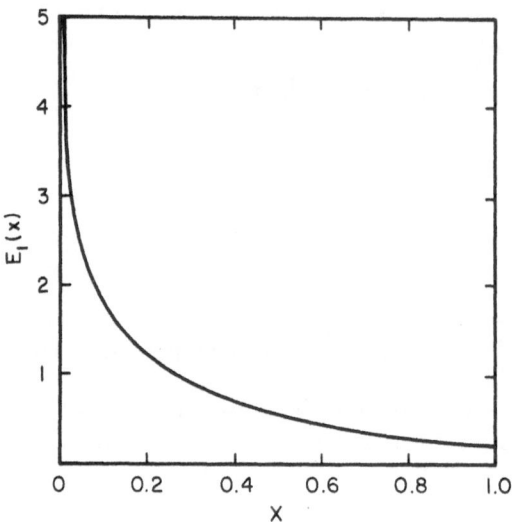

Figure 1.   First order exponential integral.

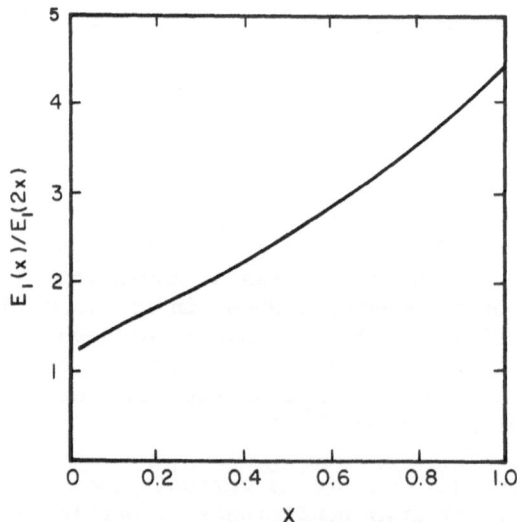

Figure 2.   Ratio of the exponential integral of argument x to that
of argument 2x.

approximated by a series expansion [7]. This is

$$E_1(x) = -\gamma - \ln(x) - \sum_{n=1}^{\infty} \frac{(-1)^n x^n}{nn!} \, , \qquad (5)$$

where γ is Euler's constant, so that the difference in temperatures at a given position in the medium at two different times t_2 and t_1, t_1 sufficiently large, is given by

$$T(r,t_2) - T(r,t_1) = \frac{q}{4\pi\lambda} \ln\left(\frac{t_2}{t_1}\right) . \qquad (6)$$

Equation (6) permits the determination of thermal conductance from the measured temperatures without reference to other properties.

EXPERIMENT AND PROCEDURE

A diagram of the experimental setup is shown in Figure 3. The line heat-source of infinite length assumed in the last section was modeled with a 0.23 mm dia. Nichrome V wire that was 16 cm long between voltage taps. This gave a length-to-diameter ratio of almost 700. This is well beyond the commonly accepted lower limit of 30 [8]. The infinite medium was approximated by a sample contained in a Teflon sample holder with dimensions of 16 x 4 x 4 cm. Prior experiments had shown that for the running times of the experiment, about 50 s, such a sample size would show no boundary effects.

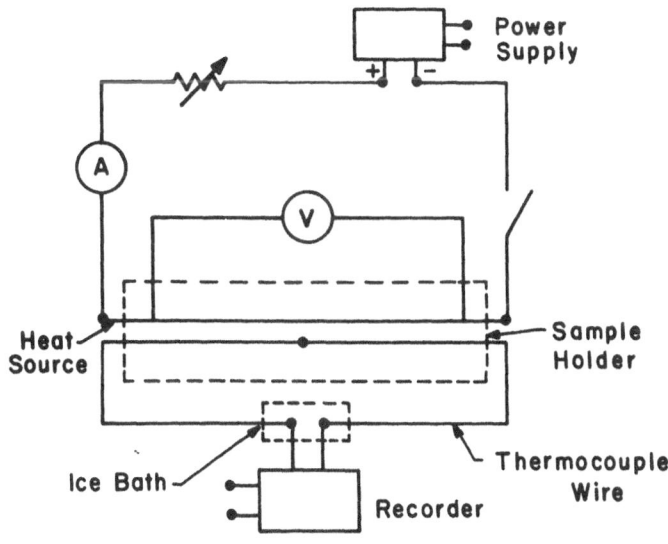

Figure 3. Diagram of experiment.

The Nichrome wire was heated by a 0.45 A current starting at time zero. The approximately 10 K temperature rise of the wire during the course of the experiment had a negligible effect on the resistance of the wire so that the heat source strength could indeed be considered constant. The temperature rise in the sample was determined with a butt-welded iron-constantan thermocouple aligned parallel to the heat source at a distance of 2.3 mm with the junction half-way between the ends of the sample holder. The other ends of the thermocouple wires were welded to copper leads and the junctions were immersed in oil-filled tubes in an ice bath. A Leeds and Northup Speedomax W recorder was used to determine the thermocouple output as a function of time. The voltage drop across the heater wire was determined with a Fluke 8000A Digital Multimeter.

Each test sample in this experiment was dried in an oven maintained at 100°C, which is well below the carbonization temperature of coal. Three different packing densities of each sample were used. These are termed loose packing, moderate packing and dense packing. The first was obtained by carefully spooning sample into the holder taking precautions to prevent the holder from being bumped or vibrated. The moderate packing was obtained by shaking the entire holder five times at both half and full packing during the sample loading process. The dense packing was made by tapping the outside of the holder with a small hammer until no further decrease in height of the sample in the holder could be noted. An additional run was made for each sample by applying a 3000 Pa loading on the surface of a sample with loose packing in an attempt to see if there was a significant effect caused by pressure effects on the contact conductance.

The sample holder was placed inside a box and surrounded on all sides with several inches of fiberglass insulation in an attempt to isolate the sample holder from environmental disturbances.

RESULTS

The chemistry of the two coal types investigated is given in the Appendix. Basically one is a medium sulfur and the other a high sulfur coal and they are of different age and reactivity. The effective thermal diffusivity and conductance are given in Table 1.

The thermal diffusivity obtained through equation (4) is essentially the same for the two samples, at least within the experimental uncertainty of the experiment which was determined to be ± 9% at the 95% confidence level. There seems to be a slight effect associated with density variation in that the thermal diffusivity increases in all cases from loose packing to dense packing. This is an indication that solid conduction effects are more important than radiative effects in the sample because, with densification, the voids between the larger particles tend to fill up with the smaller particles in the sample thereby increasing solid conduction paths and at the same time interfering with the mean free path of the radiation. The modest top loading of the sample with a pressure of 3000 Pa, which was done to simulate loading conditions that might occur in process equipment, seems to have little effect indicating a negligible change in contact conductance between the particles.

An uncertainty analysis is performed for the experiment indicated that the standard deviation of the mean was 4.5% with the major factor being the uncertainty in measuring the distance r.

As mentioned above, the effective thermal conductance was also determined from the same data sets as used for the thermal diffusivity but employing equation (6) as the working equation. These results are also shown in Table 1. Agreement with similar but limited results in the literature was good. Schack [9] gave a

Table 1. Effective Thermal Diffusivity of Powdered Coal

Sample	Thermal Diffusivity $x \ 10^{-6} m^2/s$		Thermal Conductivity (W/m-K)	Density (kg/m^3)
W.Ky.No. 9 - 10 mesh	(L)	0.11	0.10	690
	(M)	0.12	0.12	820
	(D)	0.16	0.13	880
	(LP)	0.11	0.11	
W.Ky.No. 11 - 10 mesh	(L)	0.12	0.11	680
	(M)	0.13	0.11	800
	(D)	0.15	0.14	850
	(LP)	0.11	0.11	
W.Ky.No. 9 - 200 mesh	(L)	0.13	0.10	550
	(M)	0.12	0.11	670
	(D)	0.14	0.12	800
	(LP)	0.11	0.10	
W.Ky.No. 11 - 200 mesh	(L)	0.12	0.10	540
	(M)	0.12	0.11	660
	(D)	0.13	0.12	790
	(LP)	0.10	0.10	

(L)-loose packing, (M)-moderate packing, (D)-dense packing,
(LP)-loose packing with 3×10^3 Pa top loading.

value of 0.11 W/m-K for coal characterized as pulverized and dry with a density of 730 kg/m^3. Miura et al. [3] reported a thermal conductivity of .08 W/m-K for Hunter Valley coal and 0.13 W/m-K Hongay coal. Both samples were at room temperature with a density of 850 kg/m^3.

Effects not accounted for in this study included particle separation by size that might occur during the process of densifying the samples and any changes the distance r that might occur during the loading and subsequent densification of the samples.

CONCLUSIONS

The thermal diffusivity of two samples of Western Kentucky coal showed only minor effects of particle size. The effect of densification caused by disturbing the sample produced a measurable increase in thermal diffusivity with density. There seems to be little difference in the thermal diffusivity of the two samples under similar conditions.

ACKNOWLEDGEMENT

This work was supported by the Institute for Mining and Minerals Research, Iron Works Pike, Lexington, KY 40583.

REFERENCES

1. Tye, R.P., Desjarlais, A.O. and Singer, J.M., "Thermophysical Properties of Pittsburgh Seam Coal," High Temp.-High Pres. 13, 57-68 (1981).
2. Vargha-Butler, E.I., Soulard, M.R., Hamza, H.A. and Neumann, A.W., "Determination of Specific Heats of Coal Powders by Differential Scanning Calorimetry," Fuel 61, 437-442 (1982).
3. Miura, T., Tajamia, H., Miura, K. and Ohtani, S., "Measurement of Effective Thermal Conductivity of Packed Coal during Carbonization," Heat Trans. Jap. Res. 10, 81-93 (1982).
4. Cremers, C.J., "Density, Pressure and Temperature Effects on Heat Transfer in Lunar Material," AIAA J. 9, 2180-2183 (1971).
5. Jaeger, J.C., "The Use of Complete Temperature-Time Curves for Determination of Thermal Conductivity with Particular Reference to Rocks," Aust. J. Phys. 12, 203-217 (1959).
6. Carslaw, H.S. and Jaeger, J.C., "Conduction of Heat in Solids," (2nd Ed.), Clarendon Press, Oxford (1959).
7. _____, "Handbook of Mathematical Functions," M. Abramowitz and I.A. Stegun editors, U.S. Department of Commerce, Washington, DC (1964).
8. Blackwell, J.H., "Radial-Axial Heat Flow in Regions Bounded Internally by Circular Cylinders," Can. J. Phys. 31, 472-483 (1953).
9. Schack, A., "Industrial Heat Transfer," John Wiley and Sons, New York (1965).

APPENDIX

PROPERTIES OF COAL SAMPLES

Ultimate Analysis

	Western Kentucky No. 9	Western Kentucky No. 11
Moisture	5.25 wt. %	4.83 wt. %
C	69.15	55.02
H	4.89	3.98
N	0.91	1.13
S	3.89	7.4
O	7.59	4.93
Ash	8.32	22.91
Heat Content	29082 kJ/kg	23377 kJ/kg

Proximate Analysis

	Western Kentucky No. 9	Western Kentucky No. 11
Volatiles	42.45 wt. %	34.41 wt. %
Fixed	43.79	38.05
FSI	4.5	3.5

Taken from:

O.J. Hahn, O.W. Stewart, J.C. Carter, S. Debrand, N.L. Mukherjee, and R. Wedding, "Low Btu Gas from Coal-Results From a Half-Ton per Day Moving Bed Gasification Unit of the University of Kentucky," Yearly progress report to the Institute for Mining and Minerals Research, April 1979.

SESSION Y

Poster Session

SESSION CHAIRMAN

T. Ashworth
SDSM&T
Rapid City, SD.

AN APPARATUS TO MEASURE THERMOPHYSICAL

PROPERTIES OF HEAVY CRUDES

Marcoe G. Ortiz, Jenny Montbrun, Maria I. Ramirez
and Ramon Ruilopez
Departamento De Procesos Y Sistemas
Universidad Simon Bolivar
Caracas 1080A, Venezuela

Ignacio Layrisse

Intevep S.A.
Los Teques, Venezuela

EXTENDED ABSTRACT

The main purpose of this work was to develop an apparatus to
measure thermophysical properties of a special group of "new" substances: heavy crudes from the Orinoco Belt in Venezuela. These
crudes are highly viscous and in many cases contain sand, heavy
metals and water. They have to be heated up to 80°C and more to
assist in their extraction and transportation from the well to a
location where they can be processed and utilized. Thus, it is
important to know the thermophysical properties of these substances
to design the transport systems that will handle them. Small
errors in the estimation of these properties could mean kilometers
of misused pipes.

The Hot Wire Transient Technique was chosen among many others
because of its simplicity and accuracy. The apparatus was designed
and built in accordance with its intended use for routine measurements in the laboratory.

The apparatus consists of a metallic cylindrical cell inside
a constant-temperature bath. The top cover and bottom of the cell
have electric connectors and pipe fittings to allow measuring and
constant cleaning of the cell. A frame holds the "wire" and other
important connections inside the cell. The thermal conductivity
of the sample can be measured at pressures up to 17 bars and at
temperatures up to 250°C.

Several crudes were examined in a range of temperatures between 180°C and 20°C. An important effect of volatile concentration was detected. Also the effect of moisture was measured for one of the samples. These effects are discussed in the light of existing theories.

A slight modification of the original circuit improved both the accuracy of the thermal conductivity measurement and also the ability to determine the thermal diffusivity of the sample. An effort is now being made to automate the measuring process with a microcomputer.

THE THERMAL DIFFUSIVITY/CONDUCTIVITY OF 316 STAINLESS STEEL

POWDERS IN AIR AND NITRIC ACID MIXTURES AT ROOM TEMPERATURE

K.E. Gilcrist and S.D. Preston

United Kingdom Atomic Energy Authority
Springfields Nuclear Power Development Laboratories
Salwick Preston PR4 0RR UK

SUMMARY

The thermal conductivity of AISI Type 316 stainless steel
powder, both dry and in nitric acid of 3M and 7M concentrations,
has been determined both experimentally and with the use of
theoretical models.

Using the heat pulse technique the thermal diffusivity was
measured for powders of two different particle sizes, 13 and
42 µm mean diameter. Thermal conductivity was derived from
diffusivity values using measured density data and specific heat
data from the literature. These data were compared with effective
thermal conductivity calculated from theoretical models. The
agreement was very good in all cases. Differences between the
values for the two particle sizes were small.

The thermal conductivity of solid type 316 stainless steel was
13; of type 316 powder in air, 0.2; of type 316 powder under water
3.2; of 316 powder under 3M nitric acid, 2.6; and of type 316
powder under 7M nitric acid, 2.0 W/m.°C.

INTRODUCTION

This report details the work carried out at the UKAEA
Springfields Preston England to measure the thermal diffusivity of
dry stainless steel powder and powder-fluid mixtures by the heat
pulse technique. Thermal conductivity can then be derived from
diffusivity by multiplication of the latter by volumetric specific
heat. The measured density was used to determine the void fraction
of the powders and mixtures.

These data can then be compared to theoretical models which, if successful, could be used as a means of estimating the thermal conductivity of stainless steel powders both dry and wet for a range of particle size, packing density and fluid concentration.

SAMPLES

The following samples of dry AISI Type 316 stainless steel powder were examined: (a) fine powder with a mean particle diameter of 13 μm and (b) a coarser powder with a mean particle diameter of 42 μm. The range of particle diameters were 8-20 μm and 20-65 μm respectively.

EXPERIMENTAL PROGRAMME

The aim of the experimental programme was to measure at room temperature the thermal diffusivity (α) of specimens having the following compositions:

(a) dry fine 316 powder in air
(b) dry coarse 316 powder in air
(c) fine 316 powder under water
(d) coarse 316 powder under water
(e) fine 316 powder under 3M nitric acid
(f) coarse 316 powder under 3M nitric acid
(g) fine 316 powder under 7M nitric acid
(h) coarse 316 powder under 7M nitric acid;

and, using the measured density (ρ), to derive the thermal conductivity (K) of specimens (a) through (h) from the equation, K = $\alpha \rho C_p$, where C_p is the total specific heat of the components.

THERMAL DIFFUSIVITY MEASUREMENTS

The well-characterised heat pulse, or laser flash, technique [1] was used to measure the room temperature thermal diffusivity of the samples listed above.

Basically the technique is to apply a heat pulse of approximately 1 msec duration to the front surface of a parallel-sided homogeneous sample and to monitor the temperature rise at the opposite face as a function of time. Over many years the method has been applied to solids, liquids and powders, and more recently to systems having two or three layers and to composites. Experimental data for the latter can be analysed only by using computer programs.

The samples were contained in a specially designed holder.

Initially the containers (or holders) were made of copper shim and later from sapphire optical flats 35 mm dia. and 2 mm thick sandwiched together with a pyrex spacer. This forms a vessel with internal dimensions of 25 mm diameter x 1-3 mm thickness. A top vent permits loading and unloading the powder or powder-liquid samples. Although the results obtained with the copper system were in good agreement with the data using the sapphire containers, the experiment and calculations were much more difficult to perform. Experimentally the sapphire containment vessel did not deform on loading and therefore a uniform thickness of sample could be assured. The copper containment was a three-layer system whereas the sapphire system is a 'one' layer type, i.e., the heat pulse from a laser can impinge directly on the front surface of the powder. The infra-red detector will 'see' the rear face of the powder through the sapphire disc — hence the reason for using sapphire, which has a transmission window in the far infra-red. Experimental error was normally about 1-2% for solid samples but with the powder/gas-liquid system it was about $\pm$ 5%.

TABLE 1

Experimental data: thermal conductivities of stainless steel powders in air and mixed with water or nitric acid

Material	Thickness mm	Thermal diffusivity cm^2/sec	Effective thermal conductivity W/m.°C
		$x\ 10^{-3}$	
316 ss (13 μm dia.)	3.0	1.12	0.19
316 ss (42 μm dia.)	3.0	1.48	0.23
13 μm powder + H_2O	1.0	3.09	3.21
42 μm powder + H_2O	1.0	3.17	3.26
13 μm powder + 3M HNO_3	1.0	3.17	2.63
42 μm powder + 3M HNO_3	1.0	3.17	2.60
13 μm powder + 7M HNO_3	1.0	3.00	2.02
42 μm powder + 7M HNO_3	1.0	2.95	1.96

TABLE 2

Standard room temperature data for component materials

Material	Source of the data*	Thermal diffusivitiy cm^2/sec	Thermal conductivity W/m.°C
H_2O	Ref. 11	1.46×10^{-3}	0.61
3M nitric acid	Ref. 12	1.33×10^{-3}	0.50
7M nitric acid	Ref. 12	1.06×10^{-3}	0.36
316 ss	*	3.5×10^{-2}	13.5
Air at 15 lb/in^2	Ref. 13	-	0.025

* The values quoted are taken from the literature, except those for 316 stainless steel where the thermal diffusivity quoted was measured at SNL on solid samples.

Conductivity ratios:

K_s/K_g = 540 for ss/air mixture

K_s/K_g = 27   for ss/3M nitric acid

K_s/K_g = 37.5 for ss/7M nitric acid

K_s/K_g = 22.1 for ss/H_2O

The powders were not vibropacked or mechanically pressed into the containers as this alters the packing fraction and density and in the latter case also alters the surface area of particle contact. In fact, in all cases the powders were hand tapped into the containers. When nitric acid or water was added to these powders, the addition was carried out drop by drop until a film appeared on the top of the powder, ensuring complete filling.

The data obtained experimentally are listed in Table 1, and the standard literature data used in the theoretical analysis in Table 2.

COMPARISON WITH MODELS

There are numerous powder-gas, powder-liquid models available in the literature with which the experimental data were compared. These models calculate the effective thermal conductivity of a mixture, K_e, from some or all of the following variables:

K_g - thermal conductivity of the gas/liquid phase

K_s - thermal conductivity of the solid/powder phase

ν - ratio K_s/K_g

ε - porosity of the continuous phase (% vol)

ϕ - volume fraction of one phase

K_c - thermal conductivity resulting from contact between particles

K_r - equivalent radiative heat transfer coefficient

For modelling purposes the following assumptions were made:

(a) the particles were perfectly spherical and of one size, i.e. 13 or 42 μm; spheres in a square arrangement would result in a porosity of 0.476. However, with loose hand-tapped compaction these same spheres would more likely settle into a cubic array yielding lower values of ε;

(b) no account was taken of isotherms in directions other than planes parallel to the heat flow;

(c) $K_s > K_g$;

(d) there was no convective contribution to heat flow;

(e) effects of radiation were negligible because of the low temperature of measurement;

(f) the gas/liquid was at atmospheric pressure and was the continuous medium.

Only one model took account of the contact conductivity between particles.

In order to derive thermal conductivity from the experimentally determined thermal diffusivity values, the density of the components in the solid form was required. The values of ε were derived from the weight and volume of the powders or mixtures in the cassette. For the hand-tapped compaction the values of ε for the 13 μm and 42 μm diameter particles were 0.44 and 0.40, respectively.

A comprehensive review of some of the better models available appeared in a paper by Kuzay.[2] This report compares the same seven models with the experimental data:

Imura-Takegoshi, Ref. 3

$$K_e/K_g = p + \frac{1 - p}{\phi + \frac{1 - \phi}{\nu}}$$

where $p = \dfrac{\varepsilon - \phi}{1 - \phi}$

and $\phi = 0.3 \, \varepsilon^{1.6} \, \nu^{-0.044}$

Swift, Ref. 4

$$K_e/K_g = 0.577 \, \pi \, \left[\left(\frac{1}{\nu} - 1\right)^{-1} - \left(\frac{1}{\nu} - 1\right)^{-2} \ln \frac{1}{\nu}\right] + 0.093$$

Deissler-Eian, Ref. 5,6

$$K_e/K_g = \frac{\pi}{2 \left(\frac{1}{\nu} - 1\right)^2} \left[\left(\frac{1}{\nu} - 1\right) - \ln \frac{1}{\nu}\right] + 1 - \frac{\pi}{4}$$

Kampf-Karsten, Ref. 7

$$K_e/K_g = 1 - \frac{\varepsilon' \left(\frac{1}{\nu} - 1\right)}{1 + \varepsilon'^{1/3} \left(\frac{1}{\nu} - 1\right)}$$

where $\varepsilon' = 1 - \varepsilon$

Flinta, Ref. 8

$$K_g/K_e = \frac{1}{\varepsilon'^{1/3}} \; \frac{1}{\frac{\sqrt{2}}{\varepsilon'^{1/3} \left(1 - 0.92 \, \varepsilon'^{1/3}\right)} + K_g/K_c} + \frac{1}{\nu}$$

Russell, Ref. 9

$$K_e/K_g = \frac{\varepsilon'^{2/3} + \frac{1}{\nu} \left(1 - \varepsilon'^{2/3}\right)}{\varepsilon'^{2/3} - \varepsilon' + \frac{1}{\nu} \left(1 - \varepsilon'^{2/3} + \varepsilon'\right)}$$

Eucken, Ref. 10

$$K_e/K_g = \frac{1 + 2 \varepsilon' (\nu - 1)/(2 + \nu)}{1 - \varepsilon' (\nu - 1)/(2 + \nu)}$$

Both Swift and Deissler-Eian use a similar solution for spherical particles in orthorhombic or cubical arrays where $\varepsilon = 0.40$. However it is possible to interpolate for other values of ε knowing the limiting end conditions.

DISCUSSION

Table 3 lists the effective thermal conductivity values for all the systems, powder-air, powder-water, powder-nitric acid (3M and 7M), obtained from experiment and compares them with the models.

The experimental data are in very good agreement with the data calculated from the models and show the same ordering, i.e. the thermal conductivities for the 316 stainless steel powder under H_2O, under 3M HNO_3, under 7M HNO_3, and in air are approximately 3.2, 2.6, 2.0 and 0.20 W/m.°C, respectively, from experiment, and 3.2, 2.7, 2.0 and 0.16 W/m.°C, respectively, for the models. As expected the larger particle gives the higher conductivity by theory, but experimentally this does not seem to be true for the nitric acid-powder system. However the differences in the latter between the thermal conductivities of the two particle types are only a few per cent (within experimental error on this type of sample).

The data from the models show that the highest values are obtained from the Imura-Takegoshi correlation and the lowest values from Kampf-Karsten, as found by Kuzay.[2] Also the predictions for the powder-liquid systems fit the data rather better than for the powder-gas system, although the latter fits are still good.

The best fit between the experimental and model data for the powder-gas system is from the theoretical analysis due to Swift [4] and Deissler-Eian [5,6] and that for the powder-liquid system is the one due to Flinta.[8]

As mentioned earlier, the porosity factors have been calculated from experimentally determined density values, which were all in a narrow band between 0.40 and 0.47. However any small variations in these values will not give large variations in the calculated values of K_e, as shown in Table 3.

TABLE 3

Values of effective thermal conductivity computed for mixtures
of stainless steel in air and stainless steel in liquids using
various models: comparison with experimental data

316 ss plus:	Air		H_2O		3M HNO_3		7M HNO_3	
K_s/K_g	540		22.1		27		37.5	
Mean particle diameter, μm	13	42	13	42	13	42	13	42
Thermal conductivity (W/m.°C) from:								
Experiment	0.19	0.23	3.2	3.3	2.6	2.6	2.0	2.0
Model, Ref. 3	0.25	0.30	3.5	4.0	3.1	3.5	2.5	2.8
Model, Ref. 4	0.23‡	0.25	2.7		2.3		1.9	
Model, Ref. 5,6	0.19‡	0.21	2.4		2.1		1.7	
Model, Ref. 7	0.11	0.12	2.1	2.4	1.8	2.0	1.3	1.5
Model, Ref. 8*	0.15	0.16	3.1	3.3	2.6	2.8	2.0	2.1
Model, Ref. 9	0.14	0.16	2.7	3.0	2.3	2.5	1.7	1.9
Model, Ref. 10	0.13	0.14	2.4	2.6	2.0	2.3	1.5	1.7

* Assuming $K_c = 0.40$ W/m.°C from Ref. 8

‡ value obtained by changing the particle diameter in the equation
to give $\varepsilon_{13\ \mu m} = 0.44$ from $\varepsilon_{42\ \mu m} = 0.40$.

CONCLUSIONS

The thermal conductivities of AISI 316 stainless steel powder-
gas or powder-liquid systems have been determined at room
temperature and these data have been compared with values
calculated from theoretical models.

Although two different particle sizes, 13 and 42 μm mean diameter, were tested, giving different densities, void fractions and hence volume fractions; the differences between the values of K_e obtained experimentally were less than 20% for the powder-gas and 3% for the powder-liquid systems.

The following experimental values of K_e were obtained:

(a) stainless steel powder in air 0.21 W/m.°C
(b) stainless steel powder under water 3.2 W/m°C
(c) stainless steel powder under 3M nitric acid 2.6 W/m.°C
(d) stainless steel powder under 7M nitric acid 2.0 W/m.°C.

These compare with various models from the literature and the values for the corresponding materials were:

(a) 0.11 to 0.30,
(b) 2.3 to 3.8,
(c) 1.9 to 3.3,
(d) 1.4 to 2.6 W/m°C.

The range of values from the models brackets the experimental data and certain models are in very good agreement with the latter.

It is not too surprising that the values of Imura-Takegoshi [3] are higher than the experimental values since their semi-analytical derivation was obtained for ε in the range 0.39 to 0.21. Also the Kampf-Karsten relationship [7] gives the lowest predicted values and is apparently more applicable to lower values of K_s/K_g than those in this study.

It is therefore suggested that the powder-gas system is best fitted by the models of Swift [4] and Deissler-Eian,[5,6] and the powder-liquid system by that of Flinta.[8] The corresponding values of K_e from these models for the mixture systems described above are

(a) 0.22,
(b) 3.2,
(c) 2.7,
(d) 2.0 W/m°C respectively.

In all these calculations, both experimental and theoretical, no attempt has been made to take account of the range of particle sizes or shapes other than the spherically shaped, uniformly packed particles described above.

REFERENCES

[1] Gilchrist, K. E. and Brocklehurst, J. E., "Thermophysical property measurements at RFL Springfields", Atom No. 240, 257-267, (Oct. 1976).

[2] Kuzay, T. M., "Effective thermal conductivity of porous solid gas mixtures", ASME Publication 80-WA/HT-63.

[3] Imura, S. and Takegoshi, E., "Effect of gas pressure on the effective thermal conductivity of packed beds", Heat Transfer - Japanese Research 3 (4), 13-26, (Oct. 1974).

[4] Swift, D. L., "Thermal conductivity of spherical metal powders including the effect of an oxide coating", Int. J. Heat Mass Transfer 9, 1061, (1966).

[5] Deissler, R. G. and Eian, C. S., "Investigation of effective thermal conductivities", NACA RM E52 C05, (1952).

[6] Eian, C. S. and Deissler, R. G., "Effective thermal conductivities of MgO, stainless steel and uranium oxide powders in various gases", NACA RM E53 G03, (1953).

[7] Kampf, H. and Karsten, G., "Effects of different types of void volumes on the radial temperature distribution of fuel pins", Nucl. Appl. Technol. 9, 288, (Sept. 1970).

[8] Flinta, J. E., "Thermal conductivity of uranium dioxide", Fuel Elements Conference, Paris, TID-7546, 516-525, (Mar. 1958).

[9] Russell, H. W., "Principles of heat flow in porous insulators", J. Am. Ceram. Soc. 18, 1, (1935).

[10] Eucken, A., "The thermal conductivity of ceramic refractory materials", VDI Forschung, 353 (1932).

[11] Kaye, G. W. C. and Laby, T. H., "Tables of Physical and Chemical Constants", 14th ed. (corrected), p.61, Longman, (1975).

[12] Kirk-Othmer., "Encyclopedia of Chemical Technology", 3rd ed., vol. 15, p.855, Wiley-Interscience, (1981).

[13] Kaye, G. W. C. and Laby, T. H., "Tables of Physical and Chemical Constants", 14th ed. (corrected), p.62, Longman, (1975).

THERMAL DIFFUSIVITY AND THERMAL CONDUCTIVITY

OF SOME SOLID METALS AT THE MELTING POINT

M. Lamvik* and S.O. Johansen**

*Inst. for teknisk varmelære,
NTH
N-7034 Trondheim, Norway

**Avd. for metallurgi
SINTEF
N-7034 Trondheim, Norway

ABSTRACT

This paper deals with experimental determination of thermal diffusivity and thermal conductivity of solid metals by studying the freezing process of the molten metal. Properties of pure zinc, lead and aluminium at the melting point are determined. Likewise the diffusivity and the conductivity of binary alloys of aluminium with manganese or magnesium together with some multicomponent industrial alloys of aluminium at their solidus temperature are determined. The findings are in good agreement with the results from measurements using different methods and with theoretical estimates.

INTRODUCTION

Data for thermophysical properties of materials are essential for calculation and design in thermal engineering. Statistics indicate that data for thermal conductivity, specific heat and thermal linear expansion are requested most frequently [1]. These properties are furthermore the determining properties for thermal diffusivity which is one topic of the present investigation.

Data for thermal properties are determined by measurements and/or by theoretical estimates. Which way to obtain data is a question of the rational use of available resources. Ideally, measurements and calculation should supplement each other in order to achieve the best consistency of the data.

In the past decades progress was made both in experimental techniques and in methods of estimation. However, new materials and extended application of known materials may still challenge our knowledge about their properties. Still the experimentalist may be facing the problem of either applying available techniques to the test specimen or developing a new technique or method that more properly will adapt the specimen and the given condition under which the measurements are to be made. Thus it will remain desirable to look for new methods or techniques for determining data.

Experimental determination of thermal diffusivity or thermal conductivity of solid materials near the melting point was discussed in an earlier investigation [2]. The present study applies the same method. The unidirectional solidification of a melt is further discussed and the method is applied to determination of the thermal diffusivity and thermal conductivity of solid specimens of pure zinc, lead, and aluminium together with some binary alloys of aluminium with manganese or magnesium and of some multicomponent industrial alloys of aluminium in the solid state at the melting point.

THE METHOD

The method described earlier [2] was based upon a study of unidirectional solidification of a melt of pure material. Under the assumption that the melt initially was kept at the melting point, the heat transfer at the interface during freezing was analyzed. An expression was derived for the thermal diffusivity of the solid phase

$$\alpha = \frac{\lambda}{\rho c} = \frac{i}{c} \left(\frac{\Delta x}{\Delta t}\right)^2 / \left(\frac{\Delta T}{\Delta t}\right)_s \tag{1}$$

Here

α = thermal diffusivity
λ = thermal conductivity
ρ = density
c = spec. heat of solid phase
i = heat of solidification

x = length coordinate
t = time
T = temperature
s = indicates solid phase

By simultaneously measuring temperature and time at given locations in the melt during the freezing process, the difference quotients of the expression can be evaluated, and thermal diffusivity and thermal conductivity can be evaluated when data for heat of solidification and for specific heat or density are known. The assumption for constant density during the freezing process caused the expression to give too high a value for the diffusivity, of the order of 1% for metals. This will, however, be well within the uncertainty of relevant measurements.

The expression 1) is valid when the solidification is unidirectional and the quantities time, distance, and temperature are measured with consistency, i.e. the temperature difference Δt is measured over the distance Δx over which the solidification front has propagated during the time increment Δt.

The quenching speed $\Delta T / \Delta t$, at a given location is related to the translational speed of the front by the expression above. The kinematic motion of the front is further given by the rate of thermal diffusion through the solid material and thereby of the temperature of the cooled surface of the body. The surface temperature will have a value somewhere between the solidification temperature and the temperature of the cooling fluid, which is assumed to be constant. It can be expressed by the dimensionless groups Biot number $Bi = hx/\lambda$, and the Fourier number $Fo = \alpha t/x^2$. The Biot number relates the external heat transfer through its coefficient h to the internal heat conduction through coefficient λ/x. For $Bi > 100$ it can be shown [3] that the heat flow is governed mainly by the internal heat transfer, that is by the Fourier number. This means that the surface is approaching the temperature of the fluid. The present study comprises experiments under similar conditions.

For a given Fourier number, the speed of the solidification front can be written as

$$\frac{\Delta x}{\Delta t} = \frac{\alpha}{2Fo \cdot x} \tag{2}$$

By using this expression in equation 1, the quenching speed is expressed by

$$\left|\frac{\Delta T}{\Delta t}\right| = \frac{i/c}{t} \cdot \frac{1}{4Fo} \tag{3}$$

This expression shows that the quenching speed at a location is proportional to a fictitious cooling rate i.e. the fictitious latent temperature rise, i/c, at the melting point, divided by the preceding time of freezing process. This relation was verified in the experiments.

EXPERIMENTS

The experiments reported here were primarily part of an investigation of aluminum alloys and of their microstructure by unidirectional solidification [4]. The measurements so made also fit,
however, present method for determining thermal diffusivity or
conductivity.

The solidification process was investigated by use of a vertical container with inner cross section 74 x 74 mm and with a height
of 200 mm. The side walls were from inside to outside composed of
layers of about 10 mm of Kaowool and 20 mm of asbestos. This design
ensured a relatively good insulation against heat leakage through
the walls. The bottom of the container was a copper plate, 1 mm
thick, fitting into slides in the walls for sudden removement
during the experiment. A nozzle, 12 mm in diameter, was directed
upward to the midpoint of the bottom surface through which tap
water could flow for quenching the copper plate or, directly, the
frozen end surface of the metal when the plate had been removed.

Six type K thermocouples of diameter 0.5 mm were mounted in two-
bore ceramic tubes that were kept horizontally through a side wall
of the container and with the elements located at the mid-point of
the corss section. Vertically the elements were normally located
at 40, 55, 80, 105, 130, 155 mm, respectively, from the bottom.
The electromotive force of the thermocouples was recorded with a
6-pen recorder.

The experiments were started by melting the base metal and
the alloying elements in a crucible. The melt was kept at the
desired temperature slightly above the melting point. The recorder
was started, the cooling water was turned on and the melt was poured
into the container. After ensuring that the melt next to the bottom
had frozen, the copper plate was removed to exclude any insulating
air gap between plate and metal.

The recorder traced the emf of the thermocouples as a function of
time. Zero time was defined as the moment when the melt was poured
into the container. It was recognized as an abrupt rise in the emfs.
A typical record is shown in Fig. 1, for a run with pure lead. Besides, lines are drawn in the diagram to illustrate equation 3).

For pure metal the solidification temperature is recognized
at the point where the curves change slope. The point indicates
that the thermoelement by then is just surrounded by solid metal.

For alloys the change in slope of the curves may not be as
distinct as for pure metals. There the curves bend with a certain
curvature indicating that freezing is taking place in the region
between the liquidus and the solidus temperature. The solidus and
liquidus for the different alloys were determined by differential
thermal analysis, using a Linseis DTA L-62 instrument [4]. The

726

resulting solidus temperature was marked on the curves of registered temperature versus time for the respective alloys. The propagation of the solidus mark was then studied to give measures for the diffusivity or conductivity. Thereby the assumption was incorporated that the heat of fusion had been released completely just at reaching the solidus temperature.

The metallographic investigations of the solid bodies that were produced in the experiments showed that the crystalline structure was clearly unidirectional in the region where the thermocouples were located, thus justifying the use of equation 1) for evaluation of thermal diffusivity or thermal conductivity. Only a region within 10 mm from the walls had structure that was influenced by the heat loss through the walls.

EVALUATION OF RESULTS

The evaluation of thermal diffusivity or conductivity from equation 1) needs data for the heat of solidification and for specific heat or density of the solid state at the solidus temperature.

For the pure metal the data were taken from references [5] and [6] for i or ρ, and from [7], for c. The data are given in Table I. The estimated uncertainty of the data entering the equation 1) may contribute an uncertainty of less than 3% to the evaluated thermal diffusivity or thermal conductivity.

For the investigated alloys the relevant data are not given; they are here estimated from the data of the alloying elements. The alloying elements constituted less than 7% by mass, beside aluminum. It would therefore be justified to assume, as an approximation, that the properties were nominally as for pure aluminum. The data are, however, estimated in the simple way of summing the property for each respective component according to its mass fraction. Any thermodynamic effect from the alloying process is therefore omitted. The estimate corrects the values for Al–Mn alloys upward (>2.5%), and the values for Al–Mg alloys downward (<2.2%) compared to the values for pure aluminum.

The temperature or time differences in eq. 1) are measured from the inclination of the curve at solidus temperature and their quotient is calculated.

Normally two values of the diffusivity or conductivity were computed for each run, namely from the relevant data taken at location 105 or 130 mm from bottom. The arithmetic mean of the values for the respective properties was then evaluated.

The uncertainty by which the temperature-time difference quotient was determined is estimated to be less than 4%. The uncertainty of the velocity term for the solidus was estimated to be less than 5%. However, it enters equation 1) with a double effect. The total uncertainty by which the values of thermal diffusivity or thermal conductivity are determined is then estimated to be of the order of ± 7.5%.

The resulting data is shown in Table II. For comparison, some recommended values from the literature [8] are also given in the table.

Thermal conductivity for the binary alloys are also plotted in a diagram, Fig. 2, as a function of the mass fraction of the alloying element. For comparison the recommended values are plotted for Al-Mg alloy. For the Al-Mn alloy similar data were not available.

DISCUSSION

Pure metals

The data in the literature for thermal diffusivity or for thermal conductivity are all estimated data or determined by extrapolation from experimental data at lower temperature up to the melting point [9].

By comparison with present results it is important to recognize the value for heat capacity that enters the equation 1) and likewise the crystalline structure of the specimen. The volumetric heat capacity equals the quotient of thermal conductivity to thermal diffusivity. From the data in Table II it can be derived that lead and aluminum are given higher heat capacity values by the recommended data compared to the values used by the experiments. For zinc the recommended data and experimental data are based upon the same value of heat capacity.

The recommended values for the transport properties are for metals of polycrystalline structure. That means that the crystalline grains have no predominant orientation. The present method involves conditions that the properties are coupled to the solid structure at the moment of formation. This means that the data are linked to the direction in which crystallization takes place, and before any recrystallization has taken place in the structure.

It is usually true that thermal conductivity is higher along direction parallel to than perpendicular to a grain. In the polycrystalline case the conductivity will be intermediate. The data from the present investigation would therefore be expected to be higher than the recommended values. Normally this is also the case, however, the differences may be partly due to the use of different data for heat capacity, and besides they are within the uncertainties by which the data are determined.

Alloys

The data in Table II show good agreement between experimental

728

TABLE I.   Physical Properties for Solid Materials
           at Melting Point, (5,6,7).

Element	T_s	h	c	ρ
	^{o}C	J/kg	J/kg ^{o}C	kg/m^3
Pb	321,3	24697	148,6	11013
Zn	419,4	102138	455,9	6835
Al	660,4	396784	1296	2560
Mn	1244	266648	712	5958
Mg	1380	368452	1340	1642

TABLE II.   Thermal Diffusivity and Thermal Conductivity
            at Solidus Temperature, T_s.

Metal	Run	Temp. T_s ^{o}C	Diffusivity $a \star 10^6$ m^2/s		Conductivity k w/m^{o}C	
			exp.	rec.	exp.	rec.
Pb	1	321.1	19.8	20.1	32.4	31.2
	2	"	23.4	20.1	38.3	31.2
Zn	1	419.6	38.5	32.2	120	100
	2	"	34.5	32.2	108	100
Al		660.4	69.9	68.o	230	211
Al-Mn						
.42%		660	68.0		225	
.64"		"	71.0		235	
.8 "		"	70.3		233	
1.12"		659	65.3		215	
1.5 "		"	53.7		178	
Al-Mg						
1.01%		645	58.5		204	195
3.05"		609	53.4		178	176
3.21"		606	54.9		183	173*
4.02"		593	50.9		169	165*
4.93"		580	46.9		156	156*
5.0 "		578	46.4		154	155

*interpolated

729

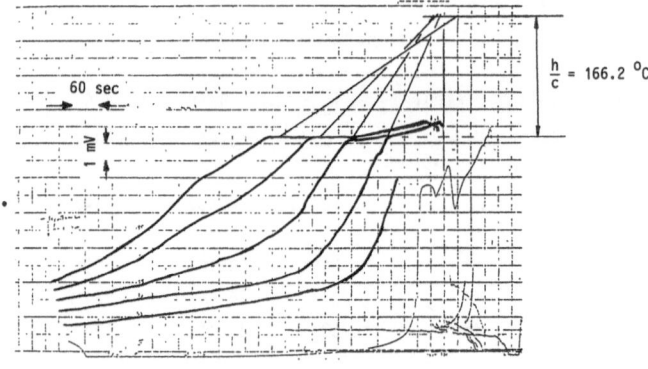

Fig. 1 Record of emf
Thermocouples by
Freezing Experi-
ment of Pure Lead.

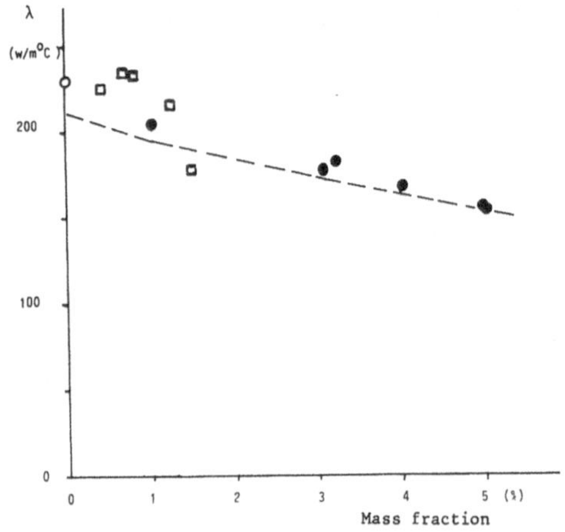

$\dfrac{h}{c}$ = 166.2 ^{o}C

60 sec

Fig. 2 Thermal Conductivity of
Binary Alloys at Solidus
Temperature.

○ Pure Aluminum, 99.996%
● Al-Mg
□ Al-Mn
--Recommended Values for
Al-Mg Alloy, [10].

Fig. 3 Thermal Conductivity of
Industrial Alloys at
Solidus Temperature, Plots
According to Table III.

-- Recommended Values for
Al-Mg Alloy, [10].

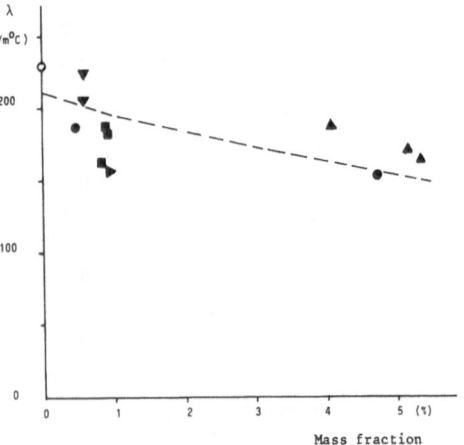

and recommended values for thermal conductivity of Al-Mg alloys. As for pure metals, no significant difference is found for polycrystalline structure and for the case ‖ to crystalline. The plots in Fig. 2 indicate a certain difference between experimental and recommended values at low Mg-content. The difference is, however, within the estimated uncertainties.

For Al-Mn alloys no reference data are available. The plots in Fig. 2 indicate a substantial decrease of the conductivity when the Mn mass fraction exceeds about 1%. For a fraction higher than about 1.5% the conductivity is lower than for an alloy with similar amount of magnesium. That manganese has a stronger influence than magnesium is consistent with the knowledge about electric conductivity for the alloy, and in view of the relation given by the Wiedemann-Franz law [11].

Fig. 3 illustrates the results from the experiments by some industrial multicomponent alloys, which analysis are given in Table III. The thermal conductivity is plotted along the ordinate and with the highest fraction of any element, next to aluminum, along the abscissa. Of special interest are the alloys which composition approximate a binary alloy. The alloy by the runs 75 and 76 approximate a binary Al-Fe alloy, the runs 77 resp. 90 and 91 have the resemblance of Al-Zn alloys and the run 84, simulates an Al-Mg alloy. From data about electric conductivity it is known that a binary alloying element in aluminum has an influence that increases in the successive order Zn, Fe and Mg. From the Wiedemann-Franz law the same relation would be suggested for thermal conductivity. The plots in Fig. 3 verify, to some extent, this suggestion. For the specimens without the predom-

TABLE III.   Analysis of Industrial Alloys.

Run	Sol. temp. °C	Cond. w/m°C	Plot Fig.3	Composition by mass percentage
65	643	181	■	98.55 Al + 0.88 Mn, 0.42 Fe, 0.1 Si
73	621	158	▶	97.57 Al + 0.99 Si, 0.73 Mg, 0.48 Mn
75	649	226	▼	99.26 Al + 0.57 Fe, 0.14 Si
76	649	209	▼	99.26 Al + 0.57 Fe, 0.14 Si
77	616	188	▲	94.00 Al + 4.08 Zn, 1.58 Mg, 0.18 Cu
84	576	155	●	94.63 Al + 4.75 Mg, 0.30 Fe, 0.12 Mn
89	643	186	■	98.55 Al + 0.87 Mn, 0.40 Fe, 0.16 Si
90	607	176	▲	93.29 Al + 5.15 Zn, 1.17 Mg, 0.18 Si
91	607	166	▲	93.17 Al + 5.27 Zn, 1.18 Mg, 0.18 Si
92	643	162	■	98.57 Al + 0.83 Mn, 0.40 Fe, 0.17 Si
94	623	189	●	98.80 Al + 0.49 Mg, 0.44 Si, 0.20 Fe

inance of any alloying element, the elements influence the conductivity in a more complex manner. As a first approximation it is reasonable that the alloying components have additive influence upon thermal conductivity. A further discussion is, however, assumed to be beyond the scope of the present investigation.

CONCLUDING REMARKS

The method applied in this investigation of thermal transport properties of metals has shown to be useful for getting data at the melting point, where other methods may present practical difficulties. The method is also unique in that it allows for the study of transport properties of a solid material while it is solidifying. By taking proper precautions in the experiments it should be possible to increase the accuracy of the data compared with the results from this investigation.

ACKNOWLEDGEMENT

Part of this investigation was funded by the Norwegian Council for Industrial and Scientific Research (NTNF), under the Project B.0520.6536.

REFERENCES

[1] Thermophysical and Electronics Newsletter, Perdue University, Vol. 8, No. 5 (1970).
[2] Lamvik, M., "Determination of Thermal Diffusivity of Solid Materials Near the Melting Point." Int. J. of Thermophysics, $\underline{3}$, 79-87 (1982).
[3] Grigull, U. and Sandner, H., "Wärmeleitung," Springer Verlag, p 82 (1979).
[4] Johansen, S.O., "Simulering av størkningsforløpet," STF34 A82027, SINTEF, Trondheim (1982).
[5] Landolt, H. and Börstein, R., "Zahlenwerte und Funktionen," 6. Aufl. IV Band, Technik, 2. Teil, Bandteil b, Springer Verlag, Berlin (1964).
[6] Van Horn, K.R., "Aluminum," Vol. 1, Am. Soc. Metals (1971)
[7] Touloukian, Y.S. and Buyco, E.H., "Thermophysical Properties of Matter," Vol. 4, IFI/Plenum, New York (1970).
[8] Touloukian, Y.S., Powell, R.W., Ho, C.Y. and Nicolaou, M.C., "Thermophysical Properties of Matter," Vol. 10, IFI/Plenum, New York (1973).
[9] Touloukian, Y.S., Powell, R.W., Ho, C.Y. and Klemens, P.G., "Thermophysical Properties of Matter," Vol. 2, IFI/Plenum, New York (1973).
[10] Ho, C.Y., Ackerman, M.W., Wu, K.Y., Oh, S.G. and Havill, T.N., "Thermal Conductivity of Ten Selected Binary Alloy Systems," J. Phys. Chem. Ref. Data, $\underline{7}$, p. 959 (1978).
[11] Eckert, E.R.G. and Drake Jr., R.M., "Analysis of Heat and Mass Transfer," McGraw-Hill, Kogkusha Ltd. (1972).

ERROR ANALYSIS AND EQUATIONS FOR THE THERMAL

CONDUCTIVITY OF COMPOSITES

J.D. Patterson

Department of Physics
South Dakota School of Mines and Technology
Rapid City, S. Dak. 57701

ABSTRACT

The overall effective thermal conductivity (K) of a composite is a function of the thermal conductivities (K_i) and volume fractions (ϕ_i) of each of the constituents. Probable errors (ΔK_i and $\Delta \phi_i$) for each of the constituents lead to a probable error (ΔK) in the overall thermal conductivity. For realistic problems, the exact function $K(K_i, \phi_i)$ is not known since it depends on the way the constituents are distributed. For this reason equations which describe the bounds on the thermal conductivity, for given K_i and ϕ_i, are very useful. Using standard calculational techniques, we plot K, K + ΔK, and K - ΔK versus volume fraction of one constituent for several cases of two-phase systems. These include the series and parallel cases as "absolute bounds" and the Maxwell equations which are the bounds for systems which are isotropic and homogeneous. We find, for common values of K_1/K_2, that the bounding equations overlap when errors are considered. This implies three things: (1) A measurement of thermal conductivity of a composite, when errors are taken into account, may not allow discrimination between cases. (2) For many cases of experimental interest a rough calculation is quite sufficient. (3) The bounds on the thermal conductivity are really not unconditional when errors are taken into account. We also consider two-phase composites where cylinders are embedded in a matrix. Complications in the calculation of the thermal conductivity include the effect of radiation, the effect of moisture and convection, and thermal resistance in the interface between phases.

733

INTRODUCTION

The term composite is sometimes restricted to mean a high-strength fiber embedded in a polymeric matrix.[1] Here the term composite is used in a general sense to include any heterogeneous system. This paper will be restricted to two-component composites but many of the ideas are easily generalized to apply to systems having more than two components. The term error analysis is used to describe the calculation of the uncertainty expected in the overall thermal conductivity given uncertainties in the thermal conductivities of the constituents and their volume fractions. Without such an analysis it is not very meaningful to claim that one equation is better than another in giving a description of the thermal conductivity of the composite. Thus error analysis relates to a central problem, namely how to estimate the thermal conductivity of a multicomponent system given the nature of the components and something about their distribution. This is a very old problem, going back to J.C. Maxwell and J.W. Rayleigh [2], and it has several equivalent statements (see appendix 2). When the thermal conductivity is not an intrinsic property of the material one might not want to define an effective thermal conductivity. This can occur in different ways. Recall, for example, the size effect in the thermal conductivity which may occur at very low temperatures.[2] It is also worth noting that Brown has rigorously shown that the distribution of the components as well as a knowledge of their volume fractions and individual dielectric constants is necessary to determine the overall dielectric constant.[2] The same remark is true for thermal conductivity.

There are several reasons why it is important to find reasonably accurate equations for the thermal conductivity of composites. One is that we often don't know exactly how the components are distributed even if we know their volume fractions. Thus the problem of determining the effective thermal conductivity is not necessarily one of straightforward, ever more detailed calculations. Equations which give bounds on or the approximate value for the thermal conductivity can be useful. This is especially so because it is not always easy or convenient to experimentally determine the thermal conductivity. For example, small laboratory samples of rocks might not be isotropic and homogeneous, but for the same type of rocks in situ it might be an acceptable approximation to so consider them. There are many materials for which calculation of the thermal conductivity is important [3]. Fiber insulations, bricks for furnace linings, building materials such as concrete, wood and fiber reinforced solids, and rocks and soils are all examples of heterogeneous materials for which we need to know the thermal conductivity. The underlying physics [4] of the heat conduction process is not considered

734

in this paper. The dominant heat carriers in dielectrics are
phonons and in metals are electrons. They are scattered by
other phonons, impurities, etc. Carriers and scattering will
need to be discussed when complications due to radiation and
interfacial boundary resistance are considered.

ERROR ANALYSIS

For small errors, we use the standard technique of estimating changes from derivatives. These can be combined in the
standard way. [5] Larger errors can be directly calculated
and again combined as for small errors. Both these methods
are discussed below and in Table 1.

Suppose the thermal conductivity of a two component composite is given by

$$K = K(K_1, K_2, \phi), \tag{1}$$

where K_1 and K_2 are the thermal conductivities of the two components and ϕ is the volume fraction of component 1 (so $1-\phi$
is the volume fraction of component 2). If ΔK, ΔK_1, ΔK_2 and
$\Delta\phi$ denote probable errors, then standard statistical results
for uncorrelated ΔK_1, ΔK_2, $\Delta\phi$ give, for small probable errors
[5],

$$\Delta K = \sqrt{\Delta K_1^2 \left(\frac{\partial K}{\partial K_1}\right)^2 + \Delta K_2^2 \left(\frac{\partial K}{\partial K_2}\right)^2 + \Delta\phi^2 \left(\frac{\partial K}{\partial \phi}\right)^2}. \tag{2}$$

Table 1 contains some examples of composite equations which may
be used with Eq. 2. When ΔK_1, ΔK_2 and $\Delta\phi$ are not relatively
small we write

$$\Delta K = \sqrt{A^2 + B^2 + C^2} \tag{3}$$

where

$$A = K(K_1 + \Delta K_1, K_2, \phi) - K(K_1, K_2, \phi) \tag{4a}$$

$$B = K(K_1, K_2 + \Delta K_2, \phi) - K(K_1, K_2, \phi) \tag{4b}$$

$$C = K(K_1, K_2, \phi + \Delta\phi) - K(K_1, K_2, \phi), \tag{4c}$$

The results given by Eqs. (3) and (4) should check the results
of Eq. (2) when Eq. (2) is valid. In Figs. 1, 2 and 3 we have
plotted $K \pm \Delta K$ vs ϕ for fixed K_2/K_1. For these figures, both
methods of obtaining ΔK give essentially the same results.

TABLE I: THERMAL CONDUCTIVITY FOR TWO-PHASE COMPOSITES.

$$\phi = \phi_1, \; 1 - \phi = \phi_2, \qquad K = \text{Thermal Conductivity}$$

$$\phi = \text{Volume Fraction}$$

Equation (Two phase)*

A. $K = K_1\phi + K_2(1-\phi)$

B. $\dfrac{1}{K} = \dfrac{\phi}{K_1} + \dfrac{1 - \phi}{K_2}$

C. $K = K_1{}^{\phi} K_2{}^{1-\phi}$

D. $K = \dfrac{K_2(1-\phi)+K_1\phi_1\left(\dfrac{3}{2 + K_1/K_2}\right)}{1 - \phi + \phi\left(\dfrac{3}{2 + K_1/K_2}\right)}$

E. $K = \dfrac{K_1\phi + K_2(1-\phi)\left(\dfrac{3}{2 + K_2/K_1}\right)}{\phi + (1 - \phi)\left(\dfrac{3}{2 + K_2/K_1}\right)}$

F. $K = K_2\,\dfrac{(K_1+K_2) + (K_1-K_2)\phi}{(K_1+K_2) - (K_1-K_2)\phi}$

* see references
  2 and 8

A = Parallel, B = Series, C = Lichtenecker, D = Maxwell (dilute
spheres K_1, K_2 continuous phase), E = Maxwell (dilute sphere
K_2, K_1 continuous phase), F = cylinders with gradient perpendicular
to axis (dilute K_1, K_2 continuous phase).

From A - F, the error ΔK can be calculated using Eq. (2) or
Eq. (3).

RESULTS AND CONCLUSIONS

Our results are presented in Figs. 1-3 and in Table II. Each figure presents the range in uncertainty in the overall thermal conductivity for two levels of uncertainty in the constituent's thermal conductivity and volume fractions. In Fig. 1 we plot the series and parallel cases. These form "absolute" bounds on the thermal conductivity.[2] Note for the 5% error case (Fig. 1a) the bounds almost overlap for all ϕ. At 10% (not shown) there iv very broad overlapping for all ϕ. For cases such as this it is very difficult to learn anything about the internal geometry from a measurement of the thermal conductivity. Alternatively it should be very easy to obtain an experimentally reasonable fit to experimental data using equations derived from almost any assumption about internal geometry. In Fig. 2 we plot the Maxwell expressions which can be reinterpreted as upper and lower (Hashin and Shtrikman) bounds for materials which are macroscopically isotropic and homogeneous. [2] Since these bounds are tighter than the absolute bounds we get significant overlapping for constituent uncertainties at the 1 to 2% level. Figure 3 just compares Lichtenecker's empirical equation with a simple equation derived for a dilute concentration of cylinders with temperature gradient perpendicular to the cylinders. Even though the equation is only valid for small ϕ we have plotted it for all ϕ. For the two cases shown, data which fell within the range for cylinders would almost equally fall within the range for Lichtenecker's equation. Table II makes a similar point. We have compared in Table II, for three levels of accuracy at $\phi = .5$, the spread in the uncertainty of the bounds relative to the difference between the upper and lower bounds ΔK_B. Whenever $\Delta K_B < \Delta K$ (upper bound) + ΔK (lower bound)or R (see table) >1, any model which fits the assumptions used to derive the bounds (such as macroscopically isotropic and homogeneous) should be good enough to fit experimental data. Three conclusions can be drawn from this work: (1) A measurement of the thermal conductivity of a composite may not discriminate between different geometries. (2) For many cases of experimental interest a rough calculation, using any reasonable model, may be about as accurate as the input data. (3) When uncertainties in input data are taken into account , bounding equations for thermal conductivity also contain considerable uncertainty.

There are additional complications in attempting to describe the thermal conductivity of a composite by the use of a simple equation. For a porous material the thermal conductivity may depend on the thermal conductivity of the material, the thermal conductivity of the gas in the pores and on radiation and convection. In addition, if there is moisture present in the pores, heat can be transferred by the latent heat of evaporation.[3] We can

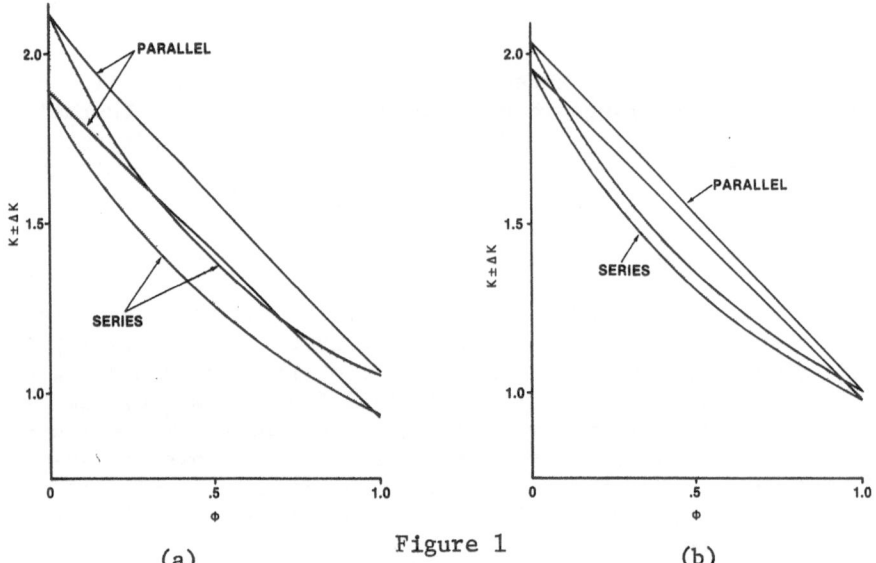

Figure 1

(a)   (b)

Thermal conductivity (K $\pm$ ΔK) vs. volume fraction (ϕ). Units
arbitrary. a. $K_1 = 1$, $K_2 = 2$, $\Delta K_1/K_1 = K_2/K_2 = \Delta\phi = .05$,
b. $K_1 = 1$, $K_2 = 2$, $\Delta K_1/K_1 = .01$, $\Delta K_2/K_2 = .02$, $\Delta\phi = .015$

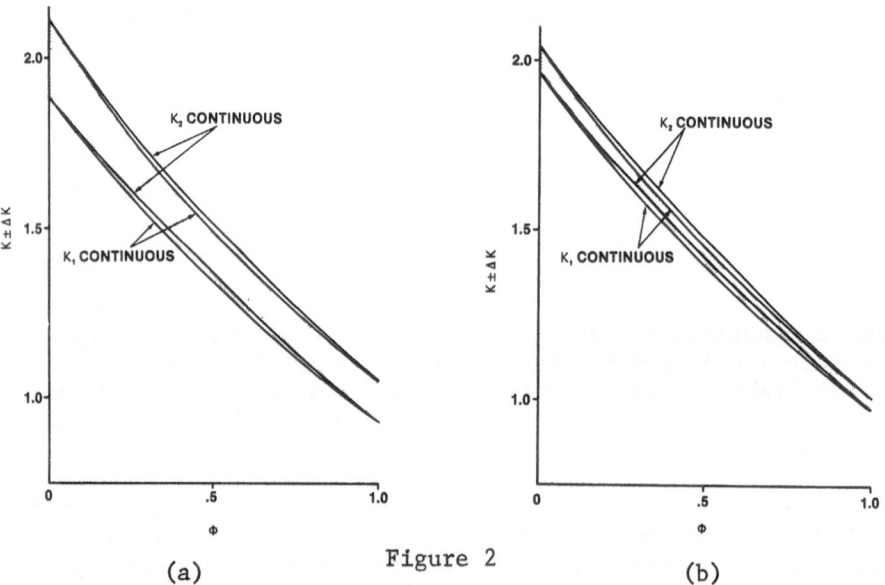

Figure 2

(a)   (b)

Thermal conductivity (K $\pm$ ΔK) vs. volume fraction (ϕ) for Maxwell
expressions. Units arbitrary.
a. and b. have same parameter values as Fig. 1.

738

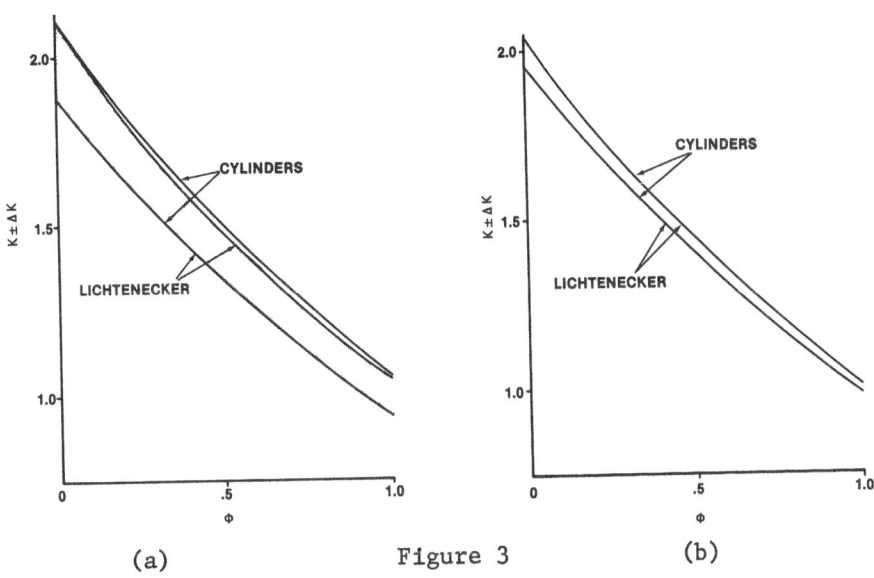

Thermal conductivity $(K \pm K)$ vs. volume fraction (ϕ) for
cylinders and Lichteneckers relation. Units arbitrary.
a. $K_1 = 1$, $K_2 = 2$, $\Delta K_1/K_1 = \Delta K_2/K_2 = \Delta\phi = .05$,
b. $K_1 = K_2 = 2$, $\Delta K_1/K_1 = .01$, $\Delta K_2/K_2 = .02$, $\Delta\phi = .015$.

Table II.   Comparison of Uncertainty in Thermal Conductivity
With Difference Between Upper and Lower Bounds

Case	I R	II R	III R
"Absolute" bounds from parallel and series cases (see Table 1).	1.66	.83	.26
Bounds for macroscopically isotropic and homogeneous materials (the Maxwell expressions, see Table 1).	10.77	5.50	1.81

$R = [\Delta K \text{ (upper bound)} + \Delta K \text{(lower bound)}]/\Delta K_B$
Volume fraction of each component equals .5, $K_2/K_1 = 2$.
ΔK = probable error in K due to the probable errors in K_1, K_2,
and ϕ (volume fraction) as listed below
ΔK_B = difference between upper and lower bounds
I. $\Delta K_1/K_1 = .1$ $\Delta K_2/K_2 = .1$ $\Delta\phi = .1$

II. $\Delta K_1/K_1 = .05$ $\Delta K_2/K_2 = .05$ $\Delta\phi = .05$

III. $\Delta K_1/K_1 = .01$ $\Delta K_2/K_2 = .02$ $\Delta\phi = .015$

expect the effects of convection to be unimportant provided the pores are sufficiently small.[2] Even for dry non-porous materials, radiative effects can be important at higher temperatures provided the material is reasonably transparent to the radiation.[2] At lower temperatures there is another complication. Acoustic mismatch can limit the flow of phonons across component boundaries and [2,6] hence give rise to an interfacial boundary resistance. One promising area for future work is the problem of including some information on how the components are arranged so as to improve the tightness of the bounds. It is possible to use information about the dielectric constant of a composite to improve the bounds on the thermal conductivity of the same composite. [7]

Appendix 1: DEFINITION OF THERMAL CONDUCTIVITY

Let T be the temperature, $K(\vec{r}) = K$ be the thermal conductivity at position $\vec{r}$, K_e be the effective overall thermal conductivity and $\vec{h}$ be the heat flux (joules/sec m^2). We consider a cylinder of length L and area A oriented along the x axis with fixed temperatures on top and bottom and with no heat flow out the sides. If we want our definition of K_e to be independent of the size, shape and orientation of the cylinder (with respect to the material from which it is out) we need the material to be macroscopically isotropic and homogeneous and we need the mean free path of the heat carriers to be much smaller than sample dimensions. The following equations define the situation.

$$\text{Let} \qquad \vec{g} = - \vec{\nabla}T \qquad \qquad (A.1)$$

$$\text{and} \qquad \vec{h} = K\vec{g} \quad \text{(Fourier's Law)} \qquad (A.2)$$

then by the steady state equation of continuity

$$\vec{\nabla} \cdot \vec{h} = 0 \qquad \qquad (A.3)$$

$$\text{Thus} \qquad \nabla \cdot (K\vec{\nabla}T) = 0 \qquad \qquad (A.4)$$

These equations are to be solved for $T(\vec{r})$ subject to the boundary conditions

$$T = T_o \quad \text{at } x = 0 \qquad \qquad (A.5)$$

$$T = T_L = T_o - g_o L \text{ at } x = L, \text{ where } g_o = \text{constant} \quad (A.6)$$

$$\text{and} \qquad \vec{g} \cdot \hat{n} = \vec{h} \cdot \hat{n} = 0 \quad \text{(on the sides)} \qquad (A.7)$$

for $\hat{n}$ being a unit vector normal to the sides of the cylinder. Eq. (A.7) implies there is no heat flow out the sides of the

740

Cylinder. For later purposes, it is convenient to define the vector

$$\vec{g} = g_o \hat{x},\qquad\text{(A.8)}$$

where g_o is defined by (A.6) and $\hat{x}$ is a unit vector in the x direction. With T determined from the above equations, the effective thermal conductivity can be defined by [7]

$$K_e = \frac{1}{Vg_o^2} \int Kg^2 dV,\qquad\text{(A.9)}$$

integrating over the volume V of the cylinder. Making some simple mathematical transformations, K_e can be shown to take the equivalent form

$$K_e = \frac{\int \vec{h} \cdot \vec{g}_o\, dV}{\int \vec{g} \cdot \vec{g}_o\, dV}.\qquad\text{(A.10)}$$

Both Eqs. (A.9) and (A.10) have been useful for theoretical analysis. Eq. (A.10) can be put in the familiar form,

$$\dot{Q} = -\left(K_e A\right)\frac{T_L - T_o}{L}\qquad\text{(A.11)}$$

where $\dot{Q}$ is the rate of heat flow along the axis of the cylinder

Appendix 2:   EQUIVALENT PROBLEMS

It is easy to see that the problem of determining the effective dielectric constant of a composite is the same as the problem of determining the effective thermal conductivity. The dielectric constant ε, the electric displacement vector $\vec{D}$ and the electric field $\vec{E}$ are related by $\vec{D} = \varepsilon\vec{E}$. In a region in which there is no free charge $\vec{\nabla} \cdot \vec{D} = 0$. The electric potential ϕ is defined by $\vec{E} = - \vec{\nabla}\phi$. Combining these three equations, we have $\vec{\nabla} \cdot (\varepsilon\vec{\nabla}\phi) = 0$. Let the composite be in the form of a right cylinder of length L with axis along the x axis. Let the boundary conditions be $\phi = \phi_o$ at x = 0 (bottom), $\phi = \phi_L$ at x = L (top) and $\partial\phi/\partial n = 0$ on the cylindrical surface where $\partial\phi/\partial n$ represents the outward normal derivative. The problem of solving $\vec{\nabla} \cdot (\varepsilon\vec{\nabla}\phi) = 0$ subject to these boundary conditions is exactly the same as the problem of finding the temperature distribution for determining the effective thermal conductivity of the composite; see Appendix 1. Just as the temperature distribution determined the effective thermal conductivity, so the potential distribution determine the effective dielectric constant. [7] Similar results hold for the magnetic case and for the electrical conductivity case. For the magnetic case $\vec{B} = \mu\vec{H}$ where $\vec{B}$ is the magnetic flux density, μ is the magnetic permeability and $\vec{H}$ is the magnetic field. In regions where there are no currents or time varying magnetic fields we can define a magnetostatic potential ϕ_m so that $\vec{H} = - \vec{\nabla}\phi_m$. Since $\vec{\nabla} \cdot \vec{B} = 0$, we then have $\vec{\nabla} \cdot (\mu\vec{\nabla}\phi_m) = 0$. For the steady state electrical conductivity case $J = \sigma\vec{E}$ where J is the electrical current density, σ is the electrical conductivity and $\vec{E}$ is the electric field. In regions in which there are no time varying magnetic fields $\vec{\nabla} \times \vec{E} = 0$ so $\vec{E} = - \vec{\nabla}\phi$ where ϕ is the electric potential. Then $\nabla \cdot \vec{J} = 0$ implies $\vec{\nabla} \cdot (\sigma\vec{\nabla}\phi) = 0$.

REFERENCES

1. DeYoung, H.G. "Plastic Composites Fight for Status", High Technology 3, No. 10, 63 (1983)

2. Hale, D.K. "The Physical Properties of Composite Materials," J. Materials Science 11, 2103 (1976)

3. Parrott, J.E. and Stuckes, A.D., "Thermal Conductivity of Solids," PION Limited, London (1975), Chapter 6

4. Slack, G.A. "The Thermal Conductivity of Nonmetallic Crystals," Solid State Physics 34, 1 (1979)

5. Taylor, J.R. "An Introduction to Error Analysis," University Science Books, Mill Valley, CA (1982. Ch. 3 and Ch. 5)

6. Chen, F.C., Choy, C.L. and Young, K., "A Theory of the Thermal Conductivity of Composite Materials," J. Phys. D. 10, 571 (1976)

7. Bergman, D.J., "The Dielectric Constant of a Composite Material - A Problem in Classical Physics," Phys. Reports 43, 377 (1978). Also Appendix 2 of this paper.

8. Mitoff, S.P. "Properties Calculations for Heterogeneous Systems," Advances in Materials Research 3, 305 (1960)

ACKNOWLEDGMENTS

Work on the thermal conductivity of rocks for RE/SPEC Inc. of Rapid City provided the original stimulus for proceeding to this more general problem. The author wishes to thank Drs. R.D. Redin, D.R. Smith, J.A. Weyland and T. Ashworth for their helpful suggestions and discussions.

TRANSIENT LINE-SOURCE APPARATUS FOR THERMAL CONDUCTIVITY

DETERMINATION OF FLUIDS AT 295-600K AND 1-700 BARS

R. C. Prasad

University of New Brunswick, Saint John, N. B. Canada

J. E. S. Venart

University of New Brunswick, Fredericton, N. B. Canada

and

E. F. Buyukbicer

University of New Brunswick, Fredericton, N. B. Canada

ABSTRACT

A transient hot-wire apparatus was developed for rapid and accurate determination of the thermal conductivity of fluids in the range of 295-600 K and up to 700 bars.

A 12.7 micron diameter platinum wire with spot-welded potential leads was used for measurements in the 295-600 K range. A computerised data acquisition system using an on-line microprocessor (LSI-11) and analog data-multiplexer was developed for fast and accurate recording of measurements and immediate data processing to obtain the thermal conductivity. Measurements obtained with this apparatus have been reported for many fluids. The reproducibility and accuracy of measurements are estimated to be within 0.5 and 2 percent. respectively.

THERMAL CONDUCTIVITY AND CONTACT CONDUCTANCE OF EXPLOSAFE

J. E. S. Venart

University of New Brunswick, Fredericton, N. B. Canada

R. C. Prasad

University of New Brunswick, Saint John, N. B. Canada

and

M. P. Callaghan

University of New Brunswick, Fredericton, N. B. Canada

ABSTRACT

The thermal conductivity of EXPLOSAFE*, an expanded aluminum foil used to augment the heat transfer in a pressure vessel, was measured over a temperature range of 5°-70° C. Measurements were obtained with different styles, orientations and thicknesses of EXPLOSAFE. In addition, the thermal contact conductance for an EXPLOSAFE matrix and wall interface was determined for one particular style as a function of pressure and a simple correlation developed.

A guarded hot-plate apparatus with an on-line Microcomputer (Hewlett Packard-85) and the Hewlett Packard-3497 Data Acquisition System was used.

INTRODUCTION

"EXPLOSAFE" is an expanded aluminum foil which is used to

---

*Vulcan Industrial Packaging Limited, Rexdale, Ont., Canada.

augment the heat transfer in a pressure vessel and minimize or eliminate the possibility of explosion [1,2]. Programs are underway [3] for evaluating the effectiveness of using EXPLOSAFE for the interior lining of railway tank cars and truck transport carrying liquid propane gas (LPG), as a protection against BLEVE (Boiling Liquid Expanding Vapor Explosion) under accident and/or fire conditions. The thermal conductivity of EXPLOSAFE and the conductance at the wall-interface are important parameters in such applications.

EXPERIMENTAL

A guarded hot plate apparatus (GHPA) was used to measure the thermal conductivity and contact conductance. The apparatus (Figure 1) consisted of a hot plate and two cold plates on either side of the hot plate. Two identical specimens of EXPLOSAFE were clamped between the hot and cold plates. A guard heater surrounded the hot plate along the edge to prevent heat loss through the edges. Cold fluid at a controlled temperature was circulated through the cold plates. Four spring scales were affixed to the apparatus to measure the interfacial contact pressure between the hot/cold plates and the EXPLOSAFE sample. The temperatures of the guard heater, the main heater and the cold plates were suitably adjusted to obtain a desired temperature difference (ΔT) or heat flux, Q, in the specimens. The steady state temperature difference across the EXPLOSAFE specimen and the heat flux were measured to determine the apparent thermal conductivity, λ_a, including the effect of contact conductance. An estimate of the true bulk thermal conductivity, λ_t, of the specimen was made (Equation 1) [4] after application of a thermal contact grease (Wakefield Thermal Compound 120-8) to the contacting surfaces. The thermal conductivity was assumed to be infinite in this case due to near perfect contact established between the specimens and the hot/cold plates.

$$Q = \lambda_t \ A \ [(\Delta T_L/\Delta X_L) + (\Delta T_R/\Delta X_R)] \tag{1}$$

Using these results, the contact conductance, h_c , was calculated (Equation 2) from the measurement in tests where thermal contact grease was not applied.

$$\frac{\Delta T_L}{(\Delta X_L/\lambda_a)} + \frac{\Delta T_R}{(\Delta X_R/\lambda_a)} = \frac{Q}{A} = \frac{\Delta T_L}{(2/h_c) + (\Delta X_L/\lambda_t)}$$

$$+ \frac{\Delta T_R}{(2/h_c) + (\Delta X_R/\lambda_t)} \tag{2}$$

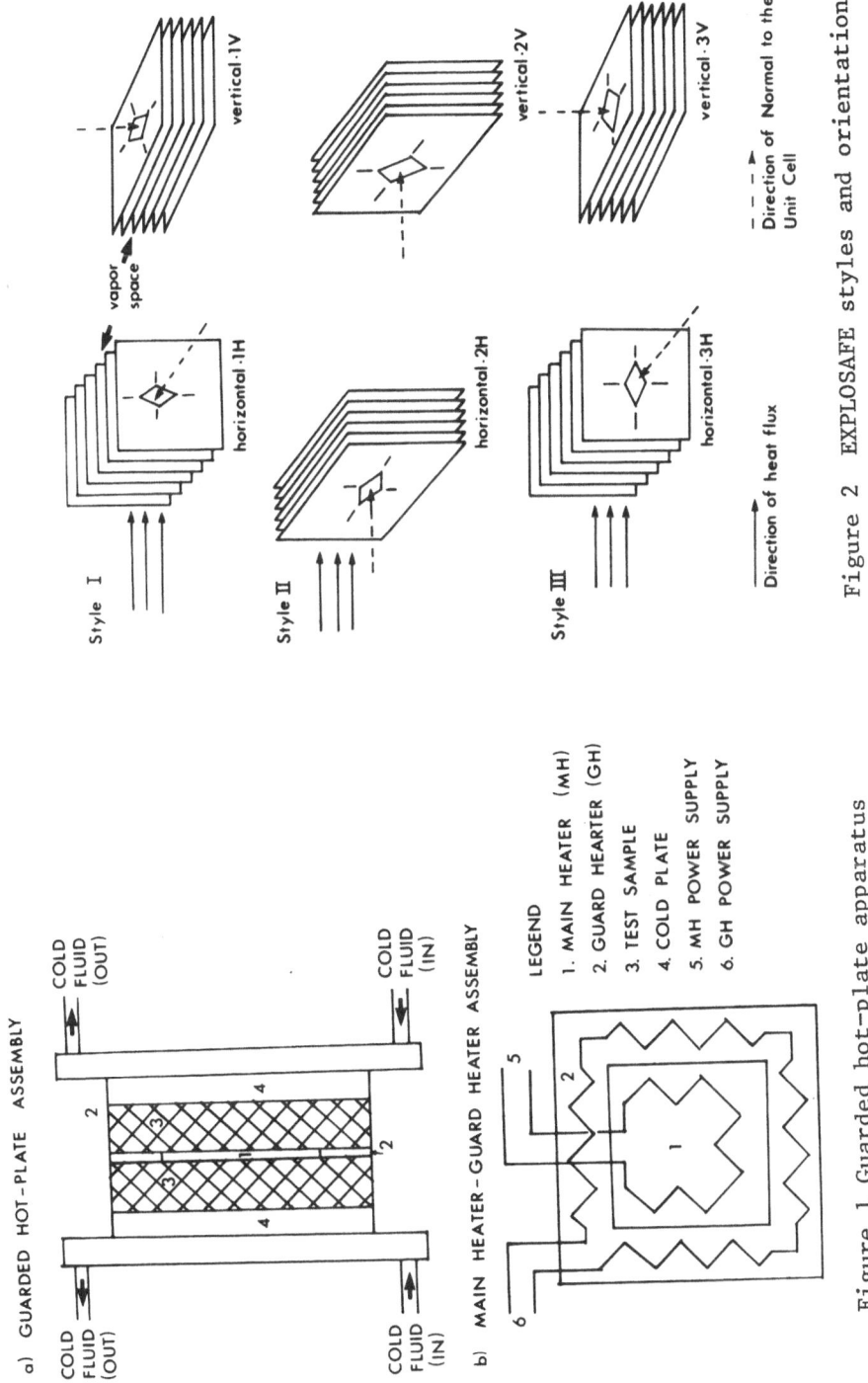

a) GUARDED HOT-PLATE ASSEMBLY

COLD FLUID (OUT)

COLD FLUID (IN)

COLD FLUID (OUT)

COLD FLUID (IN)

b) MAIN HEATER-GUARD HEATER ASSEMBLY

LEGEND
1. MAIN HEATER (MH)
2. GUARD HEARTER (GH)
3. TEST SAMPLE
4. COLD PLATE
5. MH POWER SUPPLY
6. GH POWER SUPPLY

Figure 1 Guarded hot-plate apparatus

Style I

Style II

Style III

horizontal·1H

horizontal·2H

horizontal·3H

vertical·1V

vertical·2V

vertical·3V

vapor space

Direction of heat flux

Direction of Normal to the Unit Cell

Figure 2 EXPLOSAFE styles and orientations

747

All data were acquired and processed using a Hewlett-Packard 3497 Data Acquisition System controlled by a Hewlett-Packard 85 Microcomputer.

TEST SAMPLES

EXPLOSAFE samples were supplied by the Vulcan Industrial Packaging Limited, Canada. The various styles (styles I, II, III) and orientations (V and H) in which they were tested, are shown in Figure 2. For styles I and II, V and H indicate the direction of the normal to the plane of the unit cell of the EXPLOSAFE mesh. For style III, V and H refer to the orientation of the major axis of the unit cell. Samples having thicknesses of 25 mm and 50 mm were tested.

THEORETICAL PREDICTIONS

A theoretical model was developed to predict the thermal conductivity of EXPLOSAFE for styles I and III. It was assumed that heat was transferred across the EXPLOSAFE thickness only by conduction through the aluminum foil. A theoretical value of the thermal conductivity, λ_{th}, was obtained from

$$\lambda_{th} = \lambda_{al} \frac{\Delta X}{L_p} \cdot \frac{A_f}{A_e} \tag{3a}$$

where $A_e = A_{vap} + A_1$ (3b)

The path length, L_p, and cross-sectional area, A_f, of the heat conduction path through the aluminum foil as well as the effective cross-sectional area, A_e, of EXPLOSAFE per conduction path as shown in Figure 3 were obtained from the test samples by direct measurement.

RESULT AND DISCUSSION

The apparent thermal conductivity, λ_a, of all the specimens was measured at mean temperatures of 5°-70°C (Figure 4).

The conductivity of styles II and III were lowest and highest, respectively. Also, the specimen of a given style and orientation consistently yielded an effective conductivity value lower for the 25 mm thickness than that for the 50 mm thickness of the same style and orientation. This is attributed to the effect of thermal contact resistance between test specimens and hot/cold plates. As

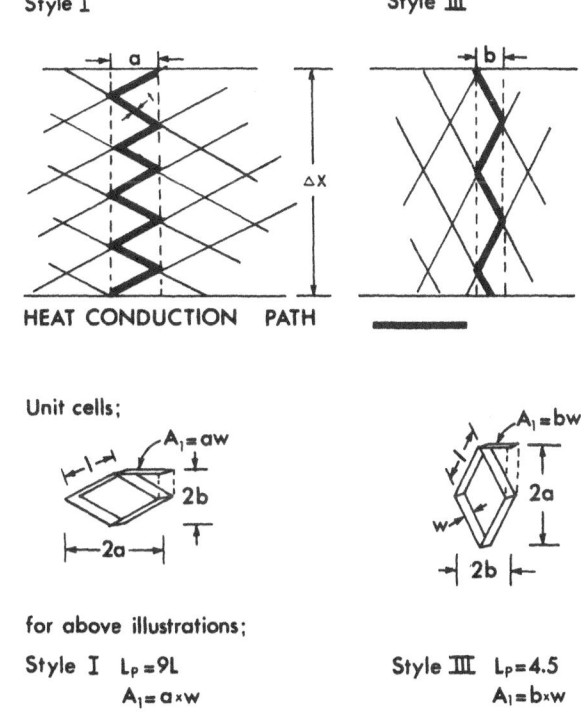

Style I

Style III

$\leftarrow$ a $\rightarrow$

$\leftarrow$ b $\leftarrow$

$\triangle$X

HEAT CONDUCTION PATH

Unit cells;

$A_1 = aw$

2b

$\leftarrow$ 2a $\rightarrow$

$A_1 = bw$

2a

w

$\leftarrow$ 2b $\leftarrow$

for above illustrations;

Style I $L_p = 9L$
$A_1 = a \times w$

Style III $L_p = 4.5$
$A_1 = b \times w$

Figure 3 Heat transfer model for EXPLOSAFE

Table 1 True thermal conductivity of EXPLOSAFE

style	λ_t, $W.m^{-1}.K^{-1}$	
	Theoretical	Experimental
style I	0.409	0.45
style II	0.106[5]	0.20
style III	2.414	2.25 H
		2.60 V

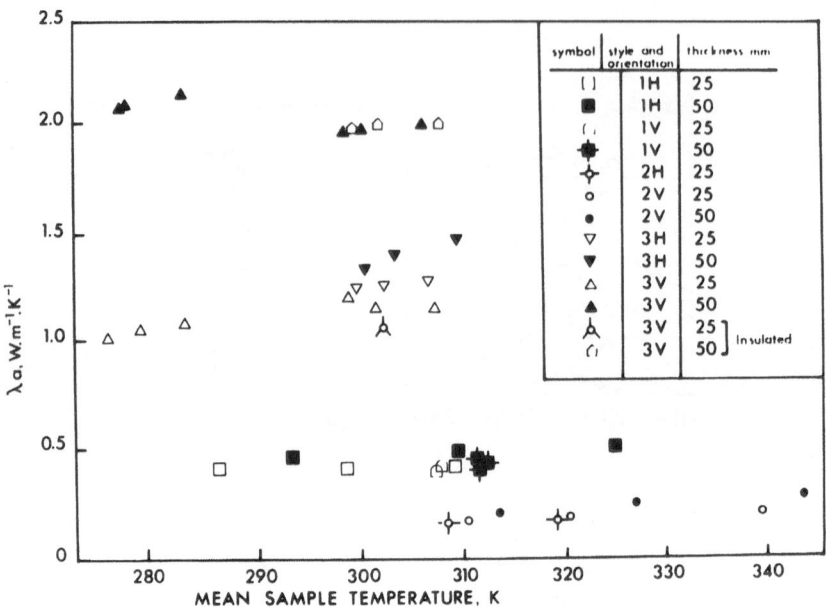

Figure 4   Apparent thermal conductivity of EXPLOSAFE

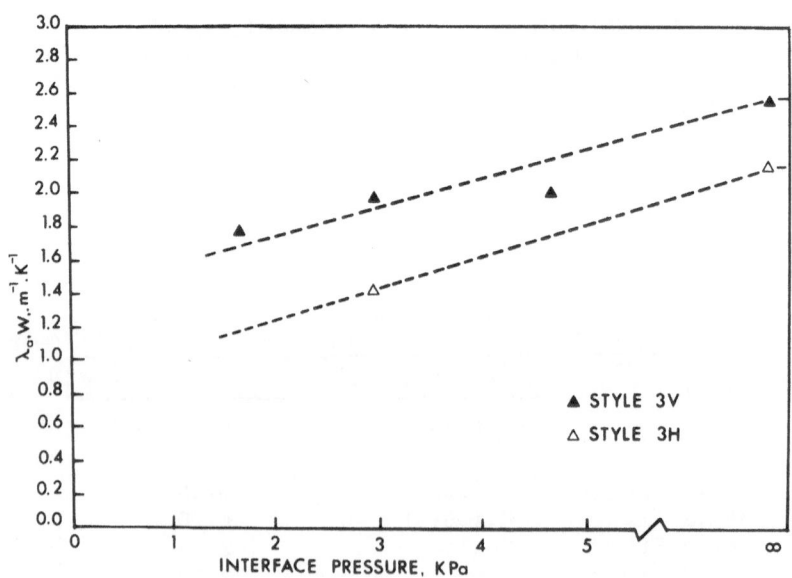

Figure 5   Apparent thermal conductivity of EXPLOSAFE as
a function of pressure.

25 mm and 50 mm samples both have the same contact resistance at constant pressure, it follows that the apparent conductivity of a thin specimen will be less than that of a thick one.

The specimens were examined for any significant physical differences. It was found that 50 mm thick samples of style 3V were approximately 10 percent more dense than the 25 mm thick samples. The apparent conductivity of all 25 mm thick samples was thus normalized (Equasion 4) to correct for the density differences.

$$\lambda_n = \frac{\rho_{3-50}}{\rho_{3-25}} \lambda_m \qquad (4)$$

True thermal conductivities for styles I and III were measured experimentally as well as estimated theoretically (Equation 3). The results are summarized in Table 1.

The apparent thermal conductivity, λ_a, as a function of pressure was measured for style III in the range of 1.69 to 4.65kPa. (Figure 5). An increase in λ_a with increasing pressure resulted due to improved contact conductance at the EXPLOSAFE-plate interface. Using the results of Figure 5 and equation (2), the thermal contact conductance was determined as a function of pressure. The following correlation was developed for h_c(P in kPa units):

$$h_c = -1.236 + 183.011\ P - 22.146\ P^2; \quad 1.69 \leq P \leq 4.65\ \text{kPa} \qquad (5)$$

CONCLUSION

Thermal conductivity of EXPLOSAFE was measured for samples of three different styles, different orientations and thicknesses. A theoretical model is developed to estimate the thermal conductivity for styles I and III. The theoretical predictions are compared with experimentally measured values.

Thermal contact conductance of EXPLOSAFE-plate interface was estimated as a function of interfacial pressure and a simple correlation is proposed.

ACKNOWLEDGEMENTS

This work was performed under a program of studies funded by the Transport Canada under contract number OSD82-00085.

NOMENCLATURE

A	Surface area of heater, m^2
A_1	Projected area of one conduction path (Figure 3), m^2
A_e	Equivalent heat flow area, m^2
A_f	Cross-sectional area of conduction path through the aluminum foil (=wt), m^2
A_{vap}	Area of vapor space between adjacent layers (Figure 2), m^2
a	Semi-major diagonal of the unit cell (Figure 3), m
b	Semi-minor diagonal of the unit cell (Figure 3), m
H	Horizontal
h_c	Thermal contact conductance, $W. \ m^{-2}, K^{-1}$
L	Length of one side of the unit cell (Figure 3), m
L_p	Path length for heat conduction through aluminum foil (Figure 3), m
Q	Heat flow rate, W
t	Thickness of aluminum foil strand (Figure 3), m
V	Vertical
w	Width of aluminum foil strand (Figure 3), m
ΔT	Temperature difference across the thickness of the test sample, deg K
ΔX	Thickness of the test sample, m
λ	Thermal conductivity, $W.m^{-1}.K^{-1}$
ρ	Density of the test sample, $Kg.m^{-3}$

Subscripts

a	apparent
al	aluminum
L	Left side
m	measured
n	normalized
R	Right side
t	true
th	theoretical
3-25	style III, 25 mm thick sample
3-50	style III, 50 mm thick sample

REFERENCES

[1] Appleyard, R. D., "Testing and Evaluation of the EXPLOSAFE System as a Method of Controlling the Boiling Liquid Expanding Vapor Explosion (BLEVE)". Report #TP2740, EXPLOSAFE Division of Vulcan Industrial Packaging Ltd., Toronto, Canada, August 1980.

[2] Financial Times-Markets, June 04, 1982.

[3]   Venart, J. E. S., "Tank Car Thermal Response Analysis",
        Interim Report for Transport Canada, Report # UNB-ME-TF18,
        August 1983.
[4]   Kreith, F., "Principles of Heat Transfer", International
        Text Book Co., Scranton, Penn., U.S.A. (1967).
[5]   Aydemir, N. V., "Thermal Stratification in a Partially
        Filled Horizontal Cylinder", M.Sc.E. Thesis in preparation,
        Univ. of New Brunswick, Canada (1984).

AUTHOR INDEX

SUBJECT INDEX